OXFORD MEDICAL PUBLICATIONS

CUNNINGHAM'S MANUAL OF PRACTICAL ANATOMY

CUNNINGHAM'S MANUAL OF PRACTICAL ANATOMY

FOURTEENTH EDITION

G. J. ROMANES, C.B.E.,
B.A., Ph.D., M.B., Ch.B., F.R.C.S.Ed., F.R.S.E.
Professor of Anatomy in the University of Edinburgh

Volume Three
Head and Neck and Brain

OXFORD
OXFORD UNIVERSITY PRESS
NEW YORK DELHI

Oxford University Press, Walton Street, Oxford OX2 6DP
LONDON GLASGOW NEW YORK TORONTO DELHI BOMBAY CALCUTTA
MADRAS KARACHI KUALA LUMPUR SINGAPORE HONG KONG TOKYO
NAIROBI DAR ES SALAAM CAPE TOWN MELBOURNE AUCKLAND
AND ASSOCIATES IN
BEIRUT BERLIN IBADAN MEXICO CITY NICOSIA
OXFORD *is a trade mark of Oxford University Press*

ISBN 0 19 263205 1

First Edition 1893
Fourteenth Edition 1979
Reprinted 1982, 1983, and 1984

Printed in Great Britain by Jarrold and Sons Ltd., Norwich

CONTENTS OF VOLUME THREE

PREFACE TO VOLUME THREE

THIS volume has been rewritten on the same principles as the other two. The order of dissection has been changed so that the median section of the skull and pharynx is carried out at an earlier phase. This allows the student to expose the entire length of the neurovascular bundle of the neck early in the dissection and to follow many structures into the walls of the nasal, oral, and pharyngeal cavities, thus giving an opportunity of relating the structures to these parts of the body which are readily examined in the living.

Accounts of the skull and cervical vertebrae are included, and all the illustrations of the bones are repeated in an atlas at the end of the book for ready reference. The development of the face, mouth, nasal cavities, and pharynx and of the central nervous system are included in a separate section. They are not placed in the appropriate parts of the book because the face, oronasal, and pharyngeal regions are best dealt with together if a proper understanding of their formation is to be obtained without undue repetition.

Tables of movements, the muscles which produce them, and the nerve supply of these muscles are also included. Several new illustrations have been added, principally in the osteology and central nervous system sections, but also to show the general structure of the head and neck and the distribution of arteries and nerves to them.

These additions, and the inclusion of a greater number of points of clinical interest, have increased the overall length of the volume in spite of the removal of a considerable amount of unnecessary detail, but it is hoped that they will improve the usefulness of the book and make it more able to meet the requirements of medical students without recourse to a number of other texts.

I am indebted to Professor E. W. Walls for reading the manuscript and making many helpful suggestions, and also to Dr. J. C. Gregory for his careful handling of the many publication problems.

Edinburgh
March 1977

G. J. Romanes

THE HEAD AND NECK

As a preliminary to the dissection of the head and neck, the dissectors should study the cervical vertebrae and skull, relating their main features to the bony points which can be felt. A sound knowledge of these structures and of those which pass through or are attached to them is necessary for an understanding of this region. The following brief accounts should be studied together with the vertebrae and a skull from which the skull-cap has been removed, so that the various points can be confirmed.

THE CERVICAL VERTEBRAE

There are seven cervical vertebrae. The third to the sixth are typical, but the first and second are modified to permit movements of the head on the neck, and the seventh shows some features of a thoracic vertebra. All seven have a foramen (foramen transversarium) in each transverse process.

Cervical vertebrae are smaller and more delicate than their thoracic and lumbar counterparts—they carry less weight—but they have a larger vertebral foramen to accommodate the cervical swelling of the spinal medulla.

In the following descriptions, individual cervical vertebrae are identified as C. 1, C. 2, C. 3, etc.

THE TYPICAL CERVICAL VERTEBRAE

Body. This is oval in plan, with its long axis transverse. The superior surface is concave from side to side, and has lateral margins that project upwards to articulate with the cut-away inferolateral margins of the body above.

Pedicles. These are short, and pass outwards and backwards from the middle of the posterolateral parts of the body to form the posteromedial wall of the foramen transversarium.

Laminae. These are long and rectangular, and almost overlap their neighbours in extension.

Spines are short and bifid.

Articular Facets. These are the obliquely cut ends of a short rod of bone, the articular process, lying at the junction of the pedicle and lamina on each side. The superior facets face upwards and backwards, the inferior pair downwards and forwards.

Vertebral Foramen. It is large and triangular.

Transverse Processes. Each is short and perforated by the foramen transversarium. Anterior to the foramen, the costal process (corresponding to a rib) projects laterally from the body to end in the anterior tubercle (attachment of scalenus anterior

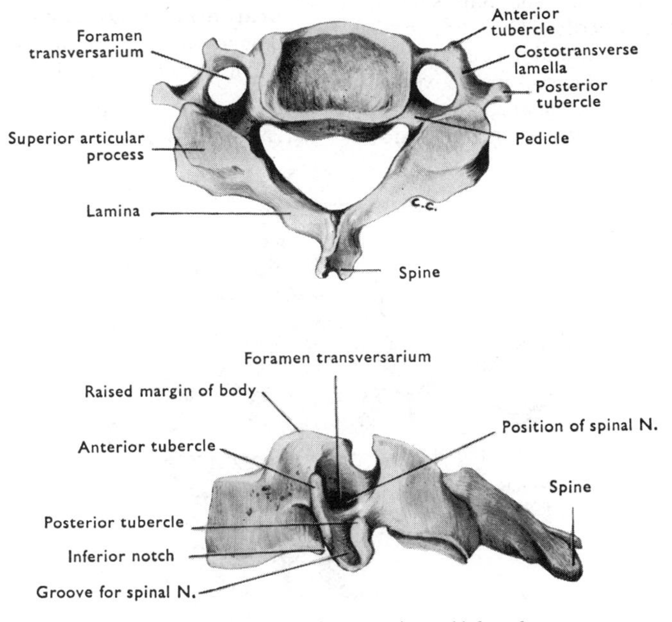

FIG. 1 The fourth cervical vertebra, superior and left surfaces.

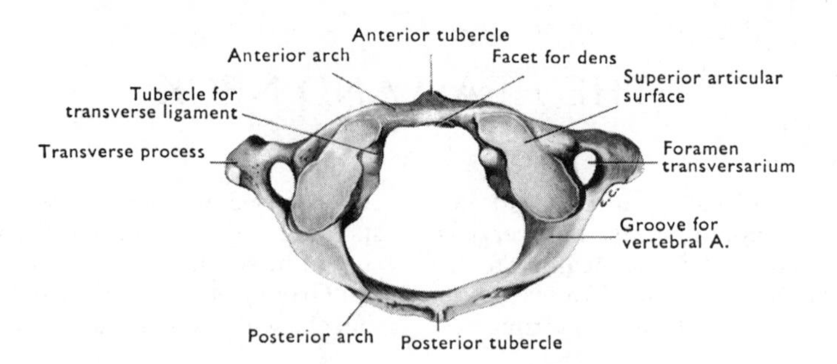

FIG. 2 The upper surface of the atlas.

and longus capitis). Behind the foramen, the true transverse process projects laterally from the junction of the pedicle and lamina to end in the posterior tubercle (attachment of scalenus medius, levator scapulae, etc.). The slip of bone which unites the tubercles and completes the foramen transversarium is concave superiorly to lodge the ventral ramus of the corresponding spinal nerve.

Foramen Transversarium. It transmits the vertebral artery, vertebral veins, and sympathetic plexus, and lies anterior to the ventral ramus of the spinal nerve.

THE ATYPICAL CERVICAL VERTEBRAE

C. 1 (the Atlas). It has no body and consists simply of two **lateral masses** united by an anterior and a posterior **arch**. The body is represented by the **dens**, a tooth-like projection from the superior surface of the body of C. 2, and each pedicle by part of the corresponding lateral mass. The **laminae** form the posterior arch which grooved on its superior surface, behind the lateral mass, by the vertebral artery and the first cervical ventral ramus. The spine is replaced by the **posterior tubercle**. The superior and inferior **facets** lie on the lateral masses anterior to the first and second cervical nerves respectively. The superior is concave and kidney-shaped; the inferior is almost circular, slightly concave, and faces downwards and medially. An inward projection from each lateral mass gives attachment to the **transverse ligament** which divides the vertebral foramen into a small anterior compartment for the dens, and a larger, oval, posterior compartment for the spinal medulla and its coverings. The **transverse process** of the atlas is long and thick, and lacks an anterior tubercle. Its **foramen transversarium** is lateral to those below.

C. 2 (the Axis). The salient feature is the **dens**. This is held against the anterior arch of C. 1 by the **transverse ligament**. There are articular facets on the arch and dens, and the posterior surface of the dens is grooved by the ligament.

On each side, the thick **pedicle** is covered by the **superior articular facet** which also lies partly on the body (lateral to the dens) and on the base of the costal process. This facet overlies the foramen transversarium, and is flatter than the inferior facet of C. 1 which it fits. The inferior facet of the axis is typical.

The **laminae** are considerably thickened for muscle attachments, and carry a massive **spine**. The transverse process has no anterior tubercle. The **foramen transversarium** turns laterally through 90 degrees under the superior articular facet so that its exit is visible from the lateral aspect.

C. 7. The **spine** is long and non-bifid, the **transverse process** lacks an anterior tubercle,

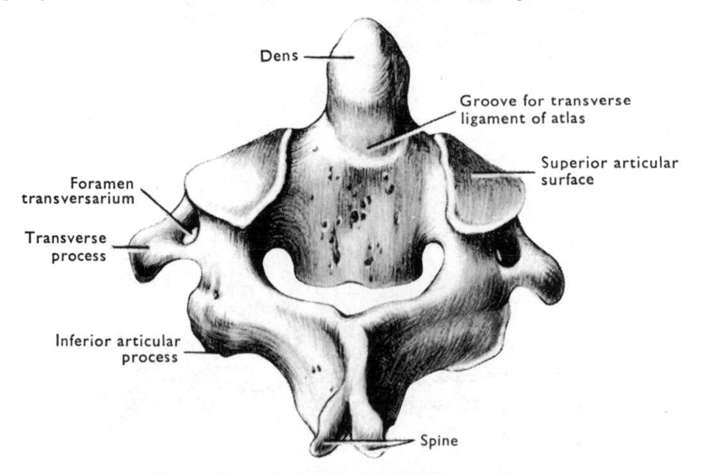

FIG. 3 The axis vertebra from behind and above.

and the **foramen transversarium** transmits only veins.

PALPABLE PARTS OF CERVICAL VERTEBRAE

The **spine of C. 2** in the nape of the neck about 5 cm below the external occipital protuberance.

The **spine of C. 7** (vertebra prominens) where the collar band crosses the posterior median line of the neck.

The **transverse process of C. 1** through the anterior border of sternocleidomastoid immediately below the tip of the mastoid process.

THE SKULL

[FIGS. 4–12]

GENERAL ARCHITECTURE OF SKULL

The skull consists of a brain-box or cranium and a facial skeleton. The **cranium** surrounds the brain and its coverings (meninges) and deepens from before backwards to accommodate them. The **facial skeleton** is slung beneath the shallow, frontal part of the cranium the anterior wall of which forms the bone of the forehead (**frontal bone**), while the floor (frontal and **ethmoid bones**) forms the roofs of the orbits (sockets for the eyeballs) and of the nasal cavities, and sends a wall (nasal septum) downwards between these cavities.

The main elements of the facial skeleton are the right and left **maxillae**. These bear the upper teeth in the curved alveolar process which projects inferiorly from each. The body of each maxilla lies below the corresponding orbit and lateral to the lower part of the nasal cavity. It contains the **maxillary air sinus**, and has the shape of a three-sided pyramid lying with its base medially. The **surfaces** are anterolateral (anterior), posterolateral (infratemporal), and superior (orbital), and the base lies in the lateral wall of the nasal cavity. The apex points laterally and is overlaid by the **zygomatic** (cheek) **bone** which extends its attachment to the maxilla by sending processes along each edge of the pyramid. The zygomatic bone forms one of the main buttresses attaching the maxilla to the cranium by extending upwards on the lateral margin of the orbit (**frontal process**) to meet the frontal bone, and backwards (**temporal process**) to join the zygomatic process of the temporal bone and complete the **zygomatic arch**. Of the extensions of the zygomatic bone on the maxilla, that between the orbital and anterior surfaces forms the lateral half of the inferior margin of the orbit, that between the anterior and infratemporal surfaces is a heavy buttress preventing upwards displacement of the maxilla, and that between the orbital and temporal surfaces is a deep flange from the frontal process of the zygoma. This flange unites with the greater wing of the sphenoid

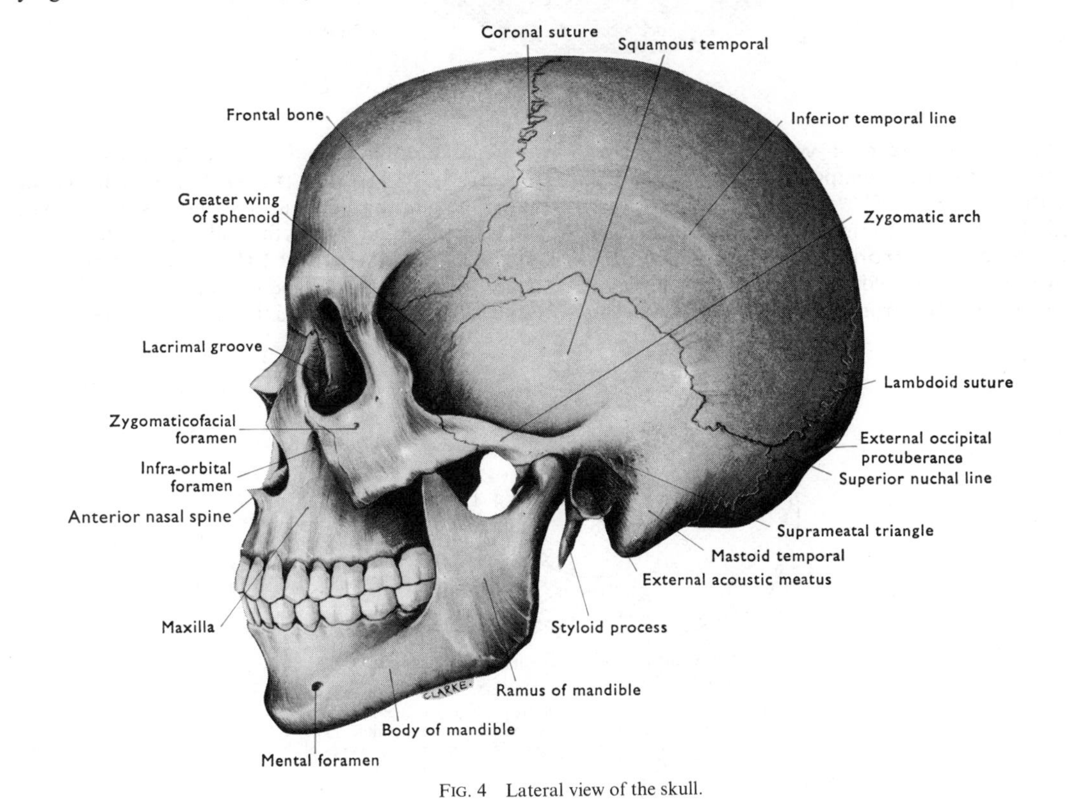

FIG. 4 Lateral view of the skull.

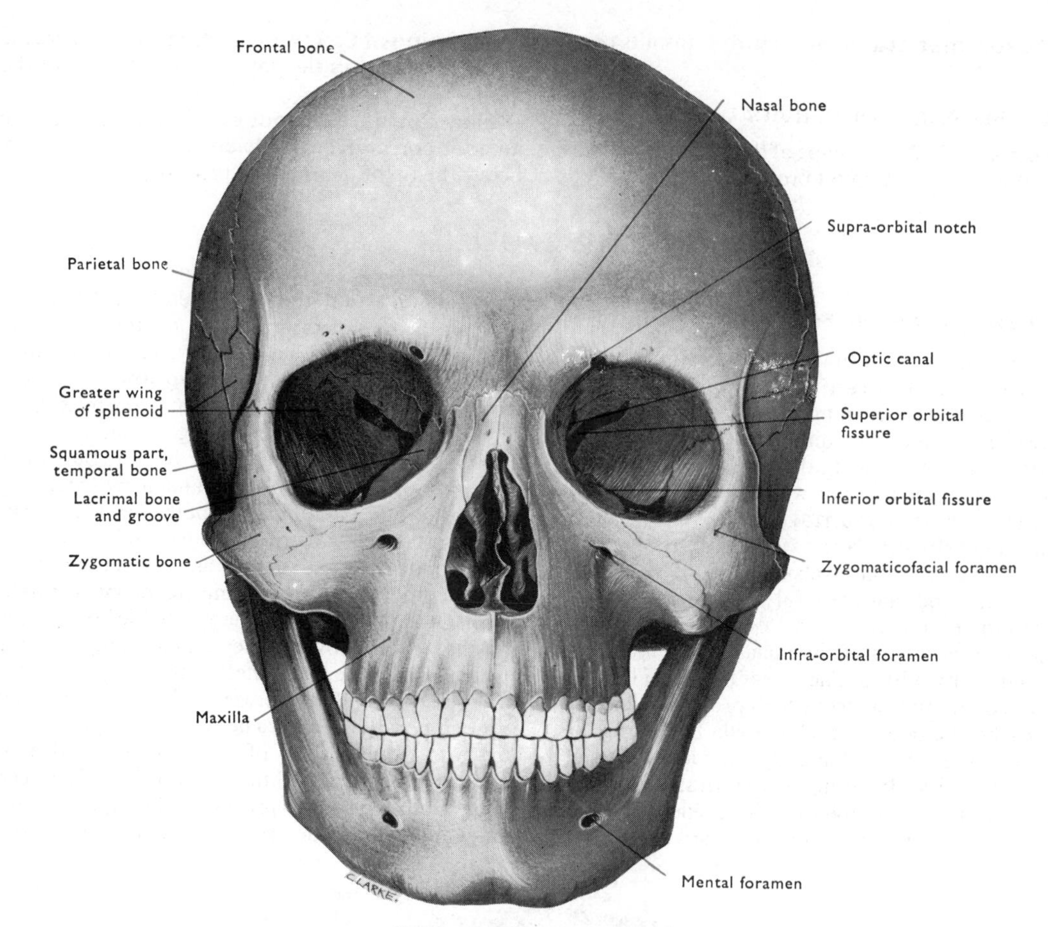

FIG. 5 Anterior view of the skull.

bone to form the **lateral wall of the orbit**, separating it from the temporal fossa deep to the zygomatic arch [FIG. 10].

Medial to the orbit, the maxilla is directly attached to the cranium by a **frontal process**. This forms the lower part of the medial margin of the orbit and articulates with the frontal bone above, with the nasal bone anteriorly, and with the lacrimal bone posteriorly to form the **fossa for the lacrimal sac**. The lacrimal bone articulates posteriorly with the orbital lamina of the ethmoid to form the greater part of the **medial wall of the orbit**, separating it from the upper part of the nasal cavity. The orbital lamina of the ethmoid and the lacrimal bones extend

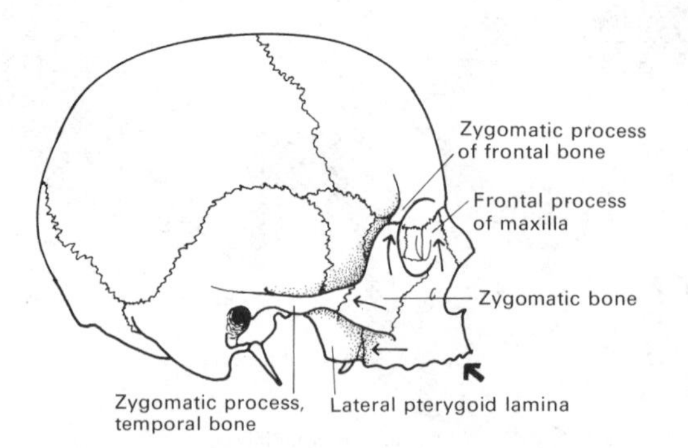

FIG. 6 Lateral view of skull to show the parts of the facial skeleton which transmit to the cranium forces applied to the maxilla. The thick arrow represents the force applied – the thin arrows the lines of transmission. These are the parts of the skull liable to fracture in blows to the face.

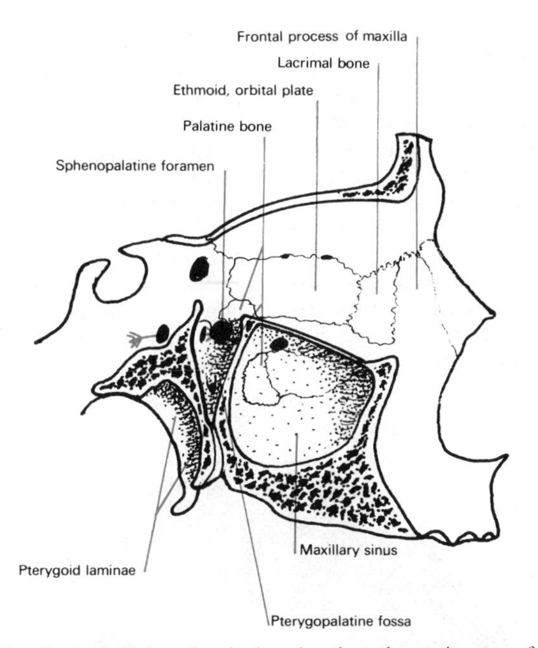

FIG. 7 Lateral view of sagittal section through anterior part of skull to show pterygopalatine fossa, and the construction of the medial wall of the orbit. Arrow in foramen rotundum.

from the orbital surface of the maxilla below to the frontal bone above, and are very thin plates of bone which add nothing to the strength of the attachment of the maxilla to the cranium. The **ethmoidal foramina** lie in the fronto-ethmoidal suture.

Posteriorly, where the infratemporal surface of the maxilla meets its base, two plates of bone (**pterygoid laminae**) extend downwards and forwards from the sphenoid bone in the base of the cranium to articulate inferiorly with the maxilla through a process of the palatine bone. They form yet another buttress attaching the maxilla to the cranium. Above, the pterygoid laminae are separated from the maxilla by the narrow **pterygomaxillary fissure** which is continuous superiorly with the **inferior orbital fissure** between the lateral wall and floor of the orbit. The importance of the buttresses which hold the maxilla to the cranium (zygomatic bone and arch, frontal process of the maxilla, and pterygoid laminae) is that they are fractured when heavy blows displace the facial skeleton.

The two maxillae are firmly united in the median plane by the articulation of the alveolar processes anteriorly and of their **palatine processes** posteriorly. These horizontal plates, which extend inwards from the upper margins of the alveolar processes, also articulate with the corresponding processes of the palatine bone posteriorly, thus completing the hard palate which separates the nasal cavities from the mouth. The **palatine bone** is an L-shaped plate. Its horizontal lamina forms the posterior one-third of the palate. The **perpendicular lamina** forms the lateral wall of the nasal cavity between the maxilla in front and the medial pterygoid lamina behind. Thus it separates the pterygomaxillary fissure from the nasal cavity except superiorly where a notch in the upper margin of the perpendicular lamina forms the **sphenopalatine foramen** with the sphenoid bone. This arrangement can be seen by looking at a skull which has been divided in the median plane.

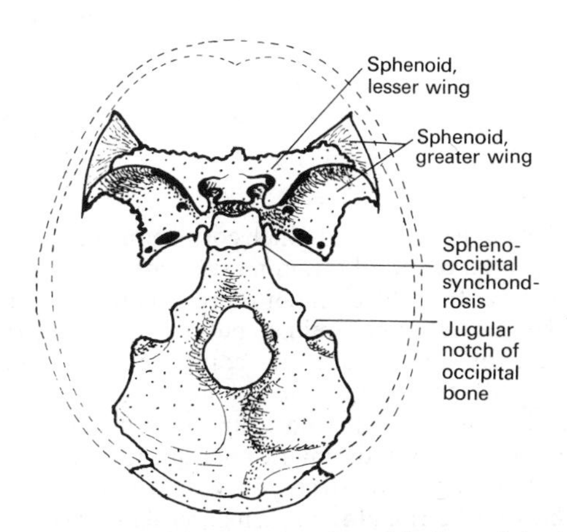

FIG. 8 The part of the internal surface of the base of the skull formed by the sphenoid and occipital bones. On each side a temporal bone is wedged between them. The frontal and ethmoid bones complete the anterior part.

THE CRANIUM

The cranium surrounds the brain and meninges much as the vertebrae surround the spinal medulla and its meninges, except that the cranium is a rigid structure composed of a number of interlocking bones which represent highly modified, fused vertebrae. A solid, median, inferior part of the cranium, which represents a number of fused vertebral bodies, is constructed from parts of two separate bones (the occipital and sphenoid [FIG. 8]) which unite in the adult. Lateral to this are several foramina which transmit nerves and blood vessels to and from the brain, and which correspond in position to the intervertebral foramina (*vide infra*).

The vault of the cranium represents the vertebral arches (pedicles and laminae) but only the **occipital bone**, by itself, forms a complete arch which surrounds the foramen magnum [FIG. 8] at the junction of the brain and spinal medulla. Like the vertebrae below it, the inferior margin of the foramen magnum has an oval, curved articular facet (**occipital condyle**) on each anterolateral surface. These articulate with the superior articular facets of the first cervical vertebra (the atlas [FIG. 2]). Lateral to each of these condyles, the occipital bone projects laterally (**jugular process**) immediately posterior to the jugular foramen. This process corresponds to a transverse process, and lying immediately above that of the atlas, separates the ventral ramus of the first cervical nerve, inferiorly, from the 9th, 10th, and 11th cranial nerves in the **jugular foramen**. The **hypoglossal canals** for the 12th cranial nerves pierce the skull immediately above the occipital condyles.

Posterior to the foramen magnum, the occipital bone forms the greater part of the inferior surface of the cranium roughened by the attachment of the muscles of the back of the neck. This area is divided transversely by an ill-defined **inferior nuchal line** [FIG. 11] and is limited posteriorly by the **external occipital protuberance** in the midline and the **superior nuchal lines** extending laterally from it. At this line, the occipital bone turns upwards, deep to the back of the scalp, as a triangular plate inset between the posterior edges of the two parietal bones with which it forms the lambdoid suture. The **parietal bones** arch forwards from the lambdoid suture and form the greater and widest part of the dome of the skull. Anterior to the occipital bone they articulate with each other in the median plane (**sagittal suture**), and with the frontal bone

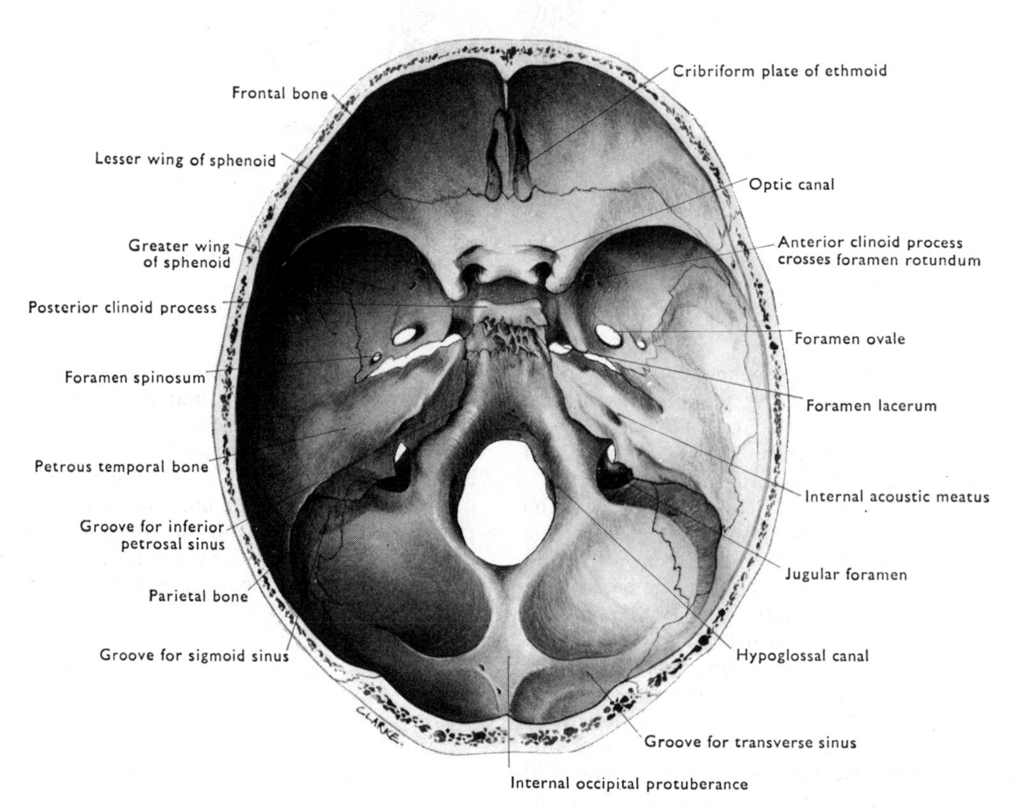

FIG. 9 Internal surface of the base of the skull.

anteriorly in the transversely placed **coronal suture** [FIG. 4].

The **frontal bone**, consisting of right and left halves which usually fuse early in life, curves antero-inferiorly to the superior margins of the orbits and the root of the nose. Here it turns backwards to form the greater part of the roof of each orbit, separated by the part of the **ethmoid bone** (**cribriform plates**) which forms the roof of the nasal cavities [FIG. 9]. The ethmoid also forms a considerable part of the septum and lateral walls of these cavities. Posteriorly, the orbital parts of the frontal bone meet the lesser wings of the sphenoid bone [FIG. 9] to complete the **roofs of the orbits** and the floor of the **anterior cranial fossa** on which the frontal lobes of the brain lie.

The **sphenoid bone** consists of a median body which is hollowed out by an extension of each nasal cavity (the **sphenoidal air sinuses**), two greater and two lesser wings, and a pair of pterygoid laminae projecting downwards from the medial part of each greater wing.

The **body** of the sphenoid articulates with the nasal septum in the midline—the ethmoid anteriorly and the vomer inferiorly. Posteriorly it is fused to the basilar part of the occipital bone, just posterior to the vomer. This fusion replaces the growth cartilage which separates the two bones and is responsible for growth of the base of the skull until this ceases (18–22 years). Superiorly, the body forms the **sella turcica**. The central portion of this is the **hypophysial fossa** which is limited posteriorly by the **dorsum sellae**—a rectangular plate of bone which projects upwards with the **posterior clinoid processes** on its superolateral corners. Anteriorly, the fossa is limited by the tuberculum sellae with the horizontal **sulcus chiasmatis** in front of it. On each side this sulcus leads to an **optic canal** which transmits the corresponding optic nerve and ophthalmic artery. The sulcus does not lodge the optic chiasma in spite of its name. The hypophysial fossa is the median part of the **middle cranial fossa**. On each side it is continuous with the expanded lateral parts of this fossa which lodge the temporal lobes of the brain.

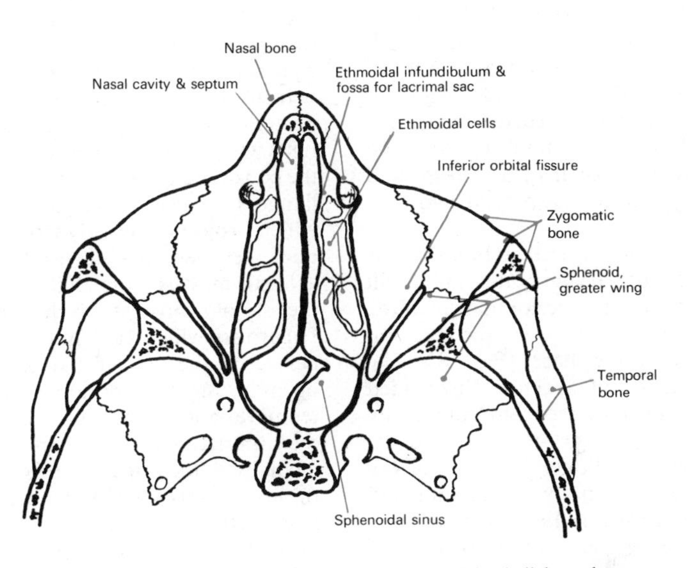

FIG. 10 Horizontal section of the anterior part of the skull through the orbits.

The **lesser wings of the sphenoid** project laterally from the anterosuperior part of the body. Each has a free, curved, posterior margin which forms the posterior limit of the anterior cranial fossa and ends medially in an **anterior clinoid process** immediately lateral to the optic canal. Laterally, the tip of each lesser wing reaches and fuses with the upturned edge of the corresponding greater wing of the sphenoid bone. This closes the lateral end of the **superior orbital fissure** which separates the two wings medially [FIG. 5] and is the route taken by structures passing between the middle cranial fossa and the orbit.

Each **greater wing** is roughly rectangular in shape with its long axis running anterolaterally, parallel to the lateral wall of the orbit. The posteromedial angle is truncated and attached to the lowest part of the lateral surface of the body of the sphenoid well below the level of the lesser wing. The posterolateral angle projects backwards and has the **spine of the sphenoid** extending downwards from it immediately posterior to the small **foramen spinosum**. The anterior part of the rectangle is bent upwards. This produces a concave cerebral surface, while the external surfaces form the posterior part of the lateral wall of the orbit and part of the medial wall of the temporal fossa; both these surfaces unite anteriorly with the zygomatic bone. The upturned end articulates superiorly with the lesser wing and with the inferior surfaces of the frontal and parietal bones. Thus this part of the bone has orbital, temporal, and cerebral surfaces [FIG. 10]. Where the greater wing is bent upwards, there is usually a well-marked **infratemporal crest** on the inferior surface. Close to the body of the sphenoid, the greater wing is pierced by the **foramen rotundum** near the medial end of the superior orbital fissure, and posterior to this by the **foramina ovale** and **spinosum**. The pterygoid laminae project below the foramen rotundum.

The temporal is the remaining cranial bone. It is as though pushed into the lateral side of the skull to fill the semicircular gap below the inferior margin of the parietal bone and the interval between the occipital and sphenoid bones in the base [FIG. 8]. The **temporal bone** consists of a dense, three-sided pyramid (**petrous part**) wedged anteromedially between the greater wing of the sphenoid and the basilar part of the occipital bone. The apex of the pyramid does not reach the body of the sphenoid bone but leaves the **foramen lacerum** between them [FIG. 9]. The anterior and posterior surfaces of the pyramid are on the internal surface of the cranium, while the rough inferior surface is on the external aspect of the base. The posterior surface of the pyramid is pierced by the cylindrical **internal acoustic meatus**, and articulates medially with the sloping cranial surface of the median parts of the sphenoid and occipital bones (the clivus) at a groove which lodges the inferior petrosal sinus. This groove leads downwards on the articulation to the jugular foramen which expands laterally into the petrous temporal bone as the **jugular fossa**. Together the clivus and the posterior surfaces of the two petrous temporal bones form the anterior wall of the **posterior cranial fossa** which lodges the cerebellum and the greater part of the brain stem. The anterior surface of the pyramid forms the posterior part of the floor of the middle cranial fossa. The base of the pyramid is the **mastoid process** on the lateral aspect of the skull [FIG. 4]. It lies in the same transverse plane as the occipital condyles.

Fused with the anterolateral part of the petrous and with the superior aspect of its mastoid process is the semicircular **squamous part** of the temporal bone. This overlaps the external surfaces of the greater wing of the sphenoid and the inferior margin of the parietal bone (**squamosal suture**).

The **zygomatic process** of the temporal bone arises as a horizontal ridge from the lower part of the lateral surface of the squamous temporal. It turns forwards to join the temporal process of the zygomatic bone in the zygomatic arch. Below the root of the zygomatic process, the inferior surface of the squamous part has two transverse notches. The larger, anterior notch, is the mandibular fossa for articulation with the head of the mandible. The smaller, posterior notch, immediately anterior to the mastoid process, is converted into a canal (**external acoustic meatus**) by a U-shaped plate of bone which forms the anterior, lower posterior, and inferior walls of the meatus, and has a sharp flange projecting inferiorly from it. This **tympanic part** of the temporal bone fuses anteriorly with the squamous part (except medially where a strip of petrous temporal intervenes) and so extends the posterior wall of the mandibular fossa. Posteriorly, it fuses with the mastoid process and inferior surface of the petrous temporal bone. The middle of the inferior flange is partly wrapped round the **styloid process** which projects inferiorly from the petrous temporal bone between the jugular fossa and the external acoustic meatus. Immediately posterior to the base of the styloid process is the small **stylomastoid foramen** which transmits the facial (7th cranial) nerve. At the medial end of the flange, a circular opening in the inferior surface of the petrous temporal bone is the **carotid canal**. This transmits the internal carotid artery which turns anteromedially in the bone and emerges from its apex into the upper part of the foramen lacerum through which it enters the cranial cavity, grooving the lateral aspect of the body of the sphenoid bone.

The **mandibular fossa** is limited anteriorly by the articular tubercle which is continuous laterally with the tubercle on the root of the zygomatic process. The smooth, articular surface of the tubercle is continuous with the mandibular fossa posteriorly—a feature which indicates the movement of the head of the mandible on to the tubercle when the mouth is opened or the jaw protruded. The articular surface for the mandible extends medially almost to the spine of the sphenoid and anteriorly beyond the root of the zygomatic process.

Features of External Surface of Base of Skull

It is important to appreciate that the posterior two-thirds of the base of the skull overlies and is continuous with structures in the neck. Anterior to this, and overlapping it laterally, are the structures passing from the skull to the jaws and orbit.

Posteriorly, the mouth and nasal cavities enter a common chamber—the **pharynx**. This is the cranial end of the gut tube which divides inferiorly into a median airway (larynx and trachea) and foodway

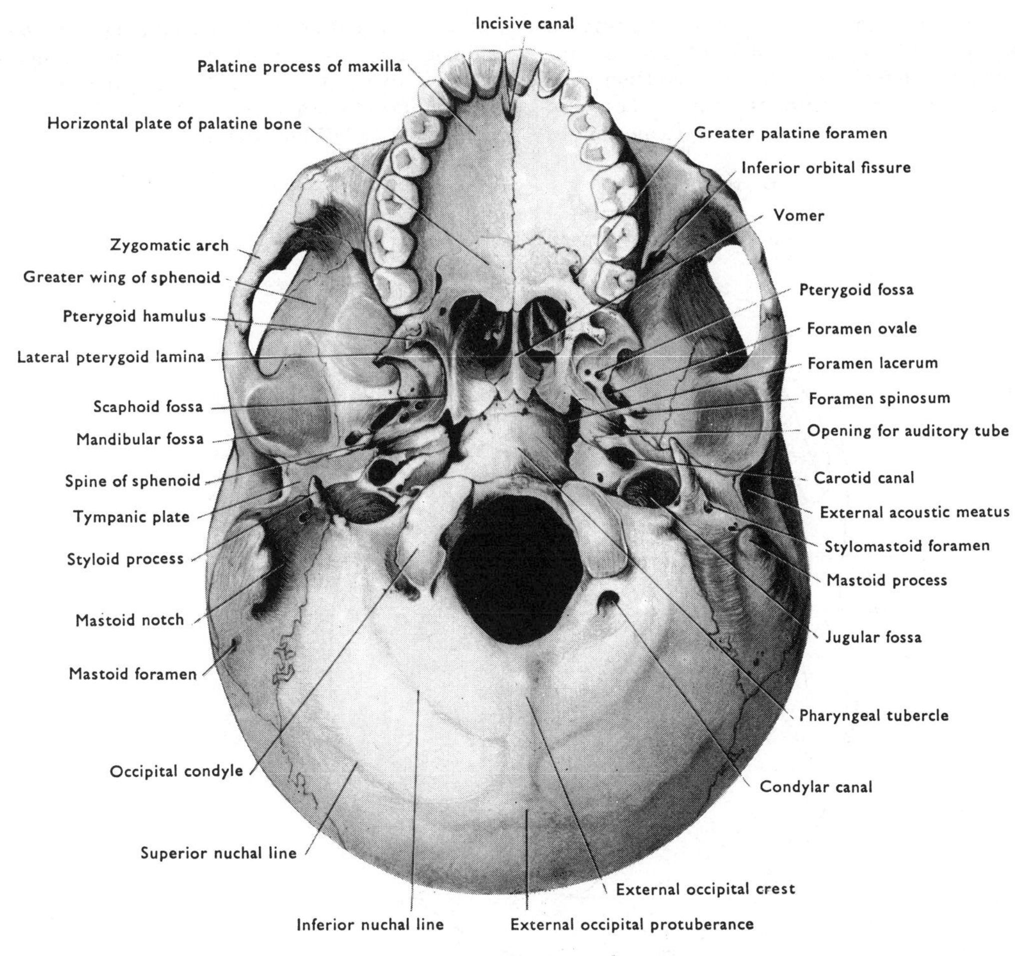

FIG. 11 The external surface of the base of the skull.

(laryngeal part of the pharynx and oesophagus) anterior to the vertebral column. The uppermost (nasal) part of the pharynx extends to the base of the skull where it is continuous anteriorly with the nasal cavities. Behind this it is closed by a musculo-fascial wall each half of which curves backwards from the posterior edge of the corresponding medial pterygoid lamina and the pterygoid hamulus [FIG. 11] to meet its fellow in a median raphe. This is attached above to the **pharyngeal tubercle** on the basilar part of the occipital bone, 1 cm anterior to the foramen magnum. Inside this curved wall is the pharyngeal part of the base of the skull [FIG. 12].

Immediately behind each medial pterygoid lamina, the upper part of the pharyngeal wall is pierced by the cartilaginous **auditory tube**. This passes posterolaterally from the pharyngeal cavity in the groove between the greater wing of the sphenoid and the petrous temporal on the inferior aspect of the skull. At the root of the spine of the sphenoid, the tube becomes bony and opens into the middle ear cavity. It maintains a pressure equilibrium between that cavity and the pharynx. Below the level of the palate, the pharynx opens into the mouth anteriorly. Here the walls of the pharnyx are continuous with the muscle (buccinator) and fasciae of the cheeks.

Posterior to the pharyngeal area on the base of the skull is the area for articulation with the vertebral column and for attachment to the pre- and post-vertebral muscles of the neck [FIG. 26].

Immediately anterior to the jugular process of the occipital bone is the **jugular foramen** with the carotid canal in front of it. These foramina lie posterolateral to the wall of the pharynx and transmit the structures which descend in this position as the neurovascular bundle of the neck—internal jugular vein, carotid artery, and vagus nerve. In its upper part this bundle also contains the 9th and 11th cranial nerves, which emerge through the jugular foramen, and the 12th cranial nerve which leaves the skull through the hypoglossal canal immediately medial to the jugular foramen.

Deep to the zygomatic arch is the **temporal fossa**. Between the infratemporal crest on the greater wing of the sphenoid and the lateral pterygoid lamina is the **infratemporal fossa**. These two fossae and the zygomatic arch give rise to most of the muscles of mastication. The mandibular nerve which supplies these muscles leaves the skull through the foramen ovale in the medial part of the roof of the infratemporal fossa, close to the auditory tube and the lateral wall of the pharynx.

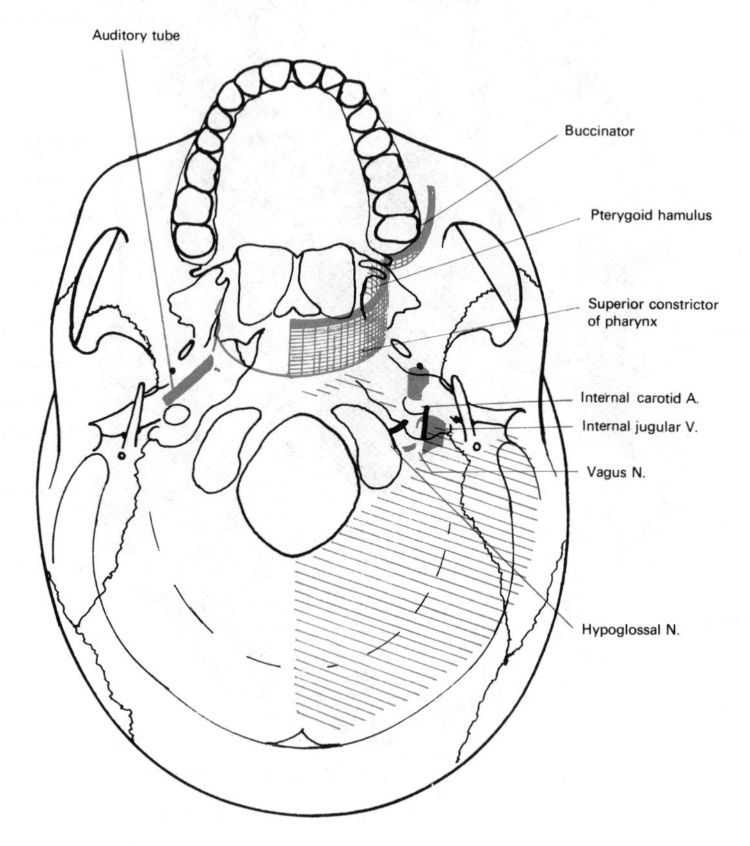

FIG. 12 External surface of base of skull to show the position of the superior constrictor, buccinator, and the auditory tube.

THE SCALP, THE TEMPLE, AND THE FACE

SURFACE ANATOMY

Begin by identifying the parts of the head mentioned in the following paragraphs by examining your own head and those of your partners.

Auricle

The main parts of the auricle are shown in FIGURE 16. It lies nearer the back of the head than the front, and is at the level of the eye and the nose. Hairs project backwards from the tragus and, particularly when they thicken after middle age, may prevent small foreign bodies from entering the external acoustic meatus.

Back and Side of Head

The **external occipital protuberance** is the knob felt in the median line where the back of the head joins the neck. From this, an indistinct, curved ridge, the **superior nuchal line**, extends laterally on each side between the scalp and the neck. Each line passes towards the corresponding **mastoid process**—a rounded bone behind the lower part of the auricle. Press your finger tip into the hollow below and in front of the mastoid process; the resistance felt is due to the transverse process of the atlas vertebra. It is overlaid by the lower part of the parotid salivary gland, the anterior border of the sternocleidomastoid muscle, the accessory nerve, and lies at the level of the lower border of the nasal orifices.

On a skull, identify the **supramastoid crest**. It is a blunt ridge which begins immediately above the external acoustic meatus and curves postero-superiorly. It is continuous superiorly with the **temporal line** which curves forwards making the upper limit of the temporal region. This line can only be felt distinctly at the lateral end of the eyebrow. The **parietal** and **frontal tubera** are the most convex parts of the corresponding bones.

Face

External Nose. The term 'nose' also includes the paired cavities which extend posteriorly from the nostrils to open into the pharynx. The mobile, anterior part of the nose consists of skin and cartilage; the rigid upper part is formed by the two **nasal bones**—the bridge of the nose—and by the two **frontal processes of the maxillae** [FIG. 5]. The skin is adherent to the cartilages, but is mobile over the bones. The part of the cavity of the nose immediately above each nostril is the **vestibule of the nose**. The vestibule is lined by hairy skin, and its lateral wall is slightly expanded to form the **ala** of the nose.

Lips, Cheeks, and Teeth

The lips and cheeks are composed chiefly of muscle and fat covered with skin and lined with mucous membrane. The space that separates the lips and cheeks from the teeth and gums is the **vestibule of the mouth**. A full set of adult **teeth** consists of 32—8 in each half jaw. From before backwards these are—2 incisors, 1 canine, 2 premolars, and 3 molars. There are 20 milk teeth, *i.e.*, 5 in each half jaw—2 incisors, 1 canine, and 2 'milk' molars. The **oral fissure**, between the lips, is opposite the biting edges of the upper teeth, the corner or angle of the mouth is opposite the first premolar tooth. The **philtrum** is the median groove on the external surface of the upper lip. The deep surface of each lip is attached to the gum by a median fold of mucous membrane, the **frenulum of the lip**.

Mandible

Identify the horizontal **body** of the mandible below the lower lip and the cheeks. Follow the lower border of the mandible backwards to its **angle**. This is the postero-inferior part of the **ramus** of the mandible—a wide, flat plate of bone which extends superiorly from the posterior part of the body and

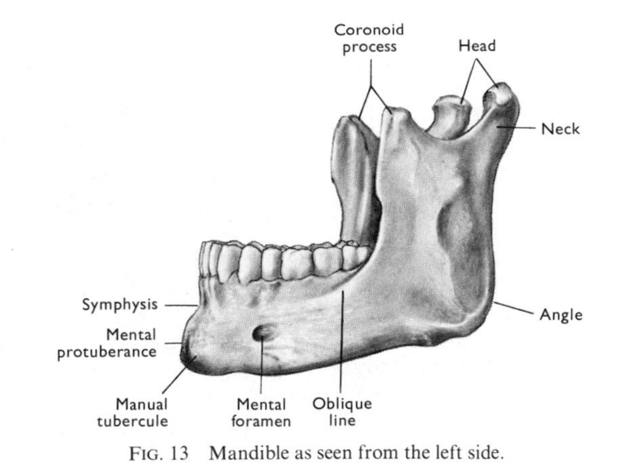

FIG. 13 Mandible as seen from the left side.

ends above in the **condylar** and **coronoid processes**. The ramus of the mandible is covered laterally by the masseter muscle so that only its posterior border is easily felt. The condylar process projects upwards from the posterior margin of the ramus and forms the **head** and **neck** of the mandible. The neck lies immediately anterior to the lobule of the auricle, the head is anterior to the tragus. Place your finger tip in front of your own tragus, and open your mouth; the head of the mandible glides downwards and forwards, leaving a fossa into which the finger slips. Note that the mouth cannot be closed while the finger remains in this fossa. Find the **mental foramen** about 4 cm from the median fusion of the two halves of the mandible (**symphysis menti**) and half way between the edge of the gum and the lower border of the mandible. It is felt in the living as a slight depression.

Zygomatic Arch

This bony bridge spans the interval between the ear and the eye. The zygomatic process of the temporal bone forms its narrow posterior part. It begins at the **tubercle** immediately anterior to the head of the mandible when the mouth is shut but above it when the mouth is open. The zygomatic bone forms the broad, anterior part of the arch in the prominent part of the cheek, lateral to and below the orbit.

Orbit

The bony structure of the orbit has been described already. Palpate the orbital margins in yourself and find (1) the **supra-orbital notch** on the highest point of the superior margin about 2·5 cm from the midline; (2) the **frontozygomatic suture** which is marked by a notch on the posterior surface [FIG. 5].

Eyebrow

This hairy skin lies above the supraorbital margin. Over its medial end is a curved ridge of bone, the superciliary arch. This is well formed only in males, and is separated from its fellow by the smooth, median **glabella**.

Eye

The white of the eye is the **sclera**, the clear front is the **cornea**. Through the cornea can be seen a dark, circular aperture, the **pupil**, surrounded by the coloured **iris**. The size of the pupil varies inversely with the degree of contraction of the iris on exposure to increasing intensity of illumination. The visible part of the sclera is covered with a moist, transparent membrane, the **conjunctiva**. This passes from the sclera on to the deep surfaces of the eyelids and is continuous with the skin at their margins. The reflexion of the conjunctiva on to the eyelids is the **fornix of the conjunctiva**, and the entire conjunctiva forms the **conjunctival sac**. This opens anteriorly between the eyelids through the palpebral fissure. It also forms the **anterior epithelium of the cornea**.

Eyelids

These folds or palpebrae protect the front of the eye and moisten it by spreading lacrimal fluid with each blink. The upper lid is larger and more mobile than the lower, and the upper conjunctival fornix is much the deeper. When the eye is closed, the **palpebral fissure** is nearly horizontal and lies opposite the lower margin of the cornea; when open, the margins of the eyelids overlap the cornea slightly, more especially the upper lid.

In the medial angle of the eye is a small, triangular space, the **lacus lacrimalis**, with a reddish elevation, the **lacrimal caruncle**, near its centre. This carries a few fine hairs which collect any debris in the conjunctival sac. Just lateral to this is a small, vertical fold of conjunctiva, the **plica semilunaris** [FIG. 14]. This corresponds to the nictitating membrane of some animals, but is not mobile in Man.

The lower eyelid is easily everted by pulling down the skin below it, and the lower fornix exposed by turning the eye upwards. The upper lid is difficult to evert because of the rigid **tarsal plate** buried in it, and once everted tends to remain so. Even when this is done the deep superior fornix is not exposed.

On the deep surface of the lids note a number of yellowish, parallel streaks produced by the **tarsal glands** [FIG. 14] the ducts of which open near the posterior margin of the flat, free edges of the lids. The eyelashes (**cilia**) project from the anterior margins. The free margins of the lids are rounded at the caruncle. Where they become flat, there is a small **lacrimal papilla** on each lid surmounted by a tiny aperture, the **lacrimal punctum**. This is the open end of a slender tube, the **lacrimal canaliculus**, which carries the lacrimal fluid (tears) to the lacrimal sac, whence it is conducted to the nose through the nasolacrimal duct. Note that the puncta face posteriorly into the conjunctival sac, and that the eyelids move medially when the eye is forcibly closed. This moves the lacrimal fluid towards the medial angle of the eye and the openings of the puncta.

Press a finger tip on the skin between the nose and the medial angle of the eye. A rounded, horizontal

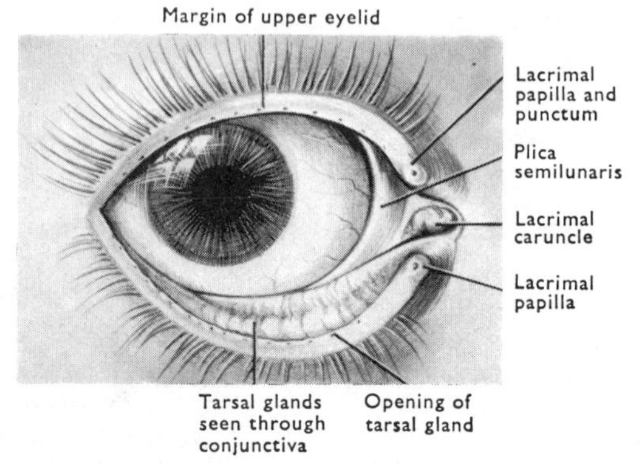

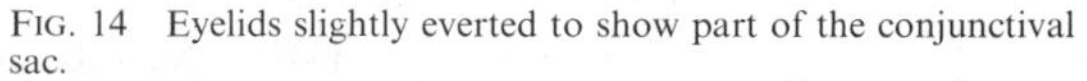

FIG. 14 Eyelids slightly everted to show part of the conjunctival sac.

cord will be felt. This is the **medial palpebral ligament** which connects both eyelids and their muscle (orbicularis oculi) to the medial margin of the orbit. If the eyelids are pulled gently in a lateral direction, the ligament is more easily felt and may show as a small skin ridge.

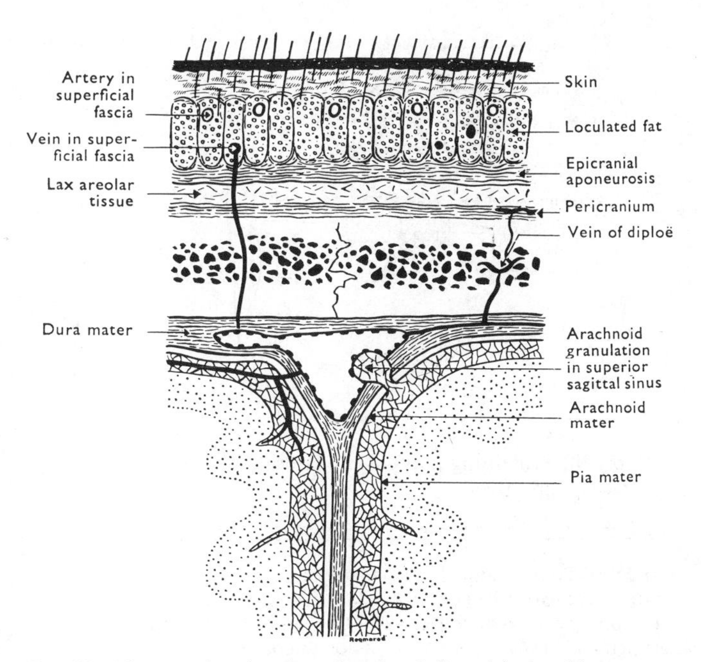

FIG. 15 Diagram of section through scalp, skull, and brain. Note venous connexions through the skull.

THE SCALP AND SUPERFICIAL PARTS OF THE TEMPLE

The scalp covers the vault of the skull and extends between the right and left temporal lines, and the eyebrows and the superior nuchal lines. It consists of skin and superficial fascia adherent to a flat, aponeurotic sheet, the **epicranial aponeurosis**. This is the tendon uniting the frontal and occipital bellies of the occipitofrontalis muscle. The skin is fixed to the aponeurosis by dense strands of fibrous tissue which traverse the subcutaneous tissue and split it into a number of separate pockets filled with fat. The blood vessels and nerves of the scalp lie in this superficial layer. Deep to the aponeurosis is a relatively avascular layer of loose areolar tissue which allows the scalp to slide freely on the periosteum **pericranium** covering the skull.

The **temple** is the area between the temporal line and the zygomatic arch. The skull is thin here and is covered by the temporalis muscle, the temporal fascia, and a thin extension of the epicranial aponeurosis from which the extrinsic auricular muscles arise.

DISSECTION. Place a block under the back of the head to raise it to a convenient angle. Make a median incision into the skin of the scalp from the root of the nose to the external occipital protuberance, and a coronal incision from the middle of the first cut to the root of each auricle. Continue this behind the auricle to the mastoid process, and in front of it to the root of the zygomatic arch. Avoid cutting deeper than the skin to preserve the vessels, nerves, and muscles in the subcutaneous tissue. Reflect the skin flaps superficial to these structures. Make use of FIGURES 21, 24, 72, and 74 to identify the positions of the main structures in the scalp so that they are not damaged.

Expose the upper part of orbicularis oculi, and follow the frontal belly of occipitofrontalis from below upwards [FIG. 17]. Find the branches of the supratrochlear and supra-orbital vessels and nerves—the supratrochlear about a finger's breadth from the median line, the supra-orbital a further finger's breadth laterally as it ascends from the supra-orbital notch. Expose the anterior part of the epicranial aponeurosis and note its extension downwards into the temple.

Find two or more **temporal branches of the facial nerve** which cross the zygomatic arch 2 cm or more in front of the auricle [FIG. 74]. Trace them upwards to the deep surface of orbicularis oculi. As the anterior part of the temporal fascia is exposed, look for the slender zygomaticotemporal nerve which pierces it a little behind the frontal process of the zygomatic bone.

Find the **superficial temporal artery** [FIG. 72] and veins and the **auriculotemporal nerve**. These cross the root of the zygomatic arch, immediately anterior to the auricle, with the small branch of the facial nerve to the superior auricular muscles. Trace these structures into the scalp, uncovering this part of the temporal fascia. The auriculotemporal nerve may be very slender and difficult to find.

Inferior and posterior to the auricle, find the great auricular and lesser occipital nerves [FIG. 21] and the posterior auricular vessels and nerve which lie immediately behind the root of the auricle. Trace the branches of these nerves.

AURICLE

The auricle consists of a thin plate of yellow, elastic, fibrocartilage covered with skin, except in the lobule and in the part between the tragus and the helix which contain no fibrocartilage [FIG. 16].

The cartilage is continuous with that in the lateral part of the external acoustic meatus which fuses to the lateral end of the bony part of the meatus. This **meatal cartilage** is incomplete above and in front, and its wall is completed by dense fibrous tissue continuous with that between the tragus and the beginning of the helix.

Vessels and Nerves. The muscles are supplied by the facial nerve which is motor to the muscles of facial expression. The great auricular nerve is sensory to both sides of the lower part, particularly the medial surface. The auriculotemporal and lesser occipital

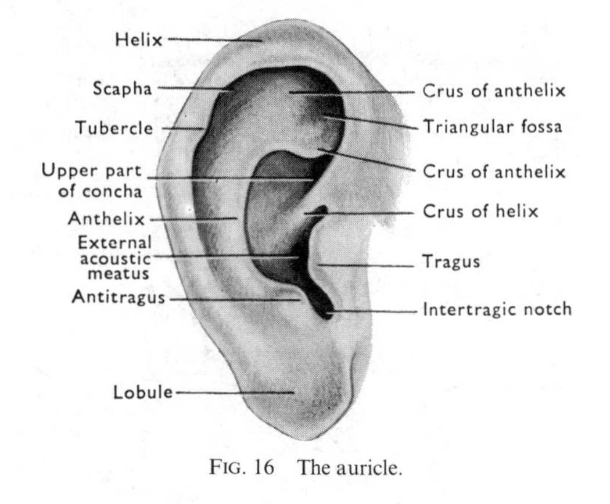

FIG. 16 The auricle.

supply the remaining parts of the lateral and medial surfaces respectively.

DISSECTION. Look for small terminal twigs of the **third occipital nerve** in the fascia over the external occipital protuberance [FIG. 24]. Cut through the dense superficial fascia over the superior nuchal line 2·5 cm lateral to the external occipital protuberance. The occipital vessels and **greater occipital nerve** pierce the deep fascia here. Find and trace them superiorly towards the vertex. Lateral to this, find the occipital belly of occipitofrontalis, and expose the posterior part of the epicranial aponeurosis. Make a small incision through the aponeurosis near the vertex. Introduce a blunt seeker through it into the loose areolar tissue beneath the aponeurosis, and expose the extent of this tissue by moving the seeker in all directions. Note that the aponeurosis is adherent to the periosteum near the temporal and nuchal lines.

Occipitofrontalis Muscle

The occipital bellies are shorter and narrower than the frontal bellies, and are widely separated by the aponeurosis. Each arises from the lateral part of the corresponding superior nuchal line.

Each frontal belly lies in the forehead and adjoining part of the scalp. It has no attachment to bone, but runs between the skin of the forehead and the epicranial aponeurosis. Because of these attachments, it both raises the eyebrows and wrinkles the forehead. The medial parts of the bellies are fused and attached to the skin of the nose which they wrinkle. **Nerve supply:** the facial nerve.

Epicranial Aponeurosis. Most of the features of this sheet have been described above. It is partly attached to the temporal lines and firmly attached to the superior nuchal lines. Elsewhere it slides freely on the pericranium because of the loose connective tissue deep to it. Traction injuries tend to tear this loose layer and the emissary veins which pass through it from the skull to the scalp. This leads to bleeding deep to the aponeurosis, and the blood spreads over a wide area raising the scalp from the skull. Alternatively, the scalp may be torn from the skull.

NERVES OF THE SCALP AND SUPERFICIAL TEMPORAL REGION

The superficial structures of this region and of the face receive a *motor* innervation from the **facial nerve,** and a *sensory* innervation from the **trigeminal nerve** and the second and third cervical spinal nerves. In addition there is a *sympathetic* innervation to blood vessels, sweat glands, and arrectores pilorum which reaches the region principally through plexuses on the arteries. Anterior to a line extending from the ear to the vertex, the sensory supply is from the trigeminal nerve, except for the skin over the postero-inferior part of the jaw and the lower part of the auricle. This, together with the antero-inferior part of the area behind the line is supplied by the **great auricular** and **lesser occipital nerves** (ventral rami of C. 2 and 3 [FIGS. 21,73]). The remaining area behind the line is supplied by the corresponding dorsal rami—the large **greater occipital** (C. 2) and the slender **third occipital** (C. 3) **nerves.** The greater occipital enters the scalp with the occipital artery by piercing trapezius and the deep fascia 2·5 cm from the external occipital protuberance. The third occipital nerve pierces trapezius 2–3 cm inferior to this.

Trigeminal Nerve (Sensory)

This is the fifth cranial nerve. It derives its name from the fact that it divides into three large nerves (ophthalmic, maxillary, and mandibular) which emerge separately from the cranial cavity. Each of the three divisions supplies sensory branches to the skin of the face and the anterior half of the head [FIG. 73].

Ophthalmic Nerve. The **supratrochlear nerve** emerges at the supra-orbital margin a finger's breadth from the median plane. It supplies the paramedian part of the forehead and the medial part of the upper eyelid. The **supra-orbital nerve** emerges through the supra-orbital notch, supplies the upper eyelid, and then divides into lateral and medial branches. Each branch sends a twig through the bone to the mucous lining of the **frontal sinus** (the cavity in the frontal bone above the nose and orbit) and together they supply the skin of the forehead and of the upper anterior part of the scalp as far as the vertex. For other cutaneous branches [FIG. 74] see later.

Maxillary Nerve. The slender **zygomaticotemporal nerve** arises from the zygomatic nerve in the orbit. It pierces the zygomatic bone and temporal fascia to supply skin of the anterior part of the temple. For other cutaneous branches, see later.

Mandibular Nerve. The **auriculotemporal nerve** emerges from the upper end of the parotid gland, close to the auricle, at the root at the zygomatic arch. It supplies the upper part of the auricle and external acoustic meatus, and skin of the side of the head.

Facial Nerve (Motor)

This seventh cranial nerve supplies the muscles of the face, scalp, and auricle. Its branches communicate freely with each other and with the branches of the trigeminal nerve through which sensory fibres are transmitted from the facial muscles. The communications add greatly to the difficulty of dissecting the nerves of the face.

Temporal branches emerge from the upper part of the parotid gland, cross the zygomatic arch obliquely, and pass to supply the frontal belly of occipitofrontalis, the upper part of orbicularis oculi, and the anterior and superior auricular muscles.

The **posterior auricular nerve** leaves the facial nerve as it emerges from the stylomastoid foramen. It curves posterosuperiorly below the root of the auricle, and runs above the superior nuchal line supplying the occipital belly of occipitofrontalis and the posterior and superior auricular muscles. For other branches of the facial nerve see FIGURE 74 and below.

ARTERIES OF THE SCALP AND SUPERFICIAL TEMPORAL REGION

This region is supplied by branches of the **external carotid artery**, except for the forehead which receives the supra-orbital and supratrochlear arteries. These run with the corresponding nerves, and originate in the orbit from the ophthalmic branch of the **internal carotid artery**.

Superficial Temporal Artery. This large, terminal branch of the external carotid begins behind the neck of the mandible in, or deep to, the parotid gland. It runs with the auriculotemporal nerve and divides into anterior and posterior branches which run towards the frontal and parietal eminences respectively. The anterior branch is frequently seen through the skin in elderly individuals, and is often very tortuous.

Other Branches [FIG. 72]. The **transverse facial** runs forwards on the masseter muscle, below the zygomatic arch. The **middle temporal** crosses the root of the zygomatic arch, pierces the temporal fascia, and runs vertically upwards, grooving the skull above the external acoustic meatus. The **zygomatico-orbital** runs anteriorly above the zygomatic arch between the two layers of temporal fascia. It anastomoses with branches of the ophthalmic artery.

Posterior Auricular Artery. This small branch of the external carotid artery arises deep to the parotid gland. It curves posterosuperiorly below and behind the root of the auricle, with the posterior auricular nerve.

Occipital Artery. This large branch arises from the external carotid deep to the angle of the mandible. It runs posterosuperiorly deep to the sternocleidomastoid and digastric muscles, and joins the greater occipital nerve as it pierces trapezius [FIG. 24]. It supplies muscles of the neck and the back of the head.

These arteries of the scalp anastomose freely with each other and with those of the opposite side. Because of this, wounds of the scalp bleed profusely, but heal rapidly. Also, if a large piece of scalp is torn downwards from the calvaria, it will survive and heal satisfactorily provided a part of the peripheral attachment containing an artery is intact.

VEINS OF THE SCALP AND SUPERFICIAL TEMPORAL REGION

Like the arteries, the veins anastomose freely and their main tributaries accompany the arteries of the scalp, but their proximal parts drain by different routes.

The **supratrochlear** and **supra-orbital veins** unite at the medial angle of the eye to form the **facial vein**. They communicate with the orbital veins.

The **superficial temporal vein** joins the **middle temporal vein** at the root of the zygomatic arch to form the retromandibular vein.

The **retromandibular vein** descends through the parotid gland and is joined by the transverse facial and maxillary veins. Inferiorly it divides. The **anterior branch** joins the facial vein which ends in the internal jugular vein. The **posterior branch** unites with the **posterior auricular vein** to form the external jugular vein on the surface of the sternocleidomastoid muscle.

The **occipital veins** run with artery in the scalp, but leave it to join the suboccipital plexus deep to the semispinalis capitis muscle (*q.v.*) at the back of the neck.

Emissary veins pierce the skull and connect this system of veins with the venous sinuses inside it. Usually one passes through each parietal foramen to the superior sagittal sinus, another through each mastoid foramen to the corresponding sigmoid sinus [FIG. 11]. These and other emissary veins, and the communications with the veins in the orbits, form routes along which infection may spread into the skull from without.

LYMPH VESSELS OF THE SCALP AND TEMPLE

These vessels cannot be demonstrated by dissection. Those from the area in front of the ear end in small **parotid lymph nodes** buried in the surface of that gland. Those from the region behind the ear terminate in lymph nodes on the upper end of trapezius (**occipital nodes**) and sternocleidomastoid (**retro-auricular nodes**).

THE SUPERFICIAL DISSECTION OF THE FACE

In this dissection the eyelids and cheeks should be stretched by packing the conjunctival sacs and the vestibule of the mouth with cloth or cotton wool

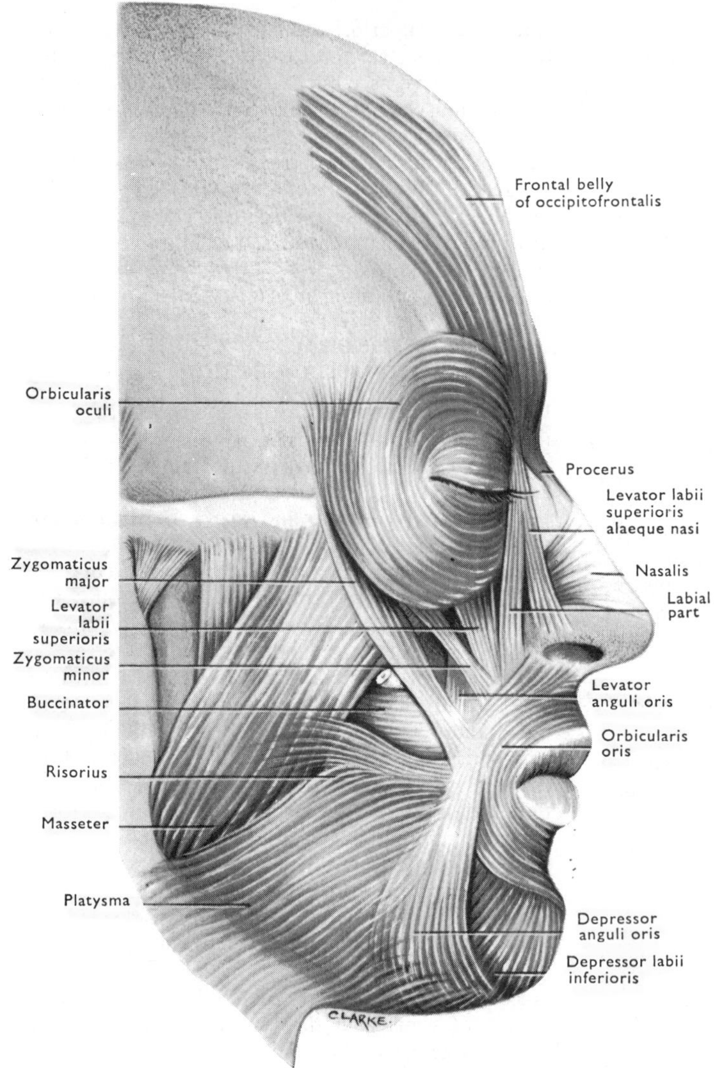

FIG. 17 The facial muscles and masseter.

soaked in preservative. This can also be done to the pharynx to prevent it drying before it is dissected. Once the skin is removed from the face, it is essential to keep it moist and to avoid drying.

The face extends from the hair margin to the chin, and from one auricle to the other. Thus the forehead is common to the face and the scalp.

DISSECTION. When the skin of the face is reflected, the attachments of the facial muscles to it are inevitably damaged. This can be minimized by keeping the knife as close to the skin as possible.

Make a median incision from the root of the nose to the point of the chin, and a horizontal incision from the angle of the mouth to the posterior border of the mandible. Reflect the lower flap downwards to the lower border of the mandible, and the upper flap backwards to the auricle.

Expose the major facial muscles [FIG. 17] taking care not to cut through them lest the major branches of the nerves and vessels are damaged.

Orbicularis Oculi. Pull the eyelids laterally and identify the medial palpebral ligament, then expose the outer part of the muscle, subsequently following its inner part to the margins of the eyelids. The small palpebral branch of the lacrimal nerve may be found entering the lateral part of the upper eyelid through the muscle.

Orbicularis Oris. This is more difficult to expose because of the large number of other facial muscles which fuse with and help to form it [FIG. 17]. At the side of the nose, find the **levator labii superioris alaeque nasi** with the facial vein lying on its surface. Try to define two slender nerves which run downwards on the side of the nose; the infratrochlear in the upper half, and the external nasal in the lower half [FIG. 74]. Trace the **facial vein** downwards till it passes deep to zygomaticus major. Expose that muscle, and then levator labii superioris, following it upwards to its origin deep to the orbicularis oculi muscle.

At the lower border of the mandible, expose the broad, thin sheet of muscle (**platysma**) which ascends over the mandible from the neck. Note that its posterior fibres curve forwards towards the angle of the mouth to form part of the **risorius muscle**. Find depressor anguli oris and depressor labii inferioris. The **buccinator muscle** lies in a deeper plane immediately external to the mucous membrane of the cheek. It is continuous with the lateral part of orbicularis oris, and will be dissected later.

FACIAL MUSCLES

These muscles, including buccinator, receive their motor supply from the facial nerve. Sensory fibres from these muscles and from the skin which overlies them, reach the brain through the trigeminal nerve [FIG. 74].

The facial muscles are known collectively as the 'muscles of facial expression'; the actions of many of them are implied by their names; the actions of others may be inferred from their positions. Because they are attached to the skin, they tend to pull open wounds of the face, which then require to be stitched. Though all these muscles are important in facial expression, the student should not burden his memory with their attachments, but should appreciate the importance of the orbiculares muscles and buccinator because of the serious consequences of their paralysis.

Orbicularis Oculi

Orbital Part. The coarse fibres of this outer part arise from the medial palpebral ligament and the

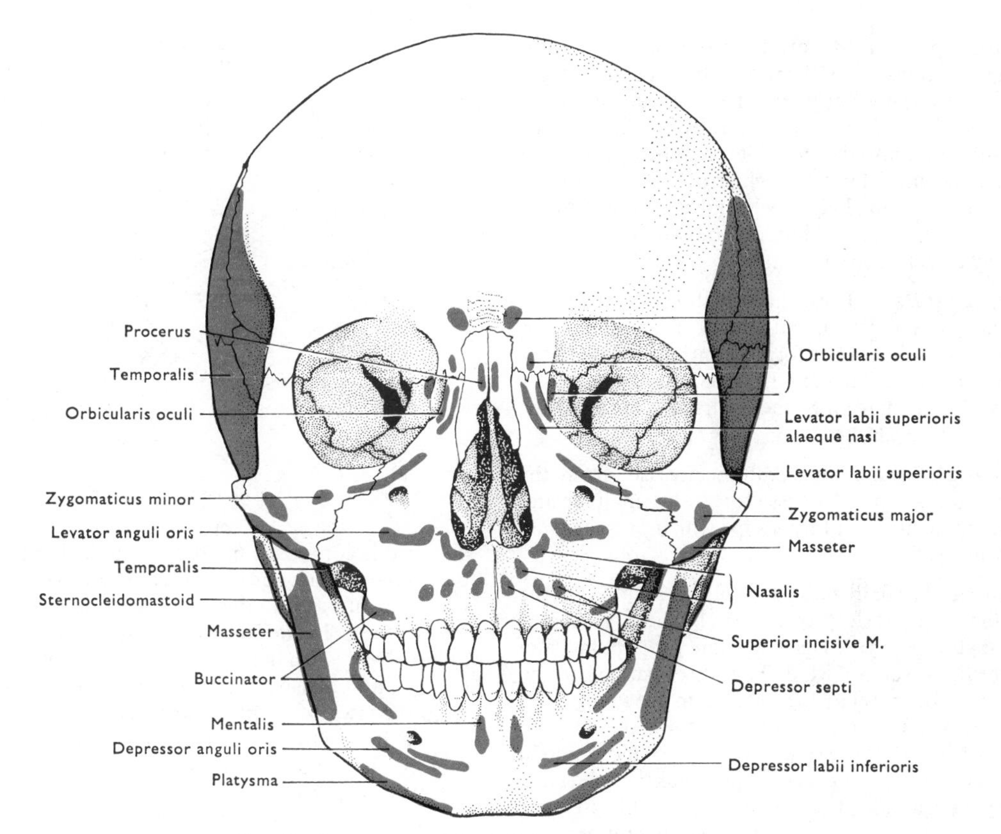

FIG. 18 Anterior view of the skull showing muscle attachments.

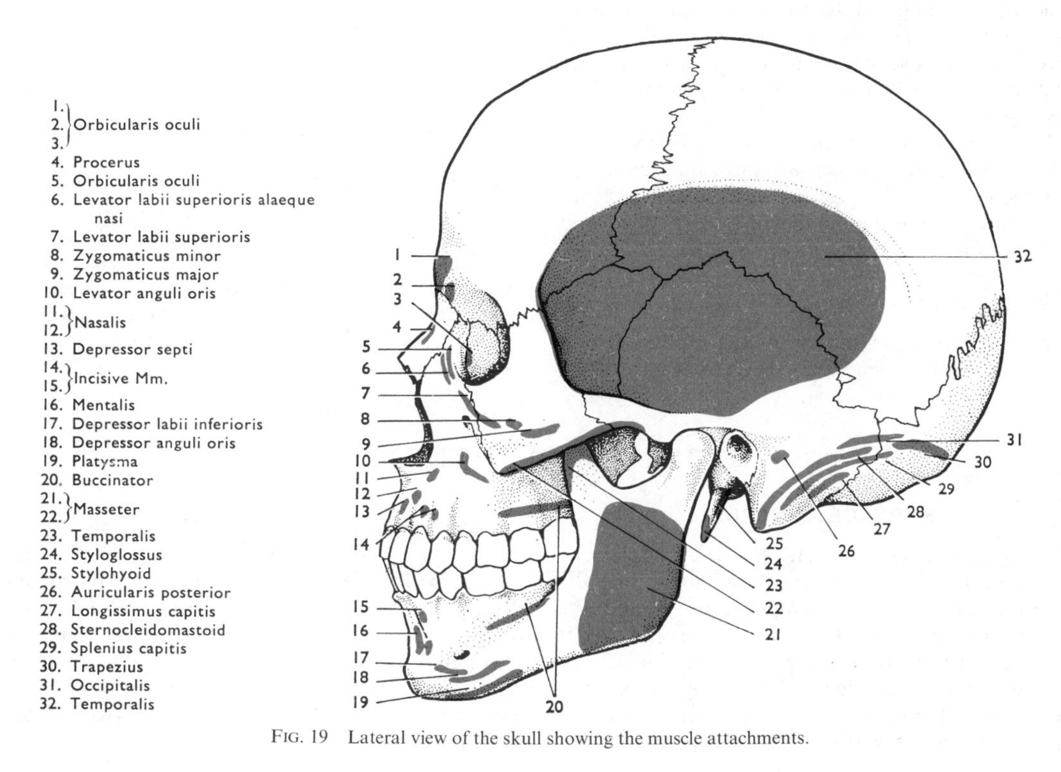

FIG. 19 Lateral view of the skull showing the muscle attachments.

adjacent part of the orbital margin. They form complete loops on and around the orbital margin which sweep superiorly into the forehead (mingling with the fibres of frontalis), laterally into the temple, and inferiorly into the cheek before returning to their point of origin. A few fibres which arise from the bone superior to the medial palpebral ligament end in the skin of the eyebrow, but the remainder are only loosely attached to the skin.

Palpebral Part. These thinner fibres arise from the medial palpebral ligament and form similar loops within the eyelids. They form a continuous layer with the orbital part, and a small, partially isolated, ciliary bundle in the margin of the eyelid, posterior to the roots of the eyelashes.

Lacrimal Part. This small sheet arises from the posterior margin of the fossa for the lacrimal sac and from the sac itself. It forms slips which pass laterally into the eyelids.

Actions. The palpebral part, acting alone, closes the eye lightly as in sleep or blinking. The orbital part screws up the eye to give partial protection from bright light, sun, or wind. The fibres passing to the eyebrows draw them together as in frowning. The orbital and palpebral parts contract together to close the eye forcibly as in protecting it from a blow, and in strong expiratory efforts such as coughing, sneezing, or crying in a child. The last action probably prevents over-distention of the orbital veins by compressing the orbital contents. When the muscle contracts it draws the skin and eyelids medially towards its bony attachments and promotes the flow of lacrimal fluid towards the lacrimal canaliculi. The lacrimal part not only draws the eyelids medially, but probably also dilates the lacrimal sac and promotes the flow of fluid through it.

Paralysis of this muscle prevents the eye from being closed, and permits the exposed cornea to dry and become sore. The lower eyelid falls away from the eyeball forming a pond in which tears collect and spill over on to the face.

Orbicularis Oris

This is the sphincter muscle of the mouth. It is a complex muscle which forms the greater part of the lips, and is composed mainly of interlacing fibres of the muscles which converge on the mouth—a feature which confers on the lips a wide variety of movements. In addition, intrinsic bundles pass obliquely between the skin and mucous membrane of the lips, and incisive slips pass laterally into the lips from the jaws adjacent to the incisor teeth.

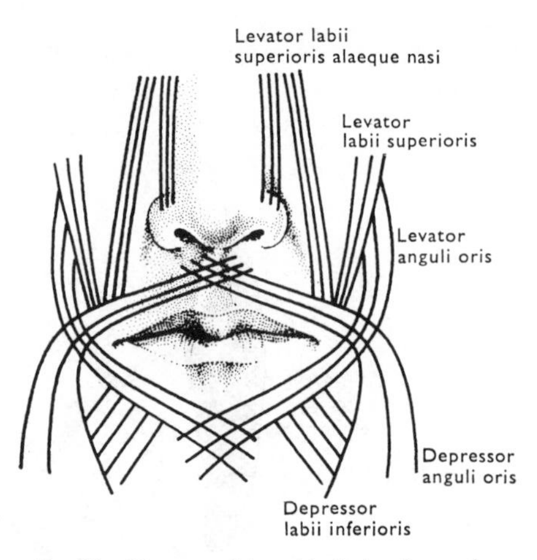

FIG. 20 Diagram of the orbicularis oris muscle.

The fibres of the various muscles converging on the mouth mingle and sweep in curves through the lips. The **buccinator** (*q.v.*) passes horizontally forwards in the cheek. Its middle fibres interlace near the corner of the mouth and sweep into the lips forming a marginal bundle. The upper and lower fibres enter the corresponding lip and interdigitate with fibres of the opposite muscle in the median plane.

Paralysis of one half of orbicularis oris prevents proper closure and movements of the lips on that side. Hence speech is slurred, and food and fluids collect in the vestibule of the mouth and escape between the lips. The lips are pulled towards the normal side by the unbalanced action of the muscles on that side, and the cheek and lips are blown out and allow air to escape on the paralysed side when the patient blows against resistance. These features, together with the inability to wrinkle the forehead on the same side (paralysis of occipitofrontalis), are diagnostic of paralysis of the muscles of facial expression as a result of injury to the facial nerve.

THE SIDE OF THE NECK

SURFACE ANATOMY

The side of the neck is bounded below by the clavicle, above by the lower border of the mandible, the mastoid process of the temporal bone, and the superior nuchal line of the occipital bone [FIG. 4]. It extends posteriorly to the anterior border of trapezius, and is divided obliquely by the sterno-cleidomastoid muscle into anterior and posterior triangles of the neck [FIGS. 21, 31]. Anteriorly, the neck extends only to the chin, but the cervical vertebrae extend to a much higher level, the upper 3–4 **cervical vertebrae** being overlapped anteriorly by the facial skeleton. Thus the first cervical vertebra lies at the level of the tip of the mastoid process—the level of the hard palate. The second lies at the level of the oral cavity, while the third lies in a plane posterior to the junction of the two halves of the mandible.

The **sternocleidomastoid muscle** extends from the manubrium of the sternum and the medial third of the clavicle to the skull behind the ear, and raises a low ridge diagonally across the side of the neck. This is made prominent when the face is turned towards the opposite side. Then the **external jugular vein** may be seen crossing the surface of sternocleidomastoid almost vertically from a point behind the angle of the mandible to the clavicle [FIG. 21].

The lesser supraclavicular fossa is a shallow depression between the sternal and clavicular parts of sternocleidomastoid. It lies above the medial part of the clavicle and is opposite the internal jugular vein. The greater supraclavicular fossa is a larger depression behind the intermediate third of the clavicle, between the lower parts of trapezius and sternocleidomastoid. It overlies the cervical part of the brachial plexus and the third part of the subclavian artery.

GENERAL ARRANGEMENT OF NECK

[FIG. 67]

The major structures of the neck are surrounded by the **investing layer of deep fascia** which encloses the trapezius and sternocleidomastoid muscles and forms the superficial covering of the posterior triangle. Internal to this layer are two compartments. (1) The larger, *posterior compartment* consists of the vertebral column and the muscles which immediately surround it. This is enclosed in a sleeve of fascia which passes anterior to the vertebral bodies, and is known as the **prevertebral fascia**. Laterally this fascia covers the muscles which form the deep surface of the posterior triangle, and then the erector spinae muscles. (2) The smaller, *anterior compartment* is composed of the pharynx, larynx, oesophagus, trachea and their associated muscles, on each side of which is the vertical *neurovascular bundle of the neck* (the carotid sheath) containing the carotid arteries, the internal jugular vein, and the vagus nerve. In the neck, this neurovascular bundle is primarily concerned with supplying the gut tube (airway and foodway) while the cervical nerves supply the skin and muscles of the neck, including an anterior, paramedian strip of muscles (strap or infrahyoid muscles) which are attached to and move the hyoid bone, the larynx, and the sternum. They are continuous above with the muscles of the tongue supplied by the hypoglossal nerve. The vascular supply to this territory is principally from the subclavian artery [FIG. 56].

The ventral rami of the **cervical nerves** emerge from the vertebrae between the muscles attached to the anterior and posterior tubercles of the transverse processes: *i.e.*, anterior to scalenus medius. Since these nerves arise within the prevertebral fascia, they either carry a sheath of this fascia outwards with them (*e.g.*, fascia of cervico-axillary canal [FIG. 23]) or remain within the fascia (phrenic nerve).

THE POSTERIOR TRIANGLE

DISSECTION. Reflect the skin of the posterior triangle carefully to avoid damage to the **supraclavicular nerves**, which lie deep to platysma in the lower part of the triangle, and to the accessory nerve which is in the investing fascia of the upper part of the triangle. Make an incision through the skin along the middle of the sternocleidomastoid muscle from the mastoid process to the sternal end of the clavicle. Do not cut into the superficial fascia or the great auricular nerve, the transverse nerve of the neck, and the external jugular vein will be divided. Extend the incision along the clavicle to its acromial end, and reflect the flap so formed to the anterior border of trapezius.

Cut through platysma along the line of the clavicle and turn it upwards and forwards, superficial to the supraclavicular nerves and the **external jugular vein** [FIG. 21]. Find this vein and trace it upwards till it is joined by the posterior auricular vein, and downwards till it pierces the deep fascia.

Find the three cutaneous nerves which pierce the deep fascia at the middle of the posterior border of sternocleidomastoid. (1) The **lesser occipital nerve** runs upwards along the posterior border of sternocleidomastoid. (2) The **great auricular nerve** crosses sternocleidomastoid obliquely towards the auricle. (3) The **transverse nerve of the neck** passes horizontally forwards across sternocleidomastoid.

Find the medial, intermediate, and lateral supraclavicular nerves either individually or by finding one branch and tracing it back to the trunk.

DEEP FASCIA OF THE POSTERIOR TRIANGLE

This fascia extends from the intermediate third of the clavicle to the superior nuchal line of the occipital bone, and is pierced by: (1) supraclavicular nerves; (2) the external jugular vein; (3) small cutaneous arteries. It is a thin sheet which splits inferiorly, the superficial layer fusing with the clavicle, the deep layer splitting to enclose the posterior belly of the omohyoid muscle [FIGS. 22, 67] and passing down behind the clavicle to its lower surface, holds that muscle in position. The gap between the two layers extends medially deep to sternocleidomastoid, and contains the terminal parts of the external jugular and transverse cervical veins, and the suprascapular vessels behind the clavicle.

The **accessory nerve** enters the triangle close to the lesser occipital nerve [FIG. 21], and runs postero-inferiorly across the triangle embedded in the investing fascia. At the sternocleidomastoid, the accessory nerve lies among lymph nodes of the posterior triangle.

The **prevertebral fascia** covers the muscles of the back of the neck, levator scapulae, and the scalene muscles, and passes in front of the cervical vertebrae and prevertebral muscles, behind the pharynx.

DISSECTION. Follow the supraclavicular nerves to their common trunk. Cut through the investing fascia

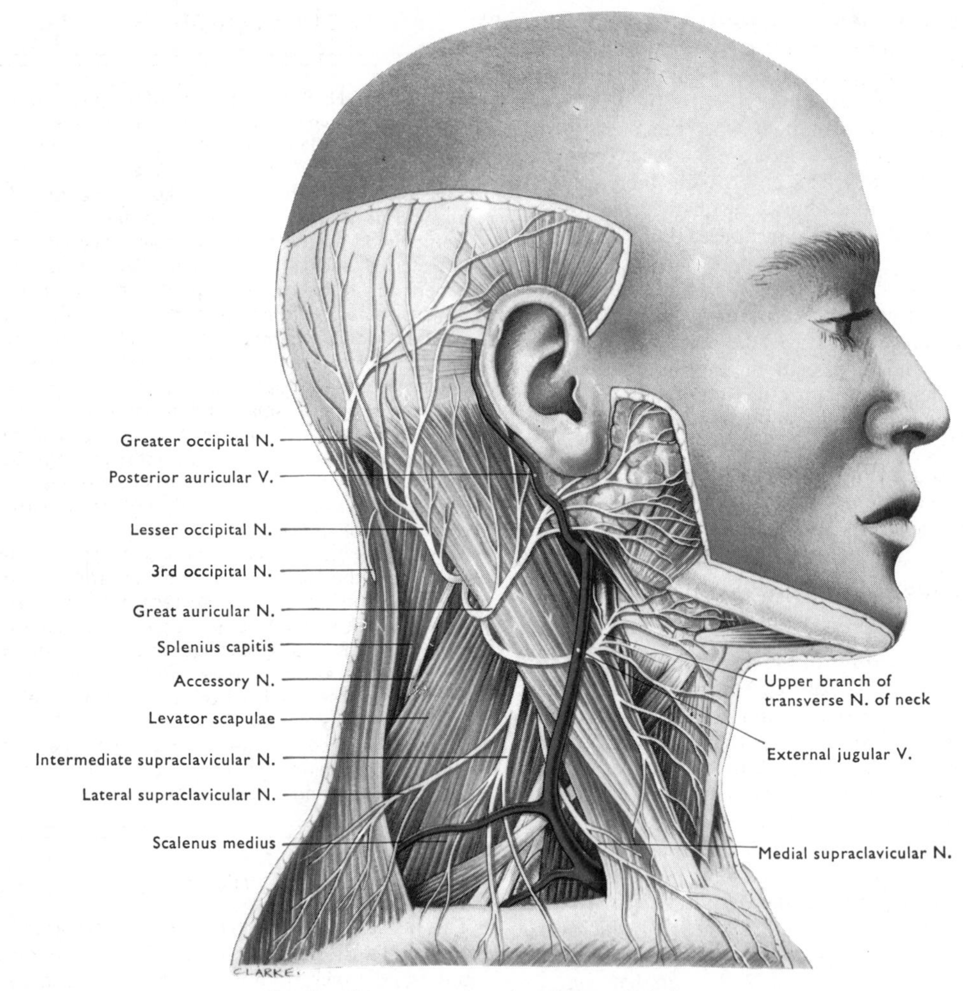

FIG. 21 The cutaneous branches of the cervical plexus.

above the clavicle and along the posterior border of sternocleidomastoid to expose its deeper layer. Find the external jugular, transverse cervical, and suprascapular veins in the space between them, and trace the veins to their junction.

Find the **nerve to subclavius** on the lateral side of the external jugular vein, and trace it in both directions. It may give an accessory branch to the phrenic nerve.

Find the entry of the anterior jugular vein into the external jugular vein, and the **suprascapular artery** deep to the clavicle. Trace the artery and its vein deep to trapezius by pushing the fat out of the way. They run to the scapular notch beside the attachment of the inferior belly of omohyoid.

Follow the cutaneous nerves emerging at the middle of the posterior border of sternocleidomastoid. Take care not to damage the accessory nerve where the lesser occipital nerve hooks round it.

CUTANEOUS BRANCHES OF THE CERVICAL PLEXUS

The cervical plexus is formed by the ventral rami of the upper four cervical nerves. It lies deep to the internal jugular vein and the sternocleidomastoid muscle in the upper part of the neck. Its cutaneous branches emerge at the middle of the posterior border of sternocleidomastoid.

Lesser Occipital Nerve (C. 2). It ascends on the posterior border of sternocleidomastoid, giving small branches to the skin of the neck. It pierces the deep fascia and sends branches to the upper part of the cranial surface of the auricle and the skin over the mastoid process. It may communicate with the greater occipital nerve.

Great Auricular Nerve (C. 2, 3). It turns round the posterior border of sternocleidomastoid, pierces the deep fascia, and runs towards the parotid gland behind and parallel to the external jugular vein. The **posterior branch** supplies skin on both surfaces of the auricle and on the mastoid process. The **anterior branch** supplies skin over the angle of the mandible and the parotid gland, and communicates with the facial and auriculotemporal nerves in that gland.

Transverse Nerve of the Neck (C. 2, 3). The nerve passes forwards towards the anterior triangle of the neck on the superficial surface of sternocleidomastoid. Its upper and lower branches supply most of the skin on the side and front of the neck. The upper branch communicates with the cervical branch of the facial nerve.

Supraclavicular Nerves (C. 3, 4). The medial, intermediate, and lateral supraclavicular nerves arise as a single trunk. Diverging, they send small branches to the skin of the neck, and pierce the deep fascia a little above the clavicle. The nerves pass over the corresponding thirds of the clavicle and supply skin on the front of the chest, down to the level of the sternal angle, and over the upper half of the deltoid muscle.

EXTERNAL JUGULAR VEIN

This vein varies greatly in size, and is often conspicuous in the neck. It begins on the surface of sternocleidomastoid, behind the angle of the mandible and below the parotid gland by the junction of the posterior branch of the retromandibular vein with the posterior auricular vein. It passes vertically downwards in the superficial fascia, deep to platysma [FIGS. 21, 22] and pierces the investing fascia on the posterior triangle at the posterior border of sternocleidomastoid, 2–3 cm above the clavicle. In the posterior triangle it descends beside sternocleidomastoid, crosses the lower roots of the brachial plexus and the third part of the subclavian artery to enter the subclavian vein behind the clavicle. It receives the transverse cervical, suprascapular, and anterior jugular [p. 31] veins.

DISSECTION. Remove the fat and fascia from the posterior triangle, starting at the apex where the occipital artery crosses it. Find the accessory nerve [FIG. 22] at the posterior border of sternocleidomastoid, and follow it and the branches from the third and fourth cervical nerves which are beside it, to trapezius. Branches of the same nerves may be found entering levator scapulae.

Remove the fascia from the inferior belly of omohyoid, and turn the muscle forwards to expose the nerve entering its deep surface near sternocleidomastoid.

Expose the upper part of the brachial plexus and trace it backwards to its roots. Avoid damage to the nerves which arise from it. (1) The **suprascapular nerve** runs postero-inferiorly immediately above the plexus, under cover of omohyoid. (2) Slightly above (1) the slender **dorsal scapular nerve** (C.5) runs postero-inferiorly deep to trapezius. (3) The roots of the **long thoracic nerve** arise from the back of the roots of the plexus (C.5, 6, 7) and descend behind it. The upper two pierce the scalenus medius muscle behind the plexus and can be found, together with the third either by turning the upper three roots forwards, or by tracing the long thoracic nerve upwards if it has been dissected in the axilla.

Find the transverse cervical artery at the upper border of omohyoid, and follow it across the posterior triangle, and back towards its origin by removing the deeper layer of the investing fascia.

Follow the nerve to subclavius to its termination, and expose the subclavian vessels and the brachial plexus posterior to them.

BOUNDARIES AND CONTENTS OF THE POSTERIOR TRIANGLE

The boundaries are the adjacent margins of the sternocleidomastoid and trapezius muscles, and the intermediate third of the clavicle.

The investing fascia covering the triangle is pierced by the external jugular vein, the supraclavicular nerves, and lymph vessels passing from superficial structures to nodes in the triangle.

From above downwards splenius capitis, levator scapulae, and scalenus medius, covered by prevertebral fascia, form the deep limit of the triangle.

Accessory Nerve (Eleventh Cranial)

The part of this nerve in the posterior triangle consists of nerve fibres which arise in the cervical part of the spinal medulla. These fibres ascend beside the spinal medulla to enter the cranium through the foramen magnum (**spinal part**). Here they join nerve fibres which arise from the medulla oblongata (**cranial part**) and together escape from the skull through the jugular foramen with the vagus (tenth cranial) nerve. The spinal part immediately separates from the remainder, and passing postero-inferiorly across the transverse process of the atlas vertebra, supplies the sternocleidomasoid muscle and pierces its deeper part.

Leaving the posterior border of sternocleidomastoid a little above its middle, the nerve runs in the investing fascia covering the posterior triangle, parallel to levator scapulae, and passes deep to trapezius about 5 cm above the clavicle. It is the only motor supply to sternocleidomastoid and trapezius, the branches from the cervical plexus (C. 2, 3, 4) being sensory in nature. Thus destruction of the spinal part of the accessory nerve paralyses these muscles, causing the neck to be flexed to the opposite side and the face turned to the paralysed side by the unapposed action of the normal sternocleidomastoid muscle—a *wry neck*.

For other branches of the cervical plexus see page 72.

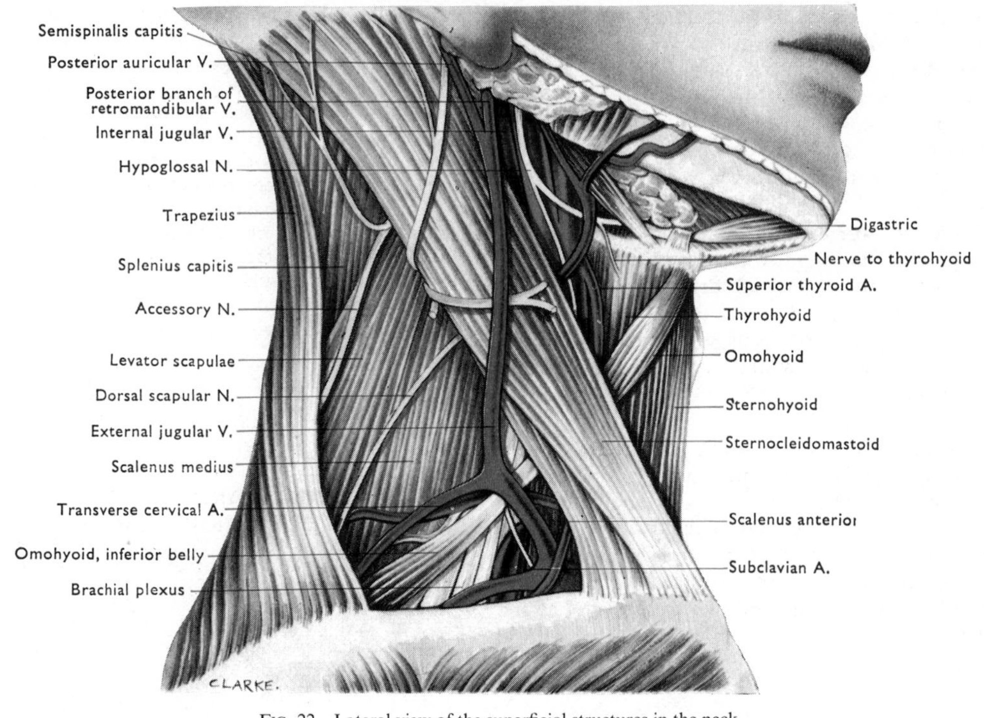

FIG. 22 Lateral view of the superficial structures in the neck.

DISSECTION. Cut through the clavicular attachment of the sternocleidomastoid and reflect this part of the muscle forwards. Remove the underlying fatty tissue to expose scalenus anterior in front of the brachial plexus. Define the borders of the muscle, and expose the omohyoid muscle, the transverse cervical artery, the anterior and internal jugular veins, and the phrenic nerve anterior to it. Depress the clavicle and expose the subclavian vein.

Scalenus Anterior [FIG. 70]

This muscle has important relations and is a key to the anatomy of the lower part of the neck. It arises from the anterior tubercles of the third to sixth cervical transverse processes and descends in front of the ventral rami of the fourth to eighth cervical nerves, the cervical pleura, and the subclavian artery, to be inserted into the medial margin of the first rib, posterior to the subclavian vein. The phrenic nerve and internal jugular vein descend obliquely in front of the muscle from its lateral to medial borders, and the muscle is also crossed more horizontally by the transverse cervical and suprascapular vessels and the anterior jugular vein [FIG. 57].

Nerve supply: the ventral rami of adjacent nerves. **Action**: it raises the first rib (inspiration) and produces lateral flexion of the neck. The lower cervical ventral rami and the subclavian artery may be compressed between it and the scalenus medius posteriorly—the scalenus anterior syndrome.

Omohyoid, Inferior Belly

This slender muscle arises from the superior transverse scapular ligament and the adjacent scapula. It crosses the posterior triangle antero-superiorly a short distance superior to the clavicle. Passing over the brachial plexus, it joins the tendon which links it to the superior belly deep to sternocleidomastoid. It can be seen contracting in thin individuals during speech.

Nerve supply: a branch of the ansa cervicalis (*q.v.*). **Action**: it helps to steady or depress the hyoid bone in speaking or swallowing.

Subclavian Artery, Third Part

The right subclavian springs from the brachio-cephalic trunk posterior to the right sternoclavicular joint. The left arises from the arch of the aorta, and enters the neck behind the left sternoclavicular joint. In the root of the neck, both arteries arch laterally in front of the cervical pleura to become the axillary arteries at the outer borders of the first ribs. The scalenus anterior is in front of the highest part of each arch—a feature which permits each artery to be divided into a first part medial to scalenus anterior, a second part behind it, and a third part lateral to it.

The third part begins about a finger's breadth above the clavicle deep to the posterior border of sternocleidomastoid. It descends from the cervical pleura to lie in the subclavian groove of the first rib anterior to the lower trunk of the brachial plexus and scalenus medius. It ends at the outer border of the first rib behind the middle of the clavicle. The suprascapular artery and the external jugular vein and its tributaries lie in front of the artery, separating it from the clavicle.

Subclavian Vein

This is the continuation of the axillary vein from the outer border of the first rib. It passes medially, antero-inferior to the third part of the subclavian artery and scalenus anterior to form the brachiocephalic vein by uniting with the internal jugular vein at the medial border of that muscle behind the sternoclavicular joint. The only tributary, the external jugular vein, joins it at the lateral border of the scalenus anterior. The subclavian vein has a single valve near its termination.

Suprascapular and Transverse Cervical Vessels

Both arteries arise with the inferior thyroid artery from the **thyrocervical trunk**—a short branch of the first part of the subclavian artery.

The **transverse cervical artery** runs laterally to the anterior border of levator scapulae, crossing scalenus anterior and the phrenic nerve, the upper and middle trunks of the brachial plexus, and the suprascapular nerve. Here it divides into a superficial branch to the deep surface of trapezius, and a deep branch which descends, deep to levator scapulae and the rhomboids, along the medial margin of the scapula. The artery may arise from the third part of the subclavian artery.

The **suprascapular artery** passes inferolaterally behind the clavicle, in front of scalenus anterior and the subclavian artery. It joins the suprascapular nerve at the postero-inferior angle of the posterior triangle, and descends with it to the scapular notch.

The corresponding veins end in the external jugular vein.

THE BRACHIAL PLEXUS

The brachial plexus should be dissected in conjunction with the dissection of the axilla. Its full exposure requires the division of the clavicle between drill holes placed in its intermediate third. This allows the shoulder to fall back and opens the cervico-axillary canal. Subsequently the clavicle should be wired together through the drill holes to replace the parts in their normal relationships.

The brachial plexus lies in the lower part of the posterior triangle, behind the clavicle, and in the axilla. The position of the plexus relative to the clavicle varies; it is higher in the erect position and lower when recumbent. The plexus is formed by the union of ventral rami at the lateral border of scalenus anterior, deep to the lower third of the posterior border of sternocleidomastoid. It ends in the axilla by dividing into the nerves of the limb.

PARTS OF THE BRACHIAL PLEXUS

Roots. These are the ventral rami of the lower four cervical and first thoracic nerves with variable contributions from the fourth cervical and second thoracic.

Trunks. The fifth and sixth cervical ventral rami, with the twig from the fourth, join to form the superior trunk. The seventh cervical ventral ramus forms the middle trunk. The eighth cervical and first thoracic ventral rami join to form the inferior trunk which contains the twig from the second thoracic ventral ramus.

Divisions. Each trunk splits into an anterior and a posterior division. The three posterior divisions unite in the **posterior cord**. The upper two anterior divisions form the **lateral cord**. The lower anterior division forms the **medial cord**. The divisions are not equal; very few fibres of the first thoracic nerve enter the posterior cord.

POSITION

The **supraclavicular part** of the plexus lies on scalenus medius with the lower trunk on the superior surface of the first rib, posterior to the subclavian artery. The roots of the long thoracic nerve are posterior to the plexus, while the external jugular vein, the inferior belly of omohyoid, and the suprascapular and transverse cervical vessels are the other main structures anterior to it.

The **infraclavicular** (axillary) **part** initially lies above and behind the first part of the axillary artery, but the cords separate to surround the second part of the artery in the positions indicated by their names.

BRANCHES IN THE NECK

The upper two roots of the brachial plexus receive **grey rami communicantes** from the middle

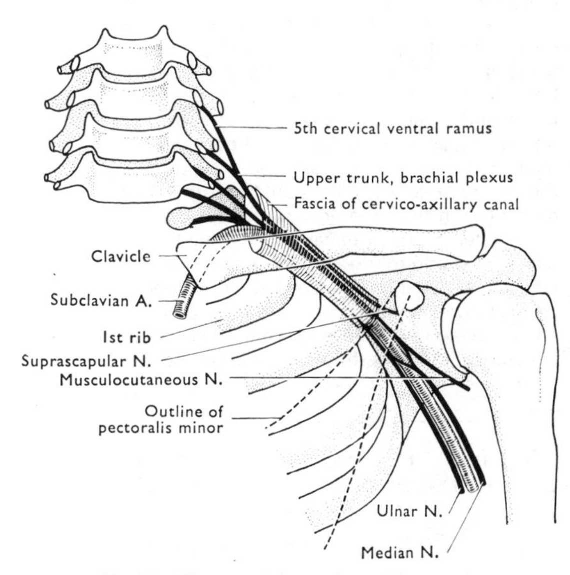

FIG. 23 Diagram of the cervico-axillary canal.

cervical ganglion of the sympathetic trunk, the others from the cervicothoracic ganglion.

All the branches in the neck are muscular and pass to the upper limb, except twigs to the scalene muscles and longus colli [FIG. 70], and the lowest root to the phrenic nerve from C. 5.

The **dorsal scapular nerve** passes backwards from the fifth cervical ventral ramus through scalenus medius, and runs inferolaterally anterior to levator scapulae and the two rhomboids, supplying all three.

The slender **nerve to subclavius** arises where the fifth and sixth cervical ventral rami unite. It descends across the brachial plexus and subclavian vessels, close to scalenus anterior, and enters the posterior surface of subclavius. It often sends a branch to the **phrenic nerve** which replaces the contribution from C. 5 to that nerve.

The **suprascapular nerve** arises from the junction of the fifth and sixth cervical ventral rami. It runs postero-inferiorly on scalenus medius lateral to the plexus, to join the suprascapular vessels and descend with them over the scapula to supply supraspinatus, infraspinatus, and the shoulder joint.

The **long thoracic nerve** arises by a branch from each of the upper three roots of the plexus. The upper two pierce scalenus medius, unite and descend on its surface to enter the axilla over the lateral surface of the first digitation of serratus anterior on the first rib. The lowest branch runs over scalenus medius to join the upper part in the axilla.

THE DISSECTION OF THE BACK

Begin by studying the surface anatomy of the back with the dissectors of the upper limb [Vol. 1, p. 24].

DISSECTION. Make a vertical median incision through the skin from the external occipital protuberance to the seventh cervical spine, and a horizontal incision laterally from this to the acromion. Reflect the skin flap and examine the posterior triangle from behind.

Expose the occipital belly of occipitofrontalis with the occipital branch of the **posterior auricular nerve** (branch of the facial nerve) running across it near its attachment. Look for the cutaneous nerves over the upper part of trapezius. They are difficult to find because of the density of the connective tissue, but the nerves from which they arise will be found when trapezius is reflected.

Find a branch of the occipital artery on the back of the scalp. A branch of the greater occipital nerve will be found beside it. Trace both downwards to the main stem of the nerve and artery. The **greater occipital nerve** pierces the deep fascia 2–3 cm lateral to the external occipital protuberance with the occipital artery. The **third occipital nerve** lies in the superficial fascia between the greater occipital nerve and the median plane. Find and follow it in both directions. Like the cutaneous branches of the other cervical dorsal rami, it pierces the trapezius and the deep fascia close to the midline.

Remove the superficial and deep fascia from the surface of trapezius.

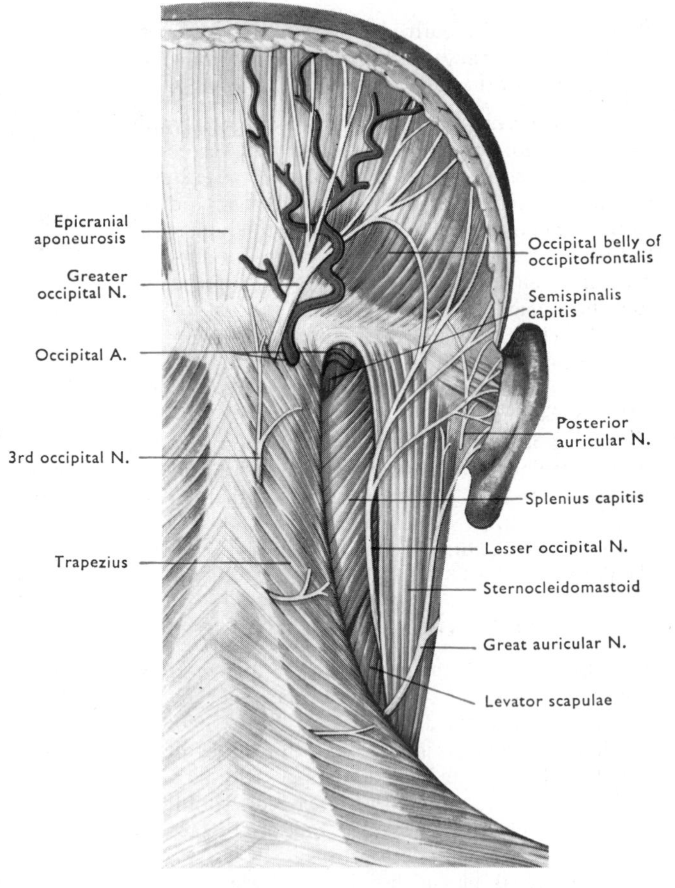

FIG. 24 Superficial structures of the back of the neck.

The parts of the occipital artery and nerves now exposed are described on pages 15 and 27.

FIRST LAYER OF MUSCLES

This consists of trapezius and latissimus dorsi.

Trapezius [FIGS. 24, 26]

The lower part of this muscle and the latissimus dorsi are dissected with the upper limb, but the head and neck dissectors should study its whole extent.

The trapezius arises from the medial third of the superior nuchal line, the external occipital protuberance, the ligamentum nuchae [p. 27], and the spines of the seventh cervical and all the thoracic vertebrae. At the cervicothoracic junction, it arises from an aponeurosis which spreads laterally into the muscle.

The upper fibres sweep downwards to the lateral third of the clavicle, forming the curve of the shoulder. The middle fibres run horizontally to the medial edge of the acromion and the adjacent part of the superior margin of the crest of the spine of the scapula. The lower fibres ascend to a small, flat tendon which slides on a bursa covering the root of the scapular spine. Its fibres end on the medial part of the upper margin of the crest.

Nerve supply: motor from the accessory nerve; sensory from the third and fourth cervical nerves. **Action**: the middle fibres pull the scapula medially, bracing the shoulder backwards. The upper fibres raise the tip of the shoulder, while the lower fibres depress the medial part of the scapula. Acting together these two parts rotate the scapula laterally, turning the glenoid cavity upwards as in raising the arm above the head. It also extends the head on the neck.

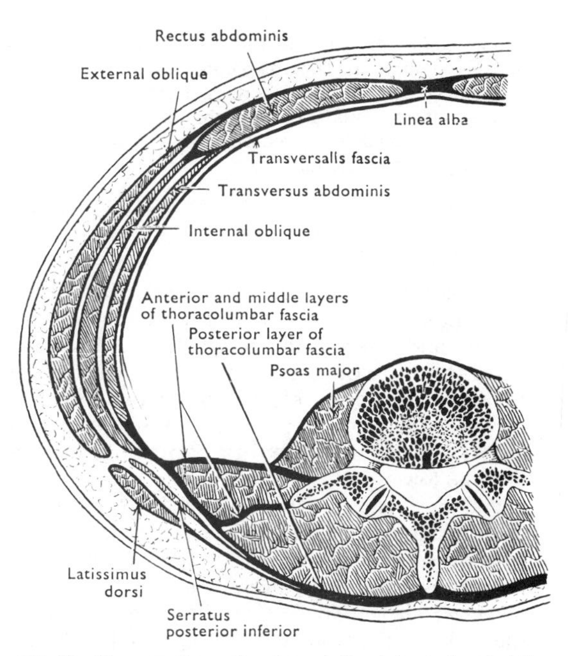

FIG. 25 Diagram of a section through the abdominal wall at the level of the second lumbar vertebra to show the arrangement of the thoracolumbar fascia.

DISSECTION. Reflect trapezius laterally by separating it from the superior nuchal line, and dividing it vertically 1 cm from the vertebral spines. Find the accessory nerve on its deep surface with the branches from the third and fourth cervical nerves and the superficial branch of the transverse cervical artery.

Define the attachments of levator scapulae and follow the deep branch of the transverse cervical artery and the dorsal scapular nerve on its deep surface.

SECOND LAYER OF MUSCLES

These are the levator scapulae, the rhomboids, the two serratus posterior muscles, and splenius.

Levator Scapulae

It arises by one slip from each of the transverse processes of the upper four cervical vertebrae. The muscle descends as two or more slips to be inserted into the medial border of the scapula from the upper angle to the spine.

Nerve supply: the ventral rami of the third and fourth cervical nerves and the fifth through the dorsal scapular nerve. **Action**: it assists in elevation and rotation of the scapula, helping to hold it steady in active movements of the upper limb.

Posterior Serrate Muscles

These are exposed when the trapezius, latissimus dorsi, and rhomboid muscles are removed and the scapula is pulled laterally. They are thin, musculo-aponeurotic sheets on the back of the thorax.

Serratus posterior superior runs inferolaterally from the seventh cervical and upper two or three thoracic spines to the second to fifth ribs.

Serratus posterior inferior passes superolaterally from the lumbar fascia and the lower two thoracic spines to the last four ribs.

Nerve supply: from the second to fourth intercostal nerves, and from the lower intercostal and subcostal nerves respectively. **Actions**: both are inspiratory muscles, the superior by elevating its ribs, the inferior by holding its ribs down against the pull of the diaphragm.

Thoracolumbar Fascia

This is a strong aponeurotic layer which extends from the ilium and dorsal surface of the sacrum to the upper thoracic region. It binds down the erector spinae group of muscles, and is particularly thick in the lumbar and sacral regions [see Vol. 2].

The thoracic part is relatively thin and transparent, and extends from the tips of the spines and supraspinous ligaments to the angles of the ribs. Superiorly it passes deep to serratus posterior superior and fades out into the neck. Inferiorly it is continuous with the posterior layer of the lumbar part deep to serratus posterior inferior.

DISSECTION. Reflect serratus posterior superior, and find the nerves entering its deep surface. Remove the thoracic part of the thoracolumbar fascia to expose the longitudinal erector spinae muscles with the splenius curving superolaterally across it above the midthoracic region. Define the attachments of splenius and find the nerves piercing it. Then separate splenius from the vertebral spines, and turn it superolaterally leaving these nerves intact. This exposes the semispinalis and longissimus muscles (erector spinae), and part of obliquus capitis superior.

Splenius

The splenius arises from the lower part of the ligamentum nuchae and the spines of the seventh cervical and upper six thoracic vertebrae. The lower part (**splenius cervicis**) runs to the back of the transverse processes of the upper two or three cervical vertebrae, deep to levator scapulae. The larger, upper part (**splenius capitis**) is inserted into the lower part of the mastoid process and the lateral part of the superior nuchal line, deep to sternocleidomastoid. Detach sternocleidomastoid from the superior nuchal line to expose the insertion.

Nerve supply: the dorsal rami of the cervical nerves which pierce it. **Action**: Together they extend the neck and the head on it. Acting on one side only the face is also turned to that side.

DEEP MUSCLES OF THE BACK

Erector Spinae

The major part of this muscle begins on the sacrum, ascends into the lumbar region, filling the region between the spines and transverse processes of the vertebrae, and splits into three columns which pass upwards into the thorax deep to the thoracolumbar fascia. Each column is inserted either into the ribs (lateral part) or the vertebrae (medial part), but fresh slips arise from the same situations and continue upwards. From lateral to medial the three columns are: iliocostalis, longissimus, and spinalis.

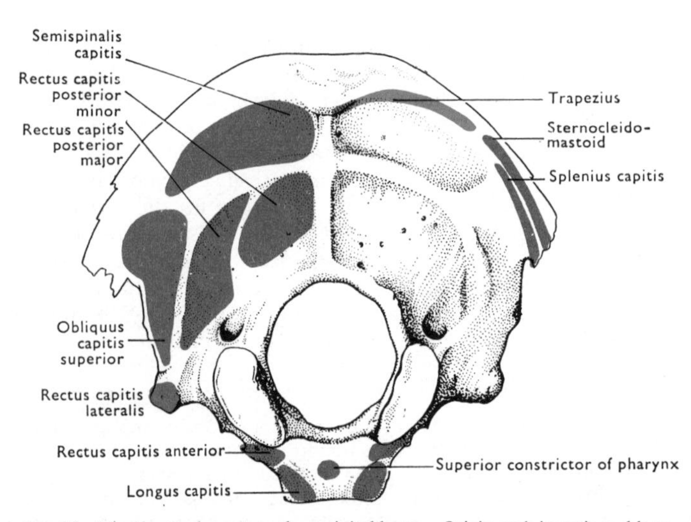

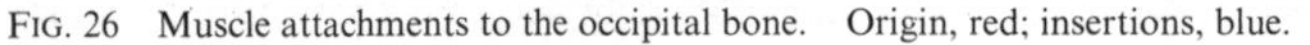
FIG. 26 Muscle attachments to the occipital bone. Origin, red; insertions, blue.

Semispinalis. This is made up of a succession of deep slips that arise from the thoracic transverse processes and are inserted into the upper thoracic spines (semispinalis thoracis) into the cervical spines (semispinalis cervicis) and into the skull (semispinalis capitis). Deeper still are the muscles which connect the sacrum with the transverse processes, laminae, and spines. They are classified with semispinalis under the general name of **transversospinal muscle**, and are **multifidus**, **rotatores**, **interspinales**, and **intertransversarii**. The details of these muscles will be found in larger textbooks, but their general arrangement should be studied as they are muscles of considerable importance in strains of the back.

Nerve supply: dorsal rami of spinal nerves. **Actions**: these are numerous; basically they extend the vertebral column, produce rotation and lateral flexion of it, and help to balance it on the pelvis. They are essential for the maintenance of the erect posture during all movements.

Levatores Costarum. These twelve small muscles on each side radiate from the transverse process of the seventh cervical and first eleven thoracic vertebrae to the posterior part of the body of the rib below. **Nerve supply**: the intercostal nerves.

DISSECTION. Remove the fascia from semispinalis capitis and from longissimus, but do not disturb the nerves which pierce semispinalis close to the median plane. Determine the attachments of the muscles.

Reflect longissimus capitis from the skull, and follow the occipital artery deep to the mastoid process over obliquus capitis superior. Trace that muscle to the transverse process of the atlas and to the skull.

Detach semispinalis capitis from the occipital bone and turn it laterally, preserving the nerves which pierce it. This exposes most of the suboccipital muscles and semispinalis cervicis with the dorsal rami running over its surface and the **deep cervical artery** ascending on it to anastomose with the occipital artery. If a branch from the first cervical dorsal ramus enters semispinalis capitis, it should be retained by cutting out a piece of the muscle with the nerve, as this makes the dissection of that ramus much easier.

Define the attachments of semispinalis cervicis.

Longissimus Capitis. This slender muscle lies under cover of splenius immediately posterior to the transverse processes [FIG. 28]. It arises from the upper thoracic transverse processes, and ascends to the back of the mastoid process under cover of splenius and sternocleidomastoid. Longissimus cervicis lies anterior to longissimus capitis. It has the same origin, but inserts into the posterior surfaces of the cervical transverse processes, except the first.

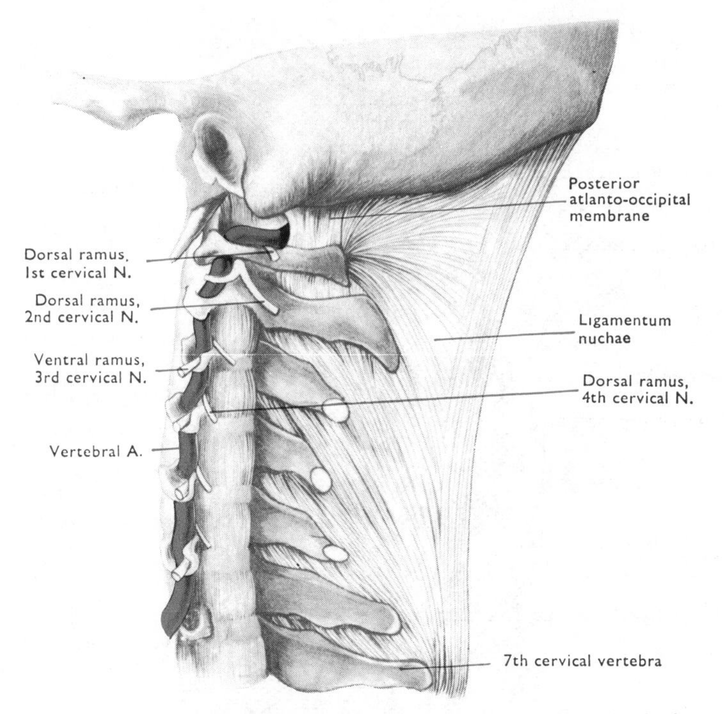

FIG. 27 Dissection of the ligamentum nuchae and of the vertebral artery in the neck. The neck is slightly flexed.

Occipital Artery

This artery arises from the external carotid artery in the front of the neck. It runs posterosuperiorly deep to the posterior belly of digastric to reach and groove the skull deep to the mastoid notch where digastric is attached. It then passes posteriorly immediately deep to the muscles attached to the superior nuchal line, crosses the apex of the posterior triangle, and pierces trapezius 2–3 cm from the midline with the greater occipital nerve. It ramifies on the back of the head, supplying the posterior half of the scalp.

Branches. In this region they are mainly muscular. A mastoid branch enters the mastoid foramen to supply the bone and dura mater. A descending branch passes deep to semispinalis capitis to anastomose with the deep cervical artery from the subclavian.

Semispinalis Capitis. This long muscle produces the rounded ridge at the side of the median furrow in the back of the neck, even though it is deep to trapezius and splenius. It has the same origin as longissimus capitis, and is inserted into the medial half of the area between the superior and inferior nuchal lines [FIG. 26]. It is separated from its fellow by the ligamentum nuchae.

Nerve supply: dorsal rami of the upper cervical nerves. **Action**: it extends the neck and the head on the neck.

Ligamentum Nuchae [FIG. 27]. This is a median fibrous septum between the muscles of the two sides of the back of the neck. It represents a powerful elastic structure in quadrupeds which helps to sustain the weight of the dependent head. In Man there is little elastic tissue, and it is a continuation of the supraspinous and interspinous ligaments from the spine of the seventh cervical vertebra to the external occipital protuberance and crest. Anteriorly it is attached to the posterior tubercle of the atlas and the spines of the other cervical vertebrae.

Deep Cervical Artery [FIG. 55]. This vessel arises from the costocervical trunk of the subclavian, and passes into the back of the neck between the seventh cervical transverse process and the neck of the first rib. It ascends deep to semispinalis capitis and anastomoses with the descending branch of the occipital artery.

Semispinalis Cervicis. This bulky muscle lies deep to semispinalis capitis and has the same origin. It passes superomedially to the spines of the second to fifth cervical vertebrae, principally the second—a feature which partly accounts for the size of the spine of that vertebra.

SUBOCCIPITAL TRIANGLE
[FIG. 28]

This small triangle is formed by the rectus capitis posterior major superomedially, the obliquus inferior below, and the obliquus superior superolaterally. It is crossed by the greater occipital nerve

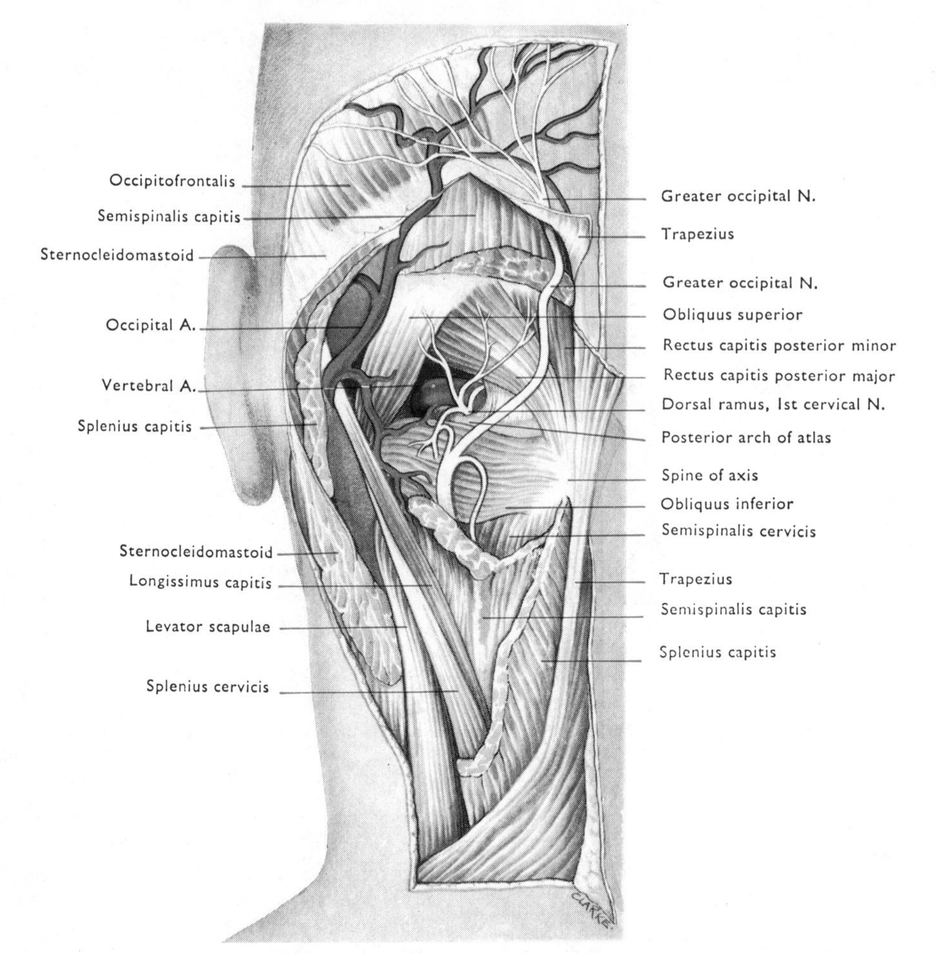

FIG. 28 The suboccipital region. Deep structures of the back of the neck.

which hooks round the lower border of obliquus inferior, and is under cover of semispinalis capitis. In the depths of the triangle are the posterior atlanto-occipital membrane, and the posterior arch of the atlas with the vertebral artery and the first cervical nerve on its superior aspect. The dorsal ramus of that nerve enters the triangle which contains the suboccipital plexus of veins draining to the vertebral and deep cervical veins.

DISSECTION. The triangle is difficult to dissect because of its content of dense fibrous tissue. If the branch of the dorsal ramus of the first cervical nerve to semispinalis capitis has been retained, it can be followed to the ramus and the other branches traced from there. Alternatively a communicating branch to the greater occipital nerve or the branch to one of the muscles of the triangle may be found and followed. The ramus emerges between the vertebral artery and the posterior arch of the atlas. Remove the fascia from the triangle and the surrounding muscles.

Dorsal Ramus of First Cervical (Suboccipital) Nerve

This nerve enters the triangle between the posterior arch of the atlas and the vertebral artery. It divides into branches to the muscles of the suboccipital triangle and semispinalis capitis, to the greater occipital nerve, and rarely a direct cutaneous branch which accompanies the occipital artery.

Vertebral Artery [FIGS. 27, 55, 71]

The third part of this artery emerges from the transverse process of the atlas, and hooks postero-medially on the posterior surface of the lateral mass of the atlas, grooving it and the lateral part of the posterior arch from which it is partly separated by the first cervical nerve. It leaves the triangle by passing anterior to the thickened lateral edge of the **posterior atlanto-occipital membrane**. This edge arches over the artery from the posterior arch of the atlas to its lateral mass and may be ossified. At this point the artery enters the vertebral canal and pierces the dura mater.

Suboccipital Plexus of Veins

This plexus lies in and around the triangle. It unites the occipital veins, the internal vertebral venous plexuses [p. 158], the emissary vein which traverses the condylar canal [FIG. 11] from the sigmoid sinus, muscular veins, the deep cervical vein, and the plexus of veins around the vertebral artery. Thus it presents a number of alternative routes for venous drainage.

Suboccipital Muscles [FIG. 28]

Rectus capitis posterior major passes superolaterally from the spine of the axis to the lateral half of the area below the inferior nuchal line of the occiput [FIG. 26].

Obliquus capitis inferior passes from the spine of the axis to the transverse process of the atlas.

Obliquus capitis superior runs posterosuperiorly from the transverse process of the atlas to the lateral half of the area between the nuchal lines.

Rectus capitis posterior minor passes from the tubercle of the posterior arch of the atlas to the medial part of the area below the inferior nuchal line, under cover of the rectus major.

Nerve supply: the dorsal ramus of C. 1. **Actions**: it is possible to ascribe specific actions to each of these muscles, but their main function is to stabilize the head on the atlas and axis, and to act as ligaments of variable length and tension for this purpose.

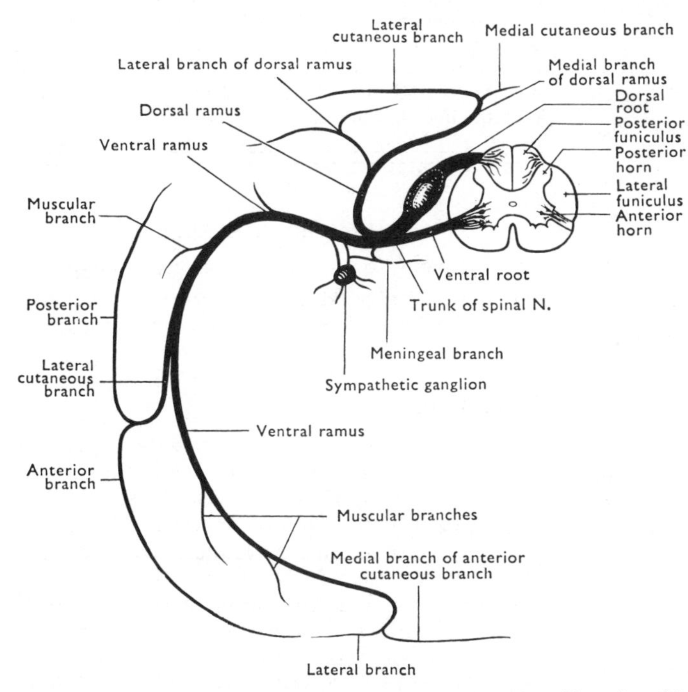

FIG. 29 Diagram of a typical spinal nerve. Both medial and lateral branches of the dorsal ramus supply muscles; the medial branch supplies skin in the upper half of the body, the lateral branch supplies it in the lower half.

DORSAL RAMI OF SPINAL NERVES

These are the nerves of the back. They supply the intervertebral joints, ligaments and muscles of the back (but not the muscles of the upper limb which extend over the back, *e.g.*, trapezius, latissimus dorsi, rhomboids) and the overlying skin.

The general arrangement and distribution of the dorsal rami follow the pattern shown in FIGURE 29, with the following exceptions.

1. The first cervical, fourth and fifth sacral, and coccygeal do not divide into medial and lateral branches.
2. The dorsal rami of the first and last two cervical (sometimes C. 5 and 6, or 6 and 7) and the last two lumbar nerves do not have cutaneous branches.
3. Above the midthoracic region the cutaneous branches arise from the medial branches of the dorsal rami, and emerge close to the midline (2–3 cm). Below this they arise from the lateral branches and emerge 8–9 cm from the median plane in line with the angles of the ribs. The cutaneous branches of the small dorsal rami of the sacral nerves emerge on a line from the posterior superior iliac spine to the tip of the coccyx.
4. The cutaneous branches of some dorsal rami are large and supply an extended area of skin. Thus C. 2 (greater occipital) supplies a large part of the scalp; T. 1 and 2 extend laterally over the scapula; L. 1, 2, and 3 extend over the buttock to the level of the greater trochanter, almost as though their territory had been pulled out by the growing head and limbs.
5. The upper cervical dorsal rami and the sacral dorsal rami form simple looped plexuses, the cervical deep to semispinalis capitis, the sacral deep to gluteus maximus.

Greater Occipital Nerve

This medial branch of the dorsal ramus of C. 2 is the thickest cutaneous nerve in the body. It emerges below the middle of the inferior oblique muscle, curves superomedially across the suboccipital triangle, and pierces semispinalis capitis. It ascends on that muscle, pierces trapezius 2–3 cm lateral to the external occipital protuberance, and ramifies in the back of the scalp, reaching as far as the vertex. Though mainly cutaneous, it also supplies semispinalis capitis.

Third Occipital Nerve

This is the small cutaneous branch of the dorsal ramus of C. 3. It pierces semispinalis capitis and trapezius and supplies skin of the nape of the neck up to the external occipital protuberance.

BLOOD VESSELS OF THE BACK

Arteries

Cervical Region. The main supply comes from the subclavian artery. The vessels are: (1) the **deep cervical** anastomosing with (2) the descending branch of the **occipital artery** deep to semispinalis capitis, and (3) the superficial branch of the **transverse cervical artery**. In addition, small twigs of the vertebral artery pass into the muscles of the back of the neck.

Thoracic Region. Posterior branches of the intercostal arteries pass to the muscles and skin together with branches from the deep branch of the transverse cervical artery.

Lumbar Region. Similar branches are derived from the lumbar arteries.

Veins

These follow the corresponding arteries. They drain into the vertebral and deep cervical veins in the neck and to the intercostal and lumbar veins in the thorax and lumbar regions. These veins are also linked to the posterior vertebral and internal vertebral venous plexuses.

Posterior Vertebral Venous Plexus. This extensive plexus lies posterior to the vertebral arches under cover of multifidus. It drains the veins of the back into the vessels indicated above.

THE ANTERIOR TRIANGLE OF THE NECK

This is the large area on the front of the neck bounded by the midline, the mandible, and the sternocleidomastoid muscle.

SURFACE ANATOMY

Draw a finger down the anterior median line of your neck from chin to sternum, and identify, in sequence, (1) the body of the hyoid bone approximately 2 cm below and 6 cm behind the chin; (2) the **laryngeal prominence** or Adam's apple (*i.e.*, the sharp protuberance of the anterior border of the thyroid cartilage) is notched superiorly; (3) the rounded arch of the **cricoid cartilage**; (4) and the rings of the trachea which are partly masked by the isthmus of the thyroid gland.

Grasp your U-shaped **hyoid bone** between finger and thumb, and trace it backwards to the greater horns. Note that their tips are close to the vertebral column and may be overlapped by the sternocleidomastoid muscles. Trace the superior border of your **thyroid cartilage** posteriorly from its notch, and note that it ends in a projection (superior horn) immediately anterior to the sternocleidomastoid.

Press the tips of your fingers into your neck along a line from the mastoid process towards the tip of the shoulder. The deep bony resistance is due to the **transverse processes of the cervical vertebrae**. Only the first can be felt at all clearly immediately antero-inferior to the tip of the mastoid process. The fourth is level with the upper border of the thyroid cartilage, the sixth with the cricoid cartilage. These points can only be felt on the living subject, especially the **isthmus of the thyroid gland** which forms a soft, cushion-like mass on the second to fourth tracheal rings, and slips upwards past the palpating finger when you swallow.

DISSECTION. Incise the skin from chin to sternum in the midline, and reflect the flap of skin inferolaterally. Do not cut deeply, but remain superficial to the fibres of platysma in the posterosuperior part.

Reflect platysma upwards, keeping close to its deep surface and dividing any nerve bundles which enter it. Find the branches of the transverse nerve of the neck [FIG. 21] as they cross sternocleidomastoid, and follow them anteriorly to their termination. Identify the cervical branch of the facial nerve as it leaves the lower border of the parotid gland, and trace it antero-inferiorly.

Find the **anterior jugular vein** near the midline. Trace it inferiorly till it pierces the deep fascia about 2 cm above the sternum. Expose the deep fascia of the triangle.

Make a transverse incision through the first layer of the deep fascia immediately above the sternum, and extend the incision 4 cm upwards along the anterior border of sternocleidomastoid. Reflect this fascia to open the **suprasternal space** between the first and second (deep) layers of fascia. Trace the anterior jugular vein and the deeper layer of the fascia laterally deep to sternocleidomastoid, and follow the fascia upwards till it fuses with the first layer (midway between the sternum and the thyroid cartilage) and downwards to its fusion with the back of the sternum.

SUPERFICIAL FASCIA

The superficial fascia in this region contains a variable amount of fat. It is only loosely connected to the skin and deep fascia, therefore the skin is freely movable.

Platysma

This thin sheet of muscle in the superficial fascia is superficial to the cutaneous branches of the cervical plexus and the external jugular vein. It extends from the face to the upper part of the chest, and lies on the superior part of the anterior triangle and the antero-inferior part of the posterior triangle.

It arises from the skin and fascia over the upper parts of pectoralis major and deltoid. The anterior fibres are either attached to the lower border of the anterior part of the mandible or meet their fellows below the chin. The posterior fibres curve upwards over the mandible to the skin of the lower face, and mingle with the muscles of the lips, helping to form risorius [FIG. 17].

Nerve supply: the cervical branch of the facial nerve. **Action**: it pulls down the corner of the mouth, elevates the skin over the upper part of the chest, and tenses the skin of the neck. It contracts most obviously in strenuous inspiratory effort, and may prevent compression of the airway by the skin. It is part of a much more extensive sheet of cutaneous muscle which is found in many animals, part of which arises in the axilla.

The **cervical branch of the facial nerve** emerges from the lower border of the parotid gland, and pierces the deep fascia. Its branches spread on the deep surface of platysma to supply it, and some cross the mandible to the muscles of the lower lip.

Anterior Jugular Vein [FIG. 30]

This is usually the smallest of the jugular veins, but it varies in width inversely with the external jugular vein. It begins below the chin, and descends in the superficial fascia about 1 cm from the median plane. Approximately 2 cm above the sternum, it pierces the first layer of the deep fascia, and runs laterally in the suprasternal space, deep to sternocleidomastoid, to join the external jugular vein. It is united to its fellow in the suprasternal space, thus forming a **jugular arch** which may be large and an important anterior relation of the trachea and thyroid gland.

The anterior jugular vein is very variable. Both may be replaced by a single median vein, or the anterior and external jugular vein may be partly or completely replaced by a single vein which descends along the anterior border of the sternocleidomastoid to the jugular arch.

DISSECTION. Remove the fat and fascia from the superficial surface and margins of the sternocleidomastoid, but retain the nerves and vessels in contact with it. Anterosuperiorly, push the parotid gland forwards and find the accessory nerve and the artery which accompanies it as they enter the anterior border of the muscle. Now lift the sternocleidomastoid, note the vessels which enter it anteriorly and the structures which lie deep to it.

STERNOCLEIDOMASTOID

This important landmark in the neck is easily defined in the living subject. When contracted, it stands out as a well-defined ridge between the anterior and posterior triangles.

It arises inferiorly by a rounded, tendinous **sternal head** from the upper part of the anterior surface of the manubrium sterni, and by a thin, fleshy **clavicular head** from the upper surface of the medial third of the clavicle. The sternal head ascends across the medial part of the sternoclavicular joint, and widening rapidly, meets and overlaps the clavicular head a short distance above the clavicle, fusing with it about half way up the neck. The thick, anterior border is inserted into the anterior surface of the mastoid process, while the posterior part, becoming thin and aponeurotic superiorly, is attached to the lateral surface of the mastoid process, and the lateral half or more of the superior nuchal line.

The deep part of the muscle is pierced by the accessory nerve, and it receives its main blood supply from the occipital and superior thyroid arteries.

Nerve supply: the accessory nerve is motor; the ventral ramus of the second cervical nerve is sensory. **Action**: acting alone, it tilts the head to its own side and rotates it so that the face is turned towards the opposite side. The two muscles acting together flex the neck, but raise the sternum and assist in forced inspiration when flexion is prevented by the postvertebral muscles.

THE MEDIAN REGION OF THE FRONT OF THE NECK

Superficial Fascia

This layer contains the anterior jugular veins, and below the chin, the decussating fibres of platysma and a few small **submental lymph nodes**. These nodes lie on the deep fascia and drain lymph from the anterior part of the floor of the mouth. They can usually be felt in the living person as small nodules when the thumb is pressed upwards behind the chin.

Deep Fascia [p. 74]

The **investing layer** is attached to the mandible, hyoid bone, and sternum. Superiorly it encloses the anterior bellies of the digastric muscles [FIG. 31]. Inferiorly it encloses the jugular arch, the sternal head of sternocleidomastoid, the lowest parts of the anterior jugular veins, and an occasional lymph node in the **suprasternal space**. The deeper **pretracheal fascia** descends from the thyroid and cricoid cartilages to invest the **thyroid gland** and cover the front and side of the trachea. It fuses with the back of the upper part of the pericardium in the thorax.

DISSECTION. Remove the deep fascia from the anterior bellies of the digastrics and from the area between them. This exposes parts of the two mylohyoid muscles which unite in a median raphe.

Continue the removal of this fascia below the hyoid. This exposes the infrahyoid muscles between the hyoid and the sternum on each side of the midline. Separate these muscles in the midline to expose the pretracheal fascia. Below the isthmus of the thyroid gland [FIG. 30] remove this fascia to expose the trachea and the inferior thyroid veins descending on it. The small thyroidea ima artery may ascend to the thyroid in this region.

At the upper border of the isthmus, look for the slender fibromuscular band (**levator glandulae thyroideae**) which may be present between the left end of the isthmus

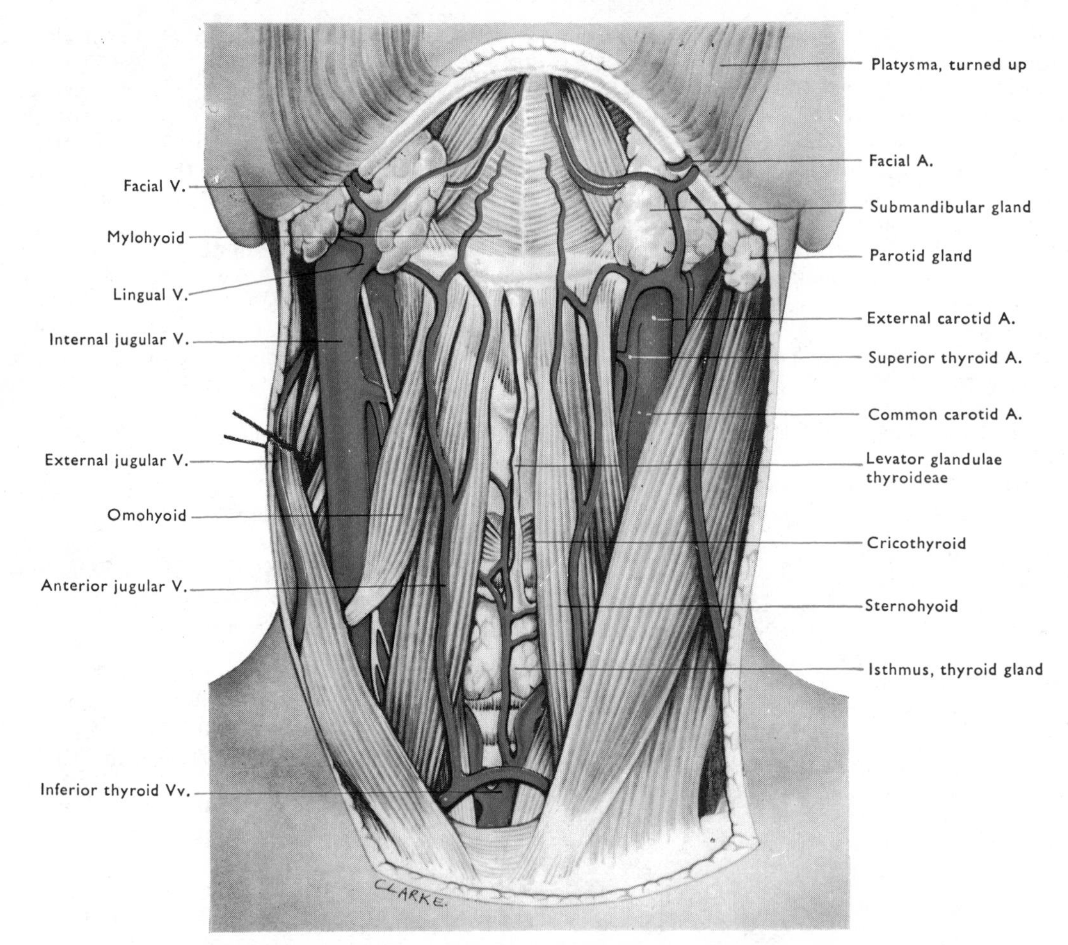

FIG. 30 Dissection of the front of the neck. The right sternocleidomastoid has been retracted posteriorly.

and the hyoid bone. Remove the pretracheal fascia from the isthmus and its vessels, and note that it can be displaced downwards when the fascia is removed. Find the **cricothyroid ligament** between the median parts of the cricoid and thyroid cartilages, with a small cricothyroid muscle covering it on each side.

Find the **median thyrohyoid ligament** with the anastomosis of the infrahyoid arteries on it, between the thyroid cartilage and the hyoid bone. Follow the ligament upwards behind the body of the hyoid bone. There is a small bursa between the ligament and the bone.

Suprahyoid Region

The **submental triangle** is in the midline, outlined by the hyoid bone and the anterior bellies of the digastric muscles. The roof of the triangle is formed by the muscular floor of the mouth, *i.e.*, the mylohyoid muscles which unite in a median fibrous raphe extending from the symphysis menti to the hyoid bone. The submental lymph nodes lie superficial to the deep fascia which forms the floor of the triangle.

Infrahyoid Region

A number of midline structures lie between the infrahyoid muscles of the two sides. From above downwards, these are:

1. The **median thyrohyoid ligament** passes from the upper border of the thyroid cartilage, behind the body of the hyoid bone, to be attached to its upper border. It is separated from the hyoid by a bursa which permits the upper edge of the thyroid cartilage to ascend inside the concavity of the hyoid bone during swallowing.
2. The **laryngeal prominence** is formed by the thyroid cartilage, and is notched on its superior margin.
3. The **cricothyroid ligament** attaches the thyroid cartilage to the arch of the cricoid. On each side of the ligament, a **cricothyroid muscle** radiates posterosuperiorly from the cricoid to the thyroid cartilage. They are important muscles of speech, and the only intrinsic laryngeal muscles to appear on the external surface of the larynx.
4. The **isthmus of the thyroid gland** usually

lies on the second to fourth tracheal rings with a rich vascular anastomosis on its surface. Occasionally a small, pointed, **pyramidal lobe** projects upwards from it, and may give rise to a slender slip of muscle (**levator glandulae thyroideae**) or fibrous tissue which attaches it to the hyoid bone. This is an embryological remnant of the tubular, epithelial downgrowth from the tongue (**thyroglossal duct**) which forms the greater part of the thyroid gland [p. 248].

5. Below the isthmus of the thyroid gland, the **jugular arch** and the **inferior thyroid veins** lie anterior to the trachea. Occasionally the left brachiocephalic vein and the brachiocephalic artery are high enough to appear in front of the trachea at the root of the neck.

THE SUBDIVISIONS OF THE ANTERIOR TRIANGLE

In addition to the submental triangle, the anterior triangle may be divided into three subsidiary triangles [FIG. 31] which are useful for descriptive purposes.

DISSECTION. Identify the facial artery and vein at the lower border of the mandible. Cut the deep fascia from the mandible and turn it downwards, exposing part of the submandibular gland. Identify both bellies of the digastric muscle at the lower border of the gland, and follow the facial vein postero-inferiorly, superficial to the gland and the posterior belly of digastric.

At the lower border of the mandible, find the **submental branch of the facial artery**. Trace it forwards by pushing the submandibular gland aside and removing the fat and lymph nodes beside it. The artery

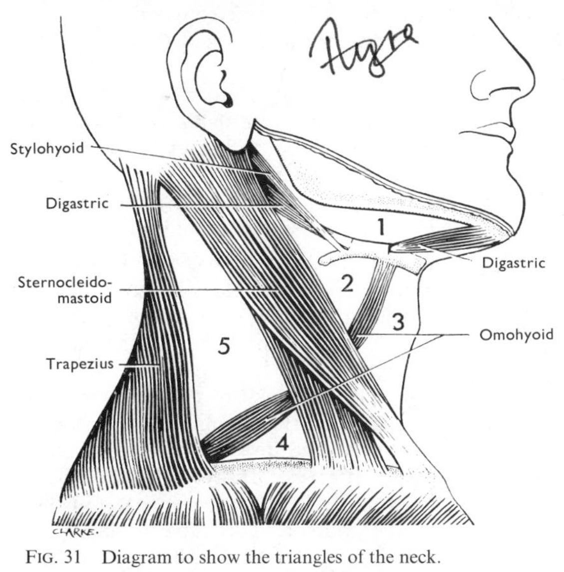

FIG. 31 Diagram to show the triangles of the neck.

1. Digastric triangle
2. Carotid triangle
3. Muscular triangle } anterior triangle
4. Subclavian (omoclavicular) triangle
5. Occipital triangle } posterior triangle

runs with the mylohyoid nerve which supplies the mylohyoid and the anterior belly of digastric.

Pull the submandibular gland laterally to expose the intermediate tendon of digastric, the fascial sling which attaches it to the hyoid, and the stylohyoid muscle which splits to embrace the intermediate tendon and lies on the upper surface of the posterior belly of digastric. Follow the stylohyoid and the posterior belly of digastric as far as the angle of the mandible.

Pull the submandibular gland backwards to expose the posterior border of mylohyoid. Expose this free border with the hypoglossal nerve and one of the veins of the tongue passing above it. The thin sheet of muscle deep to the hypoglossal nerve is the hyoglossus [FIG. 104]. Remove the fascia from the rest of mylohyoid and from the hyoglossus.

DIGASTRIC TRIANGLE

This triangle, part of the submandibular region, is bounded by the digastric, the stylohyoid, and the lower border of the mandible. Its roof is formed by the mylohyoid and hyoglossus.

The lower part of the **submandibular gland** almost fills the triangle. The **submandibular lymph nodes** lie on the surface of the gland, especially along the lower border of the mandible. They drain lymph from the side of the tongue, teeth, lips, and cheek to deep cervical lymph nodes which lie beneath sternocleidomastoid.

The **facial vein** pierces the deep fascia at the lower border of the mandible. It crosses the submandibular gland, joins the anterior branch of the retromandibular vein, and ends in the internal jugular vein. The **facial artery** curves round the lower border of the mandible and gives off the submental artery before piercing the deep fascia. The **submental artery** passes forwards with the corresponding vein and the mylohyoid nerve.

The **mylohyoid nerve** lies on the inferior surface of mylohyoid near the mandible. It supplies mylohyoid and the anterior belly of digastric. The **hypoglossal nerve** lies on a deeper plane; it passes above mylohyoid on hyoglossus. The **intermediate tendon of digastric** lies on hyoglossus immediately above the hyoid bone, and is held to the bone by a loop of fascia which forms a pulley for it.

DISSECTION. Remove the fat and fascia from the area between the posterior belly of digastric and the superior belly of omohyoid—the carotid triangle [FIG. 31]. This exposes the internal jugular vein laterally, the common and internal carotid arteries medial to it, and the external carotid anteromedial to the internal carotid.

Find the facial and lingual veins entering the internal jugular vein in the upper part of the triangle and the superior thyroid vein entering it in the lower part. Between the vein and the internal carotid artery, find the **hypoglossal nerve**. Follow it forwards across the external carotid artery, and find its branches in the triangle. (1) The superior root of the ansa cervicalis, given off where the nerve hooks round the occipital artery to enter the triangle. (2) The thyrohyoid branch given off as the hypoglossal quits the triangle [FIG. 33].

Remove the superficial part of the fascial carotid sheath which surrounds the internal jugular vein, the carotid arteries, and the vagus nerve. Avoid injury to the superior root of the **ansa cervicalis** anterior to the vein and to its inferior root lateral to the vein. The inferior root is derived from the second and third cervical ventral rami, and forms a loop communication (ansa cervicalis) with the superior root on the vein at a lower level.

Expose the **external carotid artery** and its branches [FIG. 104]. The superior thyroid is the lowest branch in the triangle [FIG. 63]. The lingual, facial, and occipital arteries arise in the upper part of the triangle, the occipital opposite the facial [FIG. 63].

Find the internal laryngeal nerve in the thyrohyoid interval [FIG. 104]. Trace it posterosuperiorly, deep to the carotid arteries, to the superior laryngeal branch of the vagus, from which the slender external laryngeal nerve can be followed downwards, deep to the superior thyroid artery.

Expose the part of hyoglossus which lies in the triangle immediately above the hyoid bone. Note the hypoglossal nerve on its surface, and the **lingual artery** passing deep to it. That artery lies on the middle constrictor of the pharynx [FIG. 32] which separates it from the mucous membrane lining the pharyngeal cavity. Elevate the posterior belly of digastric to expose the part of the middle constrictor at the hyoid bone [FIG. 32].

Find the thyrohyoid muscle in the triangle. Push the superior thyroid and carotid arteries posteriorly, and remove the fat in which the external laryngeal nerve lies. This exposes a part of the inferior constrictor muscle passing backwards from the side of the thyroid cartilage.

Separate the internal and common carotid arteries from the internal jugular vein to expose the vagus nerve in the posterior part of the carotid sheath between them. Pull the arteries anteromedially and find the sympathetic trunk posteromedial to the sheath.

CAROTID TRIANGLE

This triangle, bounded by the sternocleidomastoid, the posterior belly of digastric, and the superior belly of omohyoid, contains the vertical neurovascular bundle of the neck, mainly covered by the corresponding part of sternocleidomastoid. The medial wall of the triangle is formed by the hyoglossus and thyrohyoid muscles anteriorly and by the middle and inferior constrictor muscles posteriorly. Examine the relative positions of the contents of the triangle.

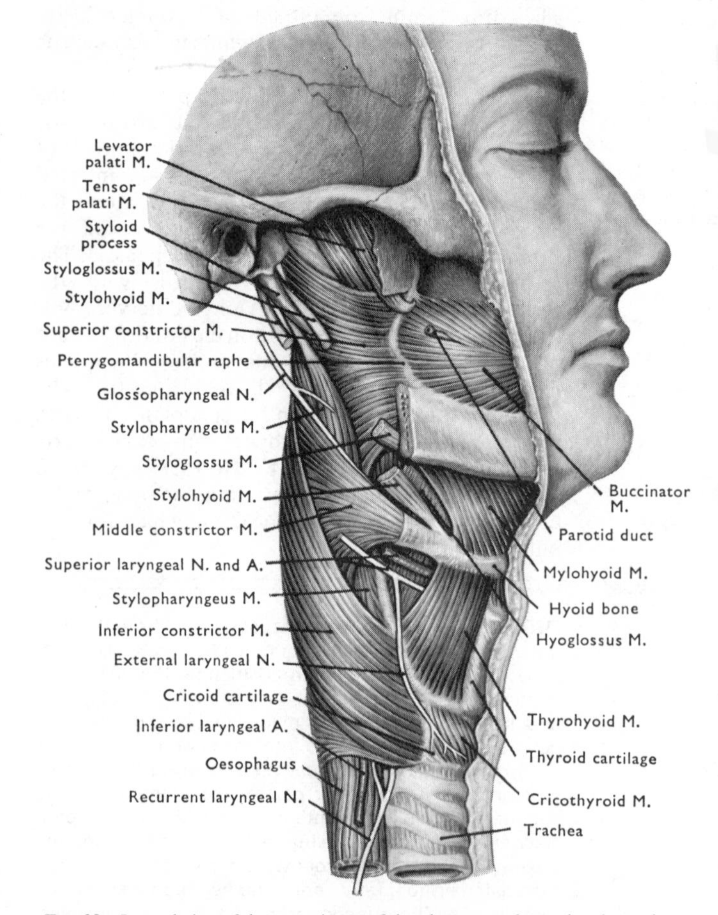

FIG. 32 Lateral view of the constrictors of the pharynx and associated muscles.

Carotid Sheath

This fascial sheath encloses the internal jugular vein, the carotid arteries, the vagus nerve, and the roots of the ansa cervicalis in the posterior part of the triangle. The sympathetic trunk is in the posteromedial wall of the sheath, and many **lymph nodes** lie on its superficial surfaces. These transmit lymph from adjacent regions and from nodes higher in the neck, to the nodes at the root of the neck.

The internal jugular vein descends vertically, first with the internal and then the common carotid artery medial to it: the common carotid divides at the level of the upper border of the thyroid cartilage into internal and external branches.

External Carotid Artery. It ascends along the side of the pharynx, anteromedial to the internal carotid, and gives off most of its branches in the carotid triangle.

1. The **superior thyroid artery** arises low in the triangle, curves antero-inferiorly, and disappears deep to omohyoid. It gives infrahyoid, sternocleidomastoid, and superior laryngeal branches in the triangle. The superior laryngeal enters the larynx with the internal laryngeal nerve.

2. The **ascending pharyngeal artery** arises from the lowest part of the external carotid, and ascends between the

internal carotid and the side of the pharynx.

3. The **lingual artery** arises behind the tip of the greater horn of the hyoid bone. It runs forwards on the middle constrictor, hooks above the tip of the greater horn, and disappears medial to hyoglossus.

4. The **facial artery** arises above the lingual, and leaves the triangle at once by ascending deep to the posterior belly of digastric.

5. The **occipital artery** runs posterosuperiorly along the lower border of the posterior belly of digastric.

Veins

The **facial vein** enters the carotid triangle over the posterior belly of digastric. It unites with the anterior branch of the retromandibular vein, but may also be joined by the lingual and superior thyroid veins as it crosses the carotid arteries to enter the internal jugular vein [FIGS. 33, 111].

The **lingual vein** is formed at the posterior border of hyoglossus by the union of veins which run with the lingual artery and the hypoglossal nerve. It crosses the carotid arteries and enters either the internal jugular or the facial vein.

The **superior thyroid vein** enters either the internal jugular or facial vein.

Nerves

The **accessory nerve** runs postero-inferiorly across the upper angle of the triangle, either superficial or deep to the internal jugular vein.

The **hypoglossal nerve** (which supplies the muscles of the tongue) appears at the lower border of the posterior belly of digastric, curves forwards round the root of the occipital artery on to the carotid arteries and the hook of the lingual artery, and disappears deep to the posterior belly of digastric. In the carotid triangle it gives off the **superior root of the ansa cervicalis** (which descends on the internal and common carotid arteries) and the thyrohyoid branch which passes forwards to its muscle. Both branches are composed of fibres of the ventral ramus of the first cervical nerve, which join the hypoglossal close to the skull and run with it [FIG. 65].

Ansa Cervicalis. Another slender nerve (the **inferior root of the ansa cervicalis**) arises from the ventral rami of the second and third cervical nerves behind the internal jugular vein. It curves forwards, usually on the lateral surface of the vein, and passing downwards out of the triangle, runs on to the common carotid artery. Here it joins the superior root from the hypoglossal nerve to form a loop called the ansa cervicalis at the level of the lower part of the larynx.

This complex consists of nerve fibres from cervical ventral rami, some of which form the **thyrohyoid nerve** (C. 1) while others may supply the **geniohyoid** through the hypoglossal [FIG. 107]. The ansa cervicalis supplies the remaining **infrahyoid muscles** (sternohyoid, sternothyroid, and omohyoid). Thus the hypoglossal nerve forms a secondary plexus with the first three cervical ventral

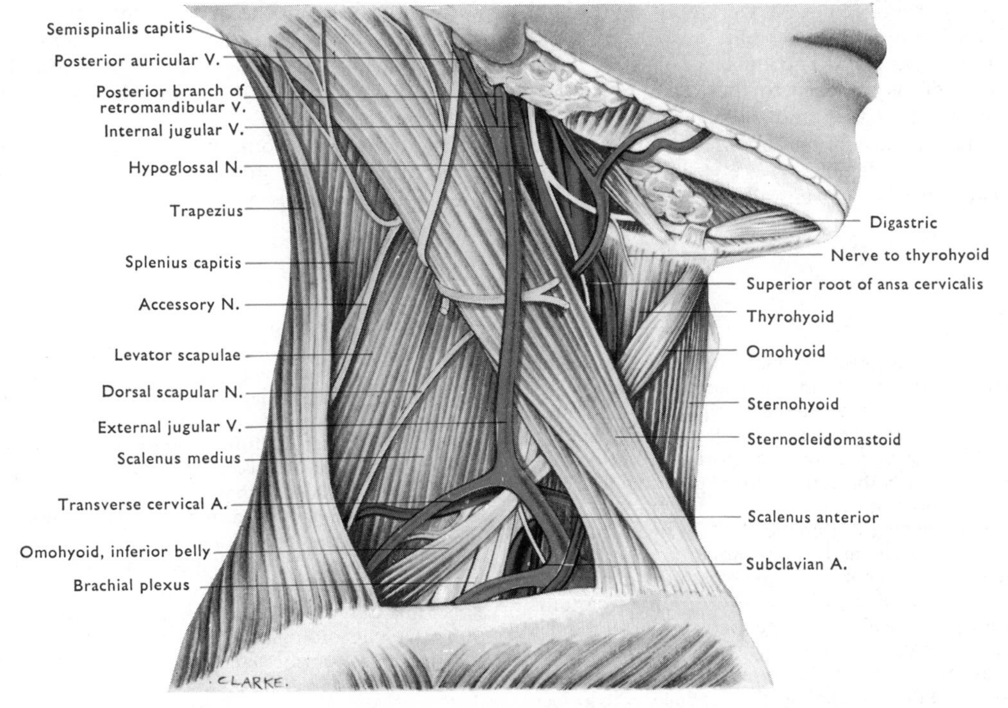

FIG. 33 Lateral view of the triangles of the neck.

rami, and with them supplies a ventral strip of muscle from the tongue to the sternum, including omohyoid. The **phrenic nerve** (C. 3, 4, 5) may be looked on as continuing this complex into the thorax for the supply of the diaphragm.

MUSCULAR TRIANGLE

This is the space bounded by the sternocleidomastoid, the superior belly of omohyoid, and the median plane. It contains the infrahyoid muscles and the structures in the midline which have been seen already [p. 32].

DISSECTION. Draw the sternal head of sternocleidomastoid aside to expose the anterior jugular vein deep to it. Find the intermediate tendon of omohyoid, and raise the superior belly to expose its nerve. Trace this nerve to the ansa cervicalis, and expose the entire ansa and its branches.

Remove the fascia from the infrahyoid muscles, but retain their nerves. Define the attachments of the sternohyoid, the sternal one by passing the handle of a knife downwards between it and the sternum. Then divide the muscle low down and, by reflecting it upwards to the hyoid, expose the other muscles and their attachments.

Infrahyoid Muscles

These ribbon-like muscles are sternohyoid, sternothyroid, thyrohyoid, and omohyoid. They lie on the trachea, thyroid gland, larynx, and thyrohyoid membrane—the sternothyroid and thyrohyoid forming a deeper layer. The thin layer of fascia which encloses them is attached to the investing cervical fascia by areolar tissue, and is thickened around the intermediate tendon of omohyoid to hold it down to the sternum and clavicle.

Nerve supply: the ventral rami of the first three cervical nerves through the thyrohyoid nerve and the ansa cervicalis. **Action**: they move the larynx and hyoid bone in speech and swallowing. They can: (1) depress the hyoid bone, or, when acting with the suprahyoid muscles, fix the hyoid to form a stable base for the tongue; (2) draw the larynx towards the hyoid (thyrohyoid) as in the early phase of swallowing; (3) depress the larynx, leaving the hyoid in position (sternothyroid) as in the last phase of swallowing [p. 120].

Sternohyoid passes from the posterior surfaces of the manubrium and medial end of the clavicle to the lower border of the hyoid bone adjacent to the midline. Its upper part is covered only by skin and fascia, but the anterior jugular vein and sternocleidomastoid overlap it inferiorly.

The superior belly of **omohyoid** passes from the inferior surface of the body and greater horn of the hyoid, immediately lateral to sternohyoid, to the intermediate tendon, which lies on the internal jugular vein under sternocleidomastoid, at the level of the cricoid cartilage. This tendon is held to the sternum and clavicle by a fascial sheath. Inferior belly, see page 22.

Sternothyroid is shorter, wider, and deeper than sternohyoid and so arises from the sternum and first costal cartilage below that muscle. It ascends to the oblique line on the lateral surface of the thyroid cartilage [FIG. 128] superficial to the attachment of the pretracheal fascia and the cricothyroid muscle. It is anterior to the large vessels of the upper thorax and root of the neck, and to the thyroid gland.

The **thyrohyoid** is the upward continuation of sternothyroid from the oblique line on the thyroid cartilage to the lower border of the greater horn of the hyoid bone. It is deep to omohyoid and sternohyoid, and covers the entry of the internal laryngeal nerve to the larynx.

The dissectors should now damp and cover the dissection already completed so as to avoid it drying and making later dissection difficult.

THE CRANIAL CAVITY

DISSECTION. Support the head on a block, and make a sagittal cut through the epicranial aponeurosis from the root of the nose to the external occipital protuberance. Pull each half laterally and detach it from the temporal lines. Strip the periosteum (**pericranium**) from the external surface of the vault of the skull down to a level below the upper attachment of the temporalis muscle [FIG. 19], detaching it from the skull. At the **sutures** of the skull it is necessary to cut through the periosteum which is continuous through them with the periosteum lining the interior of the skull (**endocranium**). This turns the scalp, periosteum, and upper parts of the temporalis muscles down over the auricles.

Remove the skull cap or **calvaria**. First make a pencil mark on the skull by encircling it horizontally with a piece of string passing 1 cm above the orbital margins and external occipital protuberance, and drawing round it on this. Make a saw cut along this line, but avoid cutting deeper than the marrow cavity (when the sawdust turns red) when the outer table of the skull has been divided. In the temporal region, the skull is very thin and there may be no **marrow cavity (diploë)**, so proceed with caution. Introduce a blunt chisel into the saw cut, and split the inner table by a series of short, sharp strokes with a mallet. Even when this is divided, the calvaria will not lift free because the endocranium, and the outer covering of the brain (dura mater) to which it is fused, adhere firmly to the interior of the skull. So introduce the thick part of the chisel into the cut and prise the skull cap upwards. Unless you have cut through the endocranium and dura, they will tear away from the skull taking the vessels of the skull with them.

Sutural Ligaments

The loosely attached pericranium is continuous with the endocranium through the sutures of the skull, forming the sutural ligaments. These hold the bones together and allow growth between them. As the skull consolidates from the third to fourth decade onwards, the adjacent bones unite by ossification of the sutural ligaments—a process which begins internally and is known as **synostosis**.

Endocranium

When the skull cap is detached, the outer surface of the endocranium is exposed. It is rough because of the fine fibrous and vascular processes which pass between it and the bones. Torn blood vessels are most numerous close to the midline. Here one of the largest intracranial venous channels lies deep to the endocranium. If a blunt instrument is pressed on the sinus, blood oozes from the numerous, small veins which have been ruptured. The endocranium is more firmly attached to the base of the cranial cavity than to the vault, the degree of adhesion varying with age and from individual to individual.

A number of branching vessels ascend on the outer surface of the endocranium towards the vertex. These branches of the **middle meningeal artery**, with the corresponding veins on their external surfaces, groove the inner table of the skull, and so stand out in relief from the surface of the endocranium. They supply the skull (particularly the red bone marrow in its diploë), the endocranium, and the dura mater which is fused to its internal surface. *These meningeal vessels play no part in supplying the pia-arachnoid* [p. 38] *or the brain itself*, but may be torn in fractures of the skull. This results in bleeding between the endocranium and the skull (*extradural haemorrhage*) producing a swelling which presses on the brain. If the fracture tears the endocranium and dura mater (the outermost of the meninges) then bleeding spreads into the space (subdural space) which separates the dura from the other meninges covering the brain—a *subdural haemorrhage*.

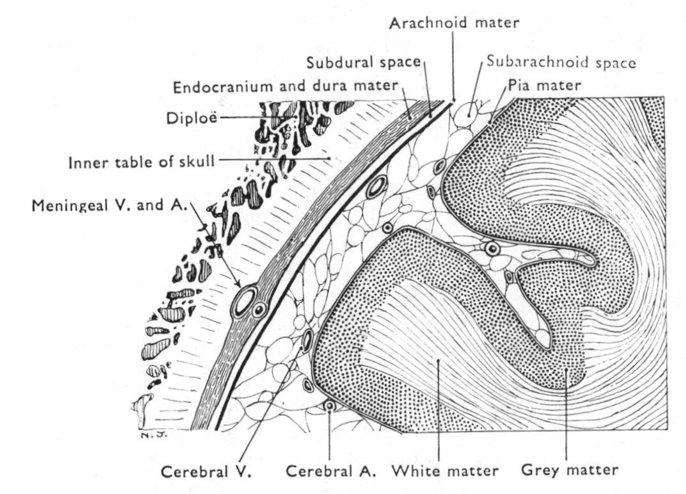

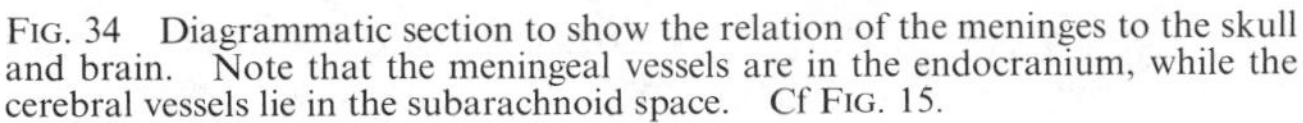

FIG. 34 Diagrammatic section to show the relation of the meninges to the skull and brain. Note that the meningeal vessels are in the endocranium, while the cerebral vessels lie in the subarachnoid space. Cf FIG. 15.

THE MENINGES

The entire central nervous system (the brain and spinal medulla) is enclosed in three membranes or meninges (the dura mater, the arachnoid, and the pia mater) which are separated from each other by two spaces (the subdural and subarachnoid spaces) containing fluid.

Dura Mater

This outermost meninx is a thick fibrous layer. Externally it fuses with the endocranium of the skull except: (1) where it forms rigid folds or partitions between the major parts of the brain, and incompletely subdividing the cranial cavity, supports the brain within it [FIGS. 37, 38]; (2) where the venous sinuses of the dura mater lie between the dura and the endocranium. The largest venous sinuses lie along the lines of attachment of the dural folds to the endocranium [FIG. 36].

The dura mater encloses the spinal medulla as a tubular sheath extending downwards from the foramen magnum to the sacrum [p. 159]. It is separated from the periosteum of the spinal canal by loose fatty connective tissue containing a plexus of veins (internal vertebral venous plexus, p. 158) which corresponds to the venous sinuses of the dura mater and is continuous with them through the foramen magnum where the dura fuses with the endocranium.

The internal surface of the dura mater is a smooth, glistening, endothelial layer. It is separated from the equally smooth external surface of the arachnoid by a capillary interval—the **subdural space**. This space surrounds the central nervous system, and is only absent where structures pierce the meninges to enter or leave that system, or where arachnoid villi (see below) are present. The space acts as a bursa which allows movement between the dura and the structures it encloses.

The Arachnoid and Subarachnoid Space

The arachnoid and pia mater develop from a single mass of loose connective tissue immediately surrounding the central nervous system, inside the dura mater. Cerebrospinal fluid from the cavities of the brain percolates into this tissue, separating it into an outer layer, the arachnoid, applied to the internal surface of the dura mater, and an inner layer, the pia mater, applied to the surface of the central nervous system, with the subarachnoid space between them. A variable number of strands (**trabeculae**) of the original tissue persist between the arachnoid and the pia.

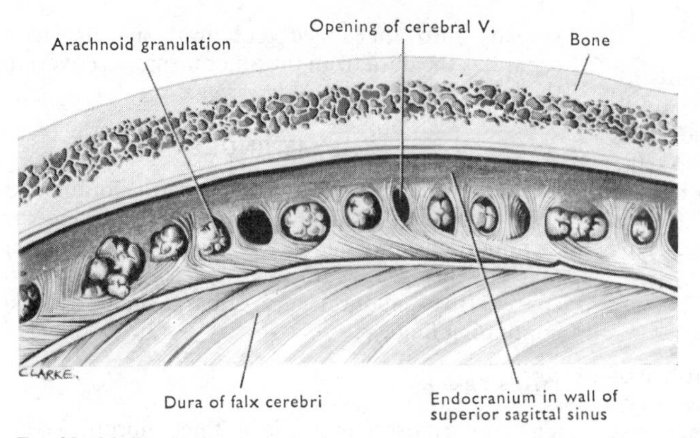

FIG. 35 Median section through the skull and superior sagittal sinus. Note arachnoid granulations protruding into the sinus.

Where the arachnoid and pia are close together, the trabeculae are numerous so that the subarachnoid space forms a fluid-filled sponge which may have some protective function for the brain. Where they are more widely separated, as where the brain is not closely fitted to the dura-arachnoid, the trabeculae are usually few and a simple fluid-filled space results. Such parts of the subarachnoid space are known as **cisterns**.

The **arachnoid** is a thin, transparent, avascular membrane covered by the endothelium of the subdural space, and lined by that of the subarachnoid space, with trabeculae passing to the pia mater from its deep surface. In a few places, notably at the superior sagittal sinus, the arachnoid pierces the dura as a number of finger-like processes (**arachnoid villi**) which form openings from the subarachnoid space into the venous sinus (see below). Like the dura, it extends along the nerves which arise from the brain and spinal medulla for a short distance before becoming continuous with the **epineurium**.

Pia Mater

This is a vascular membrane which covers the surface of the central nervous system and follows most of the irregularities of its surface. The pia mater is thicker than the arachnoid, especially on the spinal medulla and the medulla oblongata of the brain. The blood vessels of the central nervous system lie on the external surface of the pia mater. Their branches carry a sleeve of pia mater with them for a short distance into the central nervous system, together with a diverticulum of the subarachnoid space—the **perivascular space**. Also, each rootlet of nerves entering or leaving the brain carries a sleeve of pia mater as it crosses the subarachnoid space. These sleeves, surrounded by the endothelium of the space, are continuous with the perineurium of the nerves formed by the rootlets. Another modification of the pia mater is the **ligamentum denticulatum** of the spinal medulla—see page 160.

Cerebrospinal Fluid. This clear fluid is mainly derived from special vascular tufts (**choroid plexuses**) in the cavities of the brain (ventricles) from which it escapes into the subarachnoid space [p. 196]. In addition to its metabolic functions, it forms a sort of protective water bath around the brain and spinal medulla.

DISSECTION. Make a median sagittal incision through the endocranium and so open the superior sagittal sinus as far forwards and backwards as the removal of the calvaria will allow.

Structure of Venous Sinuses of Dura Mater. These venous channels are lined with endothelium applied to the endocranium externally and the dura mater internally, except in the case of the straight and inferior sagittal sinuses which are entirely enclosed in dural folds which pass between the parts of the brain. The sinuses transmit venous blood from the brain, meninges, and skull, mainly to the internal jugular vein. They also communicate with a number of extracranial veins via **emissary veins** which traverse foramina in the skull. Most of the sinuses lie in shallow grooves on the internal surface of the cranial cavity.

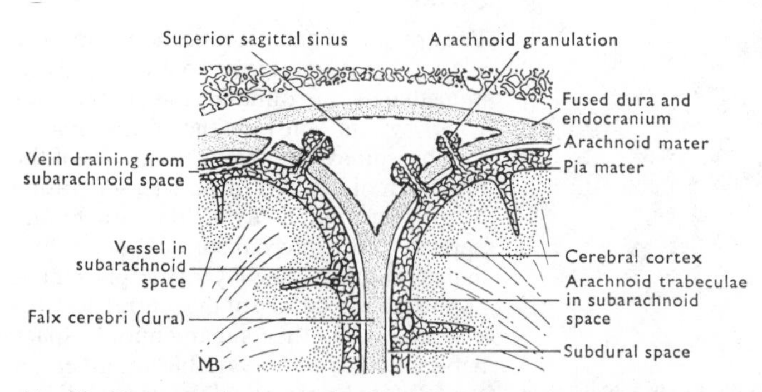

FIG. 36 Diagrammatic transverse section through the superior sagittal sinus and surrounding brain. Note the arrangement of the arachnoid granulations.

Arachnoid Granulations. These small, granular bodies lie mainly along the sides of the superior sagittal sinus. They are protrusions of arachnoid through apertures in the dura in contact with the endothelium of the venous sinus and its lateral lacunae. The granulations are normal enlargements of similar but microscopic processes of the arachnoid (arachnoid villi) which alone are present in children. In old age they may reach considerable proportions and even erode the inner table of the skull. They have also been described in most of the venous sinuses of the dura mater, and in the middle meningeal veins, but they are most numerous in the superior sagittal sinus, particularly in the highest part of its convex course. They are valvular structures which permit cerebrospinal fluid to pass into the venous system, but prevent the reflux of blood. Clotting of the blood in the superior sagittal sinus blocks these safety valves and causes a rapid rise in cerebrospinal fluid pressure.

DISSECTION. Make an incision through the dura mater and endocranium along the whole length of each side of the superior sagittal sinus. From the midpoint of each incision, make another down to the cut edge of the skull above the auricle. As these cuts are made, raise the endocranium and dura to avoid incising the arachnoid, pia, and brain. Turn the flaps down over the cut edges of the skull to avoid its sharp edges injuring the brain or your fingers during subsequent dissection. The pia-arachnoid and blood vessels on the upper parts of the cerebral hemispheres are now exposed.

The subdural space is now opened and should be examined.

Superior Cerebral Veins

These extremely thin-walled veins lie in the subarachnoid space immediately superficial to the pia mater, entirely separate from the meningeal veins. They run upwards towards the median plane, converging on the superior sagittal sinus, and can be seen through the arachnoid only when filled with blood, but are visible where they cross the subdural space to enter the sinus through the dura. Here they may be ruptured as a result of a violent blow on the skull which moves the brain within it. This produces low pressure bleeding into the subdural space. The anterior and posterior veins enter the sinus obliquely.

SUPERIOR SAGITTAL SINUS

This sinus begins anteriorly at the crista galli, where it communicates with the veins of the frontal sinus, and sometimes with the veins of the nose, through the foramen caecum. The sinus runs posteriorly, forming a median groove on the cranial vault. It becomes continuous with the right transverse sinus at the internal occipital protuberance [FIG. 48].

Lateral Lacunae. These are cleft-like lateral extensions of the sinus between the dura mater and endocranium. The largest, 2–3 cm in diameter, overlies the upper part of the motor area of the brain (*q.v.*). Many arachnoid granulations enter the lateral lacunae as well as the sinus. Lacunae and granulations both increase in size with age, and may produce shallow depressions and clean cut pits, respectively, in the skull vault on both sides of the groove for the sinus. The superior cerebral veins pass inferior to the lacunae to enter the sinus, but meningeal and diploic veins enter the lacunae.

DISSECTION. Expose the falx cerebri by dividing the superior cerebral veins on one side, and displacing the upper part of that hemisphere laterally.

FALX CEREBRI [FIGS. 37, 38]

The falx cerebri is a median, sickle-shaped fold of dura mater, which descends between the two cerebral hemispheres from its attachment to the lips of the groove for the superior sagittal sinus. Anteriorly, it is attached to the crista galli, is shallow, and often has many perforations. Posteriorly, it increases in depth,

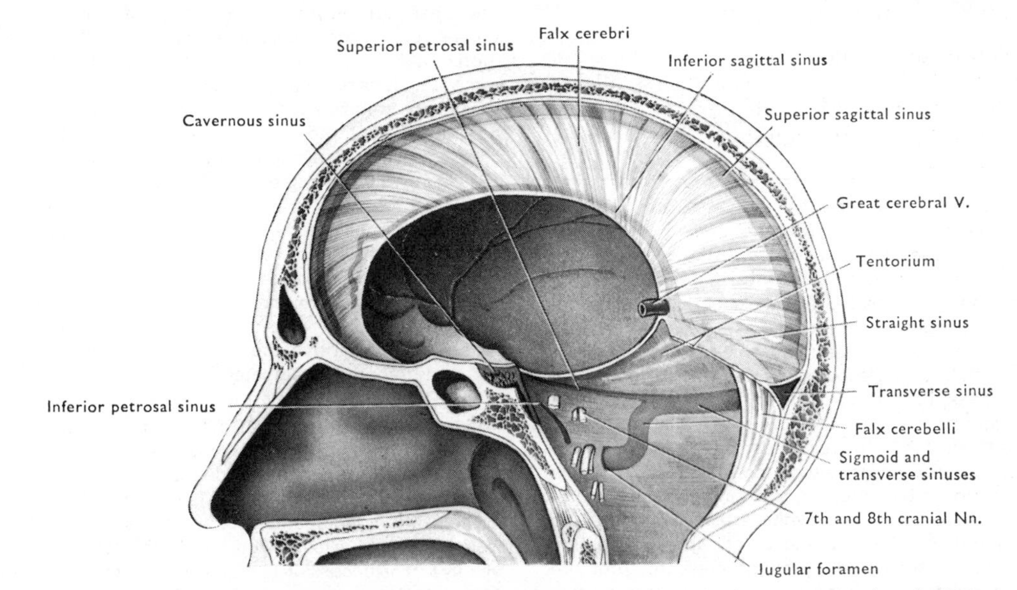

FIG. 37 Sagittal section through the skull to the left of the falx cerebri. The brain has been removed to show the folds of dura mater which incompletely partition the cranial cavity.

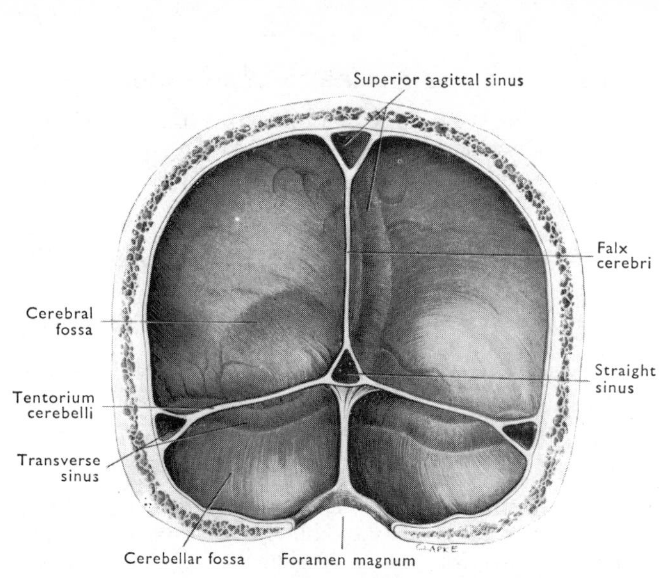

FIG. 38 Coronal section through the skull at the level of the foramen magnum. The brain has been removed to show the fossae which lodge the occipital lobes of the hemispheres and the cerebellum. The fossae are separated by the falx cerebri and tentorium cerebelli, and the cerebellar fossa is partly divided by the falx cerebelli.

and becomes continuous with the upper surface of the tentorium cerebelli—an approximately horizontal fold of dura mater which extends inwards between the posterior parts of the cerebral hemispheres above and the cerebellum below [FIGS. 37–41]. Elsewhere the lower border is free and concave, and overhangs the corpus callosum—a mass of nerve fibres which connects the two cerebral hemispheres. The **superior sagittal sinus** lies in the fixed margin of the falx; the **inferior sagittal sinus** in its free margin; and the **straight sinus** in its attachment to the tentorium cerebelli [FIGS. 37, 38].

REMOVAL OF THE BRAIN

The following dissection destroys one hemisphere of the brain, and should not be used if it is essential to retain the brain intact for future dissection. In that case, the alternative dissection should be followed.

DISSECTION. Draw the hemisphere on which the superior cerebral veins were divided away from the falx cerebri. This exposes the **corpus callosum**—a mass of white matter running between the hemispheres at a depth of approximately 5 cm. Make a median sagittal cut through the corpus callosum from its posterior end forwards, passing between the two anterior cerebral arteries which run posteriorly, one on the medial aspect of each hemisphere. Proceed carefully so as to divide no more than the corpus callosum, the posterior end of which is thicker than the remainder. As the two parts of the corpus callosum are separated, a thin layer of pia mater is exposed beneath its posterior part, while further forwards there is a thin, white, vertical sheet. This **septum pellucidum** is composed of right and left parts which can be separated by slight pressure from the handle of a knife, provided the division of the corpus callosum is strictly in the median plane.

When the two halves of the septum pellucidum and the bundle of nerve fibres on the inferior edge of each (the **fornix**) are separated, the whole extent of the layer of pia mater below the corpus callosum is exposed. Two **internal cerebral veins** may be seen lying in this pia mater, one on each side of the median plane. Posteriorly, these internal cerebral veins join to form the **great cerebral vein** which emerges from beneath the posterior end of the corpus callosum, and curving upwards over it, enters the straight sinus [FIG. 37]. Divide the pia mater between the internal cerebral veins. This opens the deep, slit-like **third ventricle of the brain** [FIGS. 246, 249].

Complete the division of the corpus callosum, and note that its anterior extremity curves posteriorly below the septum pellucidum.

Now make a horizontal slice through the mobile hemisphere about 2 cm above the level of the corpus callosum and parallel to the cut through the skull. This shows the surface layer of **grey matter** (cerebral cortex) on the cerebrum, and the deeper mass of **white matter** (nerve fibres) [FIG. 232]. Now make a slice, parallel to the first, but passing through the posterior and anterior extremities of the corpus callosum. This shows some of the internal features of the brain [FIG. 262], most of the lateral surface of the falx, and the most superior part of the tentorium. The remainder of the tentorium can be displayed by lifting the posterior part (occipital lobe) of the hemisphere upwards.

Pull the anterior part of the sliced hemisphere laterally and cut slowly downwards through the anterior wall of the third ventricle. From the antero-inferior part of the corpus callosum (rostrum) [FIG. 222] the cut passes through a rounded bundle of nerve fibres, the **anterior commissure** [FIG. 181] into a thin sheet of tissue, the **lamina terminalis**. At the inferior extremity of this lies the **optic chiasma**—the meeting of the two optic nerves. Leave the chiasma intact, and make a further 'horizontal' section which passes immediately above the optic chiasma to a point below the posterior end of the corpus callosum. This leaves only a small part of the anterior end of the hemisphere in the skull. Lift the anteromedial part of the frontal lobe gently and expose the **olfactory bulb** at the side of the crista galli, on the cribriform plate of the ethmoid [FIG. 40]. Pull away the remainder of the frontal lobe, leaving the olfactory bulb and the olfactory tract, which passes posteriorly from it, in place. This exposes the **optic nerve** passing to the chiasma, and the **internal carotid artery** lateral to the chiasma with its anterior cerebral branch passing anteromedially above the optic nerve, and its middle cerebral branch passing laterally.

Pull the occipital lobe of the hemisphere laterally and find the **posterior cerebral artery** on its medial aspect. Cut through it and try to leave its anterior part with the central part of the brain (diencephalon and midbrain) as you strip the remainder of the hemisphere forwards by pulling the occipital lobe forwards and laterally. Depending on the level of the last 'horizontal' slice, it may or may not be necessary to divide any of the brain tissue to complete the separation of the hemisphere, but it is usually possible to separate it by lifting the middle of the lower edge of the hemisphere upwards. As the lower part of the hemisphere is removed, note the tip of the temporal lobe slipping out of its recess in the anterior part of the middle cranial fossa.

The surfaces of the **falx** and **tentorium** are now exposed. Follow the free margins of both to their anterior attachments—the falx to the cristi galli, and the tentorium to the anterior clinoid process. Posteriorly, the two margins meet where the great cerebral vein pierces the dura to enter the straight sinus.

Find the **internal carotid artery** and the remains of the posterior cerebral artery. The posterior communicating artery should be seen uniting the two, and the **oculomotor nerve** lying below the posterior cerebral artery, immediately medial to the free margin of the tentorium. Follow this nerve forwards and backwards.

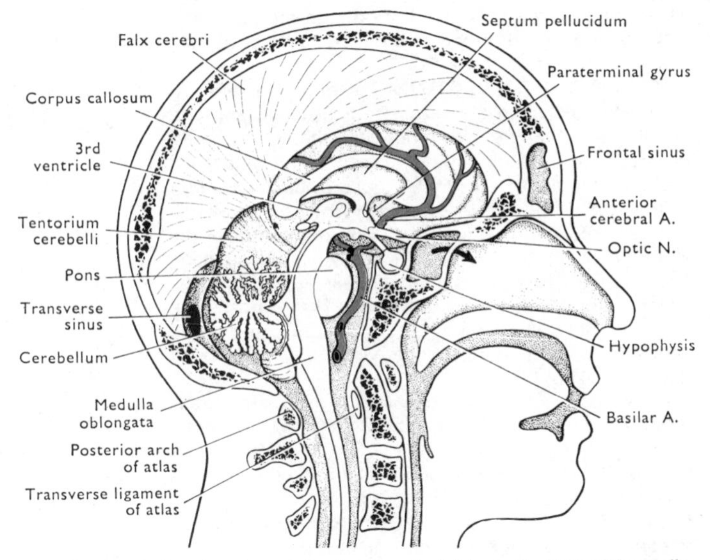

FIG. 39 Saggital section through head and neck slightly to the right of the median plane.

Examine the exposed parts of the remaining hemisphere [FIGS. 39, 222] and the base of the skull from which the hemisphere has been removed.

Note the sharp posterior margin of the **anterior cranial fossa**, which is formed by the lesser wing of the sphenoid bone. It overlies the tip of the temporal lobe which is inserted into the anterior extremity of the **middle cranial fossa**. Medially, the lesser wing of the sphenoid is continued into the anterior clinoid process, which is immediately lateral to the internal carotid artery and the optic nerve. Posteromedial to these is the **infundibulum**, a narrow funnel extending inferiorly from the floor of the third ventricle to pierce the dura mater (diaphragma sellae) and reach the hypophysis.

The remaining hemisphere may be removed intact by gently elevating it and dividing the olfactory tract, optic nerve and internal carotid artery, and then cutting transversely through that half of the midbrain (the narrow

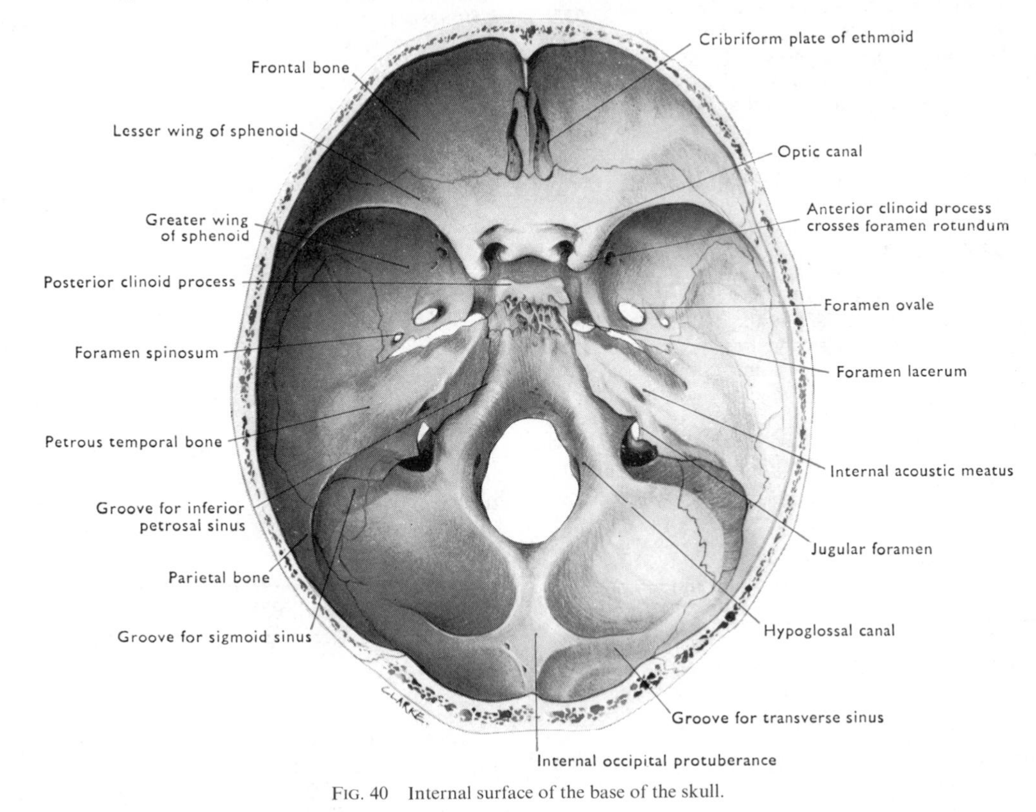

FIG. 40 Internal surface of the base of the skull.

part of the brain which passes through the aperture in the tentorium) parallel to but above the oculomotor nerve.

ALTERNATIVE DISSECTION. To ıemove the brain in one piece, detach the falx cerebri from the crista galli, and pull the falx posteriorly from between the hemispheres. Place a block under the shoulders to allow the head to fall back. This removes the frontal lobes from the anterior cranial fossae and exposes the **olfactory bulbs** which can be gently elevated from the cribriform plates of the ethmoid, tearing the olfactory nerves which pierce them. The large **optic nerves** then appear and should be divided close to the optic foramina. This exposes the internal carotid arteries with the infundibulum passing vertically to the hypophysis between them. Divide all three [FIG. 41].

Allow the brain to fall backwards as each structure is divided, but support it posteriorly so as to avoid damaging it on the skull.

Posterior to the infundibulum, identify the dorsum sellae with a posterior clinoid process at each lateral extremity [FIG. 40], and the **oculomotor nerves** passing forwards, one on each side of it [FIG. 41].

Lateral to each oculomotor nerve is the **free edge of the tentorium cerebelli**. Turn the margin of the tentorium laterally and divide the slender **trochlear nerve** which lies just under its free margin [FIG. 48].

Turn the head forcibly round to one side, raise the posterior part of the upper hemisphere with the fingers from the tentorium. Divide the **tentorium** along its attachment to the petrous temporal bone [FIGS. 40, 41], taking care not to injure the cerebellum beneath. Then turn the head to the opposite side and repeat the procedure on the other side.

Let the brain fall backwards so as to draw the brain stem away from the anterior wall of the posterior cranial fossa and bring the nerves into view.

Divide the oculomotor nerves, and allowing the brain to fall first to one side and then the other, identify and divide the trigeminal, abducent, facial, and vestibulocochlear nerves; the last two entering the internal acoustic meatus. Identify and cut the nerves passing to the jugular foramen [FIG. 48], but leave the accessory nerve on one side by detaching it from the brain. Look for the hypoglossal nerves, which are deeper and more medial, and cut them also [FIG. 47].

Press the pons [FIG. 39] further posteriorly, and identify the two vertebral arteries ascending to form the basilar artery on its surface. Pass a knife into the vertebral canal in front of the medulla oblongata, and cutting firmly from side to side, divide the spinal medulla and the vertebral arteries. Withdraw the brain from the cranial cavity, dividing any roots of the accessory nerve still attached to the spinal medulla, and those of any spinal nerves included in the specimen.

Removal of the brain in this way, without complete separation of the tentorium from the skull, ruptures the great cerebral vein where it enters the straight sinus at the junction of the free margins of the falx and tentorium.

STRUCTURES SEEN AFTER REMOVAL OF CEREBRUM

[FIGS. 40, 41]

With the help of a skull, identify the boundaries of the anterior and middle cranial fossae, and the taut but resilient tentorium cerebelli.

In the **anterior cranial fossa** note:

1. The **crista galli** with the falx attached.
2. The **cribriform plate of the ethmoid** which lodged the olfactory bulb. The olfactory nerves pass through the plate into the bulb from the nasal muscosa immediately inferior to the plate.
3. Most of the floor of the fossa is formed by the **orbital part of the frontal bone**. It is closely fitted to the irregularities of the orbital surface of the frontal lobe which it separates from the orbit.
4. The **lesser wing of the sphenoid bone**, with the sphenoparietal venous sinus running along its posterior margin towards the **anterior clinoid process**, where the free margin of the tentorium is attached.

In the **middle cranial fossa** note:

1. The recess for the tip of the temporal lobe of the brain, posterolateral to the orbit.
2. The **middle meningeal vessels** showing through the dural floor of the lateral part of the fossa, and the absence of any branches of these piercing the dura mater.
3. The raised central area (**diaphragma sellae**) formed by the dura stretched between the four clinoid processes of the sphenoid bone [FIG. 40]. This dura is perforated by:

(a) The **infundibulum** passes through the central aperture to the hypophysis.

(b) Each **oculomotor nerve** passes from the anterior surface of the midbrain, and pierces the dura medial to the free border of the tentorium [FIG. 41] to reach the lateral wall of the corresponding **cavernous sinus**. These sinuses lie lateral to the hypophysis, deep to the anterior parts of the free borders of the tentorium as they approach the anterior clinoid processes. Inferior to the dura, the right and left cavernous sinuses are united by intercavernous sinuses, anterior and posterior to the hypophysis.

(c) The **optic nerves** and **internal carotid arteries** are medial to the anterior clinoid processes. Note also the branches of the internal carotid arteries above the dura—the anterior and middle cerebral arteries, the posterior communicating artery to the posterior cerebral, and the ophthalmic artery which runs anteriorly with the optic nerve [FIGS. 43, 82].

4. Each **posterior cerebral artery** curves laterally above the corresponding oculomotor nerve and the free border of the tentorium, from the median basilar artery. It receives the posterior communicating artery [FIG. 43].
5. The **great cerebral vein** with tributaries entering it from the midbrain and cerebellum at the junction of the free edges of the falx and tentorium [FIG. 41].

TENTORIUM CEREBELLI

This wide, sloping fold of dura mater has a free margin surrounding the midbrain, and a fixed margin attached to the skull at the lip of the posterior cranial

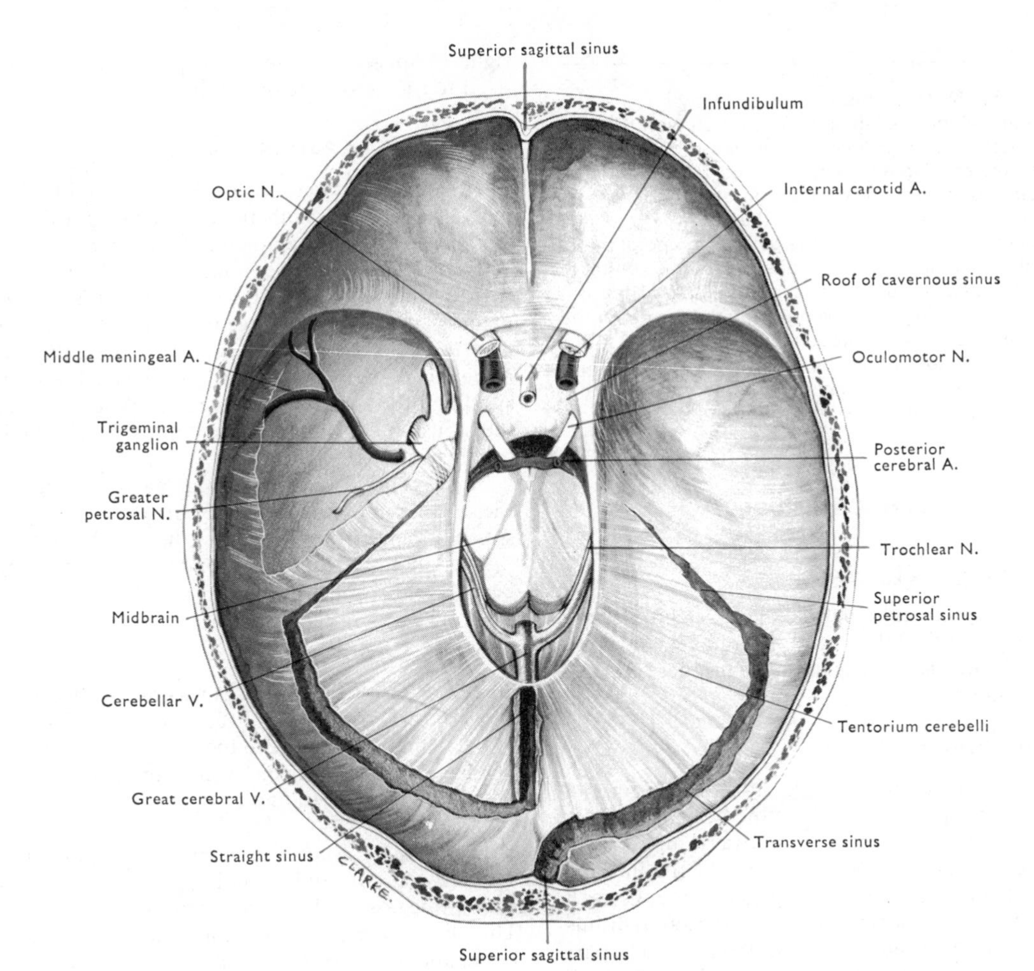

FIG. 41 Interior of the cranium after removal of the cerebrum. The transverse, straight, and superior petrosal sinuses have been opened, and the dura mater has been removed from the floor of the left middle cranial fossa.

fossa. At the attachment, the two dural layers of the fold diverge to enclose a venous sinus—the wide **transverse sinus** between the internal occipital protuberance and the base of the petrous temporal bone, and the narrow **superior petrosal sinus** on the superior margin of the petrous temporal bone [FIG. 48].

At the apex of the petrous temporal bone, the margins of the tentorium cross each other, the free margin passing forwards to the **anterior clinoid process** above the attached margin passing medially to the **posterior clinoid process**. The dura mater which unites these two margins anteromedial to the crossing, forms the roof of the cavernous sinus, while that extending downwards from the free margin into the middle cranial fossa forms the lateral wall of the cavernous sinus. The roof of the sinus is pierced by the oculomotor nerve.

The tentorium slopes upwards from its attached margin to an apex at the point of meeting of its free border with that of the falx. Here the **great cerebral vein** pierces the dura mater and joins the inferior sagittal sinus to form the **straight sinus** [FIG. 37].

The free margin of the tentorium extends almost horizontally backwards from each anterior clinoid process to the apex of the tentorium, forming the **tentorial notch**. This gap in the tentorium connects the posterior cranial fossa with the supratentorial compartments of the cranial cavity, and transmits the midbrain, the posterior cerebral arteries, and cerebrospinal fluid in the subarachnoid space.

Functions of Falx and Tentorium

These tough folds of dura mater play an important part in stabilizing the brain within the cranial cavity, and prevent this semifluid structure oscillating freely when the head is moved suddenly. When the brain does move within the cranial cavity, it carries the pia and arachnoid with it, and throws considerable stress on the very thin-walled veins which traverse the subdural space to the venous sinuses which are held fast by the dura mater.

DISSECTION. Open the straight sinus by incising the falx cerebri along the left side of its union with the tentorium. Carry this incision along the free edge of the falx to open the small inferior sagittal sinus. Open the transverse sinus by continuing the incision laterally in the fixed margin of the tentorium from the internal occipital protuberance to the base of the petrous temporal bone. Note the continuity of the transverse sinus with the sigmoid sinus at this point, and open the superior petrosal sinus by continuing the incision along the margin of the tentorium fixed to the petrous temporal bone [FIG. 48].

On the right side, follow the superior sagittal sinus into the transverse sinus, and check for continuity of all the sinuses at the internal occipital protuberance.

VENOUS SINUSES

[p. 52]

Inferior Sagittal Sinus

This narrow sinus is in the posterior two-thirds of the free margin of the falx cerebri. It drains the falx and part of the medial surface of the hemisphere into the straight sinus.

Straight Sinus

Formed by the union of the great cerebral vein and the inferior sagittal sinus where the free edges of the falx and tentorium meet, it runs postero-inferiorly in the line of union of the two folds. At the internal occipital protuberance it becomes continuous with one of the transverse sinuses, usually the left. It drains the posterior and central parts of the cerebrum, part of the cerebellum, falx and tentorium. It may communicate with the superior sagittal sinus at the internal occipital protuberance. Then four sinuses (superior sagittal, straight, and two transverse) meet in the **confluence of the sinuses**, which makes a wide, shallow impression on the bone at this point.

Transverse Sinus

On each side, the transverse sinus runs horizontally from the internal occipital protuberance to the base of the petrous temporal bone. It lies in the fixed margin of the tentorium, below the occipital lobe of the cerebrum and above the cerebellum, and grooves the occipital, parietal, and temporal bones. The transverse sinus which receives the superior sagittal sinus is larger than the other, but if the sinuses communicate they may be of equal size.

Tributaries. It receives veins from the cerebrum (occipital lobe) and cerebellum, the occipital diploic vein, and the superior petrosal sinus. Anteriorly it drains into the sigmoid sinus.

Superior Petrosal Sinus

This narrow sinus drains the posterior end of the cavernous sinus to the junction of the transverse and sigmoid sinuses. It lies in the margin of the tentorium fixed to the petrous temporal bone.

PARANASAL SINUSES

These air-filled extensions of the nasal cavities pass into a number of skull bones. Each is lined by a ciliated columnar epithelium adherent to the endosteal lining of the bone in which it lies—a **muco-endosteum**. The cilia waft the surface mucus towards the opening into the nasal cavity, and keep the sinus empty. The openings of the sinuses are all relatively narrow and easily obstructed, even by swelling of the vascular endosteum. When blocked, the sinus tends to fill with secretions which readily become infected.

Sinuses completely replace the marrow cavities of the maxilla and ethmoid, and partly of the frontal, sphenoid, and temporal bones, thereby diminishing the weight of the skull. They tend to enlarge with age because of loss of the diploë.

Frontal Air Sinuses

One or other of these sinuses was probably opened when the calvaria was removed. If not, chisel off part of the frontal bone close to the median plane till one of the sinuses is opened. Explore the cavity with a probe and attempt to find its opening into the nasal cavity.

The frontal sinuses are paired, asymmetrical cavities in the frontal bone, immediately above the root of the nose and the upper margins of the orbits. They lie between the inner and outer tables of the bone, and are separated from each other by a bony septum which is not median. Normally about 2–3 cm in height and width, they may be smaller (especially in women), absent, or very large. In addition to expanding into the forehead, they may pass posteriorly between the two tables of the roof of the orbit.

The funnel-shaped passage from each frontal sinus to the nasal cavity (the **infundibulum** = funnel) opens into the middle meatus of the nasal cavity.

THE ANTERIOR CRANIAL FOSSA

DISSECTION. Carefully remove the dura mater from the cribriform plate of the ethmoid. Attempt to find the anterior ethmoidal nerve running anteriorly on its lateral margin.

The Anterior Ethmoidal Nerve. This terminal branch of the nasociliary nerve in the orbit enters the cranial cavity with the anterior ethmoidal artery between the frontal and ethmoid bones. It appears at the lateral edge of the cribriform plate, and passing forwards, enters the nose through a small aperture at the side of the crista galli. The nerve and artery supply the mucous membrane of the upper, anterior part of the nasal cavity, and end as the external nasal.

The small **posterior ethmoidal artery** follows the same course from the orbit, but further posteriorly. It assists the anterior ethmoidal and middle meningeal arteries to supply the dura and endocranium of the anterior cranial fossa.

THE MIDDLE CRANIAL FOSSA

DISSECTION. Incise the diaphragma sellae radially, and dislodge the hypophysis from the hypophysial fossa. Examine its shape with a hand lens, and then make a median section through it.

HYPOPHYSIS OR PITUITARY GLAND

This is an exceedingly important ductless gland with a wide range of functions, including the control of the other ductless glands and of body growth.

It is a flattened ovoid lying in the hypophysial fossa, and connected to the inferior surface of the hypothalamic part of the brain by the **infundibulum**. The **posterior lobe** of the hypophysis is the expanded inferior end of the infundibulum, and is developed from the brain. The **anterior lobe** is much larger than the posterior lobe, and consists of three parts which partly surround that lobe and the infundibulum. The **distal part** forms most of the anterior lobe. It is separated from the posterior lobe by a narrow cleft and a thin sheet of glandular tissue (**intermediate part**) applied to the posterior lobe. The **infundibular part** is a narrow upward projection of the distal part. It partly encircles the infundibulum and may reach the tuber cinereum of the brain [FIG. 250]. The anterior lobe develops from the ectoderm of the roof of the primitive mouth [p. 244] and has only a vascular connexion with the brain.

Blood Supply. This is by twigs from the internal carotid and anterior cerebral arteries. The anterior lobe also receives venous blood from the hypothalamus via the **hypothalamo-hypophysial portal system** of veins which transmits releasing factors to the hypophysis which control the release of its hormones. The veins of the hypophysis drain into the cavernous sinuses.

Position. It lies posterosuperior to the sphenoidal air sinuses [p. 129], below the optic chiasma, in front of the dorsum sellae, and has a cavernous sinus on each side of it. Several venous channels (**intercavernous sinuses**) connect the two cavernous sinuses around the hypophysis.

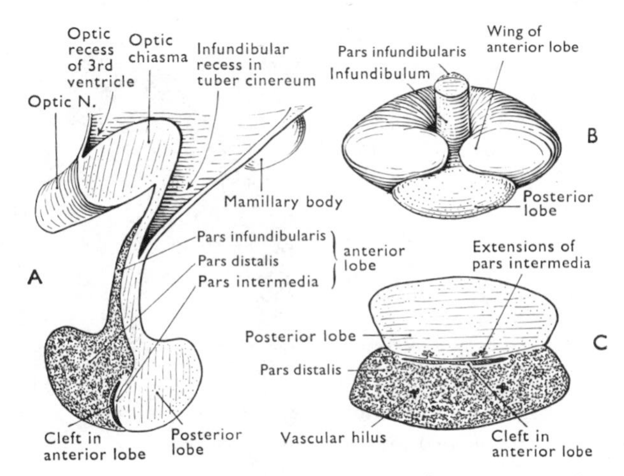

FIG. 42 Three diagrammatic views of the hypophysis. A. In median section. B. From above and behind. C. In horizontal section.

DISSECTION. *If the entire brain has not been removed*, cut through the tentorium along the line of the straight and transverse sinuses. Remove the falx, and turn each half of the tentorium anterolaterally on its attachment to the petrous temporal bone. Find the **trochlear nerve** [FIG. 43]. It arises from the dorsal surface of the lower midbrain, sweeps anteriorly on it, and pierces the inferior surface of the tentorium close to its free border, near the apex of the petrous temporal bone. Cut through the nerve before it enters the dura.

Inferior to the trochlear nerve, find the large **trigeminal nerve** [FIG. 47]. Pass a blunt seeker forwards along the line of the nerve to demonstrate the dural sac (**cavum trigeminale**) which surrounds the nerve and the greater part of its ganglion beneath the dural floor of the middle cranial fossa. Elevate the seeker so that it raises the dural floor of the middle cranial fossa and outlines the position of the nerve and ganglion [FIG. 170].

Carefully remove the dura mater from the floor of the middle cranial fossa by cutting through it on to the seeker in the cavum trigeminale, anterior to the superior petrosal sinus. Strip the dura forwards and laterally to uncover the trigeminal nerve and ganglion and the three large nerves which issue from the convex, peripheral border of the ganglion. Remove the dura from the ganglion with care, because the ganglion is a loose mass of cells which is easily destroyed. Trace the **mandibular nerve** inferolaterally to the foramen ovale, close to the entry of the middle meningeal artery. Follow the **maxillary nerve** to the foramen rotundum, and the **ophthalmic nerve** into the lateral wall of the cavernous sinus where it divides into three branches which can be traced to the superior orbital fissure [FIGS. 44, 80]. The **trochlear** and **oculomotor nerves** will also be found in the lateral wall of the cavernous sinus. When they are exposed there, pick them up as they enter the dura, pull gently on them, and thus identify their more peripheral parts. Preserve the dura where the nerves enter it, but remove it in front of that so as to follow them to the superior orbital fissure.

Remove the remains of the lateral wall of the cavernous sinus from around the nerves, and expose the **internal carotid artery** and the **abducent nerve** within the sinus. Note that the nerve passes forwards lateral to the artery. Follow the nerve forwards and backwards.

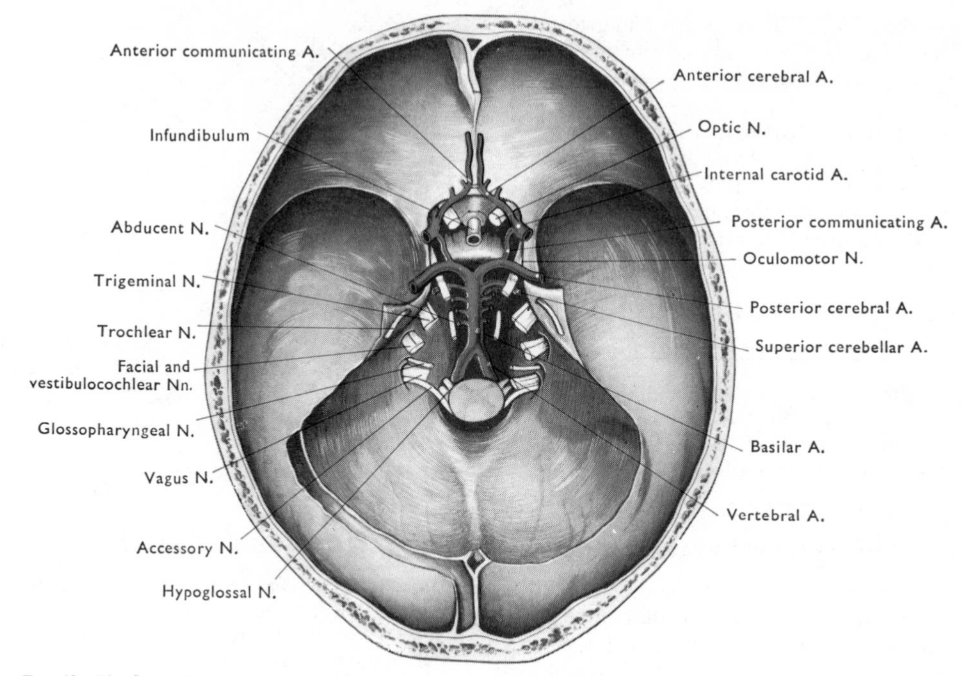

FIG. 43 The floor of the cranial cavity after removal of the brain, but with the arteries at the base of the brain *in situ*.

Carefully strip the remainder of the dura from the anterior surface of the petrous temporal bone, and look for the **petrosal nerves**. These slender nerves emerge through slits in the temporal bone and run anteromedially in shallow grooves. One pierces the skull near the foramen ovale; the other, larger and more medial, disappears under the trigeminal ganglion [FIG. 48]. Lift the trigeminal ganglion and try to identify the **motor root of the trigeminal nerve** on its inferior surface. It runs past the ganglion to the foramen ovale.

TRIGEMINAL NERVE

This is the fifth and largest of the cranial nerves. It is the principal sensory nerve of the head, supplying (1) the skin of the face and anterior half of the head; (2) the mucous membrane of the nose, air sinuses, mouth, and anterior two thirds of the tongue; (3) the teeth and temporomandibular joint; (4) the contents of the orbit, except the retina; (5) part of the dura mater.

The **motor fibres** arise in the pons, and pass only into the mandibular nerve. They supply the four muscles of mastication and mylohyoid, anterior belly of digastric, tensor palati, and tensor tympani.

Most of the **sensory fibres** are processes of the cells in the large trigeminal ganglion. Their central branches converge to form the **sensory root** which enters the pons; their peripheral branches enter the ophthalmic, maxillary, and mandibular nerves. Proprioceptive nerve fibres arise from cells in the midbrain and pons [p. 189] and pass through the ganglion into the same three nerves—a unique arrangement where primary sensory nerve cells have their cell bodies within the central nervous system.

Trigeminal Ganglion [FIG. 170]

This semilunar, sensory ganglion lies in a shallow depression on the anterior surface of the petrous temporal bone (near the apex) and on the adjacent margin of the greater wing of the sphenoid. Like other sensory ganglia, it lies in a pocket of dura mater (**cavum trigeminale**) in this case tucked forwards from the posterior cranial fossa between the dural and endocranial layers of the floor of the middle cranial fossa. Posteriorly, the narrow neck of the cavum contains the sensory and motor roots (in a sleeve of pia, arachnoid, and subarachnoid space) and grooves the upper margin of the petrous temporal bone close to the apex, inferior to the superior petrosal sinus. The distal part of the root lies above the internal carotid artery in the carotid canal. The superomedial part of the ganglion lies in the lateral wall of the cavernous sinus lateral to the artery. Sympathetic filaments run from the plexus on the internal carotid artery to join the nerve near the ganglion. These supply the arteries in the ganglion, and are distributed with its fibres.

Motor Root. This small root leaves the pons beside the sensory root, and passes under the ganglion to the foramen ovale.

Mandibular Nerve

This is the largest of the three nerves. It arises from the inferolateral part of the ganglion, gives a meningeal twig to the floor of the middle cranial fossa, and immediately enters the foramen ovale.

Maxillary Nerve

It arises from the anterior surface of the ganglion inferior to the ophthalmic nerve. It runs forwards between the dura and the lower border of the cavernous sinus, and gives off a fine meningeal branch to the middle cranial fossa before passing through the foramen rotundum [Fig. 40].

Ophthalmic Nerve

This is the smallest of the three nerves. It arises from the superomedial part of the ganglion, runs forwards in the lateral wall of the cavernous sinus, and divides into **nasociliary**, **lacrimal**, and **frontal nerves** which enter the orbit through the superior orbital fissure. Near its origin, it communicates with the oculomotor, trochlear, and abducent nerves in the lateral wall of the cavernous sinus, and gives off a small **tentorial branch** which curves back into the tentorium cerebelli. The **communications** transmit sensory fibres from the extrinsic ocular muscles to the trigeminal nerve. It receives sympathetic fibres from the internal carotid plexus.

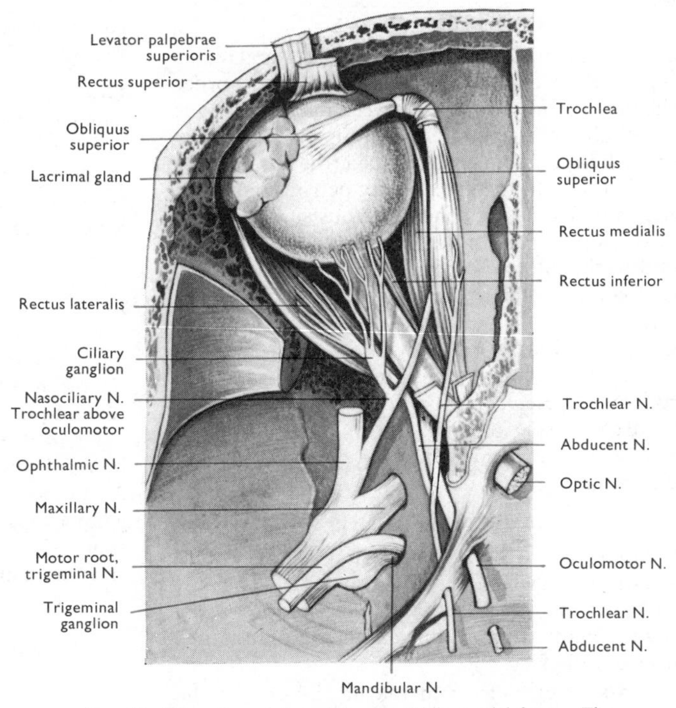

Fig. 44 Dissection of the orbit and middle cranial fossa. The trigeminal nerve and ganglion have been turned laterally to expose the motor root.

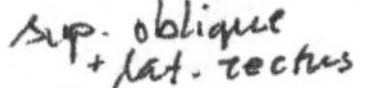

OCULOMOTOR NERVE

The third cranial nerve supplies all but two of the muscles of the orbit, and two of the three muscles in the eyeball. It emerges through the front of the midbrain, passes anterolaterally between the posterior cerebral and superior cerebellar arteries, and pierces the dura mater in the roof of the cavernous sinus. It then runs forwards in the lateral wall of the cavernous sinus (above the other nerves there) and divides into superior and inferior branches which enter the orbit through the superior orbital fissure. [Figs. 44, 45].

TROCHLEAR NERVE

This slender, fourth cranial nerve supplies only the superior oblique muscle of the eye. The nerve emerges from the dorsal surface and curves forwards on the side of the lower midbrain, between the posterior cerebral and superior cerebellar arteries. It pierces the tentorium below the free margin (near the point where that margin crosses the petrous temporal bone) and runs anterosuperiorly in the lateral wall of the cavernous sinus, crossing the lateral aspect of the oculomotor nerve to enter the orbit through the superior orbital fissure.

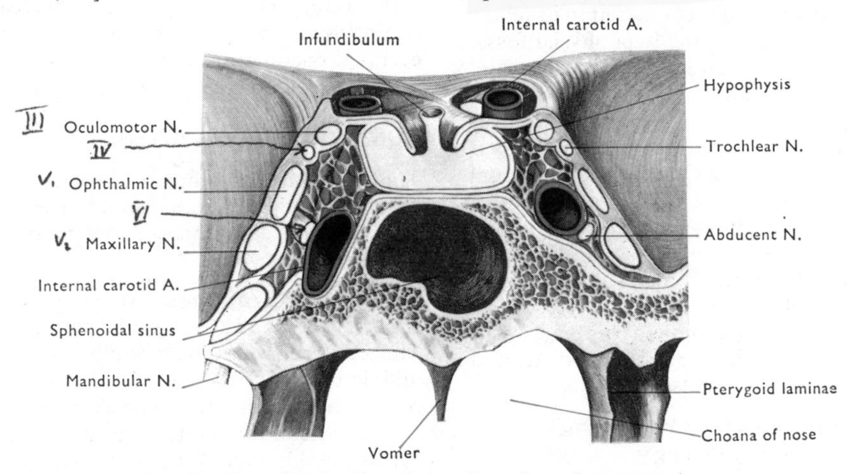

Fig. 45 Coronal section through the sella turcica and cavernous sinuses.

ABDUCENT NERVE

The sixth cranial nerve supplies only the lateral rectus muscle of the eyeball. It emerges at the lower border of the pons, bends upwards between it and the clivus of the skull, and pierces the dura mater 1 cm below the root of the dorsum sellae. Outside the dura, it runs superolaterally to the apex of the petrous temporal bone, and crossing the sphenopetrous suture, enters the cavernous sinus. In the sinus, it runs anteriorly, lateral to the internal carotid artery, and leaves the sinus to enter the orbit through the superior orbital fissure.

Communications. Apart from communications with the ophthalmic nerve, all these nerves may receive sympathetic fibres from the internal carotid plexus, partly through a well-marked bundle which has a temporary course with the abducent nerve.

CAVERNOUS SINUS

The importance of the cavernous sinus arises because of the structures which lie in and around it. They may be involved with it in infections which can spread to it along the many tributaries, *e.g.*, via the ophthalmic veins from the face.

Position. It extends from the medial end of the superior orbital fissure to the apex of the petrous temporal bone, and consists of a large number of incompletely fused venous channels surrounding the structures which lie within it.

Medially, the hypophysis with the sphenoidal air sinus below it.

Laterally, the trigeminal ganglion with its maxillary and ophthalmic nerves, and the oculomotor and trochlear nerves in its lateral wall above these.

Within the sinus, but separated from it by endothelium, the internal carotid artery and the abducent nerve.

Tributaries and Communications. *Anteriorly*, the ophthalmic veins and the sphenoparietal sinus enter it. *Posteriorly*, the superior and inferior petrosal sinuses drain it. *Medially*, it is connected to its fellow by intercavernous sinuses in the hypophysial fossa. *Superiorly*, the superficial middle cerebral vein and one or two small cerebral veins enter it. *Inferiorly*, it communicates with: (1) the pharyngeal plexus through the carotid canal; (2) the pterygoid plexus through the foramen ovale and the sphenoidal emissary foramen, if present. Thus it forms a route of communication between the veins of the face, cheek, and brain, and the internal jugular vein. The sinus may be damaged in injuries to the skull, leading to leakage of carotid blood into the sinus. This raises the pressure in the sinus and in the veins entering it, so that the eyeball may protrude (distended ophthalmic veins) and even pulsate.

INTRACRANIAL PART OF THE INTERNAL CAROTID ARTERY

The tortuous course of this artery is shown by the sinuous groove it produces on the side of the body of the sphenoid.

Course. It passes anteromedially in the carotid canal into the upper part of the foramen lacerum, and bends upwards through the endocranial floor of the cavernous sinus. It immediately turns forwards through a right angle, and runs to the root of the lesser wing of the sphenoid. Here it turns sharply upwards and backwards, and pierces the roof (dura mater and arachnoid) of the cavernous sinus, medial to the anterior clinoid process, behind the optic canal. It then bends upwards, dividing into the anterior and middle cerebral arteries on the surface of the brain [FIG. 188]. It is surrounded by the internal carotid plexus of sympathetic nerves [FIG. 64].

Branches. 1. The opthalmic artery passes forwards just above the roof of the cavernous sinus. 2. Small twigs to the hypophysis, trigeminal ganglion, and dura mater arise in the cavernous sinus. 3. The posterior communicating, anterior choroidal, middle and anterior cerebral arteries arise at the base of the brain.

MENINGEAL VESSELS OF THE MIDDLE CRANIAL FOSSA

The meningeal vessels are embedded in the endocranium, but are thicker than it and so stand out from its external surface, grooving the skull. They are distributed to the bone and dura mater, but do *not* cross the subdural space. Hence they play no part in supplying the brain or the pia-arachnoid. Their intimate relation to the bones leads to a risk of rupture of the vessels in fractures of the skull. The veins are particularly liable to injury because they lie between the arteries and the bone. If the vessels are damaged, but the dura is not torn, escaping blood collects outside the dura (*extradural haematoma*) and may compress the underlying brain. If the dura is torn, blood may pass into the subdural space and spread widely in it (*subdural haematoma*). In either case, subsequent absorption of fluid into the extravasated blood may lead to increasing brain compression causing progressive drowsiness and eventual unconsciousness, even though the original leakage of blood was insufficient to produce this.

Middle Meningeal Artery

This small artery is important because its frontal branch lies adjacent to the 'motor area' of the brain, and is the commonest source of extradural haemorrhage.

Course and Branches. From the maxillary artery it enters the skull through the foramen spinosum. Thence it runs anterolaterally on the floor of the middle cranial fossa, and divides into frontal and parietal branches on the greater wing of the sphenoid.

The parietal branch turns posteriorly, and passes across the side wall of the cranial cavity towards the apex of the occipital bone (lambda).

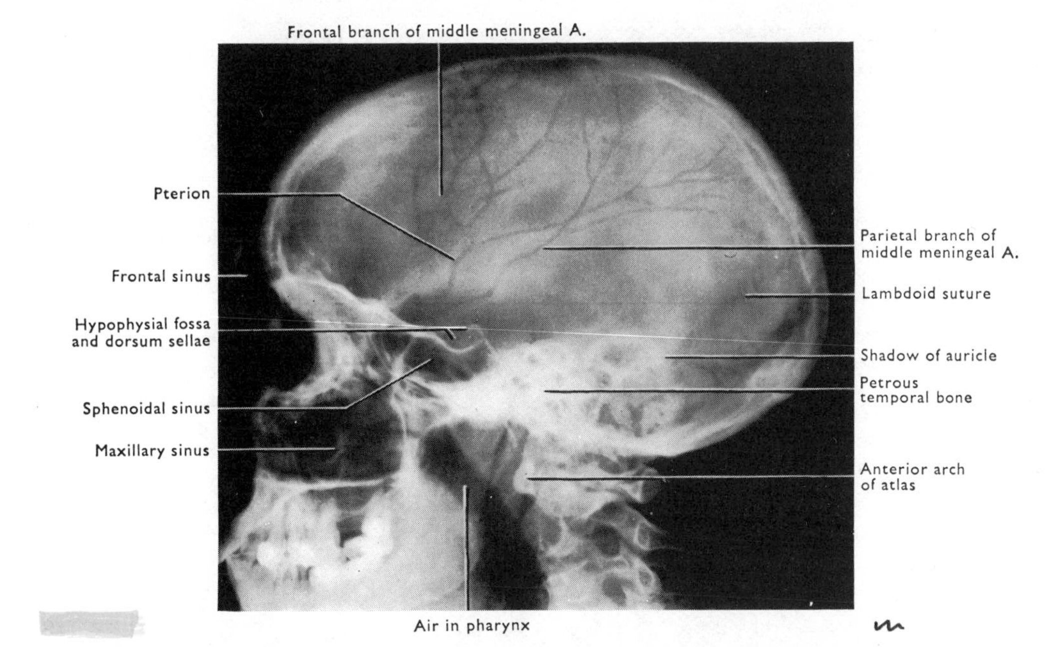

FIG. 46 Lateral radiograph of the skull. The region marked pterion indicates the position where the lesser wing of the sphenoid reaches the lateral wall of the cranium and the frontal branch of middle meningeal artery is enclosed in bone.

The larger frontal branch runs towards the lateral end of the lesser wing of the sphenoid. At this region of the skull, known as **pterion**, where the greater wing of the sphenoid meets the frontal and parietal bones, the artery grooves the skull deeply, or even disappears into a short bony tunnel. It then runs obliquely upwards and backwards, parallel to and slightly behind the coronal suture [FIG. 277], and frequently sends a large branch posteriorly on the deep surface of the parietal bone.

Check the position of the grooves for the middle meningeal vessels on the inside of the skull, and note their position on the external surface.

The middle meningeal vein accompanies the artery through the foramen spinosum to the pterygoid venous plexus.

Accessory Meningeal Vessels. A small branch of the maxillary or middle meningeal artery often traverses the foramen ovale to the trigeminal ganglion and adjacent dura. The ophthalmic and **lacrimal arteries** send twigs through the superior orbital fissure from the orbit to the middle cranial fossa. Occasionally the lacrimal branch is large and replaces the middle meningeal artery.

PETROSAL NERVES

These tiny nerves are known to transmit preganglionic parasympathetic secretory fibres to the lacrimal, nasal, palatine, nasopharyngeal, and parotid glands, and sensory fibres from taste buds on the palate.

The **greater petrosal nerve** arises from the genu of the facial nerve in the petrous temporal bone, and passes through a slit in its anterior surface into a narrow groove leading to the foramen lacerum, inferior to the trigeminal ganglion and lateral to the internal carotid artery. Here the nerve joins the **deep petrosal nerve** from the sympathetic plexus on the internal carotid artery, runs across the foramen, and traverses the pterygoid canal to the pterygopalatine ganglion. The preganglionic parasympathetic fibres of this **nerve of the pterygoid canal** synapse in the ganglion; the sympathetic fibres pass into the branches of the ganglion without synapsing in it.

The slender **lesser petrosal nerve** is formed by preganglionic parasympathetic fibres of the glossopharyngeal nerve in the middle ear. It emerges lateral to the greater petrosal nerve, and runs on the bony floor of the middle cranial fossa to leave the skull through or adjacent to the foramen spinosum (or ovale). Its fibres synapse with the cells of the otic ganglion on the medial side of the mandibular nerve [p. 103].

THE POSTERIOR CRANIAL FOSSA

DISSECTION. *If the entire brain has not been removed,* detach the tentorium cerebelli from the petrous temporal bone and remove it. Examine the superior surface of the cerebellum, the superior cerebellar artery, and the corresponding veins.

Split the cerebellum in the median plane, cutting forwards through it in a vertical plane until an aperture appears in it about the middle of the cut. This is the **fourth ventricle of the brain** which has an angular, posterior extension into the cerebellum. Extend the cut superiorly and inferiorly so as to divide the cerebellum

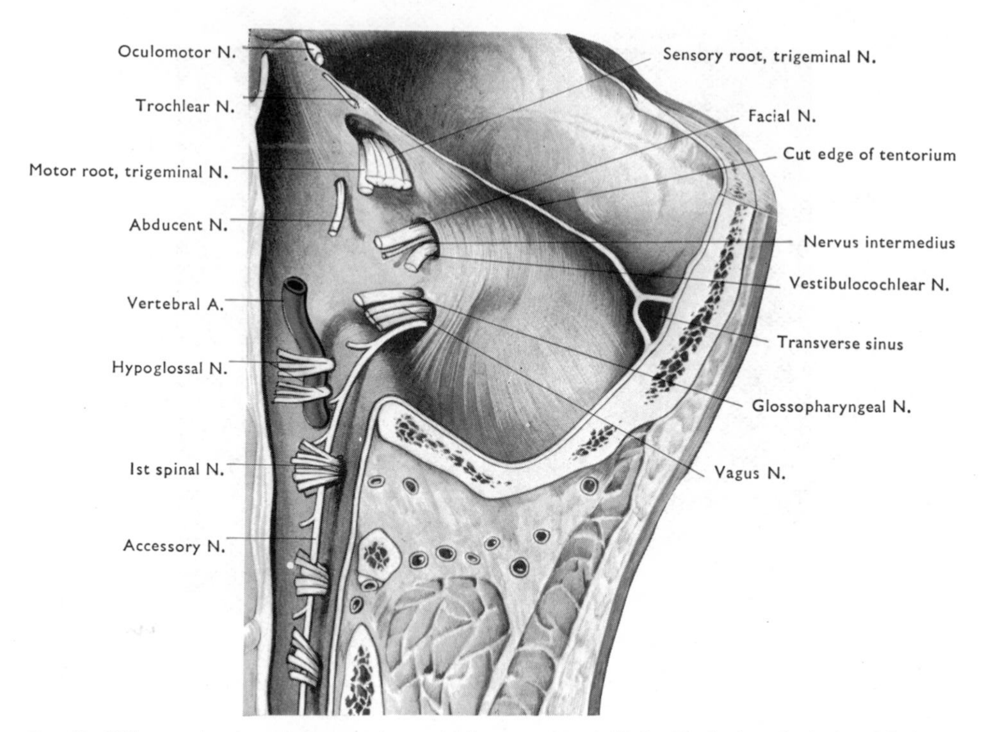

FIG. 47 Oblique section through the posterior cranial fossa seen from behind. The brain and spinal medulla have been removed.

completely, and open the full extent of the fourth ventricle, but do not cut into the midbrain.

Remove each half of the cerebellum in turn by cutting through its attachments to the brain stem (**cerebellar peduncles**) by a coronal incision parallel but posterior to the floor (anterior wall) of the fourth ventricle.

The brain stem now lies free in the posterior cranial fossa, attached only by the cranial nerves passing to their foramina and by the two vertebral arteries running on to it from below.

With the help of FIGURES 40 and 48, and a dry skull, identify the cranial nerves arising from the lateral aspect of the brain stem. (1) The **trigeminal nerve** from the pons. (2) The **facial** and **vestibulocochlear nerves** at the lower border of the pons (with the small **nervus intermedius** between them) passing to the internal acoustic meatus. (3) The **glossopharyngeal**, **vagus**, and **accessory nerves** which arise as a row of rootlets inferior to the previous two nerves. The **spinal part of the accessory** ascends beside the spinal medulla, posterior to the ligamentum denticulatum [FIG. 169], to join the vagus, and pass out through the jugular foramen, adjacent to the glossopharyngeal nerve, but in a separate dural sheath.

To expose the nerves arising from the ventral aspect of the brain stem, cut across the oculomotor and trigeminal nerves, and draw the upper brain stem posteriorly. The **abducent nerve** comes into view as it emerges below the pons, and runs upwards anterior to it. Inferior to the abducent is the **hypoglossal**. This arises as a row of rootlets which join together and pass out through the hypoglossal canal, often as two separate nerves.

Find the basilar artery on the front of the pons, and expose its continuity with the vertebral arteries on the medulla oblongata, by dividing the remaining cranial nerves (except the accessory) and drawing the brain stem further posteriorly. Separate the basilar and vertebral arteries and their main branches from the brain stem, and remove the brain stem by cutting through its junction with the spinal medulla. If the arteries are carefully removed from the brain, the continuity of the vertebral and internal carotid systems is obvious [FIG. 71].

STRUCTURES SEEN AFTER REMOVAL OF THE BRAIN STEM

The upper end of the spinal medulla is attached to the margin of the foramen magnum by the first tooth of the ligamentum denticulatum [FIG. 169 and p. 160] which has the vertebral artery anterior to it. In front of that, note the ventral rootlets of the **first cervical nerve** and, superior to this, the two parts of the **hypoglossal nerve** piercing the dura separately at the hypoglossal canal. The **accessory nerve** ascends through the foramen magnum, and turns laterally over the margin of the foramen to join the cranial part and the vagus nerve. Together they pierce the dura mater at the jugular foramen beside the separate aperture for the glossopharyngeal nerve. Find the **facial nerve**, **nervus intermedius**, and **vestibulocochlear nerve** entering the internal acoustic meatus, in that order from above down.

The **vertebral artery** pierces the spinal dura mater near the skull, and ascends through the foramen magnum anterior to the first tooth of the ligamentum denticulatum. It gives a meningeal branch to the posterior cranial fossa before piercing the dura mater, and passes superomedially between the first cervical and hypoglossal nerves.

The **falx cerebelli** is a small fold of dura mater on the internal occipital crest. It contains the occipital sinus, and fits into the posterior cerebellar notch.

DISSECTION. Slit up the falx cerebelli and look for the **occipital sinus**. Open the **sigmoid sinus** by passing a knife into it at the anterior end of the transverse sinus, and cutting along it to the base of the skull, and then forwards to the jugular foramen. Look for the mouth of the mastoid emissary vein about half way down the posterior wall of the sinus.

At the anterior end of the jugular foramen, find the end of the **inferior petrosal sinus**, and open it by cutting through the dura mater between the petrous temporal and basilar part of the occipital bone. Find the mouths of the **basilar sinuses** in the medial wall of this petrosal sinus, and trace them to the opposite side.

Pull gently on the abducent nerve, and trace it to the apex of the petrous temporal bone by slitting the dura.

Make a transverse incision through the dura mater at the anterior margin of the foramen magnum. Strip it upwards and downwards. Then incise the endocranium and try to strip it in a similar way. Note that the dura can be separated from the endocranium, but loses contact with it below the foramen magnum.

Dura Mater of Base of Skull

The dura mater of the base is firmly attached to the endocranium which is continuous round the lips of each foramen with the periosteum on the outside of the skull. At the foramen magnum the dura mater separates from the endocranium to become the spinal dura mater. At the foramina for nerves, it also separates and becomes the epineurium of the nerves.

VENOUS SINUSES OF THE POSTERIOR CRANIAL FOSSA

Sigmoid Sinus

This S-shaped sinus begins as a continuation of the transverse sinus behind the base of the petrous temporal bone. It curves downwards, grooving the mastoid and petrous parts of the temporal bone at a level posterior to the root of the auricle. The sinus is close to the **mastoid cells**, laterally; to the mastoid antrum [p. 146] and the vertical part of the facial nerve, anteriorly; to the cerebellum, medially.

On the base of the skull, the sinus curves forwards on the occipital bone, and passes through the jugular foramen to join the inferior petrosal sinus and form the internal jugular vein.

Tributaries and Connexions. The sinus receives the **posterior temporal diploic vein**. It is

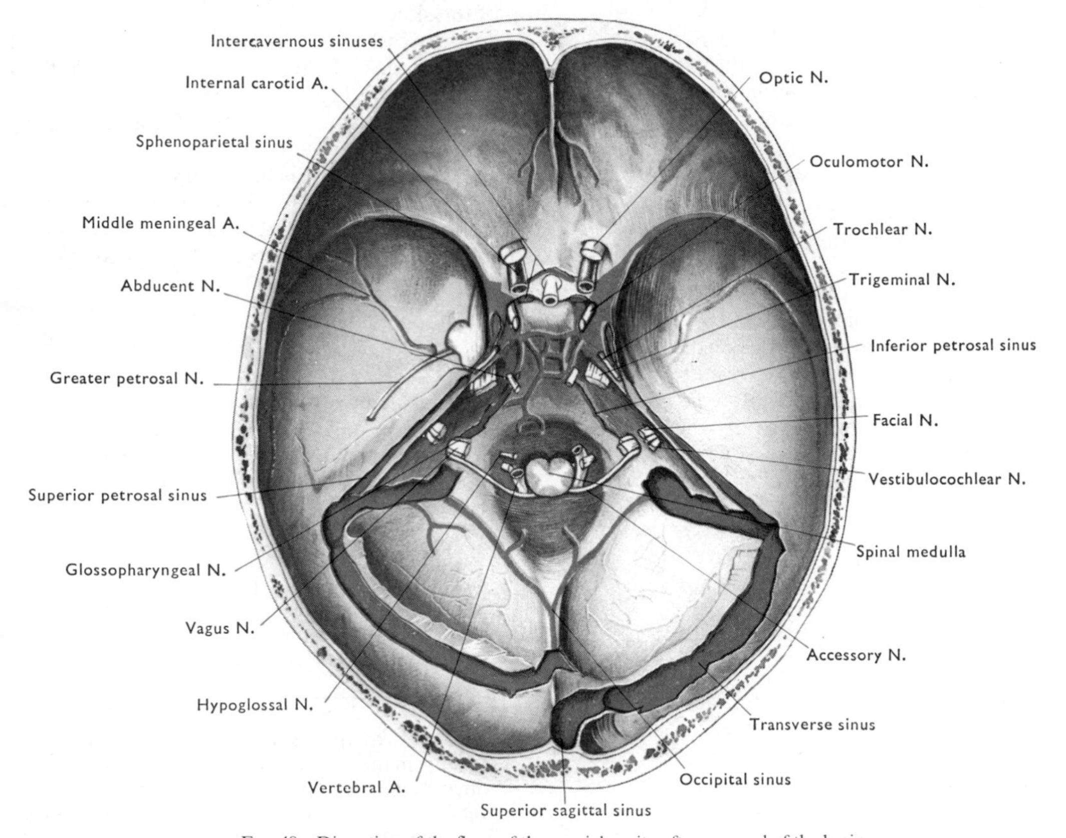

FIG. 48 Dissection of the floor of the cranial cavity after removal of the brain.

connected with: (1) the **occipital veins**, by the mastoid emissary vein in the mastoid foramen; (2) the **suboccipital plexus**, by the condylar emissary vein (when present) in the condylar canal; (3) with the beginning of the **transverse sinus**, through the occipital sinus.

Occipital Sinus [Fig. 48]

When present, this narrow sinus descends in the falx cerebelli from the beginning of the transverse sinus. It divides at the margin of the foramen magnum, and ends in the sigmoid sinuses near their terminations. It communicates through the foramen magnum with the internal vertebral venous plexus.

Inferior Petrosal Sinus

It lies in the groove between the basilar part of the occipital bone and the petrous temporal. It drains the posterior end of the cavernous sinus through the jugular foramen to the internal jugular vein.

Basilar Plexus

This network of sinuses on the clivus of the skull unites the inferior petrosal sinuses with the internal vertebral venous plexus through the foramen magnum.

DIPLOIC VEINS

These are wide venous spaces of the marrow cavities between the outer and inner tables of the flat bones of the skull. They communicate with the venous sinuses of the dura mater and with the emissary veins.

VENOUS SINUSES

These endothelial lined spaces lie either in the folds of dura mater, or between the dura mater and the endocranium. They are continuous through the foramen magnum with the corresponding veins which surround the spinal dura mater (**internal vertebral venous plexus**) and like them drain both the nervous system and the surrounding bone, and communicate with the external veins through many foramina. There are no valves anywhere in this system, so blood may flow in either direction along them according to the pressure gradients. Most of the sinuses drain eventually to the internal jugular veins, but they also drain through meningeal, diploic, and emissary veins.

The sinuses also drain cerebrospinal fluid directly from the arachnoid villi and granulations [p. 174] in the superior sagittal sinus and its lateral lacunae. Though there are other routes for the removal of cerebrospinal fluid (especially along the sheaths of cranial and spinal nerves) blockage of the superior sagittal sinus materially affects the process, causing a rapid rise in intracranial pressure.

Individual venous sinuses are described on pages 39, 43, 44, 48, and 51. Review their general arrangement, and trace the potential routes of venous drainage through them with the help of Figures 37, 38, 48, and 82.

EMISSARY VEINS

These veins connect the venous sinuses of the dura mater with veins outside the skull through cranial foramina. They have no valves. Thus blood flow may be in either direction through them, so they may transmit extracranial infections to the sinuses. Not all emissary veins are always present—the parietal, condylar, and sphenoidal are frequently absent.

The **superior sagittal sinus** may be connected to the veins of the frontal air sinus through the **foramen caecum**, and to the veins of the scalp through the **parietal emissary foramina** in the top of the skull.

The **sigmoid sinus** is connected to the occipital or posterior auricular veins through the **mastoid emissary vein** behind the auricle, and to the vertebral veins through the **condylar emissary vein**.

The **cavernous sinus** has the greatest number of such communications. (a) Through the **ophthalmic veins** to the face. (b) Through a plexus of veins along the internal carotid artery to the **pharyngeal veins**. (c) Through the foramen ovale, and occasionally through the sphenoidal emissary foramen to the **pterygoid plexus** in the infratemporal region.

Meningeal Veins. These very thin-walled veins lie between the meningeal arteries and the bone. They end either in the venous sinuses, or in veins outside the skull by passing through foramina with the corresponding arteries.

THE DEEP DISSECTION OF THE NECK

Most of the structures in this dissection are under cover of the sternocleidomastoid. Others are hidden by the infrahyoid muscles, or by the parotid gland and adjacent structures. Try to retain the sternocleidomastoid as far as possible during the dissection, for it is an important landmark in the neck.

DISSECTION. Identify again the infrahyoid muscles. Displace the sternocleidomastoid and the superior belly of omohyoid laterally. Cut through the sternothyroid near its lower end and turn it upwards to the thyroid cartilage with its nerve supply. Before removing the fat from the

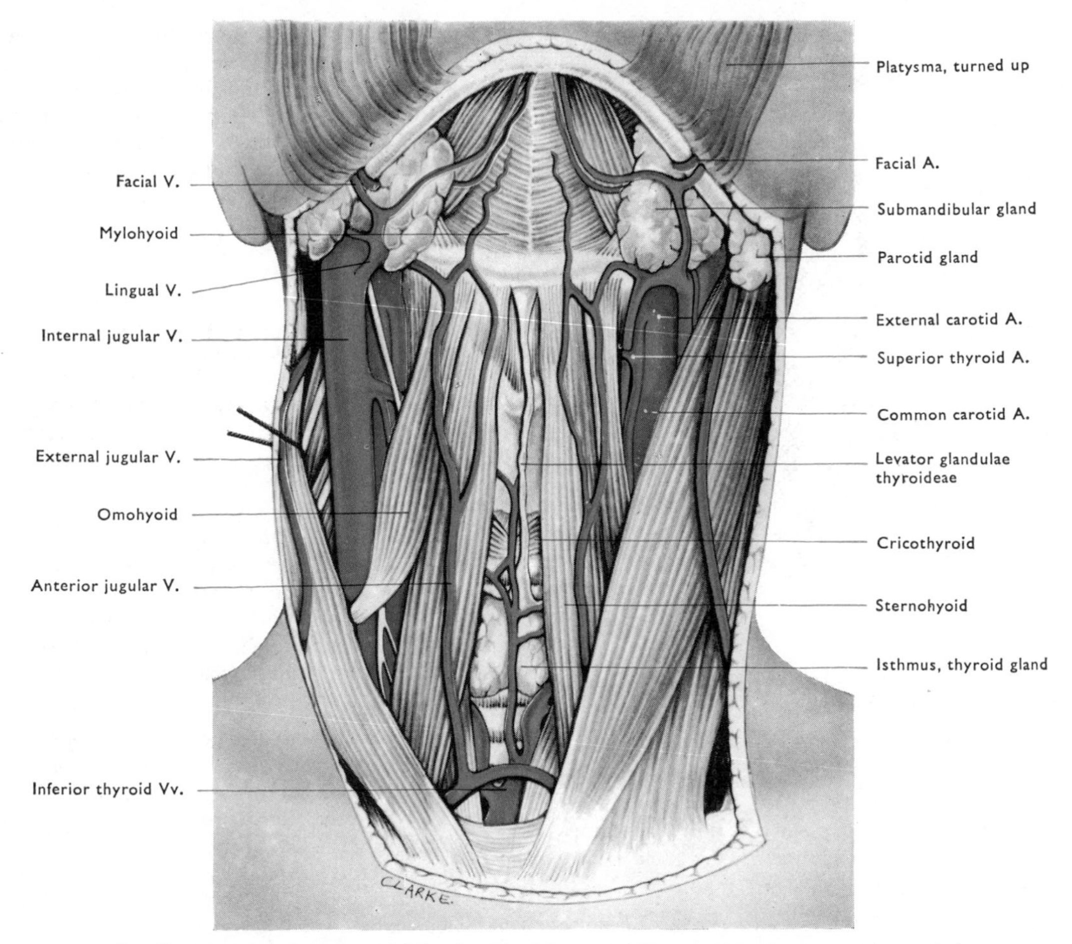

FIG. 49 Dissection of the front of the neck. The right sternocleidomastoid has been retracted posteriorly.

front of the trachea, find the inferior thyroid veins in this fat, and note if any parts of the right and left lobes of the **thymus** extend up into it. These are difficult to differentiate from the fat, but are darker and ensheathed in fascia.

Remove the fascia from the lobes of the thyroid gland, exposing its arteries and veins. Lift the lower part of the gland to expose the lateral surfaces of the trachea and oesophagus with the recurrent laryngeal nerve in the groove between them. On the left side, look for the thoracic duct on the oesophagus. If the thorax has been dissected, the duct and the recurrent laryngeal nerve can often be followed upwards. Pull the upper part of the thyroid gland laterally, and trace the external branch of the superior laryngeal nerve to the cricothyroid muscle. Find the lower part of the inferior constrictor muscle arising from a fibrous arch crossing the cricothyroid muscle and hiding its posterior part [FIG. 32].

THYMUS

This important lymph structure is concerned with the development of cellular immunity mechanisms, and is particularly large in the child. In the adult, it consists chiefly of fat and fibrous tissue. Its lobes are two slender, elongated, yellowish bodies that lie side by side, anterior to the pericardium and great vessels in the upper part of the thorax [see Vol. 2]. The superior parts may extend into the neck, anterior to the trachea, and be attached to the lower ends of the thyroid gland. This results from their *development* as lateral outgrowths of the pharynx (third pharyngeal pouches) in the neck, and their subsequent descent into the thorax attached to the pericardium [p. 248].

THYROID GLAND

This highly vascular ductless gland clasps the upper part of the trachea. It extends from the fifth or sixth tracheal ring to the oblique line on the thyroid cartilage, and is enclosed in a **sheath** of pretracheal fascia attached to that line and to the arch of the cricoid cartilage. This ensures that *the thyroid moves with the larynx in swallowing and speaking*—a feature which helps to differentiate swellings in the gland from those in adjacent structures. Deep to the sheath is the **fibrous capsule** of the gland, and between these lie the arteries of the gland and an anastomosing network of veins. The gland varies

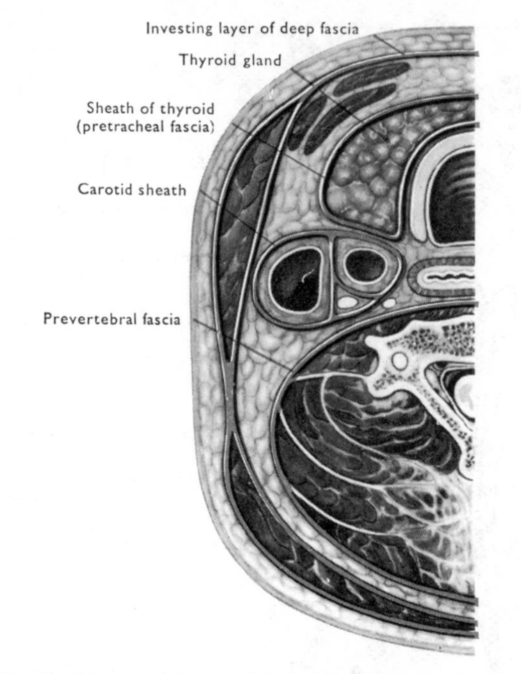

FIG. 50 Diagram of the cervical fascia (blue) in transverse section at the level of the thyroid isthmus.

greatly in size, and is relatively larger in women and children than in men. It consists of a pair of lobes joined across the median plane by a narrow isthmus.

Lobes

Each conical lobe has a convex superficial surface covered by sternohyoid, sternothyroid, and omohyoid, and overlapped by the anterior border of sternocleidomastoid [FIG. 53]. The medial surface is moulded inferiorly to the trachea and oesophagus, with the **recurrent laryngeal nerve** between them, and superiorly to the cricoid and thyroid cartilages, with the cricothyroid and inferior constrictor muscles and the **external branch of the superior laryngeal nerve** intervening. The posterior surface lies on the prevertebral fascia anterior to longus colli, and overlaps the medial part of the carotid sheath. The **parathyroid glands** are embedded in this surface.

Isthmus

This median band, of variable width, lies on the second to fourth tracheal rings, nearer the lower than the upper ends of the lobes it connects. It is covered by skin and fasciae of the neck. A slender **pyramidal lobe** frequently extends superiorly from the upper border of the isthmus, usually to the left of the median plane. It may be attached to the hyoid bone by a narrow slip of muscle (**levator glandulae thyroideae**) or by a fibrous strand. This strand is a remnant of the **thyroglossal duct**, from which the greater part of the thyroid gland is developed [p. 248]. It originates in the region of the foramen caecum of the tongue as a median epithelial downgrowth which descends in front of the body of the hyoid, hooks up posterior to it, and continues in front of the thyroid and cricoid cartilages, expanding into the thyroid gland. The suprahyoid part of this structure rarely persists.

Blood Supply

Arteries. The apex of each lobe of this very vascular gland receives a **superior thyroid artery** which divides into two or three branches. The base and deep surfaces of each lobe receive branches from the **inferior thyroid arteries**. A small **thyroidea ima** artery may arise from the brachiocephalic trunk, aortic arch, or left common carotid artery, and ascend on the anterior surface of the trachea to the isthmus. The arteries anastomose freely, especially on the posterior surface of each lobe, but there is little anastomosis across the median plane except for a branch from each superior thyroid artery which meet on the upper border of the isthmus. The thyroid arteries are arteries of the gut tube, so they also supply the larynx, the laryngeal part of the pharnyx, the trachea, and the oesophagus.

Veins. Three pairs of veins drain the thyroid gland. A **superior thyroid vein** arises near the upper end of each lobe. It either crosses the common carotid artery to the internal jugular vein, or ascends with the superior thyroid artery to the facial vein. A short **middle thyroid vein** crosses the common carotid artery to the internal jugular vein from the lower part of each lobe. **Inferior thyroid veins** pass downwards from on the front of the trachea, the network on the isthmus and lower parts of the lobes. Usually there is one on each side ending in the corresponding brachiocephalic vein close to the junction of these veins. Occasionally they unite [FIG. 51] and end in one or other of the brachiocephalic veins. They also receive tributaries from the trachea, larynx, and oesophagus.

Nerve Supply

The thyroid is supplied by branches from the cervical ganglia of the sympathetic trunk and from the cardiac and laryngeal branches of the vagus. The part they play in the function of the gland is not clear.

DISSECTION. Cut through the isthmus and the vessels of one lobe. Remove the lobe. Find the anastomostic vessel between the superior and inferior thyroid arteries on the medial part of the posterior surface, and look for the yellowish-brown parathyroid glands immediately lateral to it. The best guide to them is the small twig which the inferior thyroid artery gives to each. Make a cut through the lobe of the thyroid gland, and examine the surface with a hand lens.

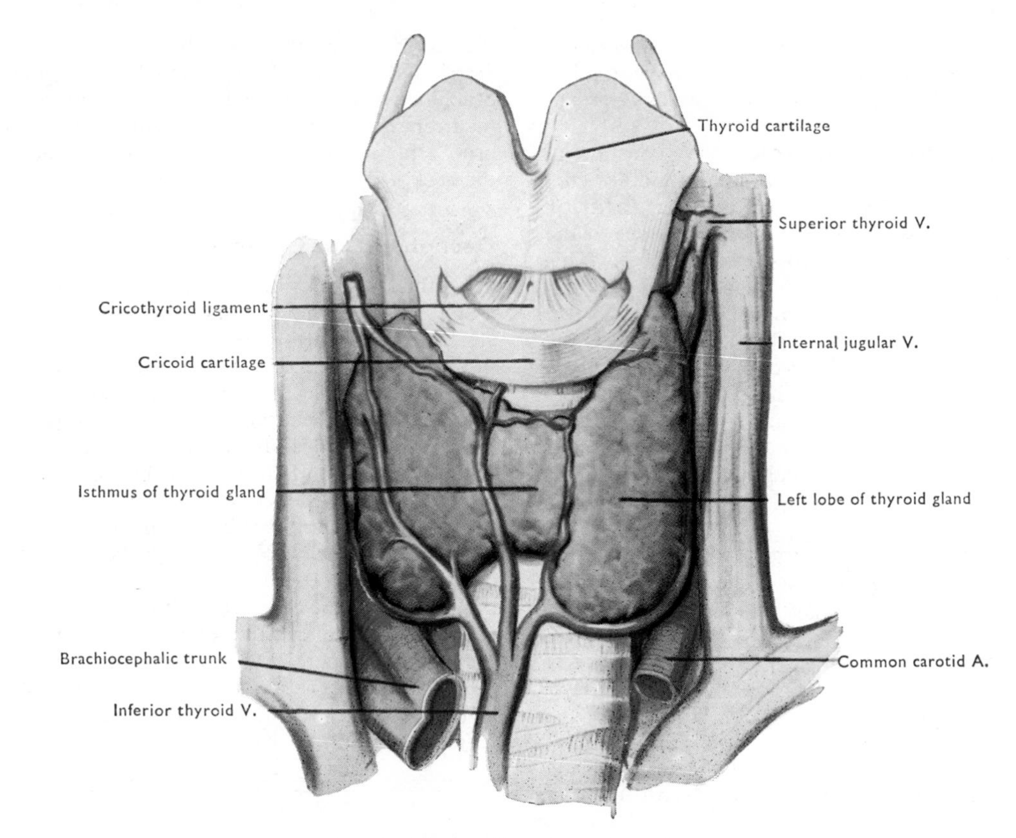

FIG. 51 The thyroid gland, anterior surface.

Structure and Function

The thyroid is composed of a large number of closed vesicles (**follicles**) held together by a stroma of delicate, vascular connective tissue. Each vesicle is filled with a semi-liquid **colloid** substance, coagulated by preservatives, which stores the active principles. The follicles vary in size up to 1 mm in diameter. The thyroglobulin of the colloid contains di-, tri-, and tetra-iodothyronine which are powerful stimulants to the metabolism of the body. Colloid accumulates in the gland which is not actively releasing these products, but disappears during excessive release into the surrounding capillaries (*e.g.*, in hyperthyroidism) by the cuboidal cells lining the vesicles. Between the follicles are occasional collections of **pale cells**. These are believed to arise from the ultimobranchial body [p. 248] and to secrete thyrocalcitonin—a hormone which antagonizes the activity of the parathyroid glands.

Lymph Vessels

These drain mostly to nodes on the carotid sheath (upper and lower deep cervical nodes).

PARATHYROID GLANDS

These two pairs of small (approximately 6 × 3 × 2 mm) yellowish-brown ductless glands are embedded in the posterior surface of the capsule of the thyroid gland: one pair in each lobe. Their important secretion stimulates the mobilization of calcium from the bones to maintain the normal blood calcium level. Thus overactivity of these glands leads to demineralization of the bones and excessive

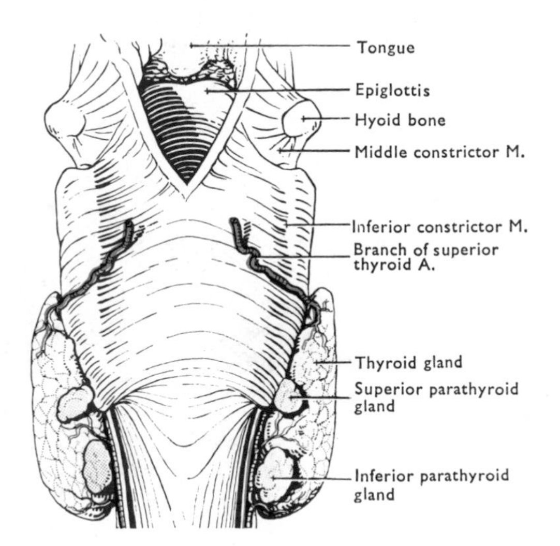

FIG. 52 Posterior surface of the thyroid gland to show the parathyroid glands.

excretion of calcium, while underactivity causes a fall in blood calcium. If severe, this leads to a condition known as tetany in which there are widespread muscle spasms.

The **superior parathyroid**, more constant in position than the inferior, lies about the middle of the posterior surface of each lobe. Each **inferior parathyroid** is close to the inferior surface of the lobe, but may lie some distance below it. This variability is the result of the development of that gland with the thymus from the third pharyngeal pouch [p. 248]. They descend together towards the thorax. The parathyroid normally leaves the thymus at the lower border of the thyroid gland, but may descend with it into the thorax.

The parathyroid glands consist of clumps and columns of small cells separated by sinusoidal capillaries.

DISSECTION. Complete the exposure of the trachea and oesophagus, and examine their relations.

TRACHEA AND OESOPHAGUS

Both these structures begin immediately below the cricoid cartilage, anterior to the sixth cervical vertebra, and descend into the thorax.

Trachea or Windpipe

This wide tube (approximately 12 cm in length) is lined by pseudostratified, columnar, ciliated epithelium containing many goblet cells. It is kept constantly patent by the U-shaped **cartilaginous bars** embedded in its walls with mucous and serous **glands** situated between them. Posteriorly, the flat wall lies against the oesophagus, and consists of smooth muscle and connective tissue uniting the ends of the bars. Superiorly the trachea is continuous with the larynx. Laterally, are the common carotid arteries, the lobes of the thyroid gland, and the brachiocephalic trunk on the right side inferiorly. A recurrent laryngeal nerve ascends on each side in the groove between the oesophagus and the trachea in the neck.

Oesophagus or Gullet

This thick, distensible, muscular tube (approximately 25 cm long) is lined by non-keratinizing stratified squamous epithelium. It has striated muscle in the upper third of its wall, smooth muscle in the lower third, and a mixture in the middle third. It extends from the pharynx to the stomach [Figs. 50, 52, 53, 58, 67] and is surrounded by loose connective tissue which permits distention. In the neck it lies between the trachea and the prevertebral fascia overlying the anterior longitudinal ligament and the longus colli muscles [Fig. 70]. On the right, it is in contact with the thyroid gland and, at the root of the neck, with the cervical pleura; on the left, with the thyroid gland, but the subclavian artery and the thoracic duct separate it from the pleura. In the neck, the oesophagus inclines to the left as it descends. Thus it is closer to the thyroid gland and is more readily accessible to the surgeon on that side.

DISSECTION. Remove the fat, lymph nodes, and carotid sheath from the common carotid artery and internal jugular vein. Separate them and find the vagus nerve. Expose the vagus, and find the **right recurrent laryngeal nerve** arising from it as it crosses the subclavian artery; follow it to the groove between the trachea and oesophagus. On the left side, find the **thoracic duct**, entering the junction of the internal jugular and subclavian veins [Fig. 54] from above and behind. Trace it backwards and downwards to where it lies behind the common carotid artery.

On both sides, expose the cervical part of the brachiocephalic vein and its tributaries. Expose the small

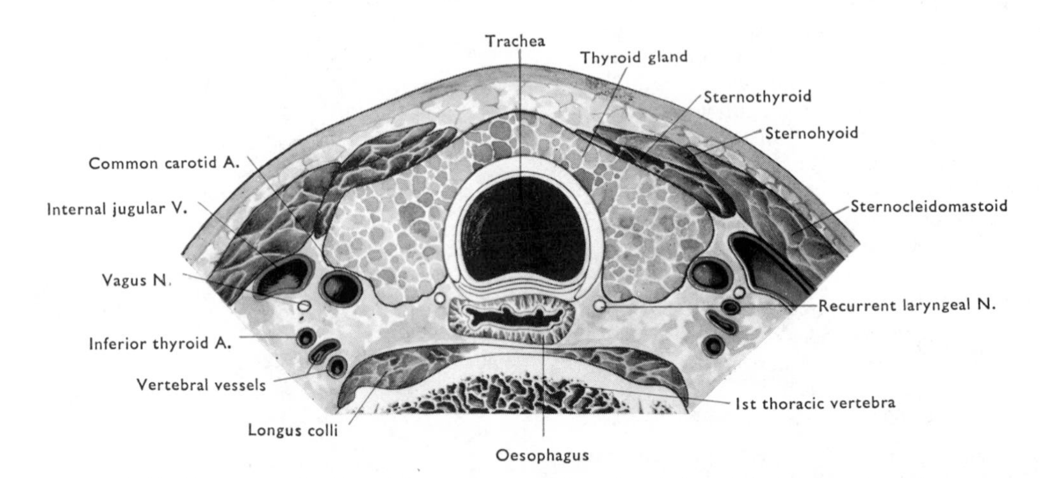

Fig. 53 Transverse section through the anterior part of the neck at the level of the first thoracic vertebra.

part of the brachiocephalic trunk which lies in the neck, on the right.

Identify the phrenic nerve behind the internal jugular vein and prevertebral fascia. Trace it to the thorax.

Displace the internal jugular vein medially to expose the subclavian artery and the cervical pleura above and below it. Avoid injury to the thoracic duct and the vertebral veins. Trace the internal thoracic artery and the thyrocervical trunk and its branches from the first part of the subclavian artery. Pull scalenus anterior laterally to expose the **costocervical trunk** arising from the subclavian artery at the medial border of the muscle. Trace the trunk over the cervical pleura to the neck of the first rib.

Separate the internal jugular vein and the common carotid artery, and, avoiding injury to the vertebral veins and thoracic duct, find the **vertebral artery** posterior to them, and trace it superiorly in front of the transverse process of the seventh cervical vertebra into that of the sixth. Remove the fat posterior to the vertebral artery to expose the ventral rami of the seventh and eighth cervical nerves, respectively above and below the seventh transverse process.

Displace the common carotid artery laterally to expose the **sympathetic trunk** posteromedial to it. Trace the trunk superiorly and inferiorly, and find the **cervico-thoracic ganglion** between the seventh cervical transverse process and the neck of the first rib, posterior to the vertebral artery. Find the small **middle cervical ganglion** on the inferior thyroid artery close to the sixth transverse process. Find the grey rami communicantes passing from these ganglia to the ventral rami of the spinal nerves.

BRACHIOCEPHALIC TRUNK

This great artery passes superiorly and to the right from the arch of the aorta. It divides into the right subclavian and common carotid arteries behind the upper margin of the right sternoclavicular joint and the sternohyoid and sternothyroid muscles. The artery lies between the trachea and the right brachiocephalic vein, and is separated from the pleura posteriorly by some fat, in which the right vagus descends obliquely towards the trachea.

SUBCLAVIAN ARTERY

This is the artery of the upper limb, but it supplies a considerable part of the neck and brain through its branches [Fig. 56].

On the left it arises from the arch of the aorta and ascends on the pleura to enter the neck behind the left sternoclavicular joint. On the right, it arises from the brachiocephalic artery behind the sternoclavicular joint. On each side it arches laterally across the anterior surface of the cervical pleura on to the first rib, posterior to scalenus anterior. It becomes the axillary artery at the outer border of the first rib.

The artery is divided into three parts, the second of which lies posterior to scalenus anterior. The third part has been described already [p. 22].

First Part

On the right, this part extends superolaterally to a point 1 cm above the level of the clavicle at the medial edge of scalenus anterior. It lies deeply, at first posterior to sternocleidomastoid, sternohyoid, and sternothyroid, to the vagus nerve (which sends the **recurrent laryngeal nerve** hooking round it), to a loop from the sympathetic trunk (**ansa subclavia**), and, occasionally, to the cardiac branches of the vagus and sympathetic trunk. Near the medial border of scalenus anterior, the internal jugular and vertebral veins are anterior to it. It lies on the

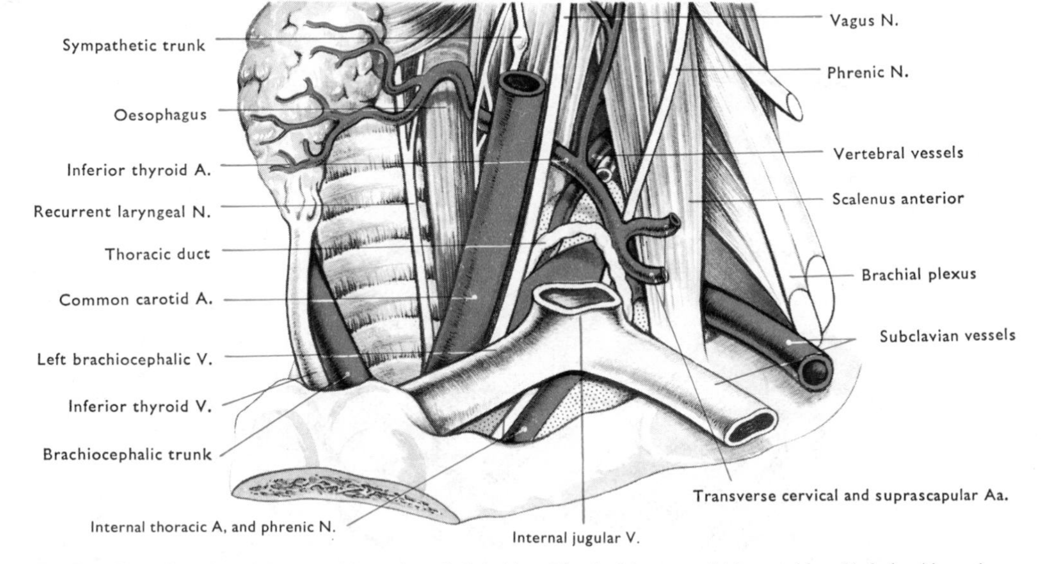

Fig. 54 Deep dissection of the root of the neck on the left side. The clavicle, sternocleidomastoid, and infrahyoid muscles have been removed, and the thyroid gland is displaced anteriorly. Pleura, blue stipple.

suprapleural membrane [p. 62] covering the anterior surface of the cervical pleura. If the lung has already been removed, investigate the position of the artery from the pleural cavity.

On the left, the first part ascends vertically from the aortic arch, with the vagus, phrenic, and cardiac nerves anterior to it, to lie behind the left brachiocephalic vein at the sternoclavicular joint. Thereafter, the course and relations are the same, except that the thoracic duct and phrenic nerve also descend anterior to it [FIG. 54]. The **left recurrent laryngeal nerve** is medial to the artery because it ascends from the aortic arch in the groove between trachea and oesophagus.

Second Part

This is the summit of the arch of the artery. It rises 1·5–2·5 cm above the level of the clavicle. Anteriorly, it is covered by the scalenus anterior with the phrenic nerve anterior to the muscle on the right side. Postero-inferiorly, it lies on the suprapleural membrane.

The only tributary of the **subclavian vein** [p. 23] is the external jugular vein. This drains blood from the transverse cervical and suprascapular veins; indeed none of the veins corresponding to the branches of the subclavian artery enters the subclavian vein directly.

Branches of the Subclavian Artery

From the first part:
- (1) Vertebral
- (2) Thyrocervical trunk { Inferior thyroid, Transverse cervical, Suprascapular
- (3) Internal thoracic

From the second part:
- Costocervical { Highest intercostal, Deep cervical

A large branch may arise from the third part of the subclavian artery. This **descending scapular artery** replaces the deep branch of the transverse cervical artery. Then only the superficial branch arises from the thyrocervical trunk and is known as the **superfical cervical artery**.

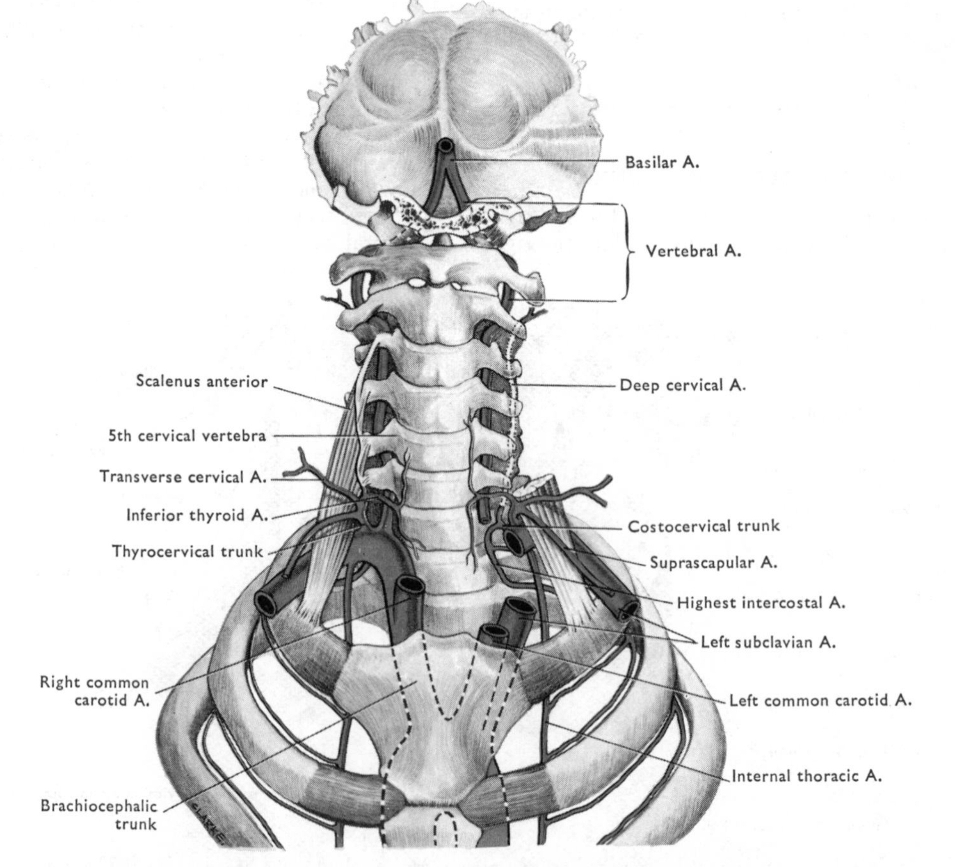

FIG. 55 Diagram of the branches of the subclavian arteries. The position of the left deep cervical artery is shown though it is completely hidden from view as it lies deep among the muscles of the back of the neck.

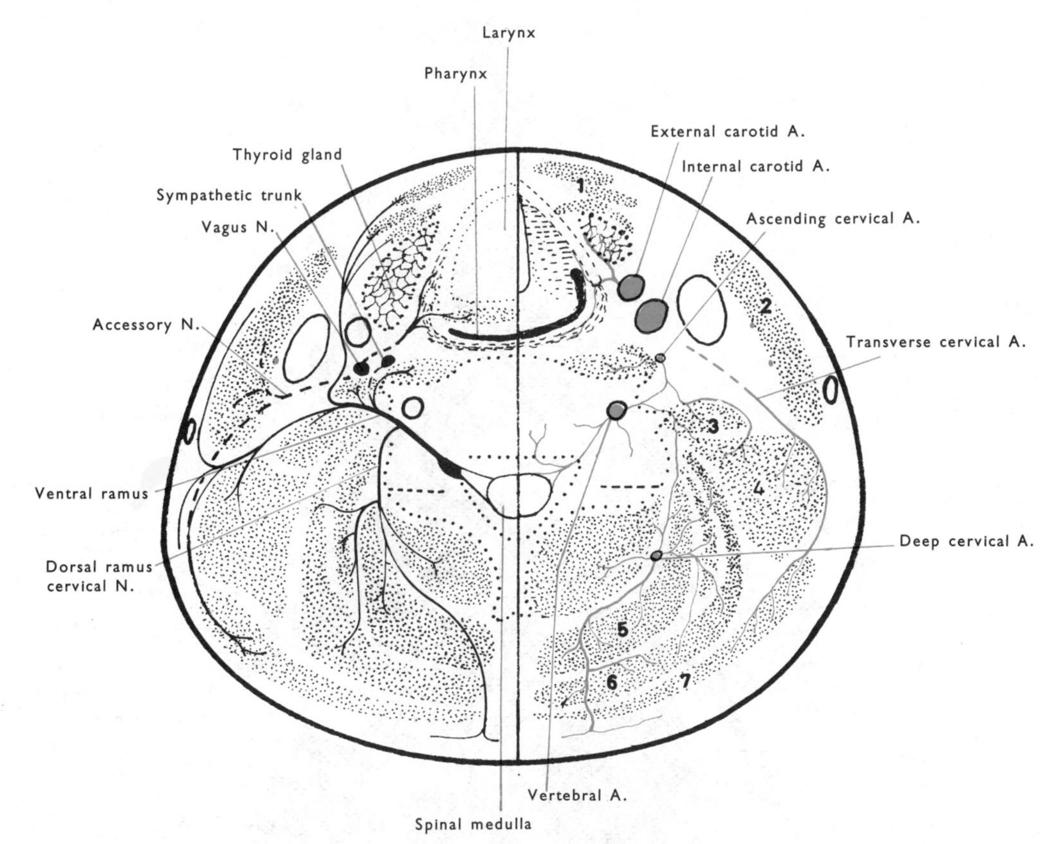

FIG. 56 Diagrammatic transverse section through the neck to show the distribution of nerves (left) and arteries (right). The diagram is constructed from a number of adjacent levels. The part of the accessory nerve shown by an interrupted line is at a still higher level. 1. Infrahyoid muscles. 2. Sternocleidomastoid. 3. Scalenus medius. 4. Levator scapulae. 5. Semispinalis capitis. 6. Splenius capitis. 7. Trapezius.

Vertebral Artery [FIGS. 55, 71]

This first branch of the subclavian is mainly distributed to the brain. It arises at the level of the upper part of the sternoclavicular joint, and ascends, between longus colli and scalenus anterior muscles, into the foramen transversarium of the sixth cervical vertebra. It is posterior to the common carotid and inferior thyroid arteries and to its own vein, and anterior to the ventral rami of the seventh and eighth cervical nerves which have the seventh transverse process between them. The sympathetic trunk lies along its medial side, and the cervicothoracic ganglion, which is partly behind the artery, sends branches to form a plexus on it. It is crossed anteriorly by the thoracic duct on the left [FIG. 54].

The vertebral vein descends posterior to the internal jugular vein from the foramen transversarium of the sixth cervical vertebra, and enters the brachiocephalic vein close to its formation.

Thyrocervical Trunk [FIGS. 54, 55]

This short vessel arises from the subclavian at the medial margin of scalenus anterior, posterior to the internal jugular vein, and between the phrenic and vagus nerves. It branches immediately into the suprascapular and transverse cervical arteries [see p. 23] and the **inferior thyroid artery** which ascends along the medial border of scalenus anterior, posterior to the internal jugular vein. Just below the transverse process of the seventh cervical vertebra, it turns medially in front of the vertebral artery and posterior to the vagus, sympathetic trunk, and common carotid artery to reach the middle of the posterior surface of the thyroid gland. Then it descends to the lower pole of the gland, and branches close to the recurrent laryngeal nerve. The main glandular branch ascends over the posterior surface of the lobe, supplies the parathyroid glands also, and anastomoses with the superior thyroid artery. No vein accompanies the artery.

The small **ascending cervical artery** ascends anterior to the cervical transverse processes. It supplies the prevertebral muscles, and sends spinal branches to the vertebral canal along the spinal nerves.

The **inferior laryngeal artery** is a small branch which accompanies the recurrent laryngeal nerve to the larynx.

Small branches also supply the trachea, oesophagus, pharynx, and adjacent muscles.

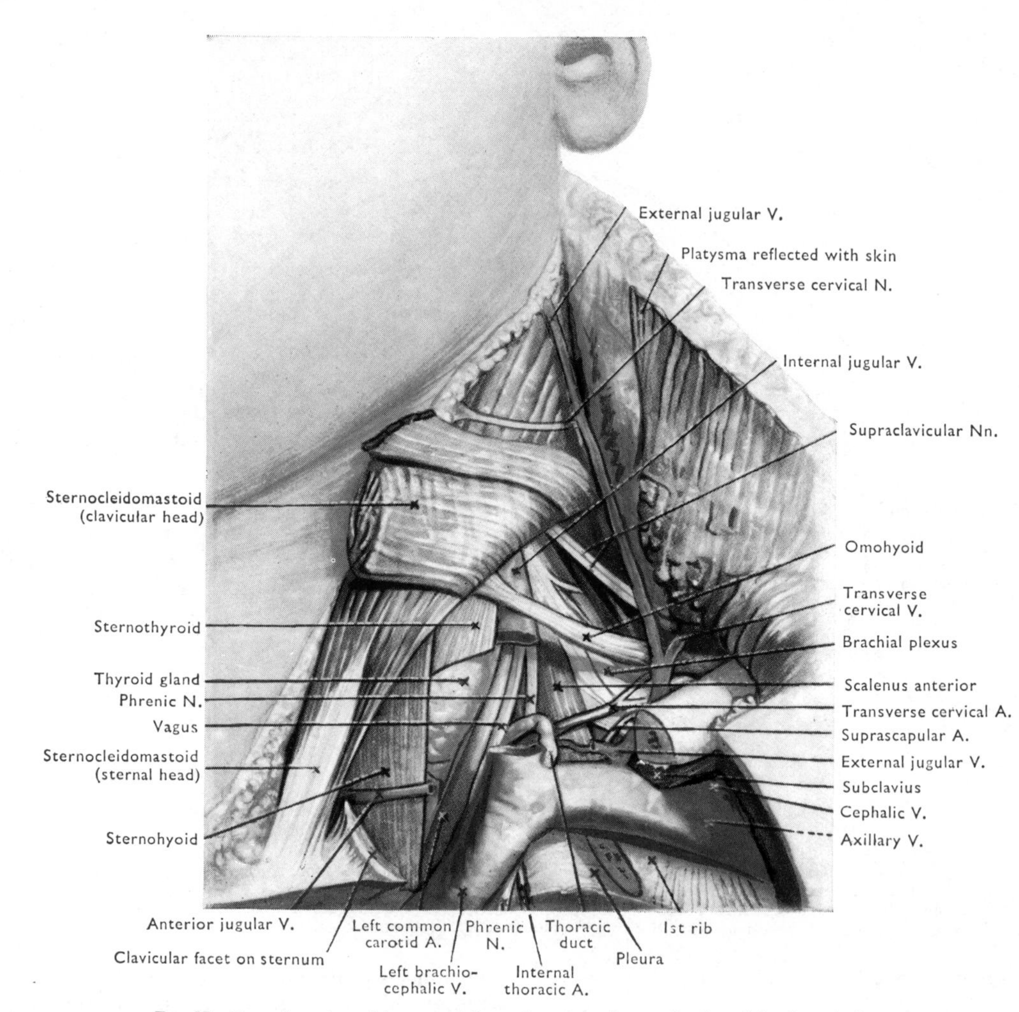

FIG. 57 Deep dissection of the root of the neck to show the termination of the thoracic duct.

Internal Thoracic Artery [FIG. 58]

This long artery arises from the inferior surface of the subclavian artery near the medial border of scalenus anterior. It passes inferomedially to the back of the first costal cartilage. In the neck it lies anterior to the pleura behind the medial part of the clavicle. The left artery runs posterior to the beginning of the left brachiocephalic vein, while the right artery passes over the lateral side of the brachiocephalic vein from posterior to anterior. The phrenic nerve passes obliquely across the artery, usually anterior to it.

The **internal thoracic vein** joins the brachiocephalic vein at the superior aperture of the thorax.

Costocervical Trunk [FIG. 55]

It arises from the posterior surface of the subclavian artery just behind scalenus anterior. It arches posteriorly over the pleura, and divides at the neck of the first rib into:

1. The **deep cervical artery** passes into the back of the neck above the neck of the first rib [p. 27]. With the descending branch of the occipital artery it is the main supply to the muscles of the back of the neck. The corresponding deep cervical vein joins the vertebral vein.

2. The **highest intercostal artery** descends anterior to the neck of the first rib, and gives posterior intercostal arteries to the first and second intercostal spaces. If the lung has been removed, examine these from the interior of the thorax.

BRACHIOCEPHALIC VEINS

These veins collect blood from the head and neck, the upper limbs, the walls of the thorax, and even from the anterior abdominal wall (internal thoracic veins). They lie in the neck and thorax, and each begins by the junction of the corresponding internal jugular and subclavian veins between the cervical pleura and the medial end of the clavicle [FIG. 58]. They end by joining to form the superior vena cava behind the lower border of the right first costal cartilage at the margin of the sternum. They have no valves.

The **right vein** descends into the thorax behind the right first costal cartilage. In the neck, it lies on the cervical pleura with the phrenic nerve and internal thoracic artery intervening. It is lateral to the brachiocephalic trunk (with the vagus behind and between them), while anterior to it is the medial end of the clavicle with the sternohyoid and sternothyroid muscles, and the costoclavicular ligament below.

The **left vein** crosses the median plane posterior to the upper part of the manubrium of the sternum to join the right vein, and is therefore much the longer. In the neck, the left vein is first posterior to the clavicle and anterior to the pleura and internal thoracic artery. It is then posterior to the sternoclavicular joint, the sternohyoid and sternothyroid muscles, and anterior to the phrenic and vagus nerves and the left subclavian and common carotid arteries [FIG. 58].

Tributaries in the Neck

Both brachiocephalic veins receive the vertebral, highest intercostal, and frequently inferior thyroid veins; also one or two lymph trunks. The thoracic duct enters the left vein.

THORACIC DUCT

This vessel drains lymph into the venous system from both sides of the body below the diaphragm (except the upper part of the right lobe of the liver and the upper half of the anterior abdominal wall), from the posterior thoracic wall on both sides, and from the left half of the body above the diaphragm. Commonly one or more of the three lymph trunks which drain the upper part of each half of the body enters the veins separately, thus altering the territory drained by the duct. The lymph in the duct has a milky appearance owing to the fat-containing lymph which enters it from the small intestine.

The thoracic duct is a slender, thin-walled vessel frequently mistaken for a vein. It ascends from the thorax along the left margin of the oesophagus, and arches anterolaterally between the carotid sheath and the cervical pleura at the level of the seventh cervical vertebra. It then turns inferiorly, anterior to the subclavian artery, and enters the left brachiocephalic vein in the angle between the internal jugular and subclavian veins [FIGS. 57, 58]. Its opening is guarded by a valve, but its terminal part is often filled with blood in the cadaver.

There are three other lymph trunks in the root of the neck on each side. (1) The **subclavian trunk** drains the upper limb. (2) The **jugular trunk** drains half of the head and neck. (3) The **bronchomediastinal trunk** drains the lung, half the mediastinum, and part of the anterior walls of the thorax and abdomen through the **internal thoracic lymph vessels**. None of these vessels is visible by dissection, and their mode of termination is very variable. (1) *On the left side* the jugular trunk may end with the other two in the thoracic duct. Commonly the subclavian trunk ends in the subclavian vein, and the bronchomediastinal trunk usually ends in the brachiocephalic vein. (2) *On the right side*, the three trunks end separately in the corresponding veins, but the jugular and subclavian trunks frequently unite in a **right lymph duct**

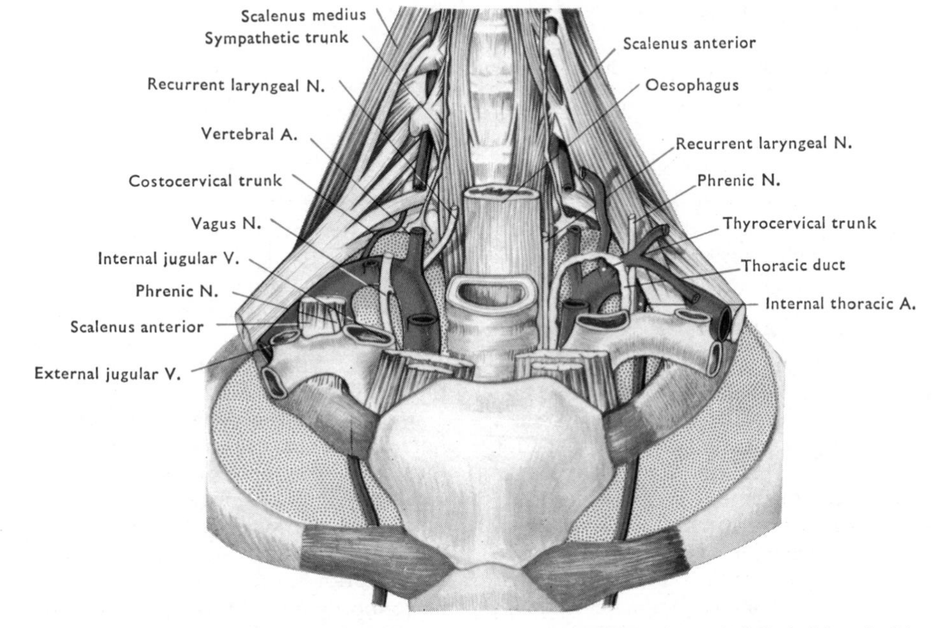

FIG. 58 Dissection of the root of the neck to show the structures adjacent to the cervical pleura (blue stipple).

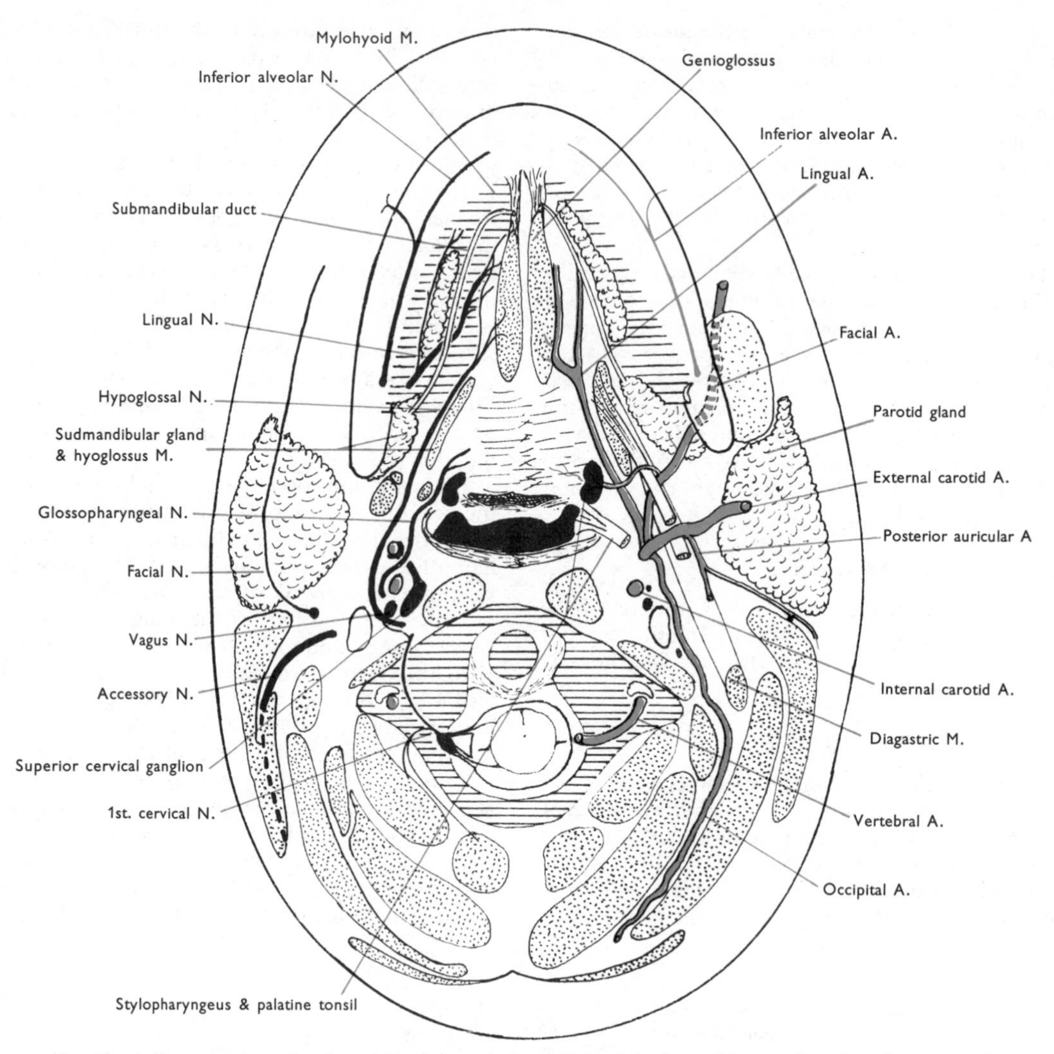

FIG. 59 A diagrammatic section through the first cervical vertebra and the floor of the mouth to show the courses of nerves (left) and arteries (right). The diagram is constructed from a number of adjacent levels. The muscle immediately superior to the facial artery is styloglossus; that inferior to it is stylohyoid.

which enters the venous system at a point corresponding to that of the thoracic duct on the left side. The bronchomediastinal trunk almost always enters the brachiocephalic vein.

CERVICAL PLEURA

On each side, the parietal pleura bulges upwards into the root of the neck to the level of the neck of the first rib, that is 2·5–5·0 cm above the sternal end of the first rib depending on the obliquity of the superior aperture of the thorax. This cervical pleura forms the dome of each pleural cavity. It is strengthened by the **suprapleural membrane** which arises from the transverse process of the seventh cervical vertebra, and fans out over the pleura to the inner margin of the first rib. The membrane may contain muscle fibres— the **scalenus minimus**.

Posterior to the cervical pleura lie the upper two ribs and the intercostal spaces and contents, and the sympathetic trunk. *Anterior* to it are the great vessels of the upper limb and head and neck. *Superiorly*, the space occupied by the cervical pleura disappears, so that the vertebral artery reaches the vertebral column, and the carotid sheath comes to lie on the prevertebral fascia and sympathetic trunk.

Medial to the cervical pleura are the vertebral bodies, oesophagus, and trachea, with the thoracic duct and recurrent laryngeal nerve on the left. *Laterally*, is the scalenus anterior with the subclavian artery and the lower trunk of the brachial plexus posterior to it. Because the subclavian artery lies on the anterior surface of the cervical pleura, its ascending (vertebral and inferior thyroid) and descending (internal thoracic) branches lie on that surface; but the costocervical trunk and its highest intercostal branch arch over the dome of the pleura from the anterior to the posterior surface.

VERTICAL NEUROVASCULAR BUNDLES OF THE NECK

One of these lies on each side of the median airway and foodway, and both are concerned principally with the supply of these structures in the neck. Each bundle, enclosed in the fascial **carotid sheath** [FIG. 61], extends from the root of the neck to the base of the skull, and lies on the sympathetic trunk and prevertebral fascia, anterior to the prevertebral muscles and the cervical transverse processes. *In the lower part of the neck*, the bundle contains the **internal jugular vein** with the **common carotid artery** medial to it, and the **vagus nerve** behind and between them. *Above the upper border of the thyroid cartilage*, the line of the common carotid artery is continued by its **internal carotid** branch. This, traced superiorly, inclines forwards to enter the carotid canal in the skull [FIG. 11] together with the **internal carotid nerve** from the superior cervical ganglion of the sympathetic trunk.

The **internal jugular vein**, which accompanies these arteries, descends vertically from the jugular foramen through which the **glossopharyngeal**, **vagus**, and **accessory nerves** emerge and descend in the bundle between the artery and vein. The **hypoglossal nerve** leaves the skull through the hypoglossal canal medial to the bundle. It spirals inferolaterally behind the artery and the other nerves to reach and adhere briefly with the lateral aspect of the vagus, descending with it in the bundle [FIG. 64]. Here the hypoglossal receives a branch from the first cervical ventral ramus [FIG. 65]. Of these nerves, only the vagus and a branch of the hypoglossal (superior root of the ansa cervicalis—C. 1 fibres) remain in the bundle in the lower part of the neck; the others leave it and pass mainly to the walls of the airway and foodway to which the vagus also sends branches [see below and p. 70].

The other branch of the common carotid artery, the **external carotid**, ascends on the pharyngeal wall giving branches (superior thyroid and lingual) mainly to the walls of the airway and foodway. It is anteromedial and parallel to the internal carotid until both have passed deep to the muscular strap (formed by the **posterior belly of digastric** and the **stylohyoid muscle**) which runs antero-inferiorly, deep to the angle of the mandible, and holds both arteries medially. Superior to this strap, the external carotid (and its facial branch) turns laterally, and, leaving the internal carotid, enters the deep surface of the parotid gland, ascending in it to the back of the neck of the mandible where it divides into superficial temporal and maxillary arteries. This lateral displacement of the external carotid [FIG. 59] places the **styloid process**, the stylopharyngeus muscle with the glossopharyngeal nerve, the styloglossus muscle, and the pharyngeal branch of the vagus between the internal and external carotid arteries, but leaves both deep to the **parotid gland** with the retromandibular vein and facial nerve buried in it. Below the digastric strap, both arteries are deep to the facial and lingual veins and to the cervical branch of the facial nerve.

Of the common, internal, and external carotid arteries on each side, only the external give branches to structures in the neck (mainly airway and foodway). The remainder of the neck (principally muscles and bone) is supplied by the subclavian arteries [FIG. 56]. The **internal carotid arteries** supply a considerable part of the brain (sharing this with the subclavian arteries—vertebral branches), the orbital contents, the forehead, the anterior part of the scalp, and parts of the external nose and of the walls of the nasal cavities. The remainder of the structures in the head are supplied by the external carotid arteries.

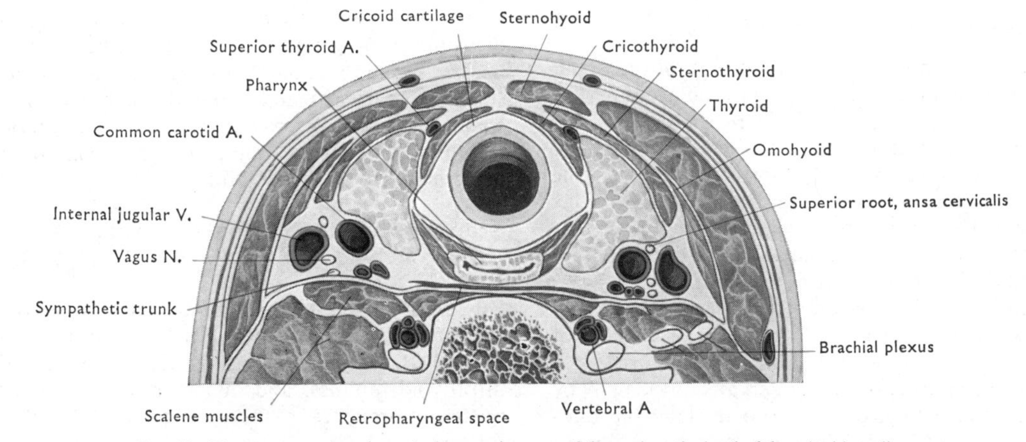

FIG. 60 Transverse section through the anterior part of the neck at the level of the cricoid cartilage.

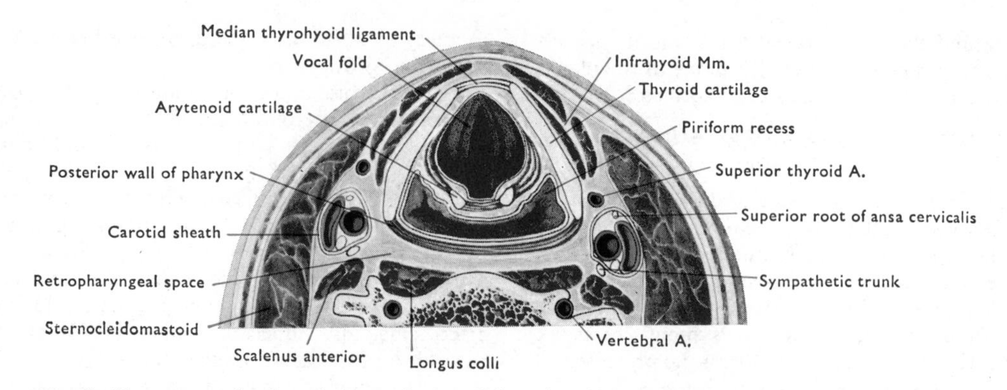

FIG. 61 Transverse section through the anterior part of the neck at the level of the upper part of the thyroid cartilage.

COMMON CAROTID ARTERY

The *right* common carotid artery arises from the brachiocephalic trunk behind the sternoclavicular joint. The *left* artery arises from the arch of the aorta, and ascends to enter the neck posterior to the left sternoclavicular joint. From the joint each artery ascends to the upper border of the thyroid cartilage, at the level of the disc between the third and fourth cervical vertebrae. Here it divides into internal and external carotid arteries—its only branches.

In the neck, the common carotid artery ascends on the subclavian and vertebral arteries to meet the prevertebral fascia on the sixth cervical transverse process (**carotid tubercle**) anterior to the inferior thyroid artery. The recurrent laryngeal nerve is posterior to the beginning of the right artery; the thoracic duct turns anterolaterally between the left artery and the vertebral vessels.

Both arteries are overlapped by the sternocleidomastoid, the infrahyoid muscles, and the thyroid gland, but not in their upper parts. Here they are covered anteriorly by deep fascia, and lie close to the thyroid cartilage. Below this they are separated from pharynx, larynx, trachea, and oesophagus by the thyroid gland—the left artery coming into contact with the trachea inferior to the gland. [FIGS. 51, 60, 61].

Carotid Sinus and Carotid Body

The carotid sinus is a slight dilatation of the upper part of the common carotid and the adjacent part of the internal carotid artery. Here the wall of the arteries is more elastic than elsewhere, and is heavily innervated. It is a pressure receptor. Distention of the wall stimulates the nerve endings and leads to a reflex slowing of the heart and a fall in blood pressure.

The carotid body is a small gland-like structure placed on the deep surface of the carotid bifurcation. It may be seen by twisting the arteries round and dissecting between them. It consists of clusters of eipthelial-like cells arranged around capillaries.

Both sinus and body are innervated principally by the **glossopharyngeal nerve** through the **carotid sinus nerve**, but they are also supplied by the vagus and sympathetic. The carotid body responds either to decreased oxygen tension or increased carbon dioxide tension in the blood. It gives rise to afferent discharges in the sensory nerve fibres, and these produce appropriate reflex changes in respiration.

EXTERNAL CAROTID ARTERY

The general course of this artery is given above. Some of its branches have been dissected already, the others will be seen shortly.

In the carotid triangle, the external carotid artery ascends on the lateral aspect of the pharynx across the branches of the superior laryngeal nerve. Superficial to it are branches of the transverse nerve of the neck and of the cervical branch of the facial nerve, the facial and lingual veins, and the hypoglossal nerve. Superior to the triangle it is overlapped by the angle of the mandible, and is crossed by the muscular strap [p. 63]. In the parotid gland, it is the deepest structure enclosed by the gland.

The **external carotid plexus** is a net of sympathetic nerve fibres on the external carotid artery. The nerve fibres arise in the cells of the superior cervical ganglion of the sympathetic trunk. Like the internal carotid nerve on the corresponding artery, this plexus is one of the main routes of distribution of sympathetic nerves to the head and neck. Similar plexuses on the vertebral and subclavian arteries and sympathetic fibres in the spinal and cranial nerves derived from the cervical sympathetic ganglia complete the sympathetic distribution to the head and neck.

Branches of the External Carotid Artery

[see also pp. 34, 67, 79, 80, 95, 99, 107, 111, 128]

Superior Thyroid Artery. This artery arises from the anterior surface of the external carotid close to its origin. It runs antero-inferiorly, deep to the infrahyoid muscles, and divides into anterior and posterior branches at the apex of the lobe of the thyroid gland.

Branches. (1) Small muscular.

(2) The small **infrahyoid artery** runs along the lower border of the hyoid bone.

(3) The **superior laryngeal artery** is larger [FIGS. 63, 104]. It enters the pharynx by piercing the thyrohyoid membrane with the internal branch of the superior laryngeal nerve. It supplies the upper parts of the larynx and the adjacent pharynx.

(4) The small **sternocleidomastoid branch** runs postero-inferiorly across the carotid sheath.

The **cricothyroid branch** arises deep to the sternothyroid muscle, and runs anteriorly across the cricothyroid muscle to anastomose with its fellow on the cricothyroid membrane.

NEUROVASCULAR BUNDLE OF THE NECK AT THE BASE OF THE SKULL

The following dissection is designed to expose the superior parts of the internal carotid artery and the internal jugular vein, together with the last four cranial nerves in the upper part of the carotid sheath, inferior to the jugular foramen. It also permits separate dissection of the two halves of the head and an approach to the cavities of the nose, mouth, pharynx, and larynx from the medial side.

Before beginning the dissection, study the base of the skull in the region of the foramen magnum and the jugular foramen [FIG. 11], also the atlas and axis vertebrae, their relation to one another and to the base of the skull, and the ligaments which unite them [pp. 165–8, FIGS. 178–9].

Identify the carotid canal, jugular foramen, and hypoglossal canal on the base of a dried skull, and note the openings of these foramina on its internal aspect [FIG. 40]. The **carotid canal** which transmits the internal carotid artery, the internal carotid nerve, and a plexus of small veins, lies furthest anteriorly. Immediately posterior to it is the **jugular foramen** which transmits the sigmoid and inferior petrosal sinuses, and the glossopharyngeal, vagus, and

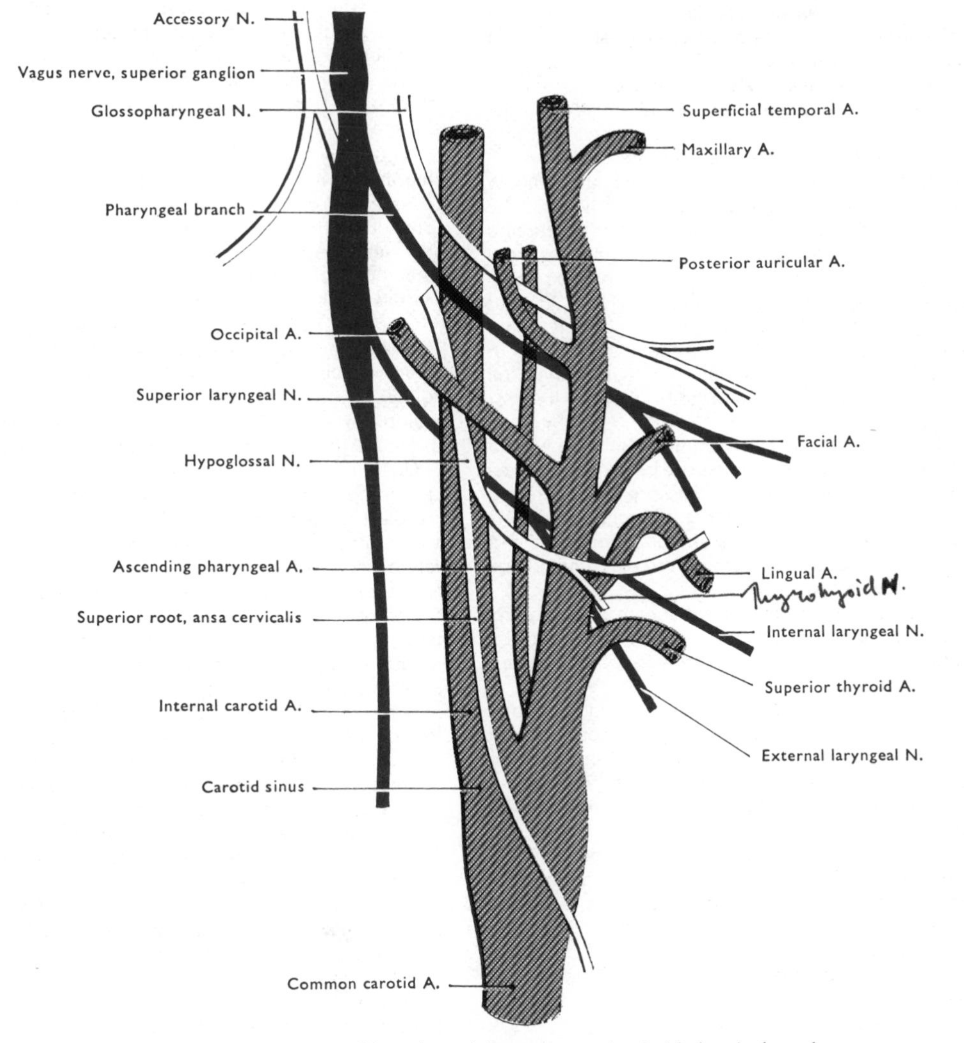

FIG. 62 Diagram of the carotid arteries and the nerves associated with them in the neck.

accessory nerves between the two veins. The **hypoglossal canal** lies posteromedial to the jugular foramen, immediately above the occipital condyle. It transmits the hypoglossal nerve which joins the other nerves between the artery and vein close to the skull. The nerves separate inferiorly [FIGS. 62, 64]. The accessory (spinal part) runs postero-inferiorly, either superficial or deep to the internal jugular vein. The glossopharyngeal passes antero-inferiorly with the stylopharyngeus muscle (*q.v.*), lateral to the internal carotid. The hypoglossal leaves the vagus at a lower level and curves anteriorly superficial to both internal and external carotid arteries. The vagus descends vertically between the artery and vein in the carotid sheath.

DISSECTION. The dissectors of the two sides should co-operate in the following dissection as certain features are best shown on one or other side.

The dissection removes from the vertebral column the right halves of the skull, mandible, and cervical viscera, and the right neurovascular bundle.

On the right side, free the greater occipital nerve from the scalp and turn it inferiorly. Divide the descending branch of the occipital artery and any muscular branches or veins which may prevent it being removed with the skull. Cut through the great auricular, lesser occipital, and transverse nerve of the neck on sternocleidomastoid. Leave sufficient nerve on both sides of the cuts to allow the ends to be identified subsequently. In the suboccipital triangle, separate the superior oblique and rectus capitis posterior major and minor muscles from their bony attachments and remove them, thus exposing the posterior atlanto-occipital membrane [FIG. 178]. Cut across this membrane close to the skull, but avoid damage to the vertebral arteries on the posterior arch of the atlas. Detach the longissimus muscle from the skull without damaging the occipital artery which may lie on or deep to the muscle.

Cut across the sternocleidomastoid 2–3 cm above the clavicle, and turn it superiorly. Expose its deep surface as far as the skull and identify the accessory nerve entering it. Cut the nerve in the posterior triangle so that its superior part remains with the sternocleidomastoid, and divide any communications which the nerve may have with the cervical plexus.

Pass a finger behind the carotid sheath and pharynx and separate them from the prevertebral fascia and the sympathetic trunk as far superiorly as the superior cervical ganglion.

Saw through the mandible in the midline, and continue this cut with a knife, splitting the tongue and epiglottis in the median plane to the hyoid bone. Cut through the hyoid bone in the midline, and extend the incision inferiorly through the larynx, pharynx, and trachea to the inferior border of the isthmus of the thyroid gland. Cut transversely through the right half of the trachea, oesophagus and neurovascular bundle (sparing the sympathetic trunk, the phrenic nerve, and scalenus anterior) at the inferior end of the sagittal cut. Separate the upper parts of the transected structures from those behind them, and cut through the inferior thyroid artery if this has not already been done.

Make a sagittal saw cut through the skull slightly to the right of the midline (so as to leave the nasal septum with the left half), and continue the cut to the foramen magnum without damage to the atlas vertebra. Within the skull, detach the dura mater and membrana tectoria [FIG. 178] from the anterior margin of the foramen magnum and turn it inferiorly, thus exposing the alar ligaments and the longitudinal fibres of the cruciate ligament [FIG. 179].

Cut across the right alar ligament, and flexing the right half of the head on the vertebral column, divide the tight, posterior part of the capsule of the atlanto-occipital joint. Now lever the occipital condyle out of its articulation, cutting any remaining parts of the capsule of the joint as they are exposed. Free the vertebral artery and the first cervical nerve from the posterior arch of the atlas, and cut across the artery where it emerges from the vertebral foramen in the atlas. Cut through the rectus capitis lateralis and anterior, the longus capitis [FIG. 70] and the anterior atlanto-occipital membrane [FIG. 178].

Complete the median division of the soft palate and of the posterior pharyngeal wall with a knife. Gently lift away the right half of the skull and attached structures, identifying and dividing the bundles of nerve fibres passing from the superior cervical ganglion to the internal carotid artery (internal carotid nerve) and to the cranial nerves in the neurovascular bundle. This leaves the ganglion and sympathetic trunk on the vertebral column.

The upper part of the cervical neurovascular bundle may now be dissected from the back on the right half of the specimen.

On both sides, find the internal laryngeal nerve at the thyrohyoid membrane. Trace it superiorly to the superior laryngeal nerve medial to the external and internal carotid arteries. On the right, this nerve can easily be followed to the vagus. Above its origin from the vagus, find the pharyngeal branch of the vagus, and trace it between the carotid arteries to the pharynx.

On the right side, find the glossopharyngeal, accessory, and hypoglossal nerves in the upper part of the neurovascular bundle and trace them distally. When following the glossopharyngeal nerve, identify the stylopharyngeus muscle on which it runs to the pharynx. Trace the muscle to its origin from the styloid process.

On the left side, cut through the posterior belly of digastric close to its origin. Turn it antero-inferiorly to expose the stylopharyngeus [FIG. 32], and avoid damage to the glossopharyngeal nerve which curves round the lateral aspect of the muscle.

The **stylopharyngeus** enters the pharynx between the superior and middle constrictor muscles of the pharynx. The upper margin of the middle constrictor can be defined and followed forwards to the hyoid bone.

Where the superior laryngeal nerve passes deep to the external carotid artery, find the **ascending pharyngeal** branch of that artery and follow it superiorly to the pharynx, medial to the internal carotid artery.

Trace a branch from the glossopharyngeal and vagus nerves downwards between the two carotid arteries to the carotid sinus. This **carotid sinus nerve** may be followed to the carotid body in the bifurcation of the common carotid artery.

On the left, lift the neurovascular bundle from the sympathetic trunk and trace the trunk upwards to the **superior cervical ganglion**. Attempt to find the branches of the ganglion to the internal carotid artery, the external carotid artery, and to the cranial nerves in the bundle. *On the right*, find the branches of the ganglion (grey rami communicantes) passing posteriorly to the upper four cervical ventral rami. Identify the longus capitis. Note its origin from the same cervical transverse processes as scalenus anterior [FIG. 70].

On the right, trace the occipital artery from the occiput to its origin. At the same time, expose the posterior belly of digastric from behind, and find the hypoglossal nerve hooking round the origin of the occipital artery, superficial to both carotid arteries.

Trace the ventral ramus of the first cervical nerve forwards with the vertebral artery, on the base of the skull. Find its branch to the hypoglossal nerve.

Occipital Artery [FIG. 59]

This artery arises from the posterior surface of the external carotid artery opposite the origin of the facial branch. It runs along the lower border of the posterior belly of digastric, under cover of sternocleidomastoid, to reach and groove the base of the skull medial to the **mastoid notch**—the attachment of the posterior belly of digastric. Its further course and branches are described on pages 15 and 27. On its way to the base of the skull, it crosses the lateral surface of the internal carotid artery, the hypoglossal nerve (which hooks anteriorly round its origin), the internal jugular vein, and the accessory nerve.

Branches. Several muscular twigs, one of which accompanies the accessory nerve to sternocleidomastoid. A meningeal branch which ascends through the jugular foramen.

Posterior Auricular Artery

This small branch arises from the posterior surface of the external carotid artery at the superior border of the posterior belly of digastric. Its proximal part may be found by displacing the digastric downwards, or it may be left till the parotid gland is dissected. The artery runs along the digastric, and passes superficial to the mastoid process with the posterior auricular nerve [p. 15]. It supplies the adjacent muscles and the parotid gland, and sends a **stylomastoid branch** superiorly into the stylomastoid foramen [FIG. 63] to supply the **facial nerve** and other structures within the temporal bone.

Ascending Pharyngeal Artery

This is the first and smallest branch of the external carotid artery. It ascends on the pharynx medial to the carotid arteries.

INTERNAL CAROTID ARTERY

The cervical part lies on the longus capitis and the sympathetic trunk. The vagus lies posterolateral to it throughout this part, while the glossopharyngeal, accessory, and hypoglossal nerves bear a similar relation to it at the base of the skull. The artery lies between the internal jugular vein and the constrictors of the pharynx [FIG. 111].

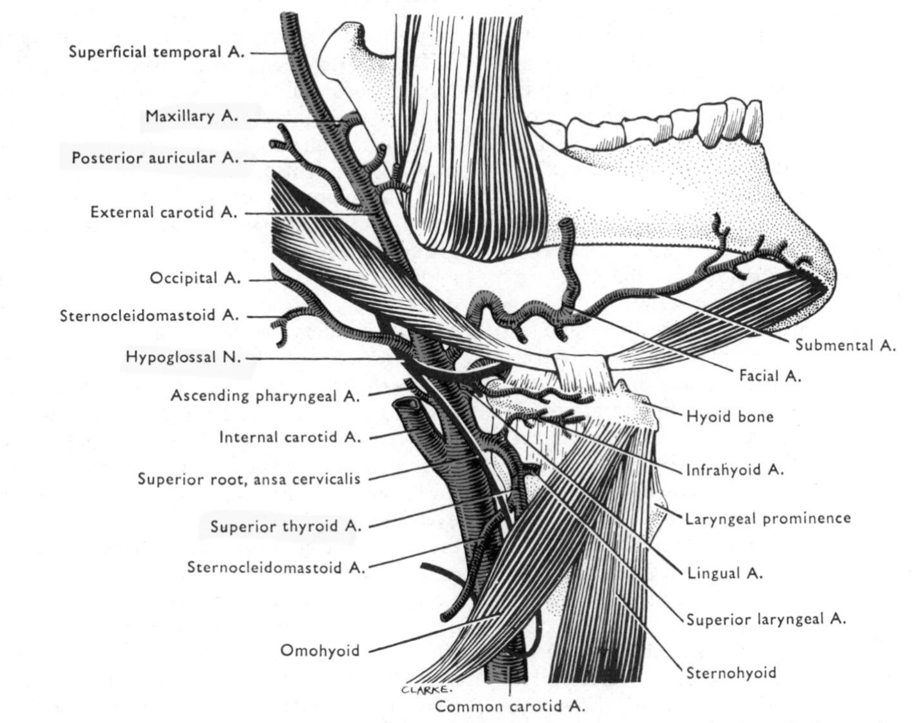

FIG. 63 Diagram of the external carotid artery and its branches. The mandible has been tilted upwards.

Inferiorly, it is deep to the sternocleidomastoid, the lingual and facial veins, the hypoglossal nerve, and the occipital artery. The superior root of the ansa cervicalis [p. 35] is anterior and the external carotid is anteromedial to it. The internal carotid artery ascends deep to the posterior belly of digastric, and the styloid process and its muscles which separate it from the parotid gland. It then passes anterior to the internal jugular vein and enters the carotid canal.

INTERNAL JUGULAR VEIN

This is usually the largest vein in the neck. It begins below the jugular foramen as the continuation of the sigmoid sinus [Fig. 64] and descends vertically applied to the lateral side of the internal and common carotid arteries, enclosed with them and the vagus nerve in the carotid sheath [Fig. 61]. It ends posterior to the medial part of the clavicle by joining the subclavian vein to form the brachiocephalic vein. In the lower part of the neck, both veins deviate to the right. The left vein therefore overlaps the left common carotid considerably, while the right vein recedes from the right common carotid artery.

The upper end of the vein dilates into the jugular fossa to form the **superior bulb**, and adheres to the fossa. Near the termination of the vein is a smaller dilatation, the **inferior bulb**, above a two- or three-cusped valve.

The *right* internal jugular vein is usually larger than the left because it normally drains blood from the superior sagittal sinus. This can be confirmed by a comparison of the sizes of the grooves for the transverse and sigmoid sinuses on the two sides of a dried skull. Note again the position of the jugular foramen, the hypoglossal canal, the occipital condyle, the styloid process, and the stylomastoid foramen on the base of a dried skull [Fig. 298]. This gives a clear picture of the relative positions of the structures at the base of the skull.

Superiorly, the internal jugular vein lies posterolateral to the internal carotid artery with the last four cranial nerves intervening. Below this, the structures in contact with the vein are the same as those in contact with the internal carotid artery above and the common carotid below: except that these arteries separate the vein from the structures medial to the arteries, and the nerves which cross the lateral surface of the internal carotid artery (glossopharyngeal, pharyngeal branch of the vagus, hypoglossal, and superior root of the ansa cervicalis) are entirely medial to the vein. Only the spinal part of the accessory nerve may run anterior to the upper part of the vein, while the inferior root of the ansa cervicalis may be lateral or medial to its lower part. Because it is lateral to the arteries, the vein is anterior to the roots of the cervical plexus and to the scalenus anterior muscle with the phrenic nerve on it; also it is lateral to the vertebral artery and the cervical transverse processes.

Tributaries. (1) The **inferior petrosal sinus** joins it at, or immediately below, the jugular foramen. (2) The **pharyngeal plexus** drains into its upper part by two or more veins. (3) The **facial**, **lingual**, and **superior thyroid veins** enter it in the carotid triangle. (4) The **middle thyroid vein** and sometimes the **jugular lymph trunk** enter it at the root of the neck.

NERVES OF THE NECK

The neck consists of two groups of structures distinguished by their different embryological origin and nerve supply.

1. The *first group* is developed from the occipital and the upper cervical spinal somites and the overlying ectoderm. It is *supplied by the hypoglossal and upper cervical spinal nerves*, and consists of the vertebral column and its associated muscles, together with the muscles of the tongue and the infrahyoid muscles.

2. The *second group* develops from the anterior extremity of the gut tube. This forms the walls of the airway (nasal cavities, pharynx, larynx, and trachea) and of the foodway (oral cavity, pharynx, and oesophagus) and their associated glands. These structures are *supplied by a group of mixed cranial nerves* which may contain (a) sensory nerve fibres, (b) motor nerve fibres to striated muscle developed from the gut tube, (c) preganglionic parasympathetic nerve fibres to glands (secretomotor) and smooth muscle in the gut tube. These nerves arise as a linear series from the lateral surface of the hindbrain—**trigeminal** (V), **facial** (VII), **glossopharyngeal** (IX), **vagus** (X), and **accessory** (XI). Their territory of supply extends from the region of the oral and nasal orifices along the gut tube into the abdomen, each supplying a region in the order of its origin. The **muscles of facial expression**, including platysma in the neck (supplied by the facial nerve) and the **sternocleidomastoid** and **trapezius muscles** (supplied by the accessory nerve) might appear to belong to the first group, but are developed from the wall of the gut tube [p. 245] and hence are innervated by its nerves.

The absence of a body cavity (coelom) in the neck, means that the derivatives of the two parts (somites and gut tube) are intermingled to an extent which does not occur in regions of the body (thorax and abdomen) where the coelom separates them. Thus the tongue and infrahyoid muscles are attached to the hyoid bone and thyroid cartilage which develop from the wall of the gut tube, and muscles such as platysma (and the other muscles of facial expression) and the sternocleidomastoid and trapezius lie in regions developed mainly from somite mesoderm. Also, where the gut tube meets the surface, the most cephalic nerve of the series, the trigeminal, extends its territory beyond the nasal and oral orifices to supply skin of the face, forehead, and scalp.

DISTRIBUTION OF THE 'GUT' NERVES

NERVE	APPROXIMATE AREA SUPPLIED IN HEAD AND NECK		
	Sensory	*Striated muscle*	*Parasympathetic*
Trigeminal	Skin and mucous membrane of oronasal region, face, forehead, scalp, dura mater	Muscles of mastication, anterior belly of digastric, mylohyoid, tensor tympani, tensor palati	Nil
Facial	Taste—ant. ⅔ of tongue and palate	Muscles of facial expression, post. belly of digastric, stylohyoid, stapedius	Submandibular, sublingual, nasal, lacrimal, palatine, oral glands
Glosso-pharyngeal	Post. ⅓ of tongue, epiglottis, palate	Stylopharyngeus	Parotid gland and glands in post ⅓ of tongue
Vagus	Pharynx, larynx, oesophagus, trachea, dura mater, ext. acoustic meatus	Of pharynx and larynx	Glands and smooth muscle of pharynx, larynx, trachea, oesophagus
Accessory cranial part	Larynx, oesophagus, trachea	Larynx	Glands of larynx, oesophagus and trachea
spinal part		Sternocleidomastoid and trapezius	

The cranial nerves which supply structures in the neck enter the neurovascular bundle through the jugular foramen, with the exception of the cervical branch of the facial nerve. They descend between the artery and the vein and either leave the bundle (XI—spinal part, and IX) or send branches from it (X with XI cranial part incorporated in it). The hypoglossal nerve also enters the bundle, but leaves it to supply the muscles of the tongue. Since the tongue muscles develop from the same mesodermal mass as the infrahyoid muscles, the nerves that supply it form a small plexus, and the hypoglossal and first cervical nerves share the innervation of the muscles on the border of their territories.

GLOSSOPHARYNGEAL NERVE

The ninth cranial nerve arises from the side of the upper medulla oblongata, and passes through the anterior part of the jugular foramen in its own dural sheath, between the sigmoid and inferior petrosal sinuses. Here it has two small **sensory ganglia** (superior and inferior). It descends between the internal jugular vein and the internal carotid artery; curves round the lateral surface of the stylopharyngeus, between the internal and external carotid arteries, and passes with the muscle into the pharynx between the superior and middle constrictor muscles [FIGS. 32, 59]. Here it lies external to the mucous membrane of the lower part of the **tonsillar fossa**, and passes forwards into the tongue, deep to hyoglossus. It supplies all forms of sensory fibres to the posterior third of the tongue and to the **vallate papillae** [FIG. 140], the palatine tonsil, part of the soft palate, and the anterior surface of the epiglottis.

Branches. The slender **tympanic nerve** enters a minute canal on the ridge of bone between the jugular foramen and the carotid canal. It ascends to the middle ear, and forms the **tympanic plexus** on its medial wall with two minute **caroticotympanic nerves** from the sympathetic plexus on the internal carotid artery [FIG. 155]. The tympanic plexus supplies the mucous membrane of the middle ear, auditory tube, mastoid antrum, and mastoid air cells [FIG. 302], and it sends preganglionic parasympathetic (secretory) nerve fibres through the **lesser petrosal nerve** to the otic ganglion. These are concerned with the innervation of the parotid gland.

The **nerve to stylopharyngeus** supplies the muscle, and passes through it to the pharyngeal mucous membrane.

The **pharyngeal branches** are: (1) One or two twigs which pass to the pharyngeal mucous membrane through the superior constrictor. (2) A larger branch which arises more proximally and accompanies the pharyngeal branch of the vagus to the pharyngeal plexus. One of its branches joins a branch of the vagus to form the **carotid sinus nerve** [p. 64] to the carotid sinus and body.

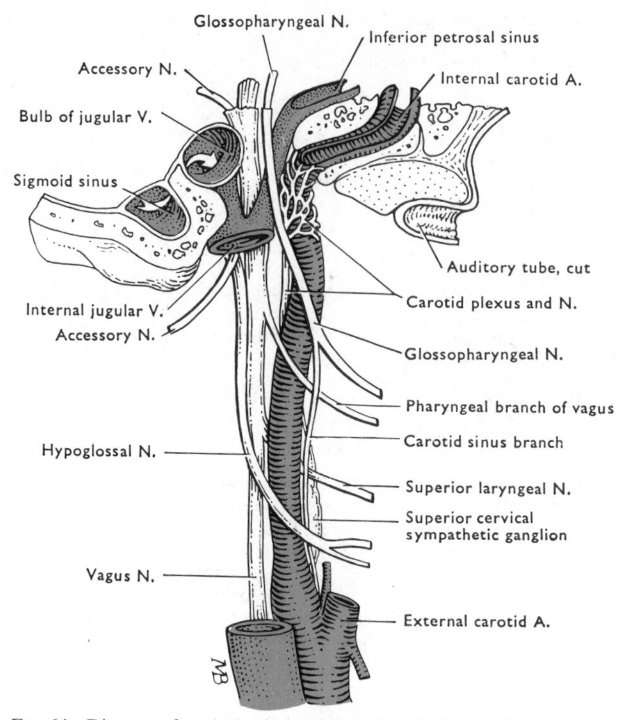

FIG. 64 Diagram of structures in and below the right jugular foramen.

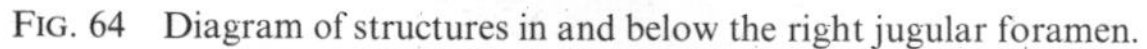

STYLOPHARYNGEUS

The longest of the three styloid muscles, stylopharyngeus arises from the medial surface of the styloid process close to its root. It runs antero-inferiorly, between the internal and external carotid arteries, to pass obliquely through the pharyngeal wall between the superior and middle constrictor muscles [FIG. 32]. Within the pharynx [FIG. 115], it blends with the anterior fibres of palatopharyngeus and is inserted partly with it and partly into the lateral aspect of the epiglottis. **Nerve supply**: the glossopharyngeal nerve. **Action**: it helps to lift the larynx and pull the base of the epiglottis upwards and backwards during swallowing and speaking.

VAGUS NERVE

Course

The vagus arises by a row of rootlets from the side of the medulla oblongata [FIG. 180] immediately inferior to the glossopharyngeal nerve. It leaves the skull with the accessory nerve through the middle compartment of the jugular foramen, and then has two **sensory ganglia** (superior and inferior) on it. The vagus descends vertically in the carotid sheath, posteromedial to the internal jugular vein and posterolateral first to the internal carotid artery and then to the common carotid artery.

At the root of the neck, each vagus crosses the anterior surface of the corresponding subclavian artery. The right vagus then descends, posterior to the brachiocephalic vessels, to the right side of the trachea in the thorax. The left vagus descends to the arch of the aorta between the subclavian and common carotid arteries, passing posterior to the left brachiocephalic vein.

Branches [FIGS. 62, 64]

It communicates with the other nerves in the jugular foramen, and is joined by the cranial part of the accessory nerve which is distributed through its branches to the larynx.

A small **meningeal branch** passes back through the jugular foramen to the dura mater of the posterior cranial fossa.

The slender **auricular branch** runs through a minute canal in the lateral wall of the jugular fossa to the tympanomastoid fissure. It pierces the cartilage of the external acoustic meatus, and supplies the cutaneous lining of its lower half and of the lower half of the tympanic membrane. Irritation of this skin may give rise to an intractable cough or cause other vagal reflexes.

The **pharyngeal branch** arises immediately below the skull. It runs antero-inferiorly between the carotid arteries, branching to form a large part of the **pharyngeal plexus** in the fascia covering the middle constrictor muscle. The plexus also receives branches from the glossopharyngeal nerve and from the superior cervical sympathetic ganglion. It supplies the pharyngeal wall.

The **superior laryngeal nerve** is larger than the pharyngeal branch, and arises inferior to it. It descends deep to both carotid arteries, and divides into:

1. The **internal branch** descends on the lateral wall of the pharynx, joins the superior laryngeal artery, and passes with it deep to the thyrohyoid muscle. They pierce the thyrohyoid membrane, and descend between it and the mucous lining of the piriform recess [FIG. 119] supplying the mucous membrane and that of the larynx down to the vocal folds.
2. The slender **external branch** descends on the side of the pharynx deep to the carotid and superior thyroid arteries, the sternothyroid muscle, and the thyroid gland. It reaches and supplies the crico-thyroid muscle [FIG. 137] (the only intrinsic laryngeal muscle not supplied by the recurrent laryngeal nerve) and the inferior constrictor muscle [FIG. 114].

Two slender **cardiac branches** arise from each vagus at variable points in the neck; one in the upper part, the other in the lower. They descend with the vagus, entering the thorax with it on the left side, but pass posterior to the subclavian artery on the right. They join the cardiac plexuses.

The right **recurrent laryngeal nerve** arises as

the vagus crosses the first part of the subclavian artery: the left as the vagus crosses the arch of the aorta in the thorax. Each nerve hooks below the corresponding artery, and ascends in the groove between the oesophagus and trachea, passing among the branches of the inferior thyroid artery deep to the lobe of the thyroid gland. The nerve enters the larynx deep to the inferior border of the inferior constrictor muscle. Within the larynx, it communicates with the internal branch of the superior laryngeal nerve, and supplies the mucous membrane below the level of the vocal folds and all the intrinsic muscles except cricothyroid. The recurrent laryngeal nerve also gives off cardiac branches near its origin, and supplies the trachea, oesophagus, and the inferior part of the pharynx.

ACCESSORY NERVE

This eleventh cranial nerve is mainly motor. The **cranial roots** emerge from the side of the medulla oblongata as a vertical row in series with those of the vagus. The **spinal roots** arise from the upper five cervical segments of the spinal medulla. They emerge between the ligamentum denticulatum [FIG. 169] and the dorsal rootlets of the spinal nerves, unite, and ascend through the foramen magnum to be joined by the cranial rootlets within the skull [FIG. 180].

Course and Termination

The nerve leaves the skull through the jugular foramen in the same dural sheath as the vagus. It is essentially the caudal part of the vagus. As it leaves the foramen the cranial part continues with the vagus, and enters its laryngeal branches. The spinal part turns postero-inferiorly, either anterior or posterior to the internal jugular vein, and crosses the tip of the transverse process of the atlas and the upper part of the carotid triangle. It then supplies and enters the deep surface of the sternocleidomastoid, emerges from the middle of its posterior border, and runs postero-inferiorly in the investing fascia of the posterior triangle to the deep surface of trapezius. The branches which the accessory nerve receives from the second to fourth cervical ventral rami are sensory to sternocleidomastoid and trapezius, the accessory being their sole motor supply.

HYPOGLOSSAL NERVE

This purely motor nerve supplies all the muscles of the tongue except palatoglossus. Some of its branches supply the geniohyoid and infrahyoid muscles, but these are principally composed of nerve fibres from the first cervical ventral ramus which join the hypoglossal nerve after it leaves the skull [FIGS. 59, 65, 104].

Origin and Course

The hypoglossal nerve arises as a row of rootlets from the anterior surface of the medulla oblongata. They form two roots which enter the hypoglossal canal in separate dural sheaths and unite externally. The nerve then enters the neurovascular bundle, adheres briefly to the lateral surface of the vagus, and descends with it to the deep surface of the posterior belly of digastric. Here it curves anteriorly on the root of the occipital artery, and enters the anterior triangle of the neck lateral to the external carotid artery [pp. 35, 110].

Branches

1. A **meningeal branch**, composed of nerve fibres of the first cervical nerve, runs back through the hypoglossal canal to supply the dura mater near the foramen magnum. 2. A descending branch, the **superior root of the ansa cervicalis** (C. 1, p. 35). 3. **Nerves to thyrohyoid** and **genio-hyoid** (C. 1, p. 35). 4. Branches to all the intrinsic and extrinsic **muscles of the tongue** except palatoglossus.

DISSECTION. Identify the cervicothoracic ganglion of the sympathetic trunk, and expose it completely by displacing the vertebral artery laterally. Find the branches of this ganglion and also of the middle and superior cervical ganglia, including the grey rami communicantes to the cervical nerves which are most easily demonstrated on the right side.

SYMPATHETIC TRUNK

In the neck, this trunk consists of **preganglionic sympathetic fibres** ascending from the upper thoracic ventral rami (white rami communicantes) to the cervical ganglia. The trunk runs almost vertically, anterior to the longus colli and longus capitis muscles on the roots of the transverse processes, and posterior to the common and internal carotid arteries. Inferiorly it is medial to the vertebral artery, and is crossed anteriorly by the thoracic duct (on the left) and posteriorly by the inferior thyroid artery.

Superiorly, the trunk ends in the superior cervical ganglion, but is continued into the skull as the **internal carotid nerve** on the internal carotid artery [FIG. 64]. Inferiorly, it enters the thorax across the neck of the first rib.

Ganglia and Rami Communicantes

All three cervical ganglia (superior, middle, and cervicothoracic) send **grey rami communicantes** between or through the prevertebral muscles to the ventral rami of the cervical nerves, but *receive no white rami from them.* Through the grey rami communicantes, and through communications with the cranial nerves and branches to the major arteries, postganglionic sympathetic fibres are distributed to the head, neck, and upper limb.

Superior Cervical Ganglion. This is the largest (2·5 cm long) ganglion of the trunk. It lies between the internal carotid artery and the longus capitis, opposite the second and third cervical vertebrae.

Branches. (1) Communicating branches to the ninth, tenth, and twelfth cranial nerves. (2) **Grey rami communicantes** pass to the upper four cervical ventral rami. (3) The **internal carotid nerve** passes on to the internal carotid artery. (4) **Laryngopharyngeal branches** pass medially to the pharyngeal plexus. (5) The **external carotid nerves** form the external carotid plexus on the external carotid artery and its branches. (6) The **superior cervical cardiac nerve** is a slender branch which descends with the common carotid artery. On the left, it ends in the superficial part of the cardiac plexus; on the right, it passes anterior or posterior to the subclavian artery, and descends behind the brachiocephalic trunk to the deep part of the cardiac plexus on the tracheal bifurcation.

Middle Cervical Ganglion. This small ganglion lies on the inferior thyroid artery at the level of the cricoid cartilage.

Branches. (1) **Grey rami communicantes** pass to the fifth and sixth cervical ventral rami. (2) **Thyroid branches** form a plexus on the inferior thyroid artery. They communicate with the recurrent and external laryngeal nerves. (3) The **middle cervical cardiac nerve** is a slender branch to the deep part of the cardiac plexus. On the left, it runs between the subclavian and common carotid arteries on to the trachea; on the right, it accompanies the superior cervical cardiac nerve. (4) The **ansa subclavia** is a slender branch which loops round the subclavian artery to the cervicothoracic ganglion. It supplies the subclavian artery and sends nerve fibres to the phrenic nerve.

Cervicothoracic Ganglion. This ganglion may consist of one or two parts. When single, it represents the fused inferior cervical and first, or first and second thoracic ganglia. When separate, the inferior cervical ganglion is small and lies behind the common carotid and vertebral arteries, anterior to the eighth cervical ventral ramus. When the ganglia are fused, the mass lies across the neck of the first rib, receives a **white ramus** from the first thoracic ventral ramus, and sends a grey ramus to it and sometimes to the second thoracic ventral ramus also. When separate, the inferior cervical ganglion has the following branches, which arise from the cervicothoracic ganglion when they are fused.

(1) **Grey rami communicantes** pass to the seventh and eighth cervical ventral rami. (2) Fine filaments pass from the ansa subclavia to form the **subclavian plexus** on the subclavian artery. (3) Larger filaments form the **vertebral plexus** on the vertebral artery. (4) The **inferior cervical cardiac nerve** passes with the middle cervical cardiac nerve to the deep part of the cardiac plexus.

All the sympathetic ganglia that send postganglionic nerve fibres to structures in the head, neck, and upper limb *receive preganglionic fibres from the central nervous system only through the white rami communicantes of the upper thoracic ventral rami.* These preganglionic fibres ascend in the trunk and reach its ganglia directly and outlying ganglia through branches of the trunk. Thus destruction of the trunk at the root of the neck, whether as a result of surgery (cervical sympathectomy) or of some pathological condition, isolates all these sympathetic ganglion cells from the central nervous system and prevents them from responding to reflex or emotional changes in the central nervous system. Thus sweating from heat or fear, vasoconstriction from cold or fright, and dilatation of the pupil as a result of darkness or terror are all lost on the side of the injury. In addition there is loss of cutaneous vasodilatation in heat, and the upper eyelid droops (**ptosis**) because of the paralysis of the smooth muscle in the elevator of that lid [p. 87].

DISSECTION. Expose the prevertebral muscles and the cervical plexus on the right side. Find the cut end of rectus capitis lateralis attached to the superior surface of the transverse process of the atlas. The ventral ramus of the first cervical nerve was medial to this muscle, and communicated with the second cervical ventral ramus. Follow the ventral rami of the cervical nerves and their branches, noting the communications between the upper five. Define the attachments of the scalene muscles [FIG. 70], and trace the lower cervical and first thoracic ventral rami as far medially as possible, detaching the scalenus anterior if necessary.

CERVICAL PLEXUS

[FIG. 65]

This plexus is formed by communications between the ventral rami of the upper four cervical nerves, which emerge superior to the corresponding vertebrae. The plexus lies posterior to the internal jugular vein and the prevertebral fascia, between the level of the root of the auricle and the superior border of the thyroid cartilage.

The **ventral rami of the cervical nerves** emerge between the corresponding intertransverse muscles (rectus capitis anterior and lateralis in the case of the first cervical) and anterior to scalenus medius (except the first cervical). They communicate serially with each other, that between the fourth and fifth uniting the cervical and brachial plexuses, and being enlarged when the brachial plexus is **prefixed**, *i.e.*, when the plexus has a higher than usual level of origin.

Branches

(1) **Communicating**. (a) A **grey ramus communicans** to each of the first four cervical ventral rami from the superior cervical ganglion. (b) A branch of the first cervical ventral ramus which

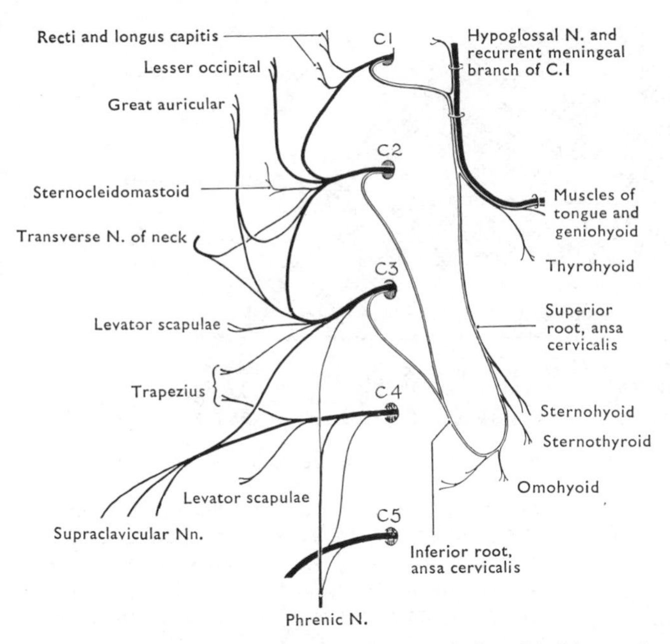

FIG. 65 A diagram of the cervical plexus and ansa cervicalis. C.1–C.5, ventral rami of the upper five cervical nerves. Note that the nerves to the geniohyoid and thyrohyoid muscles and the superior root of the ansa are derived from the first cervical ventral ramus although they appear to arise from the hypoglossal nerve.

descends to join or communicate with the **hypoglossal nerve**. This branch usually supplies geniohyoid and thyrohyoid, and forms the **superior root of the ansa cervicalis**. (c) Sensory branches from the second, third, and fourth to the **accessory nerve** for sternocleidomastoid (C. 2) and trapezius [p. 21].

(2) **Cutaneous branches**, see page 20.

(3) **Muscular branches** pass to the diaphragm (**phrenic nerve**, C. 3, 4, 5), the infrahyoid (**ansa cervicalis**, C. 1–3 [p. 35]), the prevertebral, scalene, intertransverse, and levator scapulae [p. 25] muscles.

Phrenic Nerve (C. 3, 4, 5). It arises chiefly from the fourth cervical ventral ramus at the lateral border of scalenus anterior, posterolateral to the internal jugular vein. On both sides the nerve descends with the vein obliquely across scalenus anterior, reaching the medial border of the muscle above the subclavian artery on the left. Thus it crosses the first part of the subclavian artery on the left, but lies on scalenus anterior in front of the second part of the artery on the right.

On both sides the phrenic nerve passes posterior to the end of the subclavian vein, and anterior to the internal thoracic artery. The right nerve descends into the thorax posterolateral to the right brachiocephalic vein and the superior vena cava. On the left, the nerve descends into the thorax between the common carotid artery and the pleura.

The root from the fifth cervical ventral ramus may not join the phrenic nerve on the surface of scalenus anterior, but may descend into the thorax before joining it, or reach it through a communication from the nerve to subclavius.

SCALENE MUSCLES

[FIG. 70]

These muscles form a thick mass behind the prevertebral fascia, extending from the cervical transverse processes to the first two ribs. **Nerve supply**: twigs from the ventral rami of the lower five or six cervical ventral rami. **Actions**: they elevate the first two ribs on inspiration, and produce lateral flexion of the cervical vertebrae when acting unilaterally.

Scalenus Anterior [FIGS. 54–58]

It arises from the anterior tubercles of the transverse processes of the third to sixth cervical vertebrae, and descends between the subclavian artery and vein to the scalene tubercle on the first rib [FIG. 66].

It is separated from scalenus medius posteriorly by the roots of the brachial plexus and the subclavian artery. Anteriorly, it is crossed obliquely by the internal jugular vein and the phrenic nerve, and its lowest part is separated from the clavicle by the subclavian vein. It is crossed superficially by the inferior belly of omohyoid, and by the transerve cervical and suprascapular arteries passing laterally behind the internal jugular vein, anterior to the phrenic nerve. Medially is the thyrocervical trunk [FIG. 55], suprapleural membrane, and pleura, while the vertebral artery is in contact with it superiorly.

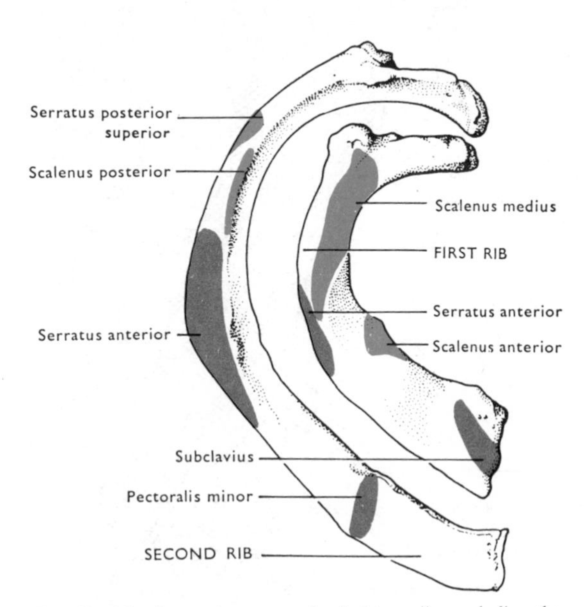

FIG. 66 Muscle attachments to the first two ribs, excluding the intercostals. Origins, red; insertions, blue.

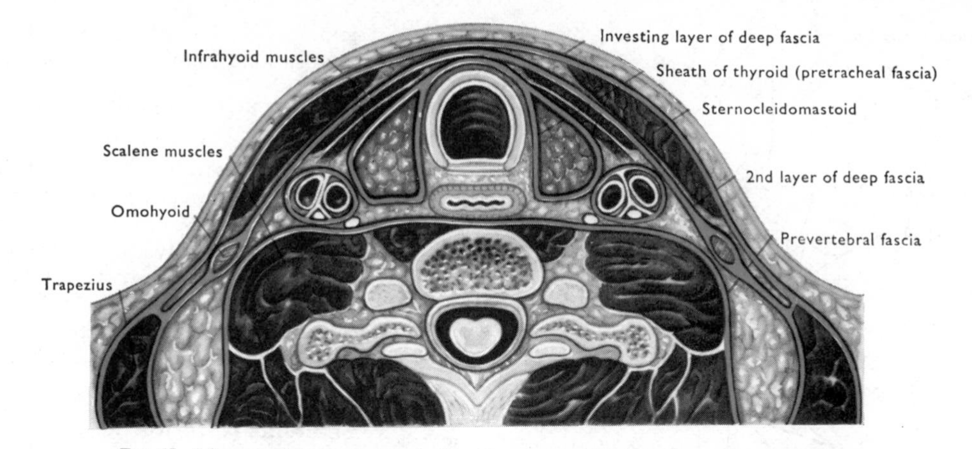

FIG. 67 Diagram of the cervical fascia (blue) in transverse section of lower part of the neck.

Scalenus Medius

This is a larger muscle than scalenus anterior. It arises from the posterior tubercles of all the cervical transverse processes, and is inserted into a rough, oval area on the superior surface of the first rib posterior to the groove for the subclavian artery and the contribution of the first thoracic nerve to the brachial plexus [FIG. 66].

Position. It lies under the prevertebral fascia in the posterior triangle of the neck immediately posterior to the roots of the cervical and brachial plexuses and the third part of the subclavian artery, and immediately anterior to levator scapulae. Scalenus medius is pierced by the **dorsal scapular nerve** and by the upper two roots of the **long thoracic nerve**. Inferiorly it is in contact with the cervical pleura.

Scalenus Posterior

This small muscle is really a part of scalenus medius which is inserted into the external surface of the second rib.

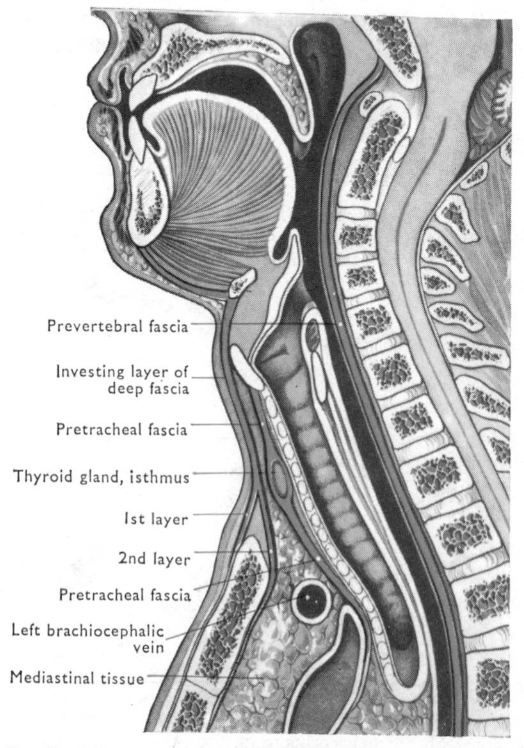

FIG. 68 Diagram of the cervical fascia (blue) in sagittal section.

CERVICAL FASCIA

The layers of cervical fascia are difficult to dissect because they are no more than laminar condensations produced by the stresses of the structures moving in the general areolar tissue. These laminae ensheathe muscles and other moving structures, and are joined to them and to other structures by loose areolar tissue. This permits movement between the sheaths and within them. The layers also form planes which tend to direct the spread of infection in the neck.

Investing Fascia. This layer encircles the neck, and splits to enclose the sternocleidomastoid and trapezius muscles, forming the lateral wall of the posterior triangle between them. *Superiorly*, it is attached to the body and greater horn of the hyoid bone, and above this it splits to enclose the submandibular gland: a thin, deep layer passing to the mylohyoid line of the mandible, a thick, superficial layer to the inferior border of the mandible [FIG. 307]. Posterior to the submandibular gland, the superficial layer is permeated by the parotid gland, its deepest part passing inwards to the styloid process—the **stylomandibular ligament**. *Inferiorly*, the investing fascia splits to surround the suprasternal space [FIG. 67, p. 31] and is attached to the clavicle, acromion, and spine of the scapula. The fascia of the infrahyoid muscles is in contact with the deep surface of the investing layer, and encloses the intermediate tendon of omohyoid, binding it down to the clavicle.

Pretracheal Fascia. This fascia covers the front and sides of the trachea, splits to enclose the **thyroid gland**, and is attached to the oblique line on the thyroid cartilage [FIG. 128] and to the arch of the cricoid cartilage anteriorly. These attachments cause the thyroid gland to rise and fall with the larynx, *e.g.*, in swallowing. Inferiorly, the pretracheal fascia surrounds the inferior thyroid veins and fuses with the pericardium.

Carotid Sheath. This tubular condensation extends from the base of the skull to the root of the neck. It surrounds the common and internal carotid arteries, the internal jugular vein, the vagus nerve, and the ansa cervicalis, and is wedged between the investing, pretracheal, and prevertebral fasciae.

Prevertebral Fascia. This layer, attached to the anterior tubercles of the cervical transverse processes, separates the prevertebral muscles from the loose areolar tissue in which the distensible oesophagus lies. The fascia extends from the base of the skull to the lower limit of longus colli (T. 3) where it fuses with the anterior longitudinal ligament. Laterally, the fascia covers the scalene muscles, levator scapulae, and splenius capitis (fascia of the medial wall of the posterior triangle), thereby enclosing the deep muscles of the back of the neck. It descends on the scalene muscles to the outer border of the first and second ribs, and is carried as a sleeve around the brachial plexus and the subclavian artery—**fascia of the cervico-axillary canal.** Medial to scalenus anterior it is continuous with the **suprapleural membrane** which extends from the seventh cervical transverse process to the inner border of the first rib [p. 62].

Buccopharyngeal Fascia. This delicate, distensible layer covers the constrictor muscles of the pharynx and buccinator, and extends from the base of the skull to the oesophagus. Together with a similar layer on the internal surfaces of these muscles (**pharyngobasilar fascia** [p. 114]), it closes the gaps in the muscular wall of the pharynx [FIG. 114].

LYMPH NODES AND LYMPH VESSELS OF THE HEAD AND NECK

Some of the lymph nodes along the carotid sheath and in the root of the neck may have been seen already, but the nodes and vessels can only be demonstrated satisfactorily by special techniques. The following account gives a brief introduction to the main features of the system in the head and neck.

The lymph nodes of the head and neck are very numerous, and may be divided into two groups—superficial and deep. The **superficial groups** are situated around the junction of the head with the

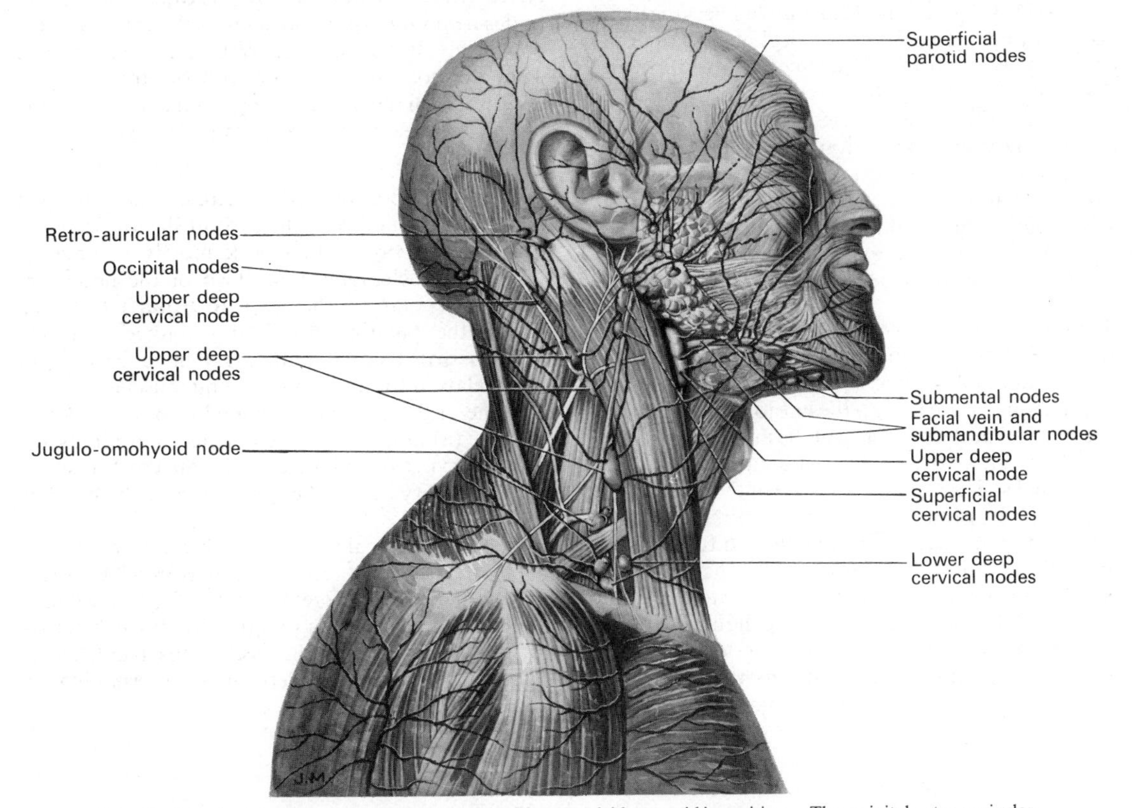

FIG. 69 Lymph nodes of the head and neck as seen with sternocleidomastoid in position. The occipital, retro-auricular, and superficial parotid nodes are inserted in accordance with descriptions. The other nodes were present in one or other of the two bodies from which the drawing was made.

neck. They drain all the superficial structures of the head and some deep parts. Most of the efferent lymph vessels from these pass to the **second group**. This consists of a vertical chain of **deep cervical nodes** arranged along the internal jugular vein from the digastric to the root of the neck, and arbitrarily divided into upper and lower groups by omohyoid. A few scattered, superficial nodes are found along the external and anterior jugular veins. These drain to the deep cervical nodes which also receive lymph from a separate group of nodes lying behind the pharynx—retropharyngeal nodes.

Superficial Lymph Nodes

Occipital Lymph Nodes. These few, small nodes lie on the upper end of trapezius and on the fascia at the apex of the posterior triangle. They drain the occipital part of the scalp and the upper part of the back of the neck to the upper deep cervical lymph nodes under cover of sternocleidomastoid.

Retro-auricular Lymph Nodes. These lie on the superior part of the sternocleidomastoid, posterior to the auricle. They drain the posterior half of the side of the head and the posterior surface of the auricle to deep nodes under sternocleidomastoid.

Parotid Lymph Nodes. Several small nodes are scattered through the parotid gland. The superficial nodes drain the area from a vertical line through the auricle forwards to an oblique line joining the medial angle of the eye to the angle of the mandible, including most of the auricle and external acoustic meatus. Deeper nodes drain the temporal and infratemporal fossae, the middle ear, auditory tube, and upper molar teeth and gums. Efferent vessels arise from the nodes at the lower pole of the parotid gland. They pass either to nodes on the external jugular vein, or to the upper deep cervical nodes.

Submandibular Lymph Nodes. These nodes lie along the submandibular gland, mainly under cover of the mandible. They receive superficial lymph vessels from the area below the line joining the medial angle of the eye and the angle of the mandible. Deeper lymph vessels drain the submandibular and sublingual salivary glands, the side of the tongue and posterior part of the floor of the mouth, most of the teeth and gums, part of the palate, and the anterior parts of the walls of the nasal cavity. Their efferents pass to deep cervical nodes under sternocleido-mastoid. Some small nodes lie along the course of the facial vein. One of these at the anterior border of masseter (mandibular node) drains the cheek and lateral parts of the lips to the submandibular nodes.

Submental Lymph Nodes. These lie on the fascia covering mylohyoid, between the anterior bellies of the digastric muscles [Fig. 30]. They drain lymph from a wedge-shaped zone which includes the incisor teeth and gums and the anterior part of the floor of the mouth. They drain to deep cervical lymph nodes, some vessels passing with the anterior jugular vein to the lower group. Though small, some of these nodes may be palpable even in healthy individuals.

Retropharyngeal Lymph Nodes. A few nodes lie in the fascia of the posterior wall of the upper pharynx, at the level of the mastoid process. They drain the oral and nasal parts of the pharynx, the palate, nose, and paranasal sinuses, the auditory tube, and the middle ear. These nodes drain postero-inferiorly to nodes in the posterior triangle.

Cervical Lymph Nodes

Superficial Cervical Nodes. 1. Three or four nodes lie along the external jugular vein. They drain the parotid nodes and the adjoining skin, either across sternocleidomastoid to deep nodes in the carotid triangle, or with the vein to deep nodes at the root of the neck. 2. A few small nodes on the anterior jugular vein drain the surrounding skin and muscles along the vein to lower deep cervical nodes.

Anterior Cervical Lymph Nodes. These small nodes lie on the front and sides of the trachea, most commonly beside the recurrent laryngeal nerves. They are continuous below with the tracheobron-chial nodes in the thorax. They drain lymph from the larynx, trachea, and thyroid gland to the lower deep cervical nodes.

Deep Cervical Lymph Nodes. These form a broad strip of nodes on the carotid sheath, from the digastric to the root of the neck, mostly under cover of the sternocleidomastoid. Two of the nodes are particularly large, the **jugulodigastric** which drains the palatine tonsil and tongue, and the **jugulo-omohyoid**. Both are named because of their relation to the corresponding muscles.

Some of the deep nodes extend into the posterior triangle (a) along the accessory nerve (draining the retropharyngeal nodes), and (b) on the transverse cervical artery across the upper part of the brachial plexus.

The deep cervical nodes are linked by afferent and efferent vessels, and receive lymph from all the other groups. Their final efferent pathway for all the lymph of the head and neck is the **jugular lymph trunk** [p. 61] at the root of the neck. This trunk enters either the thoracic duct (left) or the internal jugular vein (right).

Muscles

The prevertebral muscles are most easily seen on the right, though longus capitis, rectus capitis lateralis, and rectus capitis anterior have been cut across. The cut ends can be seen on the base of the skull, or the entire muscles exposed on the left by displacing that half of the pharynx. They are covered anteriorly by the prevertebral fascia, and are supplied by the ventral rami of the cervical nerves. As a group, they flex the neck, and the head on the neck.

Longus Colli. This is the longest and most medial of the muscles. It extends from the anterior tubercle of the atlas to the third thoracic vertebra. It is attached to the bodies of the intervening vertebrae and to the transverse processes of the third to sixth cervical [FIG. 70].

Longus Capitis. This muscle is anterolateral to longus colli and overlaps rectus capitis anterior. It passes from the anterior tubercles of the third to sixth cervical transverse processes to the base of the skull in front of rectus capitis anterior [FIG. 300].

Rectus Capitis Anterior. It passes from the anterior surface of the lateral mass of the atlas to the base of the skull, immediately anterior to the occipital condyle.

Rectus Capitis Lateralis. This muscle passes from the superior surface of the transverse process of the atlas to the jugular process of the occipital bone. It is immediately posterior to the jugular foramen, and is separated from the rectus capitis anterior by the ventral ramus of the **first cervical nerve** which supplies both muscles. The two rectus muscles act with the muscles of the suboccipital triangle to stabilize the skull on the vertebral column.

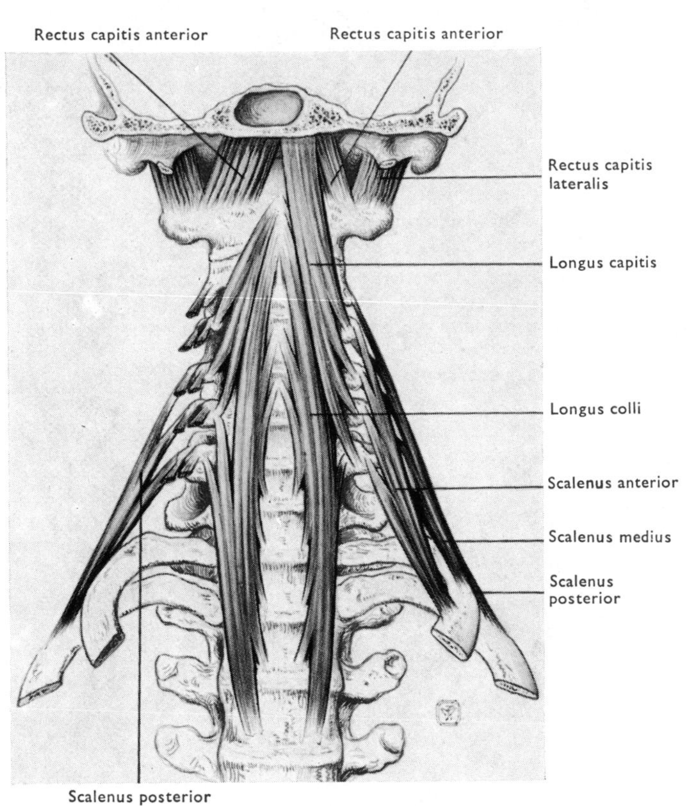

FIG. 70 The prevertebral muscles of the neck.

DISSECTION. Re-define the attachments of the scalene muscles on both sides. Remove scalenus anterior to expose the anterior and posterior **intertransverse muscles** which unite the corresponding tubercles of adjacent cervical transverse processes. They are separated by the ventral rami of the cervical nerves, and the dorsal rami pass posteriorly medial to the posterior intertransverse muscles. Rectus capitis anterior and lateralis are enlarged intertransverse muscles.

The vertebral artery may now be exposed in the intertransverse spaces by removing the anterior intertransverse muscles. Then remove the anterior tubercles and costal processes of the third to sixth cervical vertebrae to expose this part of the artery.

VERTEBRAL ARTERY

This important artery joins its fellow to form the basilar artery, and together they supply the upper part of the spinal medulla, the medulla oblongata, the pons, the cerebellum, the midbrain, and the posterior part of the cerebrum.

First Part. It begins as a branch of the first part of the subclavian artery, and ascends behind the common carotid artery to the sixth cervical transverse process [p. 59; FIGS. 58, 71].

Second Part. This part ascends through the foramina transversaria, accompanied by the **vertebral veins** and a plexus of **sympathetic nerve fibres** derived from the cervicothoracic ganglion of the sympathetic trunk. Between the transverse processes, it is anterior to the ventral rami of the cervical nerves. At first vertical, the artery turns laterally in the foramen transversarium of the axis under the superior articular facet, and bending upwards, enters the foramen transversarium of the atlas, which is placed further laterally than the others [FIGS. 55, 71].

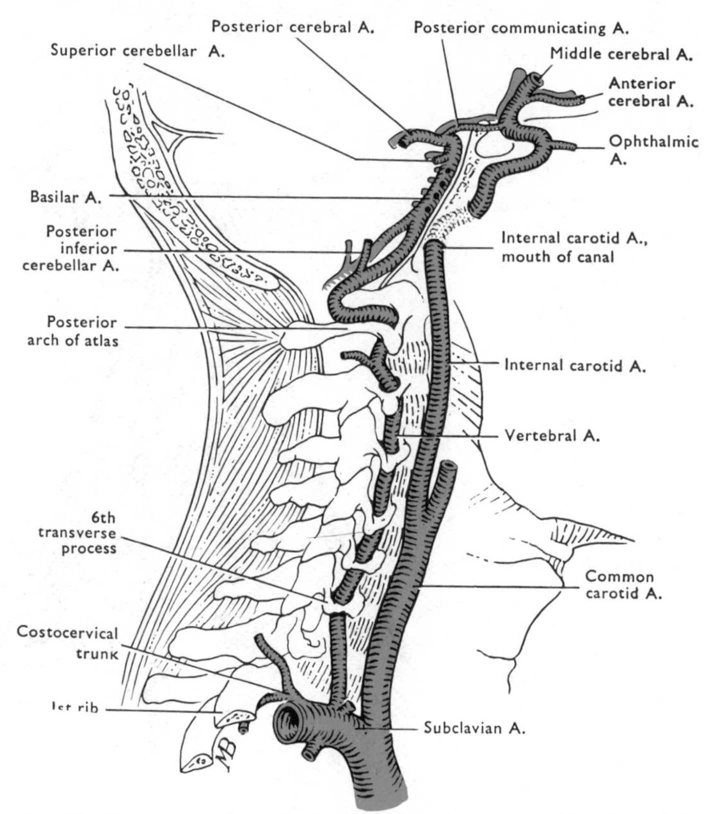

FIG. 71 The course and communications of the internal carotid and vertebral arteries.

Third Part [FIGS. 71, 288]. The artery emerges on the superior surface of the atlas between rectus capitis lateralis and the superior articular process of the atlas. It then curves horizontally over the lateral and posterior surfaces of that process with the ventral ramus of the first cervical nerve, grooving the process and the root of the posterior arch of the atlas. On the arch, it lies superior to the dorsal ramus of the first cervical nerve, in the depths of the suboccipital triangle [FIG. 28] and leaves the triangle by passing medially in front of the posterior atlanto-occipital membrane.

Fourth Part [p. 50; FIGS. 71, 186, 187]. The artery turns superiorly, pierces the dura mater and arachnoid, and enters the cranial cavity through the foramen magnum, anterior to the uppermost tooth of the ligamentum denticulatum [p. 160]. It then runs anterosuperiorly on the medulla oblongata between the rootlets of the first cervical and hypoglossal nerves. At the lower border of the pons, it forms the basilar artery by uniting with the opposite vertebral artery.

Branches

There are no branches of importance from the first part. **Spinal** and **muscular** branches arise from the second and third parts and anastomose with branches of the occipital and deep cervical arteries, and supply adjacent muscles. The fourth part gives off a meningeal branch, and a series of branches to the medulla oblongata and spinal medulla [p. 177].

VERTEBRAL VEIN

A plexus of veins is formed around the beginning of the third part of the vertebral artery by the union of veins from the internal vertebral venous plexus and the suboccipital triangle. This plexus runs with the second part of the artery. It anastomoses with the **internal vertebral venous plexus** and the **deep cervical veins**. It ends as one or two vertebral veins which descend with the artery from the sixth cervical transverse process, and passing anterior to the subclavian artery, enter the posterior surface of the brachiocephalic vein near its origin. One or two veins pass through the foramen transversarium of the seventh cervical vertebra.

In the following dissection several structures are partially exposed and will not be completely dissected until later. To avoid confusion the dissectors should refer to the illustrations as indicated.

DISSECTION. Detach the risorius and reflect it with the remains of platysma towards the corner of the mouth, avoiding injury to the underlying vessels and nerves.

Find the cut end of the great auricular nerve and trace it upwards over the lower part of the parotid gland.

Expose the facial artery and vein at the antero-inferior angle of the masseter, but do not trace them further at the moment.

Cut through the fascial covering of the **parotid gland** immediately in front of the auricle from the zygomatic arch to the angle of the mandible. Dissect the fascia carefully forwards to the margins of the gland, looking for the nerves, vessels, and the duct of the gland which emerge at the borders. The **duct** appears at the anterior border about a finger's breadth below the zygomatic arch. It is large [FIG. 90] and readily palpated in the living by rolling it against the anterior border of the clenched masseter.

Above the duct, find: (1) a small detached part of the parotid gland, the **accessory parotid**; (2) the transverse facial artery and vein [FIG. 72]; (3) the zygomatic branches of the facial nerve [FIG. 74].

Find the **branches of the facial nerve** emerging from the anterior border of the parotid gland, and trace them forwards. This is difficult because they communicate with each other and with the branches of the trigeminal nerve in the face. *The facial nerve is motor to the muscles of the face, the trigeminal is purely sensory.*

Follow the upper zygomatic branches of the facial nerve first. They pass deep to the lateral part of the orbicularis oculi where the **zygomaticofacial nerve** (sensory) may be found emerging from the zygomatic bone. Trace the lower zygomatic branches of the facial nerve forwards inferior to the orbit and deep to the zygomatic muscles. Find their communications with the **infra-orbital nerve** (sensory). Trace the branches of this nerve. Cut through zygomaticus major and minor and levator labii superioris at their origins, and turn them downwards to expose the facial artery and vein. Trace these vessels and their branches.

Find the **buccal branch of the facial nerve** at the anterior border of the parotid gland. Trace it forwards through the fat of the cheek to the buccinator muscle. Find its deep communication with the buccal (sensory) branch of the trigeminal nerve (**buccal nerve**). Follow that nerve posteriorly till it disappears deep to the ramus of the mandible.

Trace the marginal mandibular branch of the facial nerve forwards from the lower border of the parotid gland to the depressor anguli oris. Cut through that muscle, and trace the communication of the nerve with the **mental nerve** (sensory) which emerges through the mental foramen. Follow the branches of the mental nerve to the chin and lower lip.

At the upper border of the parotid gland, identify: (1) the superficial temporal artery and veins; (2) the **auriculotemporal nerve** lying behind the vessels, close to the auricle; (3) the temporal branches of the facial nerve anterior to the vessels.

At the lower border of the gland, identify again: (1) the anterior and posterior branches of the retromandibular vein; (2) the cervical branch of the facial nerve. Trace the facial vein downwards till it joins the anterior branch of the retromandibular vein [FIG. 22].

ARTERIES OF THE FACE

The face has a very rich arterial supply. The facial and transverse facial arteries anastomose freely with the various smaller arteries which accompany the branches of the trigeminal nerve into the face, and with the arteries of the opposite side, especially in the lips. Thus wounds of the face bleed freely and heal rapidly.

Transverse Facial Artery [FIG. 72]

This small vessel arises from the superficial temporal artery under cover of the parotid gland. It emerges

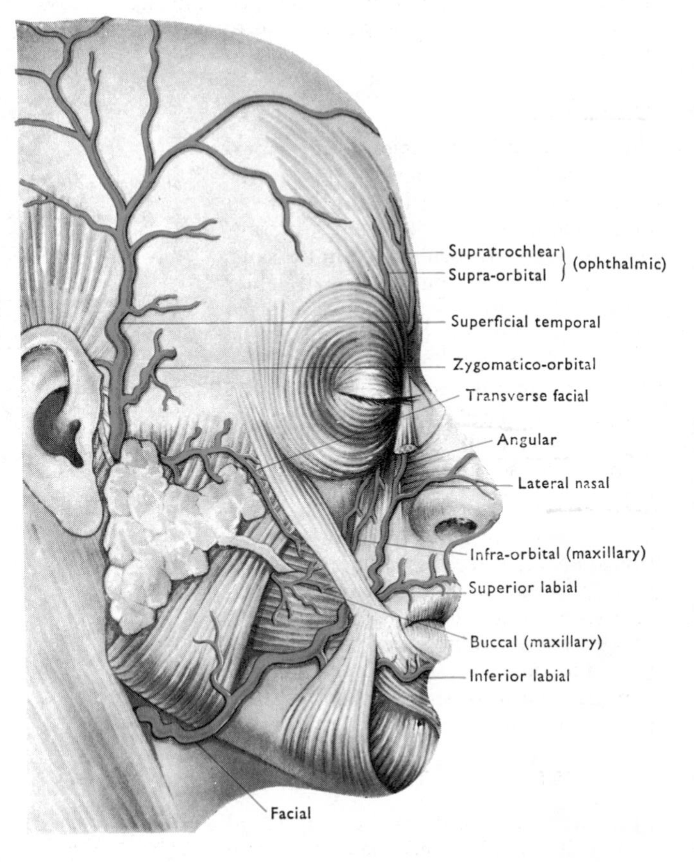

FIG. 72 The arteries of the face.

near the upper end of the gland, and runs forwards over the masseter below the zygomatic arch.

Facial Artery [FIG. 72]

This branch of the external carotid artery is the main artery of the face. It runs a sinuous course to accommodate to the mobility of the structures through which it passes. It appears on the face at the antero-inferior border of masseter (*where it can be palpated against the mandible*) by turning round the lower border of the mandible and piercing the deep fascia. On the face it runs anterosuperiorly to a point 1·5 cm from the angle of the mouth, and then ascends more vertically to end near the medial angle of the eye.

Branches. The larger branches pass to the chin, lips, and nose, but smaller branches supply the adjacent muscles. An important **anastomosis** is present at the medial angle of the eye between the facial vessels and those of the orbit. In this way venous blood may drain from the face into the orbit and skull, and the arterial blood of the facial artery (external carotid) can reach the internal carotid in the skull.

VEINS OF THE FACE

The veins anastomose freely in the face, and are drained by veins that accompany all the arteries that supply it.

Facial Vein

This is formed by the union of the supra-orbital and supratrochlear veins at the medial angle of the eye (**angular vein**). At first superficial to the artery, it runs postero-inferiorly, behind and in the same plane as the artery, but takes a straighter course, close to the anterior border of masseter, to join the artery again on the surface of the mandible. The vein then descends into the neck, pierces the deep fascia, and is joined by the anterior branch of the retromandibular vein. It crosses the carotid arteries and enters the internal jugular vein.

Tributaries. In addition to those mentioned above, it receives tributaries which correspond to the branches of the facial artery. On the surface of buccinator it gives off the **deep facial vein**, which passes medial to masseter to join the **pterygoid plexus of veins**—an important route for the spread of infection.

NERVES OF THE FACE [p. 14]

Branches of two sensory nerves (trigeminal and great auricular) and one motor nerve (facial) are found in the face.

Branches of Trigeminal Nerve in Face [FIG. 74]

The trigeminal nerve is the main sensory nerve of the face. It gives rise to three nerves (ophthalmic, maxillary, and mandibular) each of which supplies cutaneous branches to one of three roughly concentric areas of the face [FIG. 73].

Opthalmic Nerve. This nerve supplies the area of skin shown in FIGURE 73 through lacrimal, frontal, and nasociliary branches, which pass through the orbit and give rise to five nerves on the face.

1. The **palpebral branch of the lacrimal** supplies the lateral part of the upper eyelid.
2. and 3. The **supra-orbital** and **supra-trochlear** branches of the frontal nerve supply the forehead and anterior scalp.
4. The **infratrochlear** branch of the nasociliary emerges from the orbit just above the medial palpebral ligament. It supplies the medial parts of the eyelids and the root of the nose.
5. The **external nasal nerve** emerges between the nasal bone and the lateral nasal cartilage [FIG. 78], and supplies the skin of the lower half of the dorsum of the nose. It is a branch of the **anterior ethmoidal nerve** which arises from the nasociliary nerve in the orbit, and enters the wall of the nasal cavity through the cribriform plate of the ethmoid [p. 44].

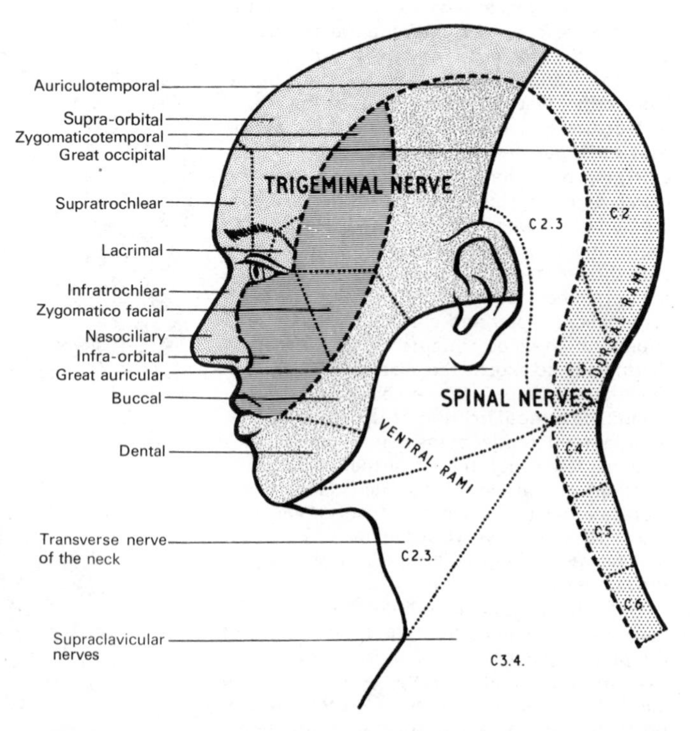

FIG. 73 Distribution of cutaneous nerves to head and neck. The ophthalmic, maxillary, and mandibular divisions of the trigeminal here are indicated by different shading.

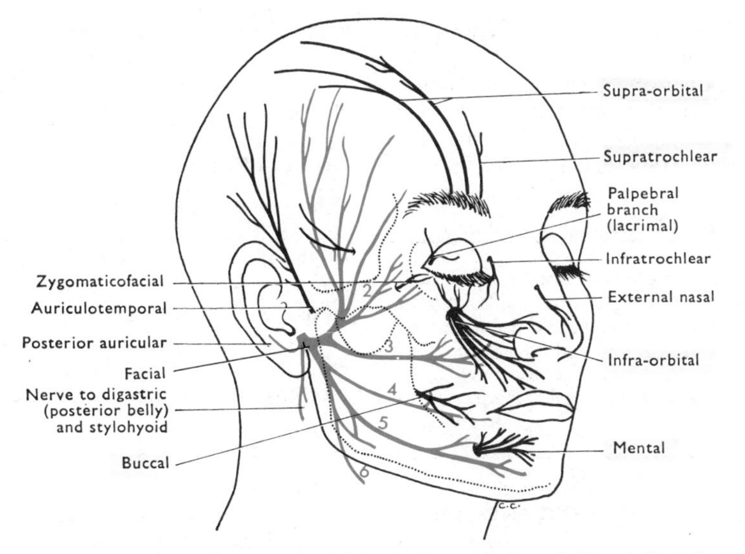

FIG. 74 The nerves of the face. The facial nerve (motor) is shown in blue, the branches of the trigeminal (sensory) in black. 1. Temporal branches of facial and zygomaticotemporal branch of trigeminal. 2 and 3. Zygomatic branches. 4. Buccal branch. 5. Marginal mandibular branch. 6. Cervical branch.

Maxillary Nerve. This nerve supplies a zone of skin inferior and lateral to the eye by three branches.

1. The **infra-orbital nerve** emerges from the infra-orbital foramen under cover of orbicularis oculi and levator labii superioris. It supplies the skin and mucous membrane of the upper lip and lower eyelid, the skin between them, and on the side of the nose, through **labial**, **palpebral**, and **nasal branches**. These nerves form a **plexus** with the zygomatic branches of the facial nerve. Such plexuses are *not* the site of union of nerve fibres in the nerves, but merely allow the nerve fibres to run together so that sensory nerve fibres from the muscles of the face can enter the brain via the trigeminal nerve.

2. The **zygomaticofacial nerve** supplies skin over the bony part of the cheek by passing from the orbit through the zygomatic bone to its facial surface [FIG. 74].

3. The **zygomaticotemporal nerve** [p. 14] also pierces the zygomatic bone, but emerges from its temporal surface. It passes through the temporal fascia, and supplies skin over the anterior part of the temple.

Mandibular Nerve. This nerve supplies a zone of skin posterior and inferior to the previous area, also by three branches.

1. The **auriculotemporal nerve** emerges from the upper end of the parotid gland beside the auricle. It supplies the upper part of the auricle, the external acoustic meatus, and the skin of the side of the head [FIG. 74].

2. The **buccal nerve** passes antero-inferiorly deep to masseter and the ramus of the mandible, to appear in the cheek on the lowest part of buccinator. It supplies the skin over buccinator, and sends branches through it to the underlying mucous membrane. It is a purely sensory nerve.

3. The **mental nerve** appears through the mental foramen from the inferior alveolar nerve in the interior of the mandible. It divides into branches under cover of depressor anguli oris. These supply the skin and mucous membrane of the lower lip, and the skin over the mandible from the symphysis to the anterior border of masseter.

Thus the trigeminal nerve supplies all the skin of the face except the area over the parotid gland and the angle of the mandible which is supplied by the **great auricular nerve**. The trigeminal cutaneous area abuts on that supplied by the ventral and dorsal rami of the second cervical nerve [FIG. 73]. *The first cervical nerve does not supply skin.*

Terminal Branches of Facial Nerve [FIG. 74]

The facial nerve is the motor nerve to the muscles of facial expression. It has five terminal branches, or groups of branches, all of which emerge from the parotid gland.

1. **Temporal branches** are described on page 15.

2. Small **zygomatic branches** run across the zygomatic arch to supply the orbicularis oculi. Larger branches run forwards below the arch to supply the muscles of the nose and those between the eye and the mouth.

3. The **buccal** branches run towards the angle of the mouth and supply the muscles of the soft part of the cheek.

4. The **marginal mandibular** branch runs forwards along the mandible, and usually curves down into the neck before running with the inferior

labial branch of the facial artery to supply the muscles of the lower lip.

5. The **cervical branch** leaves the lower border of the parotid gland, runs forwards and downwards into the neck, to supply the platysma and communicate with the transverse nerve of the neck [p. 21].

DISSECTION. Expose the levator anguli oris and the buccinator. Remove the buccal fat from the buccinator, avoiding injury to the buccal nerve, and note the small buccal glands that lie in it. Remove the fascia covering buccinator, define its attachments to the maxilla and mandible, and trace its fibres towards the angle of the mouth.

STRUCTURES IN THE CHEEKS AND LIPS

Buccinator [Figs. 17, 114, 115]

This muscle lies on the mucous membrane of the cheek. Its horizontal fibres arise from the outer surfaces of the maxilla and mandible opposite the sockets of the molar teeth. Between these two bony attachments it springs from the **pterygomandibular raphe**. The raphe is formed by the interlacing tendinous fibres of the buccinator and the superior constrictor muscle of the pharynx, where these muscles meet edge to edge. The superior constrictor will be seen later when the pharynx is dissected.

Anteriorly, the fibres of buccinator converge on the corner of the mouth, and blend with the orbicularis oris to form a large part of it. The upper and lower fibres pass directly into the corresponding lips, but the middle fibres decussate towards the angle of the mouth, so that the upper fibres enter the lower lip and vice versa. Some of the internal, postero-superior fibres pass almost vertically downwards from the maxilla to the mandible.

Nerve supply: the facial nerve. **Action**: the buccinator is used during mastication to press the cheek against the teeth and prevent food collecting in the vestibule of the mouth. It also compresses the blown out cheek and raises the intra-oral pressure.

Buccopharyngeal Fascia

This thin sheet of fascia clothes the external surface of buccinator, and continues backwards over the constrictor muscles of the pharynx. The **parotid duct** on its way to the vestibule of the mouth, pierces this fascia and the buccinator, and becomes continuous with the mucous membrane opposite the upper second molar tooth. The fascia and muscle are pierced also by the nerves and vessels of the mucous membrane.

Molar Glands and Buccal Lymph Nodes

Four to five small, molar, mucous salivary glands lie on the buccopharyngeal fascia around the parotid duct. Their ducts follow the parotid duct to the vestibule of the mouth. The buccal lymph nodes are one or two nodules found on the buccopharyngeal fascia.

Buccal Pad of Fat

This is an encapsulated mass of fat which lies on the buccopharyngeal fascia and is partly tucked between it and the masseter. It is pierced by the buccal nerve and the parotid duct, and thickens the cheek to help it resist the external pressure during sucking. This sucking pad is relatively much larger in the infant when the cheeks are not supported by teeth, and accounts for the fullness of the cheeks in infants.

Levator Anguli Oris

This muscle arises from the canine fossa below the infra-orbital foramen, under cover of orbicularis oculi, levator labii superioris, and the zygomatic muscles. It runs to the angle of the mouth, blends with orbicularis oris, and sends some fibres into the lower lip.

DISSECTION. Evert the lips and remove the mucous membrane from their deep surfaces. This exposes a number of small labial glands, and the incisive muscles near the bases of the lips opposite the sockets of the incisor teeth. Remove these from the lower lip and expose the mentalis.

Mentalis

This small muscle arises from the outer walls of the canine sockets. Its slips converge to be inserted into the skin of the chin.

Labial and Buccal Glands

These are small, closely set, mucous salivary glands that lie in the submucosa of the lips and cheeks. They are palpable as small nodules when the tongue is pressed against them. Their ducts open into the vestibule of the mouth [p. 11].

THE EYELIDS

The eyelids or palpebrae consist of the following layers from without inwards:

1. Skin and superficial fascia.
2. Orbicularis oculi [p. 16].
3. Tarsi and palpebral fascia, and joining them the tendon of levator palpebrae superioris in the upper eyelid. The tarsal glands are embedded in the tarsi.
4. Conjunctiva.

The skin of the eyelids is very thin, and its hairs (except the eyelashes) are so short that few of them appear above the surface. The superficial fascia is thin, loose, and devoid of fat. It allows the skin to move freely over the lid, and can become greatly swollen with fluid or blood after an injury.

DISSECTION. Separate the palpebral part of orbicularis oculi from the remainder by a circular incision, and turn the palpebral part towards the palpebral fissure, avoiding injury to the palpebral fascia, vessels, and nerves. At the medial angle of the eye, this dissection will expose the medial palpebral ligament [p. 13]: elsewhere it uncovers the palpebral fascia and tarsi.

Tarsi

The tarsi are two thin plates of condensed fibrous tissue which lie close to the free margins of the eyelids, and stiffen them.

The **inferior tarsus** is a narrow strip attached to the inferior orbital margin by the palpebral fascia.

The **superior tarsus** is much larger, and can be felt if the upper lid is pinched sideways between finger and thumb. Its deep surface is adherent to the palpebral conjunctiva. The palpebral fascia is attached to its anterior surface some distance below its upper border, together with the greater part of the expanded tendon of **levator palpebrae superioris**, which is attached to the deep surface of the palpebral fascia [FIG. 75]. As a result, when the upper lid is everted to expose the conjunctival surface of the superior tarsus, it tends to remain in this position and is not immediately replaced by contraction of the levator. The lower edge of the tarsus is adherent to the skin of the margin of the lid.

The **tarsal glands** lie in furrows on the deep surfaces of the tarsi, and can be seen through the conjunctiva as closely placed, parallel, yellow streaks that run at right angles to the margin of the lid. Their ducts open on the margin behind the eyelashes.

The **ciliary glands** are arranged in several rows immediately behind the roots of the eyelashes. Their ducts open on the margin close to the lashes. They are too small to be seen by dissection, but when infected, they produce a red swelling of the margin of the lid known as a 'stye'.

The **palpebral fascia** is a thin fibrous membrane which connects the tarsi to the orbital margins, and forms an **orbital septum** with them. Medially it passes posterior to the lacrimal sac and is attached to the posterior margin of the groove which lodges the sac. In the lower lid it is attached to the inferior margin of the tarsus, but in the upper lid, it is attached to its anterior surface (see above). It is pierced by the nerves and vessels which pass from the orbit to the exterior.

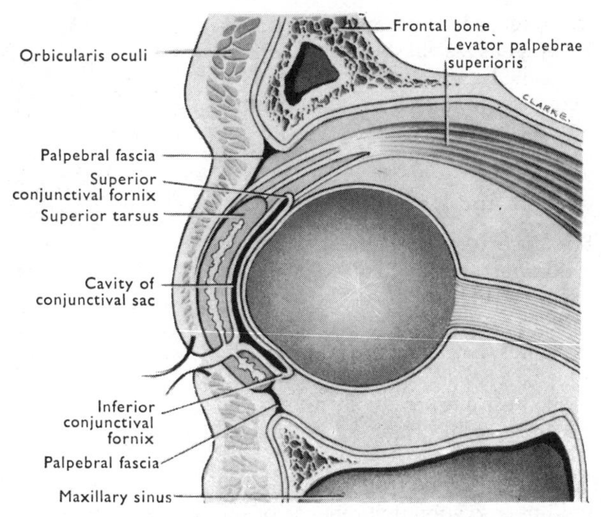

FIG. 75 Diagram of the structure of the eyelids as seen in section.

Palpebral Ligaments

The **medial** palpebral ligament is a strong fibrous band that connects the two tarsi with the medial margin of the orbit. It lies under cover of the skin, anterior to the lacrimal sac, and gives origin to many of the fibres of orbicularis oculi.

The **lateral** palpebral ligament is a slender fibrous band which connects the two tarsi with a small tubercle on the lateral orbital margin. It lies posterior to the palpebral fascia and is separated from it by a little fat.

Levator Palpebrae Superioris

The tendon of this muscle expands into a wide, thin sheet which enters the upper eyelid from the orbit, and blends with the deep surface of the palpebral fascia. A few of the fibres of the tendon pass through orbicularis oculi to be inserted into the skin; others

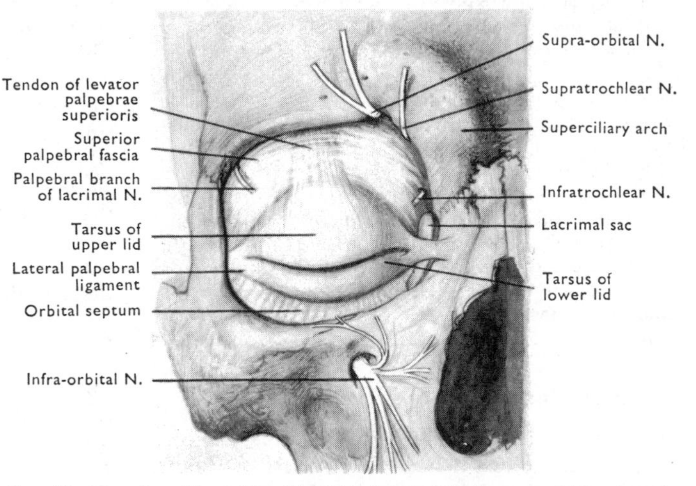

FIG. 76 Dissection of the right eyelids. Orbicularis oculi has been completely removed.

are attached to the front of the tarsus. Through its fascial sheath, the tendon is attached to the superior fornix of the conjunctiva, which is, therefore, raised with the eyelid [FIG. 75].

Vessels and Nerves of Eyelids

The **arteries** are derived from the ophthalmic, and pierce the palpebral fascia to enter the lids. They anastomose to form arches near the margins of the lids, between the tarsus and the orbicularis oculi.

The **veins** run medially and end in the supratrochlear and facial veins.

The **motor nerves** to the orbicularis oculi come from the temporal and upper zygomatic branches of the facial nerve. The **sensory nerves** to the upper lid are the palpebral branch of the lacrimal nerve and twigs from the supra-orbital, supratrochlear, and infratrochlear nerves. The infra-orbital and infratrochlear nerves supply the lower lid.

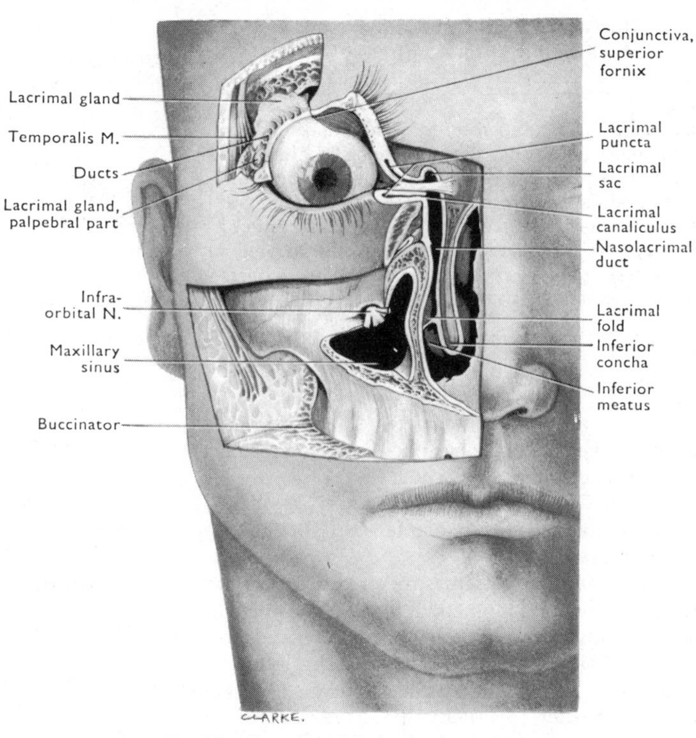

FIG. 77 Dissection of the lacrimal apparatus.

THE LACRIMAL APPARATUS

Lacrimal fluid, produced by the lacrimal gland, passes through numerous ducts into the superolateral part of the conjunctival sac. Thence it flows to the medial angle of the eye, aided by the contraction of the orbicularis oculi, and enters the lacrimal canaliculi through the puncta [FIG. 14]. These canaliculi discharge into the lacrimal sac, which transmits the fluid into the nose through the nasolacrimal duct.

DISSECTION. Cut through the superolateral part of the palpebral fascia and expose the lacrimal gland. Raise the gland, and moving the points of a fine pair of forceps up and down in the loose tissue below the gland, find its ducts.

Find the lacrimal papillae at the medial ends of the eyelids, and attempt to pass a fine bristle along the lacrimal canaliculi through the puncta.

Identify the medial palpebral ligament and the lacrimal sac which lies posterior to it. Note the lacrimal part of the orbicularis oculi passing round the lateral side of the sac. Make an opening into the sac, and passing a probe into it, explore its extent, and then pass the probe downwards into the nose through the nasolacrimal duct. Confirm its orifice in the nasal cavity.

Lacrimal Gland

This gland [FIGS. 77, 79, 80, 111] lies mainly in the orbit, but a part of it (the **palpebral process**) extends into the upper eyelid lateral to the tarsus, between the conjunctiva and the palpebral fascia. When the lid is everted it may be seen bulging the conjunctiva.

The **ducts** of the lacrimal gland (ten or less) are short and slender, and open into the conjunctiva near the superior fornix.

Conjunctiva

This membrane is covered with moist, stratified, squamous epithelium, and it lines the deep surfaces of the eyelids and covers the exposed surface of the eyeball. Over the cornea it is known as the anterior epithelium of the cornea and is firmly attached to it. The **fornices** are produced where the thick, vascular, palpebral part is reflected from the roots of the eyelids on to the eyeball. Here it is thin and transparent and is loosely attached to the sclera. As a whole it is known as the conjunctival sac, which is closed when the eye is shut. The cavity is no more than a capillary interval which is moistened by lacrimal fluid. It opens to the exterior through the **palpebral fissure** and into the lacrimal sac through the canaliculi [FIG. 77].

Lacrimal Canaliculi

These two slender tubes are about 1 cm in length. Each begins as a tiny hole (the **lacrimal punctum**) which is situated in the anterior wall of the closed conjunctival sac on the summit of a lacrimal papilla. Thence it runs medially in the margin of the eyelid, and opens into the lacrimal sac posterior to the medial palpebral ligament.

Lacrimal Sac

This sac lies posterior to the medial palpebral ligament, lodged in the lacrimal groove just posterior to the medial orbital margin. The sac is approximately 1 cm long by 0·5 cm wide. Its upper end is blind; its lower end is continuous with the nasolacrimal duct.

Nasolacrimal Duct [FIGS. 77, 126]

This tube is about 1·5 cm long and 0·5 cm wide. It begins at the anteromedial corner of the floor of the orbit, and passes downwards through the nasolacrimal canal, to end in the inferior meatus of the nasal cavity. The mucous membrane at the medial side of its opening is raised up as the **lacrimal fold**. This acts as a flap valve which prevents air and secretions being blown up the nasolacrimal duct when the intranasal pressure is raised, *e.g.*, in blowing the nose or sneezing.

Lacrimal Fluid. This fluid flows downwards over the eyeball, and most of it evaporates in ordinary circumstances. It is also carried medially by the frequent involuntary contractions of orbicularis oculi which move the eyelids medially owing to the attachment to the medial palpebral ligament. Thus, in paralysis of the orbicularis oculi, the fluid is not spread over the permanently open eye and the cornea dries. Also the sagging lower lid forms a pond in which tears collect and spill over on to the face.

Excessive secretion by the gland, due to irritation of the conjunctiva or other causes, floods the conjunctiva and the fluid overflows as tears.

Accessory Lacrimal Glands. These are minute glands that lie near the fornices; they can be effective in moistening the conjunctiva if the lacrimal gland is removed.

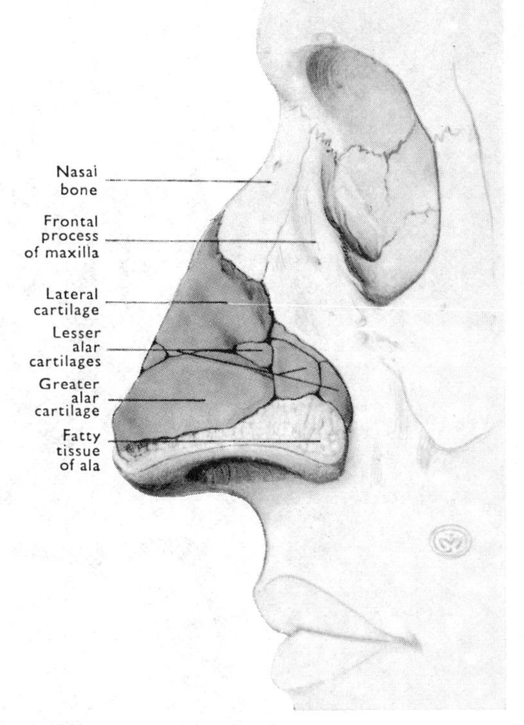

FIG. 78 Cartilages of the external nose.

DISSECTION. Strip the muscles and skin from the nose, and define the cartilages. Details of these are available in the larger textbooks.

THE ORBIT

DISSECTION. To remove the roof of the orbit, strip the periosteum from the floor of the anterior cranial fossa, except over the cribriform plate of the ethmoid. With a gentle tap from a mallet on a chisel, crack the orbital roof and lever up the broken pieces of bone from the underlying orbital periosteum. Extend the opening with bone forceps, keeping outside the orbital periosteum, until all but the anterior margin of the orbital roof is removed. Occasionally the **frontal air sinus** extends into the roof of the orbit; in this case, two layers of bone and the contained sinus must be removed.

The **orbital part of the frontal bone** forms most of the roof of the orbit, and extends medially over the ethmoidal air cells, roofing them also. Between the ethmoid and this part of the frontal bone lie the ethmoidal vessels and nerves. Chip away this part of the frontal bone, avoiding injury to the vessels and nerves, and expose the mucoperiosteal lining of the **ethmoidal sinuses**. Open some of them.

Remove the remains of the lesser wing of the sphenoid, but leave the margin of the optic canal intact. The superior orbital fissure is now opened, and the nerves which have been followed from the wall of the cavernous sinus can be traced into the orbit.

Ethmoidal Sinuses

These multiple, thin-walled cavities (**cells**) occupy the whole of the ethmoidal labyrinth, between the orbit and the upper part of the cavity of the nose [FIGS. 82, 123]. They form three groups, of which the anterior and middle cells open into the middle meatus of the nose [FIG. 124], while the posterior open into the superior meatus [FIG. 125]. Explore them with a blunt probe, and attempt to find their openings into the nasal cavity.

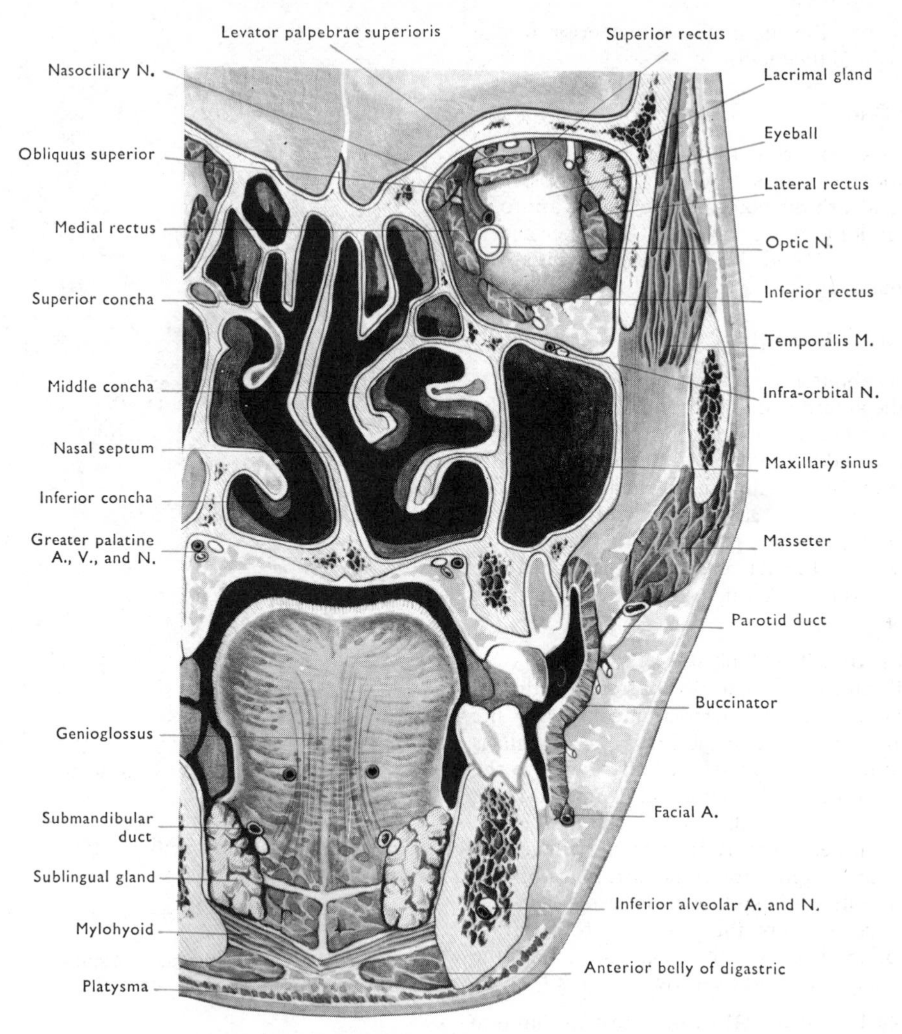

FIG. 79 Coronal section through the head at the level of the second molar teeth.

Orbital Periosteum

The orbital periosteum forms a funnel-shaped sheath which is loosely attached to the bony walls, and encloses all the contents of the orbit except the zygomatic nerve and the infra-orbital nerve and vessels. It is continuous with the endocranium through the optic canal and the superior orbital fissure.

DISSECTION. Divide the periosteum of the orbital roof transversely, close to the anterior margin of the orbit, and then anteroposteriorly along the middle line of the orbit.

Take care not to injure the nerves which pass through the superior orbital fissure. The **trochlear nerve** which lies immediately beneath the periosteum is most likely to be damaged. Find the trochlear nerve and trace it forwards and medially to the superior oblique muscle in the upper medial part of the orbit. In the midline of the orbit, the frontal nerve will be found lying on levator palpebrae superioris. Trace the nerve forwards to its division into supra-orbital and supratrochlear nerves; each runs with the corresponding artery. Follow the supratrochlear to the medial angle of the orbit, where it passes above the pulley (trochlea) of the superior oblique muscle.

Expose the superior oblique muscle, and follow its tendon through the pulley at the superomedial angle of the orbit, then posterolaterally to disappear beneath the levator palpebrae superioris and the superior rectus.

Raise levator palpebrae superioris and identify beneath it the superior rectus muscle and the branch of the oculomotor nerve which pierces that muscle to enter the levator.

Find the lacrimal nerve and artery which lie in the fat along the superolateral part of the orbit. Trace them to the lacrimal gland and define it.

THE STRUCTURES IN THE ORBIT

Frontal Nerve [FIG. 80]

This is the direct continuation of the ophthalmic nerve. It enters the orbit through the superior orbital fissure, and runs forwards, under the roof of the orbit, on levator palpebrae superioris, to divide into supra-orbital and supratrochlear branches at a variable point.

Supratrochlear Nerve. This, the medial and smaller branch, runs towards the pulley of the superior oblique muscle, pierces the palpebral fascia above it, and leaves the orbit to run upwards into the forehead [p. 14]. In the orbit it communicates with the infratrochlear nerve.

Supra-orbital Nerve. This branch continues in the line of the parent stem, passes through the supra-orbital notch or foramen, and turns upwards into the forehead [p. 14]. Normally it divides into two in the scalp, but if this occurs in the orbit, the larger, lateral part occupies the supra-orbital notch.

Lacrimal Nerve

This smallest branch of the ophthalmic passes through the lateral part of the superior orbital fissure [FIG. 86] and runs forwards above the lateral rectus muscle. At the anterior part of the orbit it receives a filament of postganglionic parasympathetic fibres from the **zygomatic nerve**, sends numerous twigs to the lacrimal gland, and one branch below it to the upper eyelid, the **palpebral branch**.

Trochlear Nerve

This fourth cranial nerve supplies only the superior oblique muscle. Having entered the orbit through the superior orbital fissure, it passes forwards and medially just under the roof of the orbit, and enters the superior oblique. The remaining nerves are seen later.

Lacrimal Gland [FIGS. 77, 79, 111]

The lacrimal gland is a lobulated structure situated in the hollow on the medial side of the zygomatic process of the frontal bone at the superolateral angle of the orbit. It is mostly hidden by the orbital margin, and is bound to it by short fibrous strands. The concave, medial surface rests on levator palpebrae superioris and the lateral rectus, which separate it from the eyeball. The **palpebral part** projects down into the upper lid between the palpebral fascia and the conjunctiva.

Nine to ten slender ducts open on the deep surface of the upper eyelid near the conjunctival fornix [FIG. 77].

The parasympathetic secretory nerve fibres arise in the **pterygopalatine ganglion** and reach the gland through the orbital branches of the ganglion, more especially through the branch from the zygomatic nerve.

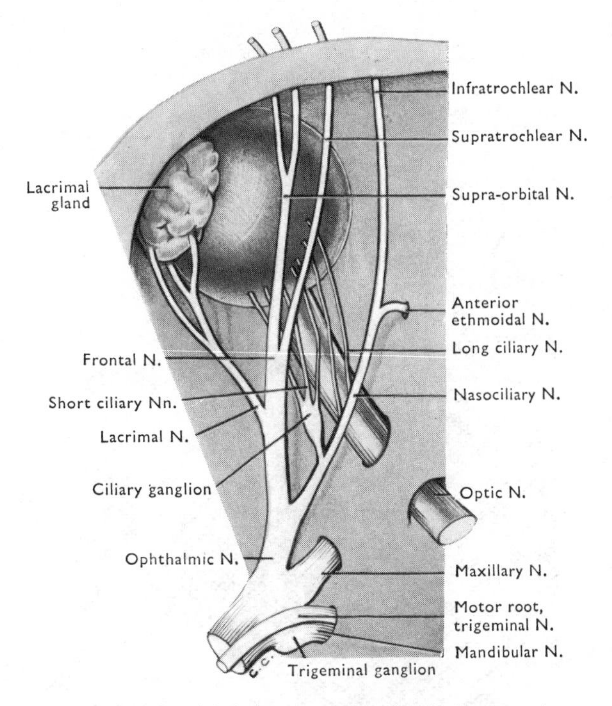

FIG. 80 The left ophthalmic nerve. The trigeminal nerve and ganglion have been turned laterally.

oculomotor nerve

Levator Palpebrae Superioris [FIG. 75]

It arises from the roof of the orbit immediately anterior to the optic canal, and passes forwards above rectus superior. Anteriorly, it widens into a broad, membranous expansion which is inserted into the skin of the eyelid, the anterior surface of the superior tarsus, and the superior conjunctival fornix. In addition there is a layer of **involuntary** (smooth) **muscle** which arises from the aponeurosis and is attached to the superior tarsus. This smooth muscle is supplied by nerve fibres from the cervical sympathetic and its denervation leads to drooping of the eyelid (**ptosis**).

Nerve supply: oculomotor nerve, superior division, and cervical sympathetic via the carotid plexus. **Action**: it opens the eye by raising the upper eyelid and the superior conjunctival fornix.

DISSECTION. Divide the frontal nerve and the levator palpebrae superioris, and turn them out of the way. If possible, inflate the eyeball through a small cut in the white part (sclera).

Posterior to the eyeball, note a quantity of loose tissue. This is the fascial sheath of the eyeball. Pick up and cut out a small portion. Insert a blunt seeker between it and the eyeball, and gauge the extent of the sheath, defining the extensions it gives over the muscles at their attachments to the eyeball.

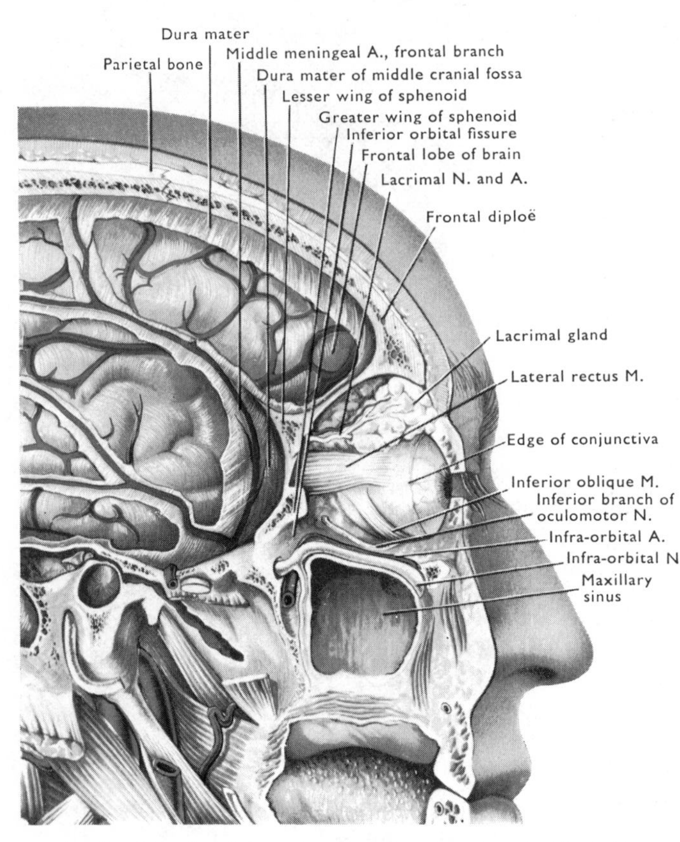

FIG. 81 Dissection of the orbit and maxillary sinus from the lateral side.

Rectus Superior

It arises from the upper margin of the optic canal, passes anterolaterally above the optic nerve, and is inserted into the sclera about 6 mm posterior to the sclero-corneal junction.

Nerve supply: superior division of the oculomotor nerve. **Action**: the actions of the extrinsic ocular muscles are complicated by the fact that the axis of the orbit passes anterolaterally, and the rectus superior and inferior run in this direction. The visual axis is, however, anteroposterior, and so the superior and inferior recti are not simple elevators and depressors of the cornea respectively, except when the eye is turned laterally and the two axes correspond. In more medial positions of the cornea, they tend to turn the cornea further medially, and this becomes more pronounced the further the cornea is turned medially [FIG. 88].

Obliquus Superior [FIG. 83]

It arises from the roof of the orbit immediately anteromedial to the optic canal, and passes anteriorly along the upper part of the medial orbital wall. Anteriorly, it ends in a slender tendon which enters the trochlea, and at once turns posterolaterally to pass between the superior rectus and the eyeball. Lateral to the superior rectus, the tendon flattens out and is inserted into the sclera midway between the entrance of the optic nerve and the cornea. **Nerve supply**: the trochlear nerve. **Action**: it turns the cornea downwards when it is already turned medially; a position in which the inferior rectus is ineffective as a depressor of the cornea [p. 92].

The **trochlea** is a small fibro-cartilaginous ring attached by fibrous tissue to the trochlear fossa on the frontal bone. It is lined with a synovial sheath which allows the tendon to slide freely in it.

DISSECTION. Cut through and reflect the superior rectus. Remove the fat beneath it and expose the optic nerve. Posteriorly, three structures cross the optic nerve; the nasociliary nerve, the ophthalmic artery, and the superior ophthalmic vein. Follow these and their branches. Two thread-like branches from the nasociliary nerve (long ciliary nerves) pass along the optic nerve to the eyeball. The short ciliary branches are much more numerous, and run forwards in the fat around the optic nerve. Select one of these and follow it posteriorly to a small swelling (the ciliary ganglion) which lies between the optic nerve and the lateral rectus. With care, the branches to the ciliary ganglion from the nasociliary and oculomotor (inferior division) nerves can be found, and even the sympathetic branch from the internal carotid plexus may appear.

Remove the fat lateral to the ganglion, and expose the abducent nerve on the medial side of the lateral rectus muscle. Expose the optic nerve in its sheaths.

Optic Nerve

This nerve enters the orbit through the optic canal, and carries with it sheaths of dura mater, arachnoid, and pia mater which enclose extensions of the subdural and subarachnoid spaces as far as the eyeball. The nerve runs anterolaterally and slightly downwards, and pierces the sclera a short distance medial to the centre of its posterior surface. The nasociliary nerve, ophthalmic artery and vein cross above it, and the ciliary nerves and vessels surround it near the eyeball. The nerve is slightly longer than the distance it has to run, so that it does not restrict the movements of the eyeball. It is a sensory nerve, and the great majority of its fibres originate in the retina (the light sensitive layer of the eye [p. 155]), though there are efferent fibres passing from the brain to the retina in the nerve.

Nasociliary Nerve

The nasociliary nerve arises from the ophthalmic nerve in the anterior part of the cavernous sinus. It passes through the superior orbital fissure and between the two heads of the lateral rectus muscle [FIG. 86]. It runs anteromedially above the optic nerve to the medial wall of the orbit. Here it continues forwards between the superior oblique and medial rectus muscles, and ends by dividing into infratrochlear and anterior ethmoidal nerves. It also gives off: (1) a branch to the ciliary ganglion; (2) the long ciliary nerves; (3) the posterior ethmoidal nerve.

The **communicating branch to the ciliary ganglion** runs along the lateral side of the optic nerve to reach the ganglion.

Two **long ciliary nerves** pass along the medial side of the optic nerve and pierce the sclera in this position. Their nerve fibres are sensory to the eyeball except the retina, but they also transmit some postganglionic sympathetic fibres, which enter the nasociliary nerve from the internal carotid plexus, and pass to supply the dilator of the pupil.

The **posterior ethmoidal nerve** arises at the medial wall of the orbit, passes through the posterior ethmoidal foramen, and supplies the mucous membrane of the ethmoid and sphenoid sinuses.

Infratrochlear Nerve. This is the smaller terminal branch, and it runs forwards to leave the orbit below the trochlea. It appears on the face above the medial angle of the eye, and supplies the skin of the eyelids and upper half of the external nose.

Anterior Ethmoidal Nerve. This nerve leaves the orbit by the anterior ethmoidal foramen, crosses above the ethmoidal sinuses [FIG. 82], and appears at the lateral margin of the cribriform plate of the ethmoid. Here it turns forwards under the dura mater, and descends into the walls of the nasal cavity through a slit-like aperture at the side of the crista galli. It gives **internal nasal branches** to the adjacent mucous membrane of the lateral and septal walls, and running on the deep surface of the nasal bone, emerges between it and the lateral nasal cartilage (**external nasal branch**) to supply the skin of the lower half of the nose.

Ciliary Ganglion

This small collection of parasympathetic nerve cells, about the size of a large pin head, lies in fatty tissue between the optic nerve and the lateral rectus muscle.

Connexions. (1) The long, slender filament from the nasociliary nerve consists of **sensory** fibres to

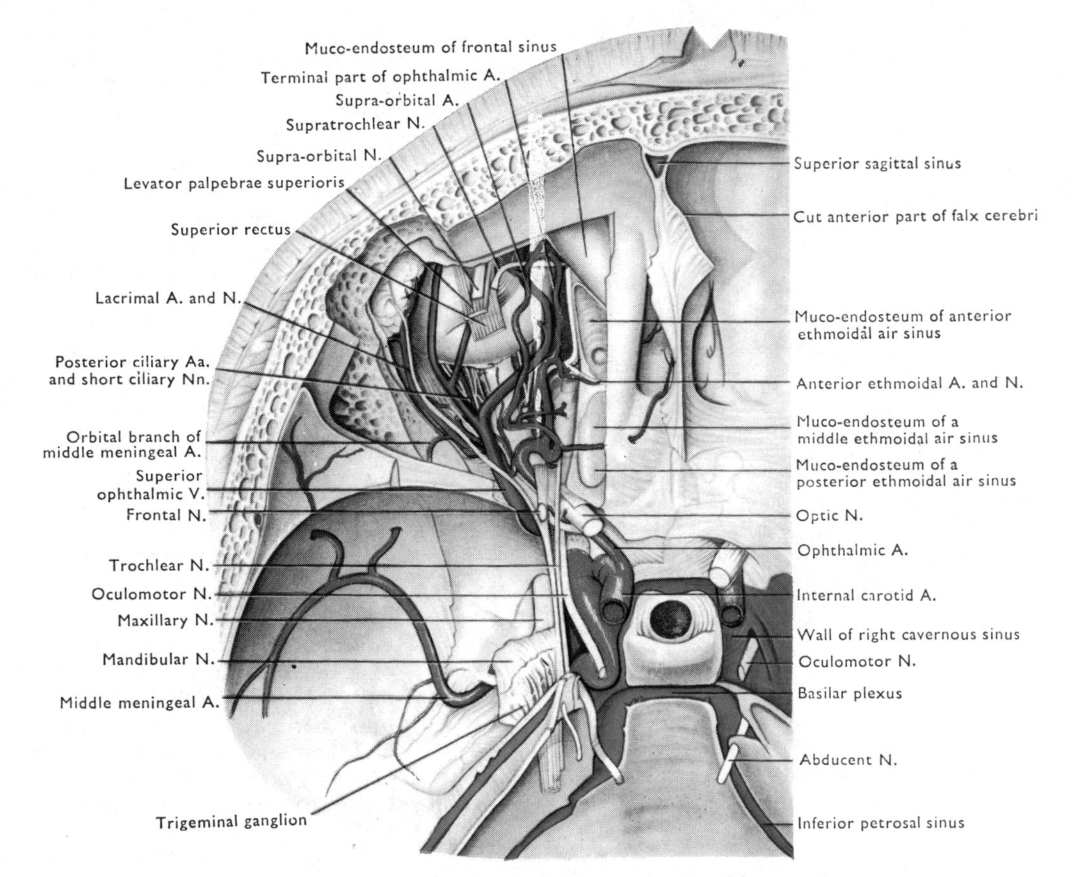

FIG. 82 Dissection of the orbit and middle cranial fossa.

the eyeball, and postganglionic **sympathetic** fibres which enter the ophthalmic nerve from the internal carotid plexus and supply the dilator of the pupil and blood vessels in the outer coats of the eye. Both of these pass straight through the ganglion to the eyeball in the short ciliary nerves. (2) The oculomotor root is a short, stout branch from the nerve to the inferior oblique muscle. It enters the ganglion from below, and contains preganglionic **parasympathetic** fibres from the oculomotor nerve which synapse with the cells of the ganglion. The postganglionic fibres, which arise from these cells, pass with the short ciliary nerves to the eyeball, and innervate the sphincter of the pupil and the **ciliary muscle**. Increasing tension in the ciliary muscle decreases the focal length of the lens in the eye and allows focusing (accommodation) on near objects [p. 157].

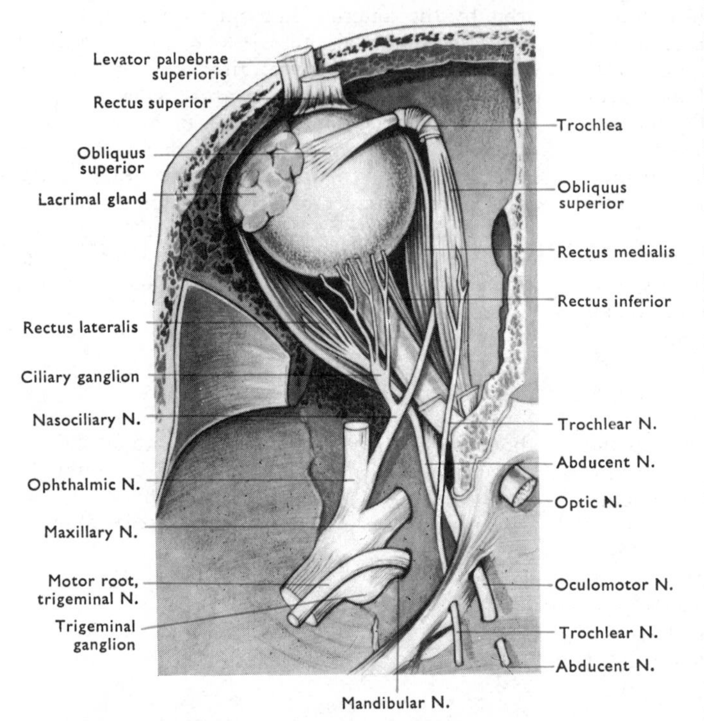

FIG. 83 Dissection of the orbit and middle cranial fossa. The trigeminal nerve and ganglion have been turned laterally.

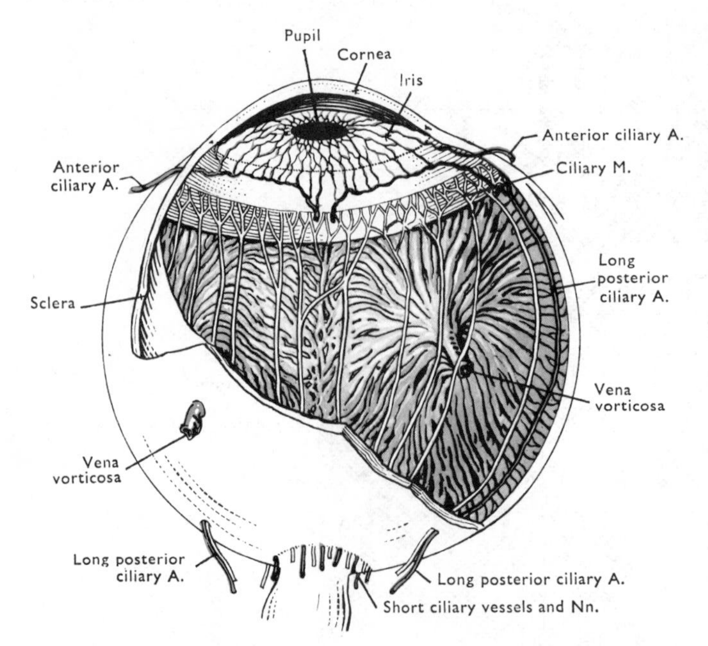

FIG. 84 Dissection of the eyeball to show the vascular coat and the arrangement of the ciliary nerves and vessels.

Approximately six **short ciliary nerves** pass beside the optic nerve and divide, so that twelve to twenty pierce the sclera around the entrance of the optic nerve.

Parasympathetic Ganglia

The four small ganglia associated with the branches of the trigeminal nerve (ciliary, pterygopalatine, otic, and submandibular) belong to the parasympathetic division of the autonomic nervous system. They receive **preganglionic fibres** either from the oculomotor, facial, or glossopharyngeal nerves, and the cells they contain give rise to **postganglionic nerve fibres**. These are distributed mainly through the branches of the trigeminal nerve, to glands of the orbit, nose, nasal part of the pharynx, and mouth, and to the sphincter of the pupil and ciliary muscle of the eyeball. Postganglionic fibres of the sympathetic system and fibres of the trigeminal nerve frequently run through these ganglia, but have no functional connexion with them. These parasympathetic ganglia lie closer to the structures they supply than the cervical sympathetic ganglia which send their postganglionic fibres to the same structures, often through plexuses on the arteries which supply them.

Ophthalmic Artery

This artery arises from the internal carotid artery as soon as it pierces the arachnoid [FIG. 82].

Course. It enters the orbit through the optic canal, inside the arachnoid sheath of the optic nerve. At first it lies below the optic nerve, but gradually pierces its arachnoid and dural sheaths, and winding round the lateral side of the nerve, crosses above it to reach the medial wall of the orbit. Thence the artery runs forwards below the superior oblique muscle, and ends by dividing into the supratrochlear and dorsal nasal arteries near the front of the orbit.

Branches. These are numerous and difficult to display, but correspond to the nerves in the orbit [FIG. 85]. They supply all the contents of the orbit and extend beyond this to the eyelids (**palpebral**), the forehead and anterior scalp (**supra-orbital** and **supratrochlear**), to the ethmoidal cells and the walls of the upper anterior parts of the nasal cavity (**anterior** and **posterior ethmoidal**), and to the skin of the nose (**dorsal** and **external nasal**). The most important branch is the **central artery of the retina**. It enters the optic nerve in the optic canal, and runs in it as the only direct arterial supply to the retina. *If it is occluded, blindness of the eye results.* The **cutaneous supply** is also important because it is possible to use the rate of reheating of the cooled forehead as a measure of the efficiency of the internal carotid circulation to the ophthalmic artery at least, provided the major branches of the external carotid artery to the scalp are occluded by pressure. The lacrimal artery communicates with the middle meningeal artery through the superior orbital fissure [p. 49].

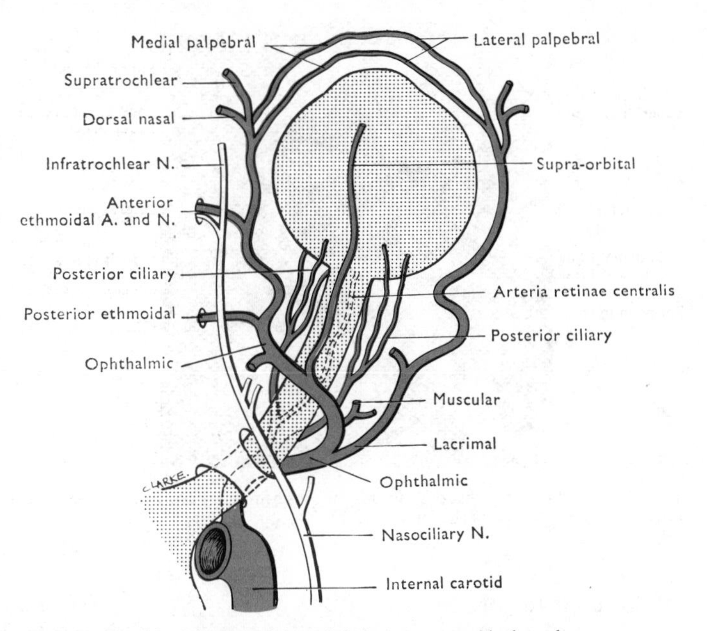

FIG. 85 Diagram of the ophthalmic artery and its branches.

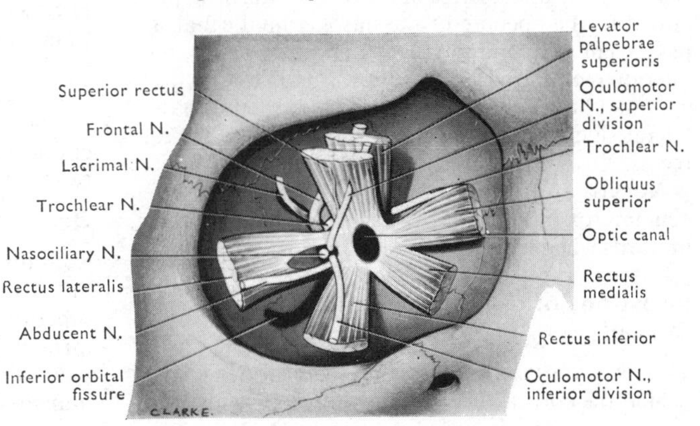

FIG. 86 Diagram of the origin of the ocular muscles and the routes of entry of nerves into the right orbit.

Ophthalmic Veins

The **superior** ophthalmic vein begins in the anterior part of the orbit close to the artery, and communicates with the supra-orbital and supratrochlear tributaries of the facial vein. It runs with the ophthalmic artery in the orbit.

The **inferior** ophthalmic vein is smaller, and lies below the optic nerve. Both veins receive numerous tributaries in the orbit, and passing through the superior orbital fissure, open into the cavernous sinus either separately or by a common trunk.

Origins of the Muscles that Move the Eyeball

The four rectus muscles arise from a common tendinous ring which surrounds the orbital end of the optic canal, and encloses the medial part of the superior orbital fissure. The superior, inferior, and medial recti arise from the ring above, medial to, and below the optic canal respectively. The lateral rectus arises by two heads from the lateral part of the ring and the adjoining margins of the orbital fissure. The two divisions of the oculomotor nerve, the nasociliary nerve, the abducent nerve, and the ophthalmic veins pass between the two heads of the muscle.

The superior oblique arises from the body of the sphenoid between the superior and medial recti. The inferior oblique is entirely separate from the others, and arises from the floor of the orbit far forward on the medial side.

DISSECTION. To display the attachments of the muscles of the eyeball, divide the optic nerve close to the canal and turn it forwards with the eyeball. Expose the two heads of lateral rectus but do not damage the structures between them.

Replace the eyeball, and make an incision through the inferior fornix of the conjunctiva and the palpebral fascia. Raise the eyeball slightly and dissect away the fat and loose connective tissue to expose the inferior oblique muscle.

Expose the rectus muscles and trace the nerves into the inferior oblique, and inferior and medial recti from the inferior division of the oculomotor nerve.

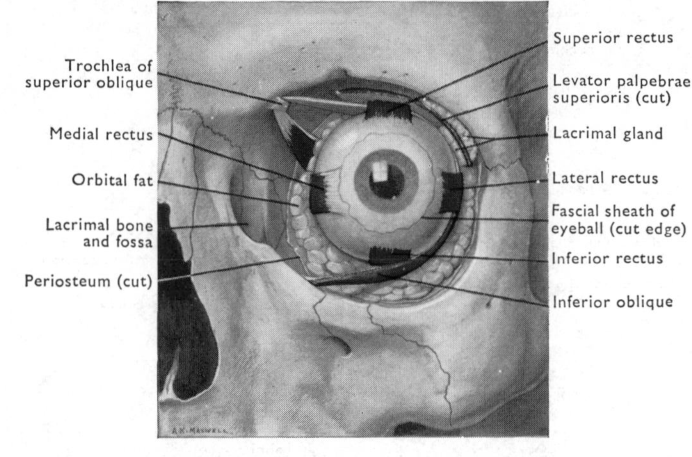

FIG. 87 Dissection of the left orbit to show the insertions of the muscles of the eyeball after removal of the fascial sheath.

The eye muscles may be used in any combination to produce intermediate movements, but it should be appreciated that very delicate control of eye movements is essential for normal vision. Thus failure to train both eyes exactly on to an object leads to *double vision*, because the images thrown on the two retinae do not fall on exactly corresponding parts. To test this, hold up your finger about 30 cm from your nose and look past it at a distant object. In this situation the finger appears double and out of focus. Now slowly change your direction of vision on to the finger; the two images move together and come into focus as a single object. This is due to convergence of the eyes throwing the image of the finger on to the corresponding parts of the two retinae, and the automatic focusing process which is linked to convergence (**convergence-accommodation reflex**) coming into play. Double vision is always found in paralysis of one or more of the ocular muscles, but only in those movements which the paralysed muscle or muscles produce.

Once again trace the nerves forwards from the cavernous sinus into the orbit, and note the relative positions which they take up in the fissure [FIG. 86].

Obliquus Inferior

It arises from a small area of the floor of the orbit just lateral to the opening of the nasolacrimal canal. It passes laterally and slightly backwards below the inferior rectus, and curving upwards [FIGS. 81, 87] ends in a flattened tendon which is inserted into the lateral side of the sclera under cover of the lateral rectus, close to the insertion of the superior oblique.

Nerve supply: the inferior division of the oculomotor nerve. **Action**: it turns the cornea upwards when the cornea is already turned medially.

Insertions of the Muscles that Move the Eyeball

The recti are inserted into the sclera about 6 mm behind the cornea. The oblique muscles are inserted much further back, their attachments to the sclera lying behind the equator of the eyeball, lateral to its sagittal meridian.

Actions of the Ocular Muscles

The **medial** and **lateral recti** have the simple function of rotating the eye around a vertical axis, thus turning the cornea medially and laterally. The **superior** and **inferior recti** produce simple elevation and depression of the cornea only when the eye is turned laterally so that the **visual axis** corresponds with that of these muscles [FIG. 88]. As the cornea is turned medially, they become progressively less effective in elevating and depressing the cornea, but tend to turn it further medially. At the same time the visual axis is moving progressively nearer to that of the **oblique muscles**, and they become increasingly effective in elevation and depression of the cornea as movement continues; their tendency to turn the cornea laterally helping to offset the medial movement produced by the superior and inferior recti.

Oculomotor Nerve

Both divisions of this nerve enter the orbit between the two heads of the lateral rectus muscle. The superior division passes to the rectus superior and, through it, to the levator palpebrae superioris. The inferior division is larger. It divides into three branches to supply the inferior and medial recti and the inferior oblique. The branches to the recti enter their internal surfaces. The branch to the inferior oblique runs forwards between the inferior and lateral rectus muscles, and gives the **parasympathetic root to the ciliary ganglion.**

Abducent Nerve

This nerve enters the orbit between the two heads of the lateral rectus, and continues forwards closely applied to the medial surface of that muscle. It supplies only the lateral rectus.

Fascial Sheath of the Eyeball

The connexions of this sheath cannot be demonstrated satisfactorily in an ordinary dissection, but some of the following points can be confirmed.

Relation of Sheath to Eyeball. It forms a membranous socket for the eyeball, but is deficient in front over the cornea. It is separated from the eyeball by some soft, semifluid areolar tissue which allows the eyeball to slide freely in the sheath. The free anterior margin of the sheath fuses with the ocular

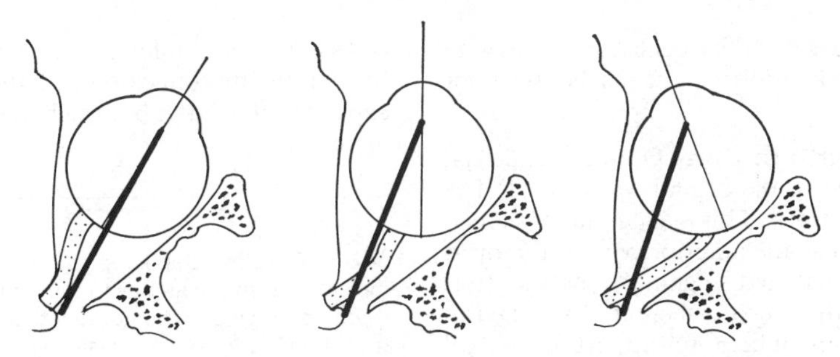

FIG. 88 Three diagrams to show the changing relationship and actions of the superior and inferior rectus muscles (thick line) as the cornea is turned from lateral to medial. When the cornea is turned laterally the two muscles elevate and depress it respectively. As the cornea is turned further medially they become progressively less effective in this action, tending instead to turn the cornea further medially. It is in this position that the oblique muscles become effective elevators (inferior oblique) and depressors (superior oblique) of the cornea.

conjunctiva close to the margin of the cornea, while posteriorly, it is adherent to the dural sheath of the optic nerve. The sheath is loosely attached to the orbital fat.

Relation of Sheath to Extrinsic Ocular Muscles. Each of these muscles pierces the fascial sheath at the equator of the eyeball, and receives a covering sleeve from it which fades out posteriorly in continuity with the epimysium. Each of these sleeves is strengthened by a slip of fibrous tissue which passes to the bony wall of the orbit, and makes the sleeve act as a pulley which prevents the muscle compressing the eyeball when it contracts. The sheath of the

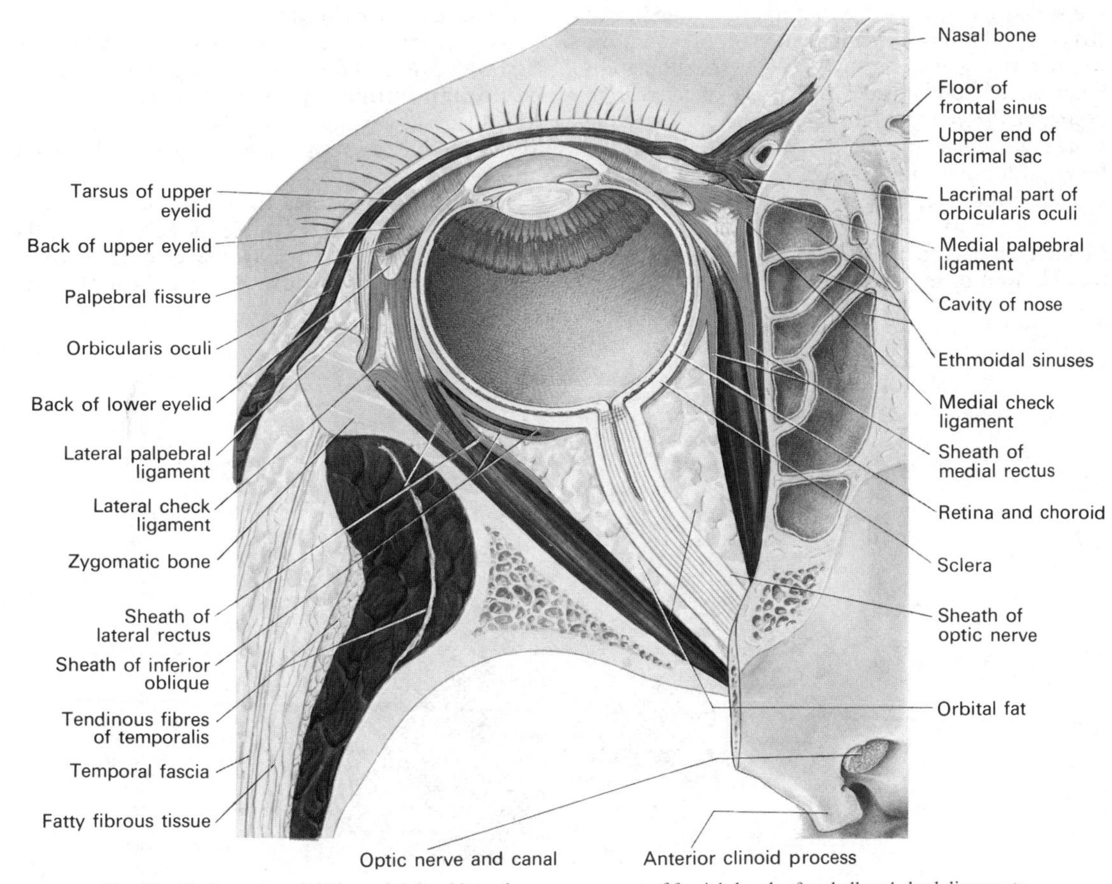

FIG. 89 Horizontal section through left orbit to show arrangement of fascial sheath of eyeball and check ligaments.

superior oblique passes to the trochlea and fuses with it, while that of the inferior oblique reaches the floor of the orbit.

Relation of Sheath to Bony Orbit. The fascial sheath is connected to the orbital walls by: (1) The **suspensory ligament**. This is a hammock-shaped sling stretched across the anterior part of the orbit between the lacrimal and zygomatic bones. It is broadest beneath the eyeball where it is attached to the fascial sheath, and it helps to support the eyeball and steady the fascial sheath. (2) The **check ligaments**. These are strong bands which pass from the sheaths around the lateral and medial rectus muscles to be attached to the zygomatic and lacrimal bones close to the attachments of the suspensory ligament. These ligaments limit the amount of movement which the lateral and medial rectus muscles can produce. Similar checks are present on the superior and inferior rectus muscles, the former by an intimate connexion with the levator palpebrae superioris, the latter by attachment to the suspensory ligament.

DISSECTION. Separate the eyelids completely by incisions from the angles to the orbital margins, and pull them apart. Divide the conjunctiva immediately beyond the cornea by a circular incision which will also divide the fascial sheath. Turn the conjunctiva and fascial sheath outwards, thus exposing the openings for the passage of the tendons through the sheath.

There remain only the zygomatic and infra-orbital nerves to be dissected, but as these lie outside the orbital periosteum they are most conveniently dissected later with the maxillary sinus.

THE PAROTID REGION

PAROTID GLAND

[FIGS. 72, 90–94]

This is the largest of the salivary glands, and is predominantly serous in type. It has a very irregular shape because it is wedged in among a number of structures. During development it does not grow within a well-defined capsule—as does the submandibular gland—but invades the fascia of its region, becoming disseminated through it and enclosing the structures which traverse that fascia. It lies in the fossa posterior to the ramus of the mandible, and extends from the external acoustic meatus above, to the upper part of the carotid triangle below. Medially, it extends to the styloid process (close to the side wall of the pharynx) and wraps round the neck of the mandible. Posteriorly, it overlaps sternocleidomastoid, and extends anteriorly over masseter for a variable distance; a portion of this superficial part is often detached from the rest, the **accessory parotid gland**.

A part of the cervical fascia (in which it is embedded) is thickened deep to the gland to form the **stylomandibular ligament**. This passes from the styloid process to the posterior border of the ramus of the mandible, and separates the parotid from the submandibular gland.

Branches of the facial, great auricular, and auriculotemporal nerves pierce the substance of the parotid gland, and so may the external carotid, superficial temporal and transverse facial arteries,

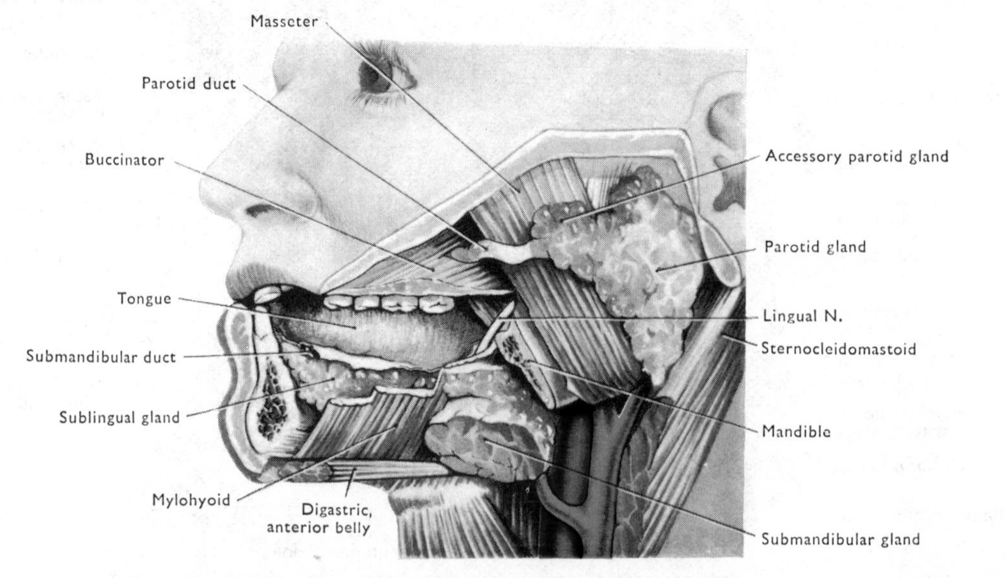

FIG. 90 Dissection of the parotid, submandibular, and sublingual glands.

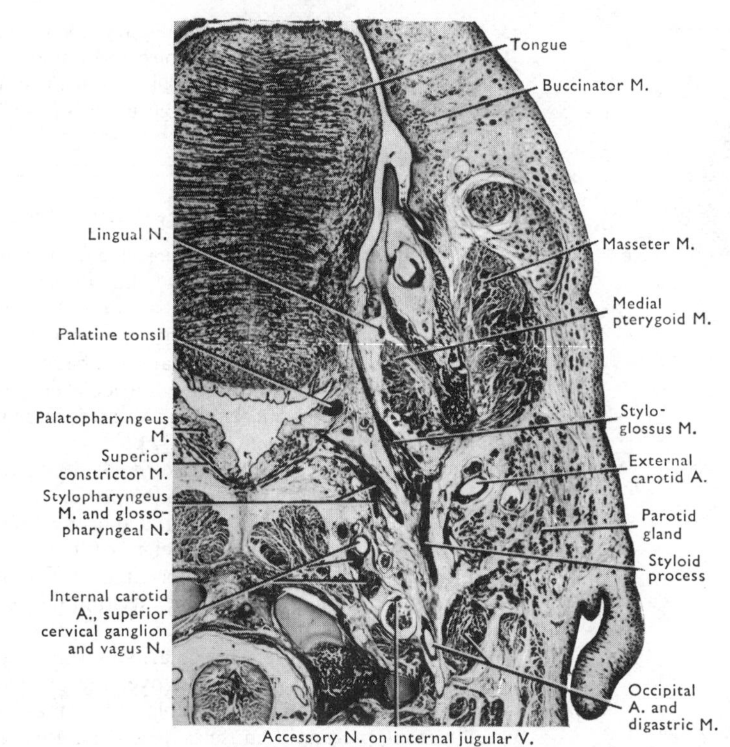

FIG. 91 Horizontal section through part of the head of a seven-month human foetus.

and the retromandibular vein. All of these have been seen already and their position should be reviewed [pp. 14, 15, 20, 79, 81].

Parotid Duct

This is a thick-walled tube formed within the gland by the union of the ductules which drain its lobules. It appears at the anterior border of the gland on the surface of masseter a finger's breadth, or less, below the zygomatic arch. It runs anteriorly across the masseter, below the accessory parotid gland, in company with the zygomatic branches of the facial nerve. The duct hooks medially over the anterior border of the masseter, and can be felt there by rolling it against the muscle with the jaw clenched. The duct pierces the buccal pad of fat, the buccopharyngeal fascia, and the buccinator muscle to open into the vestibule of the mouth on a small papilla opposite the second upper molar tooth.

Structures within the Parotid Gland

1. The **external carotid artery** enters and leaves the gland on its deep surface. It gives off the **posterior auricular artery** immediately before entering the gland, and divides into the **maxillary** and **superficial temporal arteries** where it emerges behind the neck of the mandible. Deep to the gland, the superficial temporal gives off the **transverse facial** and **middle temporal arteries**.

2. The **retromandibular vein** is joined by the superficial temporal, middle temporal, maxillary and transverse facial veins, and divides into anterior and posterior branches at the lower end of the gland.

3. The **facial nerve** enters the deep surface of the gland close to the stylomastoid foramen, and divides into five terminal branches in the gland. These radiate forwards, superficial to the artery and the vein, and branching repeatedly, appear at the borders of the gland as numerous smaller branches. Within the gland they receive communicating branches from the great auricular and auriculotemporal nerves.

4. The **parotid lymph nodes** are embedded in the gland, especially near its superficial surface. The superficial nodes drain the auricle, the anterior part of the scalp, and the upper part of the face; the deeper nodes receive lymph from the external acoustic meatus, middle ear, auditory tube, nose, palate, and deeper parts of the cheek. Both groups drain to the cervical lymph nodes [p. 76].

DISSECTION. Clean the surface of the parotid gland, and follow its duct to the buccinator muscle. Follow one of the branches of the facial nerve back through the gland to the trunk of the nerve, and then trace the others out through it, looking for the communicating branches from the auriculotemporal nerve which are relatively large. Trace the trunk of the facial nerve to the stylomastoid foramen, and find its posterior auricular branch and the branch to the posterior belly of digastric and the stylohyoid which are given off deep to the gland. Find and trace the posterior auricular artery.

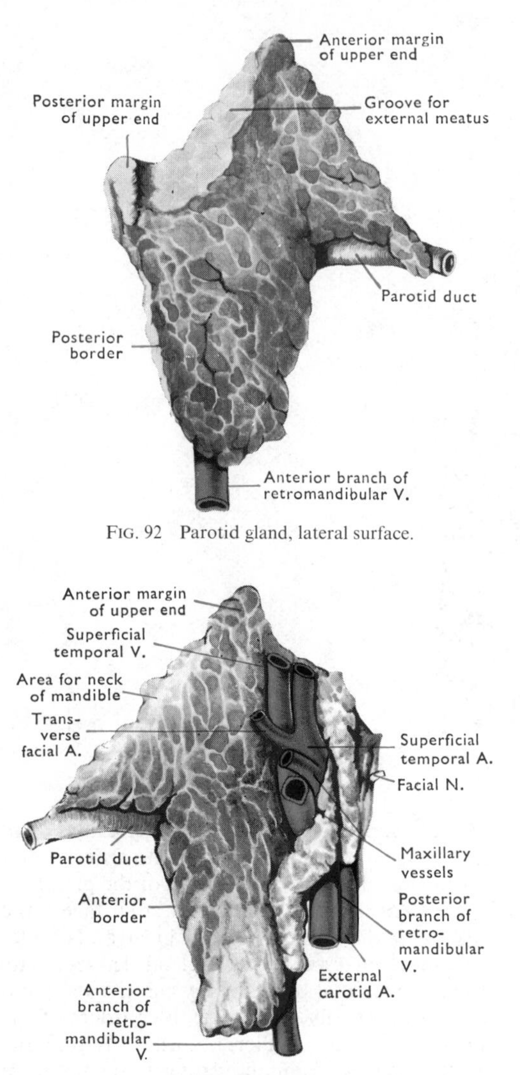

FIG. 92 Parotid gland, lateral surface.

FIG. 93 Parotid gland, anteromedial surface.

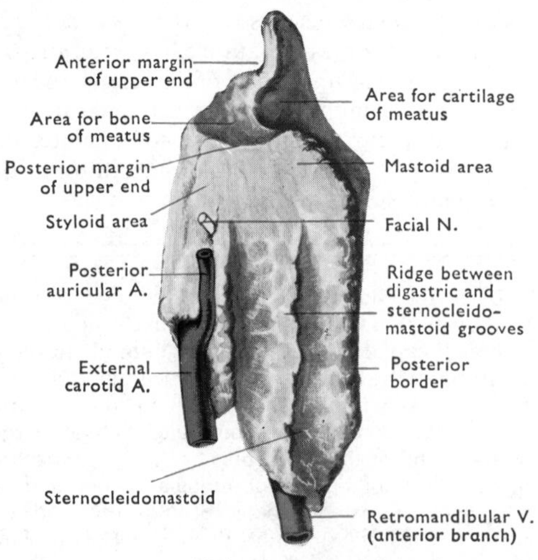

FIG. 94 Parotid gland, posteromedial surface.

Expose the retromandibular vein and the external carotid artery by removing more of the gland. Remove the remainder of the gland piecemeal and expose the structures which surround it, retaining, as far as possible, the structures which pass through it.

Shape and Position of the Parotid Gland

Compare FIGURES. 92–94 with the space from which the gland was removed and note that its shape is determined by the surrounding structures [FIG. 95].

The upper end is grooved by the external acoustic meatus and is wedged between it and the back of the temporomandibular joint; the auriculotemporal nerve and the superficial temporal vessels pierce the gland just behind the joint [FIG. 72].

The lower end lies between sternocleidomastoid and the angle of the mandible, on the posterior belly of digastric. It is pierced by the cervical branch of the facial nerve and the two branches of the retromandibular vein, and is separated from the submandibular gland by the stylomandibular ligament.

The deep surface is very irregular. Anteromedially it is deeply concave where it fits over the posterior margin of the ramus of the mandible and the masseter and medial pterygoid muscles [FIG. 91]. Posteromedially it is grooved to fit the mastoid process, sternocleidomastoid muscle, and the posterior belly of digastric. The stylohyoid muscle and styloid process are medial.

Vessels and Nerves

The vascular supply is from the adjoining vessels. Sensory nerve fibres reach it through the **great auricular** and **auriculotemporal** nerves, and postganglionic parasympathetic secretory fibres come from the otic ganglion through the auriculotemporal nerve. The preganglionic parasympathetic fibres reach the otic ganglion from the **glossopharyngeal nerve**. Sympathetic postganglionic fibres enter the gland from the plexus on the external carotid or middle meningeal arteries [FIG. 102].

FACIAL NERVE

This cranial nerve leaves the brain at the lower border of the pons [FIG. 180], passes with the vestibulocochlear (eighth cranial) nerve into the internal acoustic meatus, and emerges from the temporal bone at the stylomastoid foramen after a complicated course through that bone [p. 148]. The facial nerve then curves anteriorly and enters the posteromedial surface of the parotid gland. Before entering the gland, it gives off the posterior auricular nerve (to the occipital belly of occipitofrontalis and the auricular muscles), and a small branch which divides to supply the posterior belly of digastric and the stylohyoid muscle. In the gland it divides into its main branches [FIG. 74] and of these, the temporal and zygomatic can sometimes be felt on the zygomatic arch.

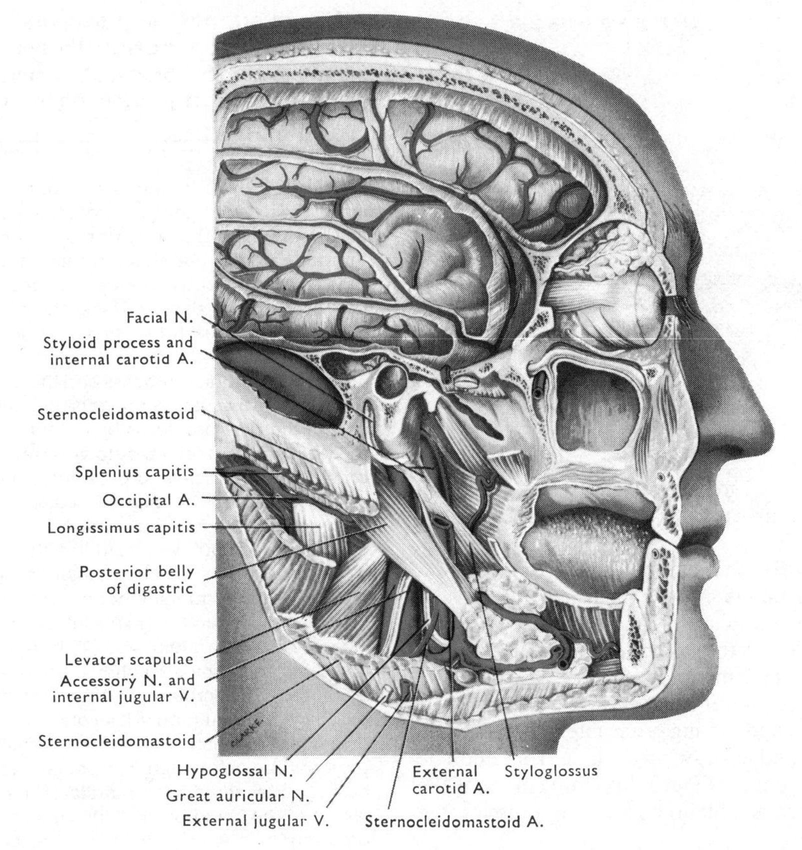

FIG. 95 Dissection of the head to show the structures deep to the parotid gland.

THE TEMPORAL AND INFRATEMPORAL REGIONS

Begin by revising the superficial part of the temporal region [pp. 13–15].

Temporal Fascia

This strong, glistening membrane is stretched over the temporal fossa and the temporalis muscle. Superiorly, it is attached to the upper temporal line. Inferiorly, it splits into two layers, of which the superficial is attached to the upper margin of the zygomatic arch, while the deep layer, separated from the superficial by a little fat, passes medial to the arch to become continuous with the fascia deep to masseter.

Masseter [FIGS. 17, 19]

This thick, quadrate muscle covers the lateral aspect of the ramus and coronoid process of the mandible, but leaves its head and neck uncovered and therefore palpable. It arises from the inferior margin and deep surface of the zygomatic arch, from the tubercle at its root posteriorly, to the junction with the zygomatic process of the maxilla anteriorly. It is inserted into the lateral surface of the ramus and coronoid process

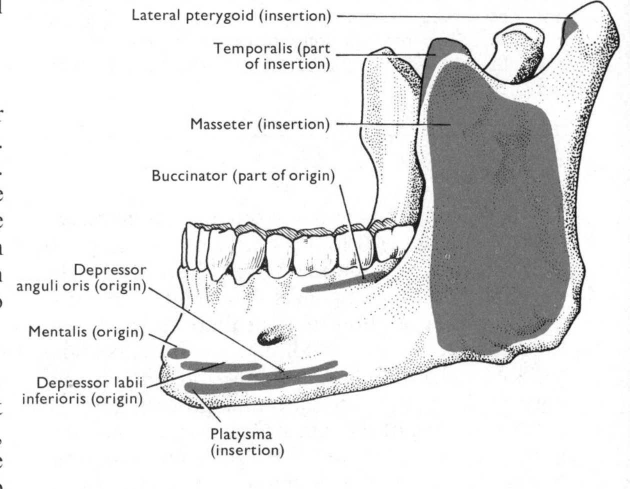

FIG. 96 Muscle attachments to superficial surface of mandible.

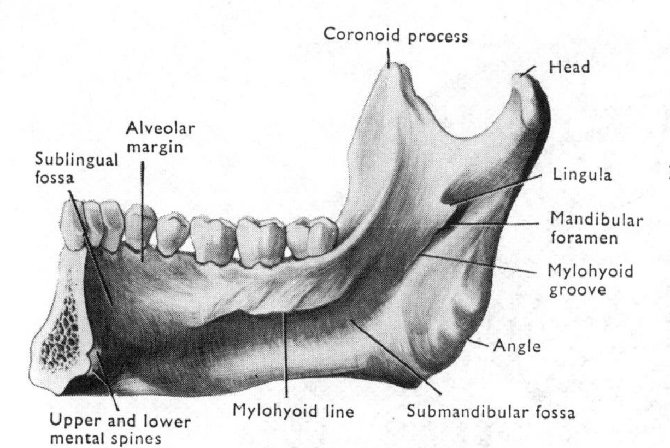

FIG. 97 Medial surface of the mandible.

of the mandible [FIG. 96]; its deep fibres passing vertically, its superficial fibres running postero-inferiorly.

Nerve supply: the mandibular nerve through a branch which enters its deep surface by passing through the mandibular notch immediately anterior to the capsule of the temporomandibular joint. **Action**: it raises the mandible, clenches the teeth. Its superficial fibres, running obliquely, help to protract the mandible, and acting alternately in the two muscles (right and left), swing the chin from one side to the other, producing a grinding movement of the teeth which is assisted by the pterygoid muscles (see below).

DISSECTION. Divide the zygomatic arch anterior and posterior to the attachment of masseter, and turn it down with that muscle, dividing the neurovascular bundle (which enters the deep surface of the muscle immediately anterior to the temporomandibular joint) and any fibres of the temporalis that may join the masseter. Strip the masseter from the surface of the mandible as far as the angle, but leave it attached there. Expose temporalis.

Temporalis

This fan-shaped muscle arises from the medial wall of the temporal fossa [FIG. 19] and from the temporal fascia. It converges on the coronoid process of the mandible, the anterior fibres descending vertically, and the posterior fibres running almost horizontally forwards. A tendon is formed on its superficial surface, and this is inserted into the summit and anterior margin of the coronoid process and the anterior margin of the ramus. The deeper, muscular fibres are attached to the medial side of the coronoid process and, becoming tendinous, reach down to the junction of the anterior border of the ramus with the body of mandible behind the third molar tooth [FIG. 98]. Some of the superficial fibres may join masseter and pass with it to the mandible.

Nerve supply: deep temporal branches of the mandibular nerve. **Action**: the temporalis raises the mandible and its horizontal, posterior fibres retract the mandible after protraction [p. 101].

DISSECTION. Separate the coronoid process from the mandible by an oblique cut from the mandibular notch to the point where the anterior margin of the ramus meets the body of the mandible. Be particularly careful at the lower end of the cut, for here the **buccal nerve** and artery lie either on the deep surface of, or embedded in, the lowest, tendinous fibres of temporalis, and are easily destroyed.

Turn the coronoid process and the attached temporalis upwards, and separate the muscle fibres from the lower part of the temporal fossa by blunt dissection to expose the **deep temporal vessels and nerves** which ascend between the muscle and the bone. The middle temporal artery [p. 9] will be exposed passing upwards on the squamous part of the temporal bone. Follow the zygomaticotemporal nerve, if it is still intact, to the small foramen through which it emerges from the temporal surface of the zygomatic bone.

The deeper structures in the infratemporal fossa may be exposed from the lateral side (1) by removal of part of the mandible, or (2) by dissection from the medial aspect at a later phase. A combination of the two methods gives the best view of this region. Make one horizontal cut through the neck of the mandible, and another immediately above the mandibular foramen. The position of the latter cut can be found by sliding the handle of a knife between the ramus of the mandible and the subjacent soft parts, and pressing it inferiorly till it is arrested by the inferior alveolar nerve and vessels entering the foramen. Cut half way through the bone with a saw along the lower border of the knife handle, and complete the division with bone forceps, avoiding injury to the underlying structures. Remove the pieces of bone and expose the underlying muscles, vessels, and nerves. Remove the parts of the pterygoid plexus of veins which obscure your view.

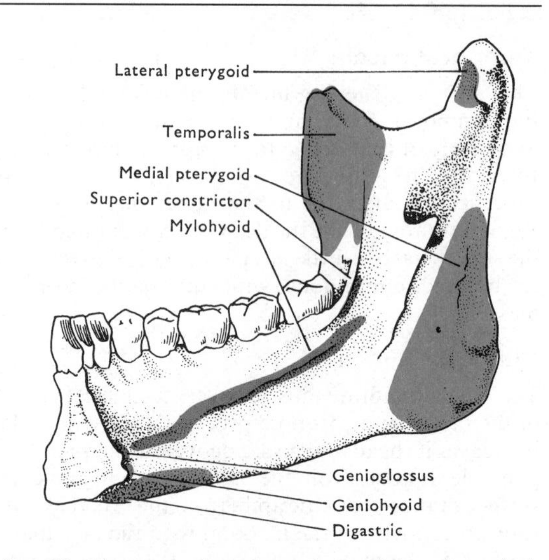

FIG. 98 Muscle attachments to the medial surface of the mandible. Origins, red; insertions, blue.

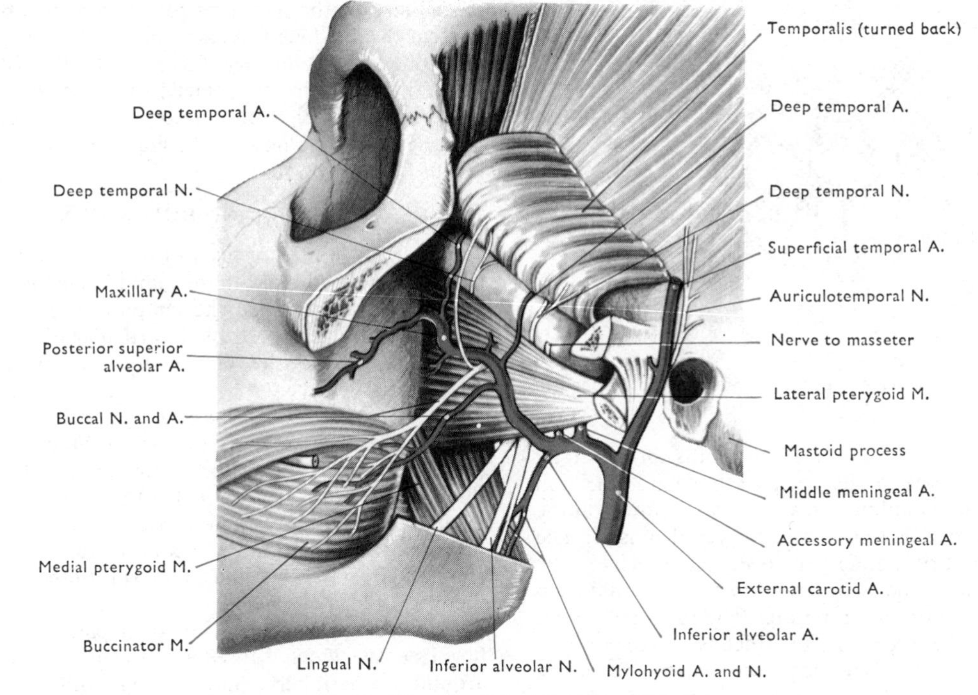

FIG. 99 Dissection of the infratemporal fossa.

THE SUPERFICIAL CONTENTS OF THE INFRATEMPORAL FOSSA [FIG. 99]

When the fatty tissue is cleared away, the pterygoid muscles are exposed, as are parts of the maxillary vessels and the branches of the mandibular nerve related to them.

DISSECTION. Follow the maxillary artery antero-superiorly till it disappears medially. Carefully remove the fat from the region just superior to this. The maxillary nerve may be seen passing towards the inferior orbital fissure to become the infra-orbital nerve [FIG. 103]. It gives two branches in this region: (1) the **zygomatic nerve** passing through the inferior orbital fissure, and (2) the **posterior superior alveolar nerve**, which divides into branches that descend and disappear into small holes in the posterior surface of the maxilla.

Lateral Pterygoid Muscle [FIGS. 98, 99]

This muscle arises by two heads. The smaller, **upper head** springs from the infratemporal ridge and infratemporal surface of the greater wing of the sphenoid medial to it. The **lower head** arises from the lateral surface of the lateral pterygoid plate. The muscle narrows as it passes posteriorly, and is inserted into the front of the neck by the mandible and the articular disc through the capsule of the temporomandibular joint.

Nerve supply: the mandibular nerve. **Action**: acting together the two muscles protrude the mandible and depress the chin, drawing the head of the mandible and the disc forwards on to the articular tubercle. When one muscle acts alone, the head of the mandible on that side is drawn forwards, the mandible pivots around the opposite joint so that the chin is slewed towards the opposite side.

Medial Pterygoid Muscle

This muscle also has two heads of origin, and they embrace the lower head of the lateral pterygoid. The **superficial head** is a small slip arising from the maxillary tuberosity. The **deep head** forms nearly the whole muscle, and arises deep to the lateral pterygoid from the medial surface of the lateral pterygoid plate. The two heads unite inferior to the anterior part of the lateral pterygoid, and passing downwards, backwards, and laterally, are inserted into a rough area between the mandibular foramen and the angle of the mandible. Its fibres are nearly parallel to the superficial fibres of masseter.

Nerve supply: the mandibular nerve. **Action**: it raises the mandible, assists protrusion, and slews the chin to the opposite side. The two medial pterygoid muscles acting alternately produce a grinding movement similar to the action of the superficial fibres of masseter.

Maxillary Artery [FIG. 99]

This artery arises posterior to the neck of the mandible as the larger terminal branch of the external

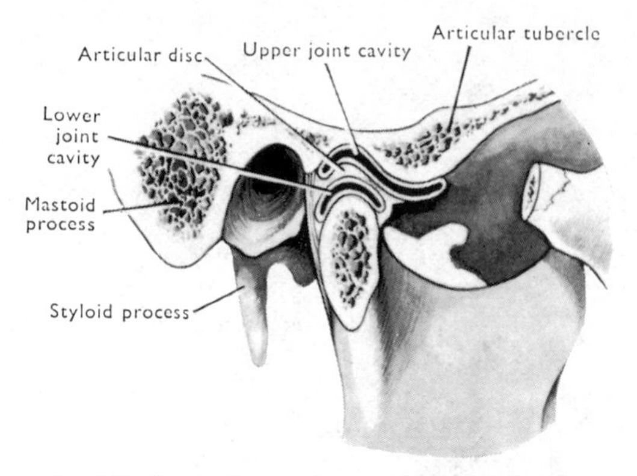

FIG. 100 Section through the temporomandibular joint.

carotid artery. The first part runs horizontally forwards between the neck of the mandible and the sphenomandibular ligament (see below), on the lower border of the lateral pterygoid muscle. The second part runs anterosuperiorly, superficial to the lower head of the lateral pterygoid muscle and deep to the insertion of temporalis. The third part turns medially, between the two heads of the lateral pterygoid, and ends in the pterygopalatine fossa in a number of branches which are distributed over the surfaces of the maxilla. The second part of the artery often lies between the two pterygoid muscles, and bends laterally between the heads of the lateral pterygoid muscle before entering the pterygopalatine fossa [FIG. 7].

During its course the artery gives **branches** to the external acoustic meatus and the middle ear, the muscles of the region, the skull bones and dura mater (especially by the middle meningeal [p. 48]), and branches which accompany the nerves traversing the infratemporal and pterygopalatine fossae.

The **inferior alveolar artery** descends with the inferior alveolar nerve, and entering the mandibular foramen, courses through the mandibular canal to supply the teeth, gums and mandible, and the skin over the chin and lip through the **mental artery**. It sends the mylohyoid artery with the mylohyoid nerve. The corresponding veins drain into the pterygoid venous plexus.

Pterygoid Plexus and Maxillary Vein.

The numerous veins of the infratemporal fossa are difficult to dissect, since they form a dense pterygoid plexus around the lateral pterygoid muscle. Veins corresponding to the branches of the maxillary artery open into this network, which is drained posteriorly by one or two short, wide maxillary veins. These pass to the parotid gland, and drain into the retromandibular vein posterior to the neck of the mandible.

Communications of Pterygoid Venous Plexus. This plexus has widespread communications with all the surrounding veins, but particularly with: (1) the **cavernous sinus** by an emissary vein which traverses the foramen ovale or the sphenoidal emissary foramen; (2) the **facial vein**, through the deep facial vein which passes posteriorly on the buccinator muscle, deep to masseter and the ramus of the mandible, to join the plexus.

TEMPOROMANDIBULAR JOINT

This synovial joint is formed by the articulation of the head of the mandible with the mandibular fossa and the articular tubercle of the temporal bone. These two bones are separated by an articular disc which completely divides the joint cavity into upper and lower parts.

The **fibrous capsule** is attached to the margins of the articular area on the temporal bone and around the neck of the mandible. Laterally it is thickened to form the **lateral ligament**. This is a triangular band attached by its base to the zygomatic process of the temporal bone and the tubercle at its root, and by its apex to the lateral side of the neck of the mandible.

The **articular disc** is an oval plate of dense fibrous tissue which is fused with the fibrous capsule around its periphery, and, through this, is more tightly bound to the mandible than to the temporal bone. The upper surface of the disc is concavo-convex to fit the articular tubercle and the mandibular fossa; its inferior surface limits the smaller of the two cavities of the joint, and is concave to fit the head of mandible.

Remove the lateral ligament and expose the disc and the two separate synovial cavities.

The fibrous capsule and its thickened lateral ligament are the only proper ligaments of the joint. The sphenomandibular and stylomandibular [p. 94] ligaments also connect the mandible to the skull, but add little if anything to the strength of the joint, which is maintained principally by the muscles of mastication. As with any joint maintained mainly by muscles, the temporomandibular joint is relatively readily dislocated.

The **sphenomandibular ligament** is a long, membranous ribbon that passes from the spine of the sphenoid, superficial to the medial pterygoid muscle, to reach the lingula and lower margin of the mandibular foramen. At the latter point it is pierced by the mylohyoid vessels and nerve. The ligament is of considerable developmental interest, as it is the remnant of part of the first branchial arch cartilage, the superior part of which gives rise to the malleus (one of the middle ear ossicles) while the inferior part is fused to the medial aspect of the mandible which develops on its lateral side [p. 246].

Movements [FIG. 101]

When the chin is depressed to open the mouth, the articular disc and the head of the mandible move forwards on the upper articular surface until the head of the mandible lies inferior to the articular tubercle. At the same time the head of the mandible rotates on the lower surface of the disc in the inferior part of the joint. The latter movement alone is capable of permitting simple chewing movements over a small range, but if the mouth is opened wide the former element is added also. This can be confirmed by placing a forefinger on the head of the mandible immediately in front of the tragus. When small chewing movements are made without separating the lips, the head of the mandible moves in the mandibular fossa. When the mouth is opened wide, the head of the mandible swings forwards and downwards and the finger slips into the mandibular fossa vacated by the head. Note that the mouth cannot be shut while the finger remains in the fossa. The *axis* of the two movements is different. In small movements, the axis is through the head of the mandible; in the wider range of movement, the axis passes approximately through the mandibular foramen. Thus the vessels and nerves entering the mandible are not stretched when the mouth is opened wide.

In **protraction** (pushing the jaw forwards) the head of the mandible and the articular disc slide forwards on the temporal articular surface on both sides, but the mouth is not opened. The reverse movement is **retraction**. When these movements alternate on the two sides, a grinding movement is produced as the chin swings from side to side.

The muscles which are active in these movements are: (1) depressors of the chin—the lateral pterygoid and the digastric acting through a fixed hyoid bone (infrahyoid muscles); (2) elevators—temporalis, masseter, and medial pterygoid; (3) protractors—lateral pterygoid and, less effectively, medial pterygoid and the superficial fibres of masseter which also prevent the lateral pterygoid opening the mouth; (4) retractor—the posterior fibres of temporalis; (5) side to side, grinding movements are produced by the muscles of opposite sides acting alternately.

DISSECTION. Separate the two heads of the lateral pterygoid muscle, avoiding injury to the buccal nerve between them. Carefully remove the upper head by detaching it from the capsule of the temporomandibular joint and removing it piecemeal from the infratemporal fossa. In the latter position, take care not to damage the deep temporal nerves which pass between it and the skull. Separate the lower head of the lateral pterygoid from the lateral pterygoid lamina, and strip it posteriorly from the underlying structures, leaving the buccal nerve intact. Disarticulate the head of the mandible from the articular disc and remove it with the lower head of the lateral pterygoid, taking care not to damage the **auriculotemporal nerve** which curves round the medial and posterior surfaces of the joint capsule. Expose the underlying structures.

Trace the middle meningeal artery to the foramen spinosum, and note the two roots of the auriculotemporal nerve which surround the artery close to the skull.

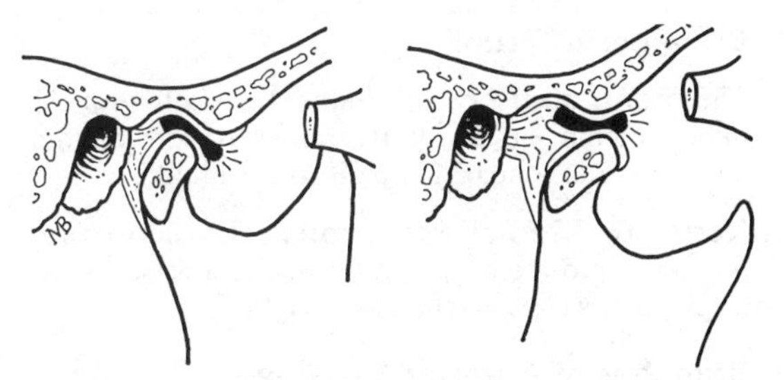

FIG. 101 Two diagrams to show the changing relationship between the head of the mandible and the temporal bone when the mouth is opened (right illustration). Note that the head of the mandible, together with the articular disc (black), slides forwards on to the articular tubercle, while the head of the mandible rotates on the disc.

Identify the origin of the nerve from the mandibular nerve and trace it posteriorly. Expose the other branches of the mandibular nerve [FIG. 102] and identify the chorda tympani nerve entering the posterior surface of the lingual nerve. Trace the chorda tympani towards the spine of the sphenoid.

THE DEEPER CONTENTS OF THE INFRATEMPORAL FOSSA

MIDDLE MENINGEAL ARTERY

This branch arises from the maxillary artery at the lower border of the lateral pterygoid muscle, and ascending between the roots of the auriculotemporal nerve, enters the skull through the foramen spinosum. It is posterolateral to the mandibular nerve, and lies on the lateral surface of the tensor palati muscle, which separates it from the auditory tube. Its intracranial course is described on page 48.

The **accessory meningeal** and **anterior tympanic arteries** arise either from the middle meningeal artery or from the maxillary. The accessory meningeal artery runs anterosuperiorly through the foramen ovale with the mandibular nerve; the anterior tympanic passes postero-superiorly to enter the middle ear through the petrotympanic fissure, close to the chorda tympani nerve.

MANDIBULAR NERVE

The mandibular branch of the trigeminal nerve arises from the trigeminal ganglion in the cranium, and enters the infratemporal fossa through the foramen ovale in the plane of the lateral pterygoid plate. In the foramen ovale it is joined by the entire motor root of the trigeminal nerve, and emerges from the skull as a mixed nerve.

Immediately below the skull it lies between the lateral pterygoid muscle and the tensor palati which separates it from the auditory tube [FIG. 12]. It divides almost immediately into anterior (predominantly motor) and posterior (predominantly sensory) divisions.

Branches of Trunk

The **meningeal branch** enters the skull with the middle meningeal artery. It supplies the dura mater and skull and sends a filament to the middle ear.

Nerve to Medial Pterygoid. This nerve runs forwards into the deep surface of the muscle. At its origin, it lies close to the otic ganglion.

Branches of Anterior Division

Buccal Nerve. This is the largest branch of the division. It passes between the two heads of the lateral pterygoid, and runs antero-inferiorly to the surface of the buccinator muscle, often piercing the lowest fibres of insertion of temporalis on the deep surface of the ramus of the mandible [FIG. 98]. On buccinator, it forms a plexus with the buccal branch of the facial nerve (which is motor to buccinator) and supplies the skin and mucous membrane on the lateral and medial surfaces of buccinator.

Nerve to Lateral Pterygoid. It arises with the buccal nerve, and enters the muscle as that nerve passes between its heads.

Deep Temporal Nerves. These two nerves, anterior and posterior, pass into the temporal fossa between the skull and the lateral pterygoid, and grooving the bone, enter the deep surface of the temporalis. The anterior deep temporal nerve may be replaced by a branch of the buccal nerve which ascends lateral to the upper head of the lateral pterygoid [FIG. 99]. Apart from this occasional branch, the buccal nerve is purely sensory.

Nerve to Masseter. It arises with the posterior deep temporal nerve, runs laterally between the skull and the lateral pterygoid muscle (immediately anterior to the capsule of the temporomandibular joint) and enters the deep surface of masseter by passing through the mandibular notch. It gives one or two twigs to the temporomandibular joint.

Branches of Posterior Division

Auriculotemporal Nerve [FIGS. 102, 120]. It arises by two sensory roots. Each receives a bundle of postganglionic parasympathetic, secretomotor fibres for the parotid gland from the cells of the otic ganglion. Preganglionic fibres reach that ganglion from the glossopharyngeal nerve. The roots surround the middle meningeal artery, unite posterior to it, and run backwards, lateral to the spine of the sphenoid, to hook round the posterior surface of the neck of the mandible. Here the nerve turns superiorly in contact with the parotid gland, and crossing the root of the zygomatic process of the temporal bone with the superficial temporal artery,

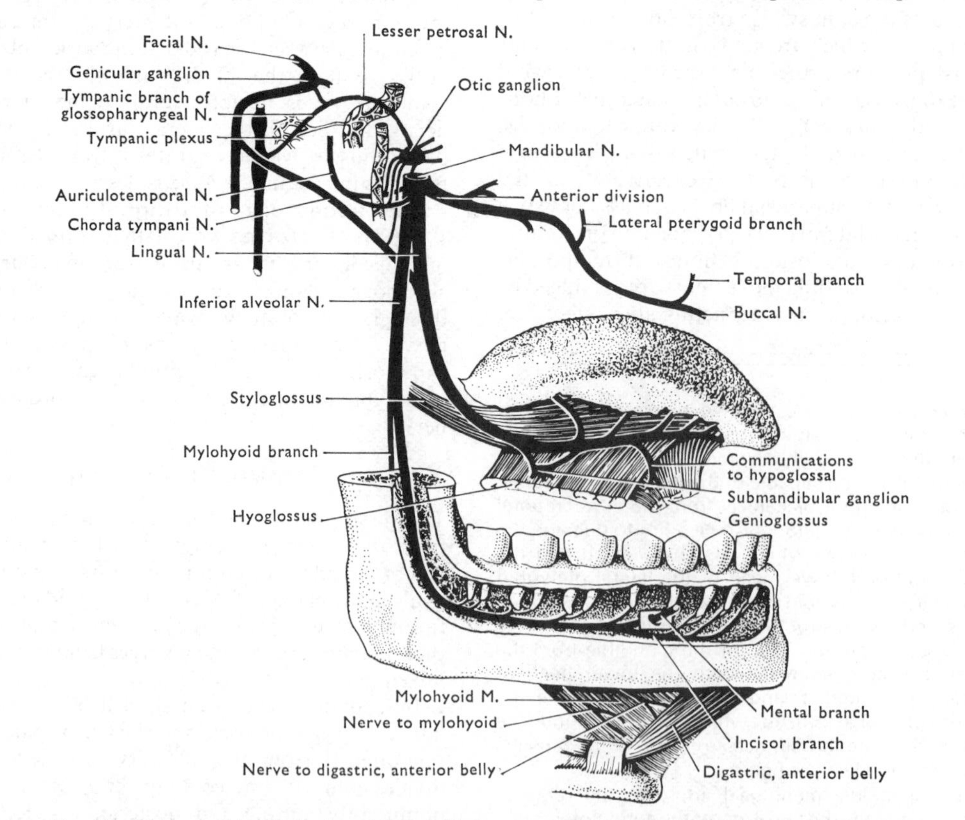

FIG. 102 Diagram of the distribution of the mandibular nerve. Note the sympathetic filaments to the otic ganglion and the tympanic plexus from the plexuses on the middle meningeal and internal carotid arteries.

breaks up into its terminal branches on the temple [FIG. 74].

Branches. (1) A few slender filaments to the posterior part of the capsule of the temporomandibular joint. (2) One or two thick branches which enter and supply the parotid gland, and mingle with the branches of the facial nerve in its substance. (3) Cutaneous branches to the auricle and the temple.

Inferior Alveolar Nerve. This is the largest branch of the posterior division. It runs vertically downwards with the inferior alveolar artery, medial to the lateral pterygoid muscle and lateral to the sphenomandibular ligament and the medial pterygoid muscle. The nerve and artery each give off a mylohyoid branch, and then enter the mandibular foramen. In the body of the mandible, the nerve and artery give branches to the teeth and gums, and each sends a branch (**mental nerve** and **artery**) through the mental foramen to supply the skin of the chin and the mucous membrane and skin of the lower lip.

The **mylohyoid nerve** contains the only motor fibres in the posterior division. It pierces the sphenomandibular ligament and runs antero-inferiorly in a groove on the medial aspect of the mandible to the digastric triangle, inferior to the mylohyoid muscle. In the triangle, it is joined by the submental artery, and supplies the mylohyoid muscle and the anterior belly of digastric [FIG. 104].

Lingual Nerve. All the trigeminal fibres in this nerve are sensory to the mucous membrane of the anterior two-thirds of the tongue and to the adjacent part of the floor of the mouth and gum. The lingual nerve curves antero-inferiorly on the medial pterygoid muscle to reach the medial surface of the mandible in front of the inferior alveolar nerve. *Here it passes between the mandible and the mucous membrane covering it just inferior to the last molar tooth* and above the posterior fibres of the mylohyoid muscle [FIGS. 98, 105]. Its further course will be followed later.

The nerve gives no branches in the infratemporal fossa, but is joined by the chorda tympani branch of the facial nerve deep to the lateral pterygoid muscle, and may send a communicating branch to the inferior alveolar nerve.

Chorda Tympani [FIG. 102]

This slender branch of the facial nerve contains sensory and preganglionic parasympathetic fibres. It supplies the submandibular and sublingual glands through the submandibular ganglion, and sends sensory fibres to the taste buds on the anterior two-thirds of the tongue. It arises from the facial nerve posterior to the middle ear cavity, runs anteriorly across the lateral wall of that cavity (tympanic membrane), and escapes from it through the petrotympanic fissure. It then grooves the medial side of the spine of the sphenoid, running antero-inferiorly to join the posterior surface of the lingual nerve.

DISSECTION. Lift the mandibular nerve laterally, and attempt to find the otic ganglion medial to it. If it cannot be found easily, do not dissect further as it can be seen more readily from the medial side later.

Expose as much as possible of the tensor palati medial to the middle meningeal artery and the mandibular nerve. The muscle will be seen more clearly from the medial surface later.

Otic Ganglion

This minute collection of parasympathetic nerve cells lies between the mandibular nerve and the tensor palati, immediately below the foramen ovale. It is on the origin of the nerve to the medial pterygoid muscle [FIG. 102].

A number of different fibres pass to the otic ganglion, but only the preganglionic parasympathetic fibres, which reach it in the lesser petrosal nerve [p. 49], synapse with the cells of the ganglion. The remainder traverse the ganglion and have no functional connexion with it. Postganglionic fibres arise from the cells of the ganglion and pass in the auriculotemporal nerve as secretomotor fibres to the parotid gland. The nerve fibres which traverse the ganglion are: (1) Motor fibres to the tensor palati and tensor tympani muscles from the nerve to the medial pterygoid muscle. (2) Sympathetic fibres from the plexus on the middle meningeal artery, for distribution through the branches of the ganglion. (3) Sensory fibres from the glossopharyngeal and trigeminal nerves which are distributed through the branches of the ganglion.

Tensor Palati (Tensor Veli Palatini) [FIGS. 114, 115, 120, 154]

This thin, triangular muscle arises from the scaphoid fossa at the root of the medial pterygoid lamina, and from the posteromedial margin of the greater wing of the sphenoid as far posteriorly as the spine of the sphenoid [FIGS. 11, 12]. Thus it lies between the groove for the auditory tube medially, and the foramina ovale and spinosum laterally, in the uppermost part of the lateral wall of the pharynx. It runs antero-inferiorly to a small tendon which hooks round a bursa on the base of the pterygoid hamulus—a protrusion from the inferior margin of the medial pterygoid lamina. The tendon spreads medially into the soft palate to form the palatal aponeurosis with the tendon of the opposite side. **Nerve supply**: the mandibular nerve. **Action**: with the muscle of the opposite side it tenses the anterior part of the soft palate.

The maxillary nerve will be completely dissected at a later stage, but its course and some of its branches can be seen now.

DISSECTION. Lift the contents of the orbit upwards and medially, or remove them if they are too unyielding.

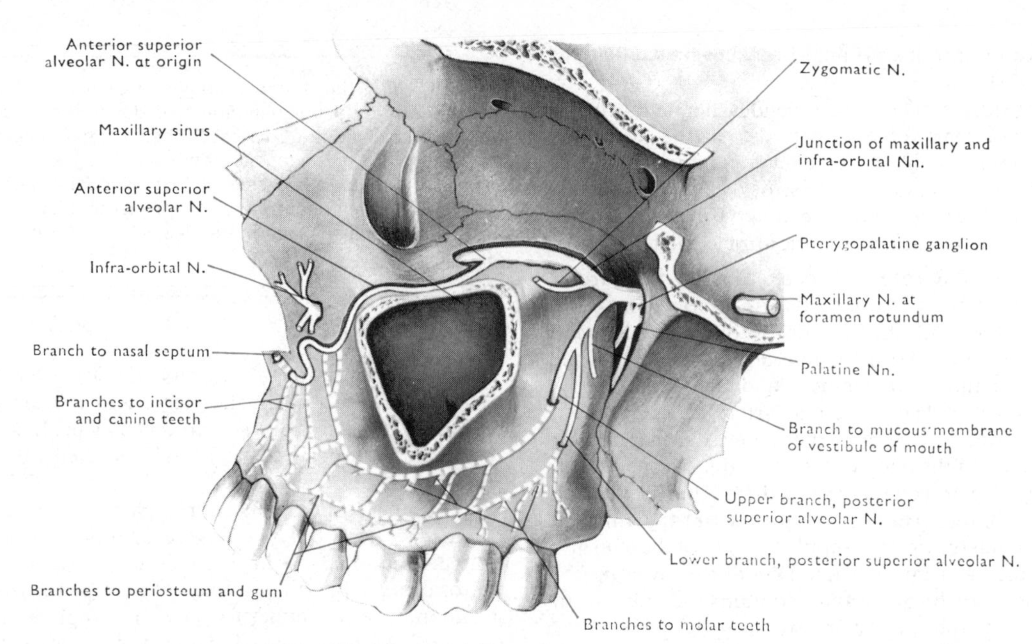

FIG. 103 Diagram of the maxillary nerve. Cf. FIG. 126.

Find the **zygomatic nerve** near the angle between the floor and lateral wall of the orbit, and trace the nerve forwards till it enters the bone.

Remove the periosteum from the orbital floor, and identify the infra-orbital groove and the infra-orbital vessels and nerve lying in it.

MAXILLARY NERVE

The maxillary nerve is the second of the three divisions of the trigeminal nerve. It arises from the trigeminal ganglion [p. 46] and passes forwards on the side of the body of the sphenoid, at the lower border of the cavernous sinus, to the **foramen rotundum**. It emerges from the foramen rotundum into the upper part of the pterygopalatine fossa, and curves laterally through the pterygomaxillary fissure to the infratemporal fossa. Here it turns sharply forwards into the infra-orbital groove as the **infra-orbital nerve** [FIGS. 103, 126]. Confirm this course by passing a bristle through the foramen rotundum into the infra-orbital groove in a dried skull.

Branches. (1) A **meningeal** branch arises near the origin of the nerve. (2) Two **ganglionic** branches pass inferiorly in the pterygopalatine fossa to join the pterygopalatine ganglion. (3) The **posterior superior alveolar nerve** arises in the infra-temporal fossa, and divides into two branches which descend over the posterior surface of the maxilla. They supply filaments to the gum and to the mucous membrane of the cheek, and then enter canals in the bone with corresponding branches of the maxillary artery. In the canals they run forwards above the tooth sockets and form a plexus with the superior alveolar branches of the infra-orbital nerve. They give dental branches to the molar teeth. (4) The **zygomatic nerve** arises close to the foramen rotundum and enters the orbit through the inferior orbital fissure. It gives a delicate filament to the lacrimal nerve, pierces the periosteum, and divides into **zygomaticotemporal** and **zygomatico-facial** branches. These pass forwards in the periosteum of the lateral wall of the orbit, and pierce the zygomatic bone to reach the skin of the temple and face [pp. 14, 81]. (5) The **infra-orbital nerve** is the continuation of the maxillary nerve in the infra-orbital groove and canal. It is accompanied by the corresponding vessels, and emerges through the infra-orbital foramen on to the face, deep to the orbicularis oculi. It divides into sensory branches to the upper lip, nose, and lower eyelid.

About the middle of the floor of the orbit the infra-orbital nerve gives off the **anterior superior alveolar nerve** which descends through the anterior wall of the maxilla, and joining the plexus formed by the posterior superior alveolar nerves, supplies branches to the upper first molar and to the premolar, canine, and incisor teeth. It also supplies the adjacent gum, and sends branches to the maxillary sinus and to the mucous membrane of the antero-inferior part of the nose.

A middle superior alveolar nerve is described in 30 per cent of individuals.

DISSECTION. Expose the mandibular canal by removing the outer table of bone with chisel and bone forceps.

STRUCTURES WITHIN THE MANDIBULAR CANAL

The canal is traversed by the inferior alveolar vessels and nerve, which give dental branches to the roots of the lower teeth, and gingival branches to the adjacent gum.

The mental nerve and artery arise from the inferior alveolar nerve and artery in the canal, and emerge through the mental foramen.

THE SUBMANDIBULAR REGION

The submandibular region lies between the body of the mandible and the hyoid bone. The superficial part includes the submental and digastric triangles which have been dissected already. Its deeper parts, now to be dissected, include the root of the tongue and the floor of the mouth.

DISSECTION. To expose this region, turn the airway and foodway medially on the right: on the left, extend the neck and flex the head to the right. Cut across the facial artery and vein at the lower border of the mandible, and detach the anterior belly of digastric from the mandible. Turn the mandible upwards and fix it with hooks.

Complete the exposure of the posterior belly of digastric and of the stylohyoid muscles.

Digastric Muscle [FIGS. 33, 63, 98]

The **anterior belly** of the digastric springs from the lower border of the mandible, close to the symphysis. The **posterior belly** arises from the floor of the mastoid notch on the medial side of the mastoid process. The two bellies are united by an intermediate tendon, which passes through a short, strong loop of fibrous tissue binding it down to the upper border of the hyoid bone. This fibrous pulley is attached at the junction of the body with the greater horn of the hyoid bone, and allows the tendon to slide backwards and forwards through it, in a synovial sheath [FIG. 141].

Nerve supply: the posterior belly by the facial nerve, the anterior belly by the mylohyoid nerve. **Action**: if the hyoid bone is fixed, it helps to depress the mandible. The chief action of both bellies, acting together, is to raise the hyoid bone (and hence the tongue) in the action of swallowing; an action which cannot be carried out when the mouth is open because the muscle is already shortened. Acting with the infrahyoid muscles, it fixes the hyoid bone, thus forming a stable platform on which the tongue can move.

The **anterior belly** is covered by deep fascia, and is overlapped by the submandibular gland; its deep surface is in contact with the mylohyoid muscle.

The **posterior belly** begins deep to the mastoid process and superficial to the accessory nerve. It

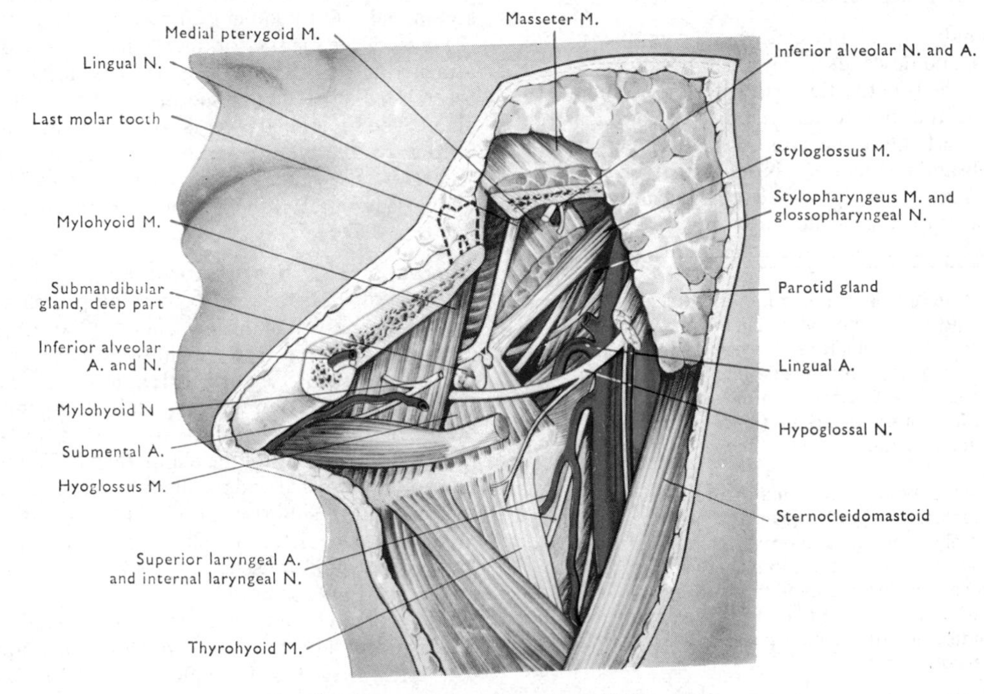

FIG. 104 Deep dissection of the submandibular region.

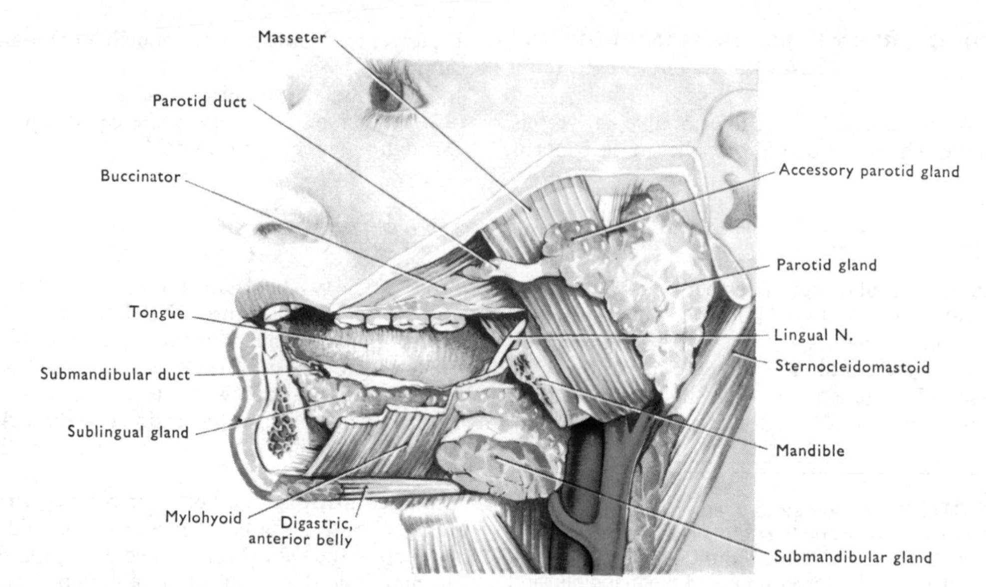

FIG. 105 Dissection of the parotid, submandibular, and sublingual glands.

passes antero-inferiorly, deep to the lower part of the parotid gland and the angle of the mandible. It is crossed by the facial vein, and overlapped by the submandibular gland, and runs obliquely across the neurovascular bundle of the neck—the internal jugular vein, internal and external carotid arteries, and the hypoglossal nerve. The occipital artery runs posterosuperiorly on its deep surface and grooves the temporal bone just medial to the origin of the muscle [FIG. 59].

Stylohyoid Muscle

This small slip of muscle arises from the styloid process, and descends along the upper border of the posterior belly of the digastric. Inferiorly, it divides to surround the intermediate tendon of the digastric, and is inserted into the hyoid bone at the junction of the body and greater horn. **Nerve supply**: the facial nerve. **Action**: it helps to pull the hyoid upwards and backwards during swallowing.

DISSECTION. Turn the submandibular gland posteriorly, and expose the mylohyoid muscle. Note the deep part of the gland which hooks round the free posterior border of the mylohyoid muscle, and runs forwards on its superior surface. Dissect out the facial artery from the deep surface of the gland, and trace its branches in this region. Identify the mylohyoid nerve on the mylohyoid muscle.

Turn the submandibular gland anteriorly and identify the hypoglossal nerve lying on the hyoglossus muscle immediately superior to the greater horn of the hyoid bone. At a higher level, the lingual nerve crosses the same muscle, and has the submandibular ganglion suspended from its lower margin. Identify this ganglion and the submandibular duct, which passes forwards from the deep part of the gland.

Submandibular Gland

This salivary gland is about half the size of the parotid gland [p. 94]. Its superficial part is wedged between the body of the mandible and the mylohyoid muscle, and reaches superiorly to the mylohyoid line on the medial surface of the mandible. Posterior to the free margin of the mylohyoid, the mucous membrane of the mouth lies superior to it, while inferiorly it is limited by the bellies of the digastric. It extends posteriorly to the angle of the mandible, where it is separated from the parotid gland by the stylomandibular ligament, and anteriorly it reaches the level of the mental foramen. The gland is loosely attached to a capsule of deep cervical fascia which ascends from the hyoid bone and splits to enclose the gland. The superficial layer is attached to the inferior border of the mandible; the thinner deep layer separates the gland from the mylohyoid and hyoglossus muscles, and is attached to the mylohyoid line [FIG. 97].

Surfaces. The **inferolateral** surface is covered with: (1) superficial fascia containing platysma and the cervical branch of the facial nerve; (2) deep fascia; (3) deep to this the facial vein and a few **submandibular lymph nodes**, most of which lie in the groove between the submandibular gland and the mandible. The **facial artery** grooves the posterosuperior part of the gland, then loops antero-inferiorly between it and the medial pterygoid muscle to appear at the inferior margin of the mandible.

The **lateral** surface is related, posteriorly, to the medial pterygoid muscle and the facial artery, and anteriorly, to the mandible.

The **medial** surface extends from the bellies of the digastric muscle to the mylohyoid line, and is applied to the mylohyoid and hyoglossus muscles. On the latter, it lies on the hypoglossal and lingual nerves

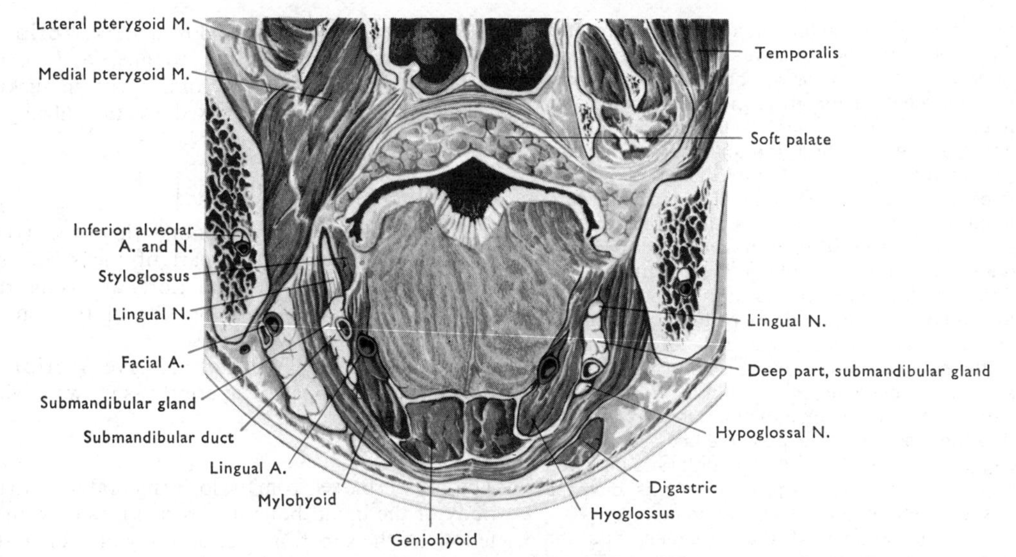

FIG. 106 Coronal section through the mouth posterior to the molar teeth. Cf. FIG. 110.

and the submandibular ganglion, while further posteriorly it lies in contact with the pharynx and is indented by the facial artery.

The duct arises from the medial surface, and passes anteriorly in the angle between the side of the tongue and mylohyoid. The **deep part of the gland** is a thin, flat process which extends anteriorly on the lateral side of the duct [FIGS. 59, 106] superior to mylohyoid.

Nerves and Vessels. The nerve supply is derived from the *parasympathetic* nerve cells in the **submandibular ganglion**, and from the *sympathetic* plexus on the facial artery. The preganglionic parasympathetic fibres, which synapse in the ganglion, come from the **chorda tympani** via the lingual nerve, and the latter sends *sensory* fibres to the gland. The arterial supply comes from several small branches of the facial and submental arteries.

Facial Artery [FIGS. 63, 72, 106]

This artery arises from the external carotid immediately superior to the tip of the greater horn of the hyoid bone, in the carotid triangle. It passes vertically upwards on the middle and superior constrictor muscles of the pharynx, passing close to the palatine tonsil, and lying under cover of the angle of the mandible and the digastric and stylohyoid muscles. Turning forwards above these muscles, it hooks round the posterosuperior part of the submandibular gland, and the lower border of the mandible to appear on the face at the antero-inferior angle of the masseter.

Branches in the Neck. (1) **Ascending palatine** (*q.v.*). (2) The **tonsillar artery** ascends superficial to the styloglossus, and pierces the superior constrictor to reach the palatine tonsil [FIG. 59]. (3) Small **glandular branches** pass to the submandibular gland. (4) The **submental artery** arises between the submandibular gland and the mandible and runs forwards to the inferior surface of the mylohyoid muscle. It supplies the submandibular and sublingual glands and the adjacent muscles and skin.

DISSECTION. Displace the submandibular gland and the submental vessels posteriorly. Cut the mylohyoid nerve and turn the anterior belly of digastric downwards. Examine the attachments of the mylohyoid muscle.

Mylohyoid Muscle [FIGS. 98, 104, 109]

This thin sheet arises from the whole length of the mylohyoid line on the mandible. The fibres of the right and left muscles pass downwards, medially, and forwards to meet in a median fibrous raphe which extends from the symphysis of the mandible to the body of the hyoid bone, into which a few of the most posterior muscle fibres are inserted. The two muscles form a supporting sling [FIG. 110] under the tongue, and separate it from the submandibular region. Posteriorly each muscle has a free border around which the superficial and deep parts of the submandibular gland are continuous; the deep part and the duct thus lie above mylohyoid in the floor of the mouth.

Nerve supply: the mylohyoid nerve. **Action**: it helps to raise the hyoid bone and tongue in swallowing, and forms the muscular floor of the mouth.

DISSECTION. Identify the cut edge of the mylohyoid muscle on the medial surface [FIG. 116], and separate it from the overlying geniohyoid by sliding a blunt instrument backwards and forwards between them. Find the plane of separation between geniohyoid and genioglossus in the same way, and note the fibres of genioglossus radiating from the upper mental spine. Pull the tongue laterally and cut carefully through the mucous membrane between it and the mandible. Strip the mucous membrane from the floor of the mouth and the mandible to expose the sublingual gland, the submandibular duct, the lingual nerve, the submandibular ganglion, and the **hypoglossal nerve** [FIG. 107] with the **deep lingual vein.** Follow the **lingual nerve** posteriorly and find its branches to the submandibular ganglion and the sublingual gland. Note the proximity of the nerve to the last molar tooth, and identify the inferior edge of the superior constrictor muscle of the pharynx below which the nerve enters the mouth. This is above the posterior margin of mylohyoid [FIGS. 115, 121]. Medial to these structures, expose the lateral surface of the hyoglossus muscle with the styloglossus mingling with it posteriorly. The **lingual artery** is deep to hyoglossus, but appears at its anterior margin.

On the lateral surface of the specimen, find the hypoglossal nerve and the lingual artery and vein. Confirm their continuity with the structures found lateral to the tongue. Trace the styloglossus back to the styloid process, and find the stylohyoid ligament and the stylopharyngeus attached to it. Try to trace the ligament towards the lesser cornu of the hyoid bone, and follow the stylopharyngeus downwards. The **glossopharyngeal nerve** curves round the **stylopharyngeus** and passes forwards deep to hyoglossus. Below this, confirm the superior border of the middle constrictor muscle, and follow it to the hyoid bone and stylohyoid ligament, lateral to stylopharyngeus.

Hyoglossus [FIGS. 107, 109]

The hyoglossus is a flat, quadrate muscle which arises from the greater horn and body of the hyoid bone [FIG. 109]. Its fibres pass superiorly into the posterior half of the side of the tongue, interdigitating with the fibres of styloglossus.

Nerve supply: the hypoglossal nerve. **Action**: it depresses the side of the tongue, and assists genioglossus to enlarge the oral cavity in sucking when the hyoid bone is fixed by the infrahyoid muscles.

Styloglossus [FIGS. 107, 114]

This elongated slip arises from the tip of the styloid process and the adjacent part of the stylohyoid ligament. It passes antero-inferiorly and is inserted into the whole length of the side of the tongue, intermingling with hyoglossus.

Nerve supply: hypoglossal nerve. **Action**: it pulls the tongue posterosuperiorly as in swallowing.

Geniohyoid

This muscle passes from the lower mental spine to the body of the hyoid bone. It lies side by side with its fellow on the superior surface of mylohyoid, and is almost horizontal in the resting position.

Nerve supply: the first cervical ventral ramus through the hypoglossal nerve. **Action**: it pulls the elevated hyoid bone directly forwards, sliding the pulley on the taut intermediate tendon of digastric. This increases the anteroposterior diameter of the pharynx to receive the bolus in swallowing.

Genioglossus [FIGS. 141, 143]

This flat, fan-shaped muscle is in contact with its fellow in the median plane. It arises from the upper mental spine, and its fibres spread out into the tongue in a vertical plane. It is inserted into the paramedian part of the dorsum of the tongue from the tip almost to the hyoid bone.

Nerve supply: hypoglossal nerve. **Action**: the postero-inferior fibres of the two muscles, acting together, protrude the tongue. If one muscle only is active or the other is paralysed, the tip of the tongue deviates towards the inactive side. The muscles also

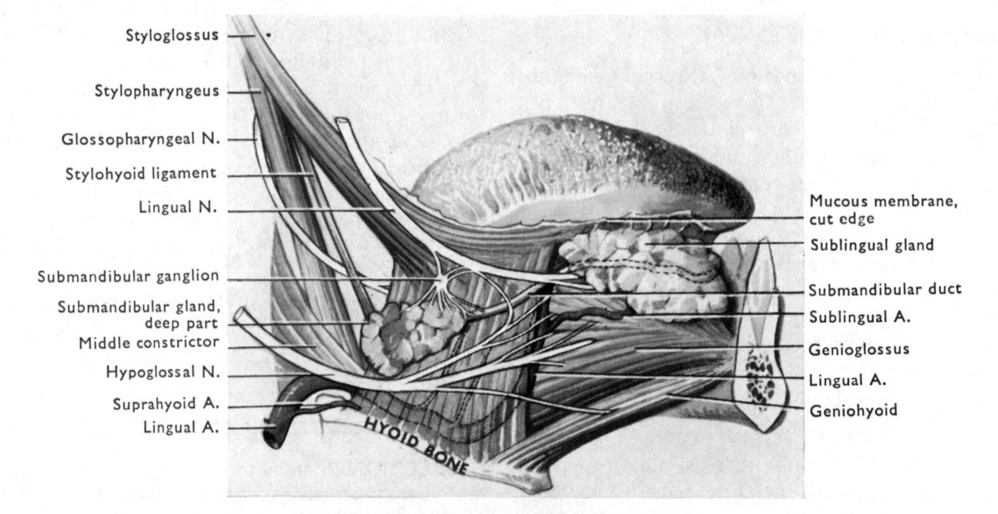

FIG. 107 Dissection of the submandibular region.

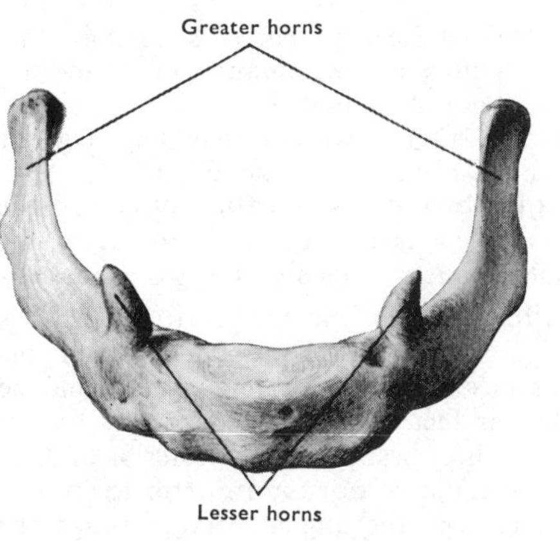

FIG. 108 Anterior view of the hyoid bone.

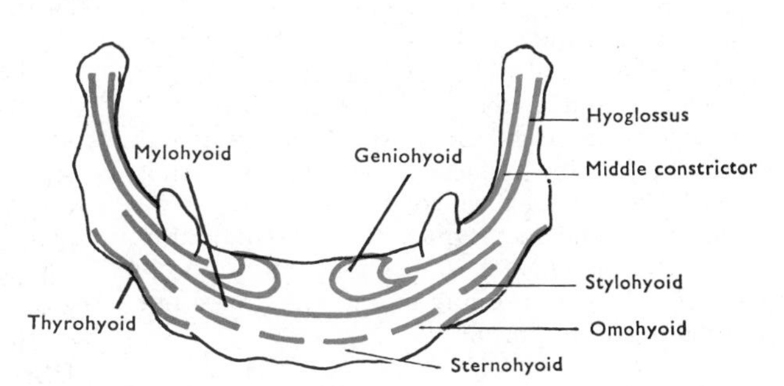

FIG. 109 Anterior view of the hyoid bone to show muscle attachments.

depress the central part of the tongue and increase the volume of the mouth, *e.g.*, in sucking. The anterior fibres depress the tip of the tongue and retract it.

Hyoid Bone [FIG. 108]

This is a U-shaped structure, deficient posteriorly, which lies between the root of the tongue and the thyroid cartilage. It forms a movable base for the tongue, and is held in position by a large number of muscles which connect it to the mandible (mylohyoid, geniohyoid, and anterior belly of digastric); to the skull (stylohyoid and posterior belly of digastric); to the thyroid cartilage (thyrohyoid); to the sternum (sternohyoid); and to the scapula (omohyoid). It can be raised by the contraction of both bellies of the digastric; pulled forwards (and upwards) by the geniohyoid; moved posterosuperiorly by the stylohyoid; depressed by the infrahyoid muscles [p. 36], and held fixed by the contraction of these opposing muscles so that the tongue may be moved on it. The hyoid is also attached to the skull by the stylohyoid ligament (*q.v.*), to the thyroid cartilage by the thyrohyoid membrane, and is in continuity with the pharyngeal wall through the attachment of the middle constrictor muscle of the pharynx to it [FIG. 114].

Submandibular Duct [FIGS. 107, 110, 121]

It emerges from the medial surface of the submandibular gland and passes anterosuperiorly in the angle between the mylohyoid and the side of the tongue. It opens on the floor of the mouth, beneath the anterior part of the tongue, on the summit of the sublingual papilla, an elevation at the anterior end of

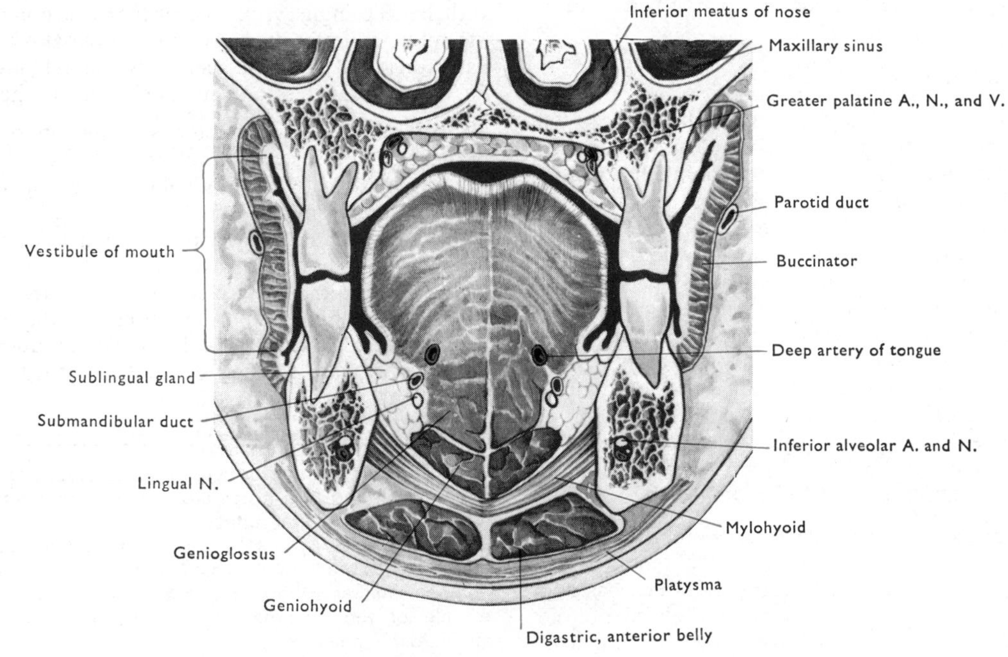

FIG. 110 Coronal section through the mouth at the level of the second molar teeth.

the sublingual fold [FIG. 112]. Examine these structures in the floor of the living mouth.

At first the duct lies on the hyoglossus, between the lingual nerve above, the hypoglossal nerve below, separated from the mylohyoid muscle by the deep part of the gland. Next it passes on to the genioglossus, and has the sublingual gland between it and mylohyoid. Here the **lingual nerve** hooks round the inferior surface of the duct, and runs upwards into the tongue medial to the duct.

The wall of the duct is much thinner than that of the parotid gland, and is therefore not so easily palpated. Make an opening into it, and pass a fine probe along it into the mouth.

Sublingual Gland

This gland lies on the mylohyoid between the mandible and the genioglossus muscle. It raises the mucous membrane covering its superior surface to form the sublingual fold [FIGS. 110, 112]. It is the smallest of the large salivary glands (3 cm long) and touches its fellow in the median plane above the origin of the genioglossus. The lingual nerve and submandibular duct lie medial to it.

Ducts. Numerous small ducts (eight to twenty) open into the mouth on the summit of the sublingual fold. It is predominantly a mucous gland, in contrast to the serous parotid and the mixed submandibular glands.

Vessels and Nerves. The arterial supply is from the sublingual branch of the lingual artery, and its nerves are branches of the lingual nerve. The latter contain postganglionic parasympathetic fibres from the cells of the submandibular ganglion, as well as sensory and postganglionic sympathetic fibres [FIG. 121].

Lingual Nerve

The lingual nerve descends between the ramus of the mandible and the medial pterygoid muscle. It then inclines fowards and enters the mouth by passing inferior to the lower border of the superior constrictor muscle of the pharynx at its attachment near the posterior end of the mylohyoid line [FIGS. 98, 121]. It continues antero-inferiorly between the mucous membrane of the mouth and the body of the mandible, postero-inferior to the last molar tooth [FIG. 105]. In this position it is liable to be injured by the clumsy extraction of the adjacent tooth, and is accessible to local anaesthetics. In its further course, the nerve lies close to the side of the tongue, crosses the styloglossus and the upper part of hyoglossus, and hooks beneath the submandibular duct.

Branches.

Twigs of Communication.
- (1) Two or more to the submandibular ganglion.
- (2) One or two which descend along the anterior border of the hyoglossus muscle to unite with the hypoglossal nerve.

Branches of Distribution.
- (1) Slender filaments to the mucous membrane of the mouth and gums.
- (2) A few twigs to the sublingual gland.
- (3) Branches to the tongue.

The **lingual branches** pierce the substance of the tongue, and then incline superiorly to supply the mucous membrane over its anterior two-thirds.

Submandibular Ganglion [FIGS. 107, 121]. This small ganglion lies on the upper part of the hyoglossus muscle, between the lingual nerve and the submandibular duct. It is suspended from the lingual nerve by two short branches. The posterior of these carries preganglionic **parasympathetic nerve fibres** which join the lingual nerve through the chorda tympani branch of the facial nerve and synapse on the cells of the ganglion. They may be accompanied by some **sensory fibres** which traverse the ganglion and are distributed through its branches. The anterior branch carries axons of the ganglion cells (postganglionic fibres) to the lingual nerve for distribution to the sublingual gland and to the glands in the anterior two-thirds of the tongue. From the inferior border of the ganglion, several minute branches carry postganglionic fibres to the submandibular gland and duct, and to the mucous membrane of the mouth.

Postganglionic **sympathetic fibres** reach this territory from the plexuses on the facial and lingual arteries, and supply the glands.

Hypoglossal Nerve

This nerve has already been traced to the posterior margin of the mylohyoid muscle [p. 33]. It can now be followed anteriorly across the hyoglossus muscle, with the deep lingual vein, between the hyoid bone and the submandibular duct. Anterior to hyoglossus, it passes on to and pierces the genioglossus muscle, breaking up into branches to the muscles of the tongue.

Branches. Through numerous purely muscular branches it supplies: (1) styloglossus; (2) hyoglossus; (3) genioglossus; (4) geniohyoid; (5) the intrinsic muscles of the tongue.

It communicates freely with the lingual nerve on the lateral surface of hyoglossus and in the substance of the tongue, and these communications transmit sensory nerve fibres from the muscles to the brain stem via the trigeminal nerve. For the distribution of nerve fibres from the first cervical nerve with the hypoglossal nerve, see page 35.

DISSECTION. Detach the hyoglossus carefully from the hyoid bone, and turn it upwards without dividing the structures on its lateral surface.

This exposes (1) the lingual artery and its dorsal branches; (2) the lingual veins; (3) the posterior part of genioglossus, and the origin of the middle constrictor muscle of the pharynx; (4) the attachment of the stylohyoid ligament.

Lingual Artery [FIG. 107]

It springs from the front of the external carotid artery opposite the tip of the greater horn of the hyoid bone, and hooking on to the top of it, runs anteriorly under cover of hyoglossus. It turns superiorly along the anterior border of hyoglossus, and ends by becoming the deep artery of the tongue.

Initially it lies deep to skin, fasciae, and the hypoglossal nerve, and gives off the **suprahyoid branch**, which runs along the superior border of the hyoid bone, lateral to hyoglossus.

Deep to hyoglossus, it gives off two or more **dorsal lingual branches**, which run postero-superiorly to supply the muscular substance of the tongue, and end in the mucous membrane of its pharyngeal surface and in the palatine tonsil.

At the anterior border of hyoglossus it is crossed by the branches of the hypoglossal nerve, by the submandibular duct, and by the lingual nerve. Here it gives off the **sublingual artery**, which runs anterosuperiorly to supply the sublingual gland and the neighbouring muscles.

It is continued as the **deep artery of the tongue**, which enters the tongue about its middle and runs forwards to the tip, separated only by the deep vein from the mucous membrane of the lower surface of the tongue near the frenulum linguae. It is a tortuous vessel to allow for the elongation of the tongue when protruded, and it sends numerous branches into the substance of the tongue.

Veins of the Tongue

The arrangement of these veins is variable. Two venae comitantes run with the lingual artery, and these are joined by the dorsal lingual veins. The **deep vein** is the principal vein. It begins at the tip and runs posteriorly near the median plane. It lies immediately deep to the mucous membrane on the inferior surface of the tongue and can be seen through it in life. It descends along the anterior margin of hyoglossus, and crosses the superficial surface of that muscle below the hypoglossal nerve. All the veins unite at the posterior border of hyoglossus to form the **lingual vein**. This either joins the facial vein, or, crossing the external and internal carotid arteries, joins the internal jugular vein.

Stylohyoid Ligament [FIGS. 107, 115]

This fibrous cord passes from the tip of the styloid process to the lesser horn of the hyoid bone. It may be stout or slender, and it is not uncommon to find it partly cartilaginous or ossified. Occasionally it may contain muscle fibres.

It is of embryological interest for it represents the remnant of part of the cartilage of the second pharyngeal arch of the embryo. The remainder forms the styloid process and the lesser horn and upper part of the body of the hyoid bone. It also forms the stapes, one of the three ear ossicles, and it corresponds to the sphenomandibular ligament [p. 100] in the first pharyngeal arch [p. 246].

THE MOUTH AND THE PHARYNX

THE MOUTH

The mouth is the first part of the digestive tube. It is separated by the teeth and gums into the smaller, external, vestibule, deep to the lips and cheeks, and the larger mouth proper within the dental arch.

Vestibule of Mouth

The vestibule is a cleft into which the ducts of the parotid glands and the mucous glands of the lips and cheeks open. Superiorly and inferiorly, it is bounded by the reflexion of the mucous membrane from the lips and cheeks on to the maxillae and mandible. Posteriorly, it communicates, on each side, with the cavity of the mouth proper through the interval between the last molar teeth and the ramus of the mandible. In paralysis of the facial muscles, the lips and cheeks fall away from the teeth and gums, and food is apt to lodge in the vestibule.

Lips

The superficial structures in the lips have been examined already [p. 82]. Each lip is a flexible structure which consists of a sheet of muscle covered externally by skin and internally by mucous membrane and submucosa. The mucous membrane is continuous with the skin at the margins of the lips, and here the stratified squamous epithelium changes from a non-keratinizing to a keratinizing type. At its reflexion on to the jaws, the mucous membrane forms a median, raised fold, the **frenulum** of the lip. The chief bulk of the lips is the muscular layer which is formed by orbicularis oris and the various facial muscles which converge on it [p. 18].

The submucous layer consists of areolar tissue which binds the mucous membrane to the muscular layer. It contains numerous mucous **labial glands** with ducts which pierce the mucous membrane and open into the vestibule. In each lip there is an **arterial arch** formed by the labial branches of the facial arteries [FIG. 72].

The **lymph vessels** of the lower lip pass to the submandibular nodes; those of the upper lip pass to the submandibular and superficial parotid nodes.

Cheeks

The cheeks are directly continuous with the lips and have the same general structure. **Buccinator** forms the muscle layer, and it follows the line of the teeth, passing deep to masseter posteriorly, to become continuous with the superior constrictor muscle of the pharynx at the pterygomandibular raphe [FIG. 114]. It lies between the **buccopharyngeal**

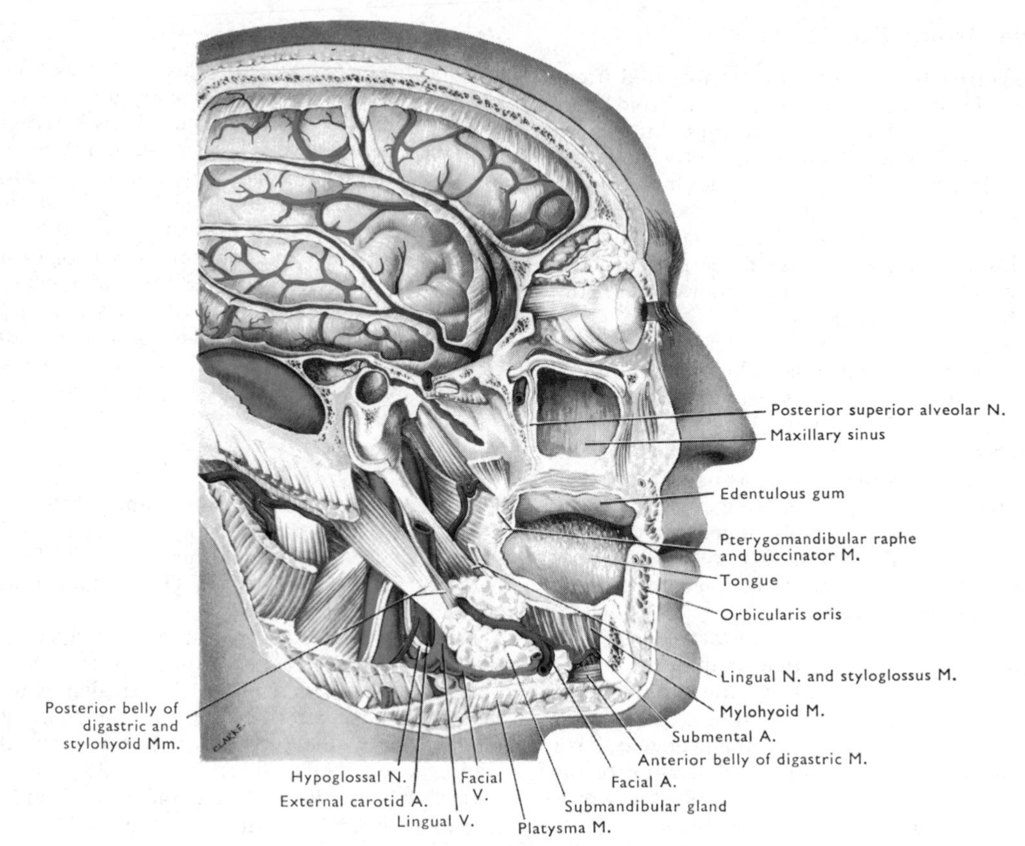

FIG. 111 Deep dissection of the face and adjoining parts.

fascia externally and the **pharyngobasilar fascia** internally. There is a considerable quantity of subcutaneous fat in the cheek, particularly in infants where the encapsulated **buccal pad of fat** is especially well formed, and assists the sucking process by increasing the rigidity of the cheek. Numerous **buccal glands** lie in the submucosa, and four or five larger mucous, **molar glands** lie external to the buccinator around the entry of the parotid duct. The **parotid duct** passes above the buccal pad of fat, pierces buccinator and its fasciae, and enters the mouth opposite the second upper molar tooth.

Gums and Teeth

The gums are composed of dense fibrous tissue covered with a smooth, vascular mucous membrane. They are attached to the alveolar margins of the jaws and the necks of the teeth. Here the fibrous tissue is continuous with the **periodontal membrane** which attaches the teeth to their sockets. The epithelium is adherent to the tooth surfaces.

In the adult there are sixteen **permanent teeth** in each jaw. These consist, on each side from the median plane, of 2 incisors, 1 canine, 2 premolars, and 3 molars. The complete **milk dentition** consists of ten teeth in each jaw—on each side 2 incisors, 1 canine, and 2 molars. The first of these (lower central incisors) erupts at approximately six months after birth, the last (second milk molars) at approximately two years. The first permanent tooth to erupt is the first molar at approximately 6 years.

Floor of Mouth

The surface of this floor is formed by the mucous membrane which connects the tongue to the mandible. Laterally the mucous membrane passes from the side of the tongue on to the mandible. Anteriorly it stretches from one half of the mandible to the other, beneath the free anterior part of the tongue. Here it forms a median fold between the floor of the mouth and the inferior surface of the tongue, the **frenulum of the tongue**. On both sides of the frenulum, the sublingual gland bulges upwards forming the **sublingual fold** which ends anteriorly in the **sublingual papilla**. On the apex of the papilla is the opening of the submandibular duct. The sublingual ducts form a series of minute apertures on the summit of the sublingual fold [p. 110].

Roof of Mouth

This is formed by the vaulted palate [FIGS. 113, 116]. The anterior two-thirds is the bony or **hard palate**, the posterior third is the **soft palate**. The soft palate is attached to the posterior border of the hard palate, and has a free posterior margin from the middle of which the **uvula** hangs down and rests on the dorsum of the tongue. A poorly marked median raphe runs forwards from the uvula, and ends anteriorly, below the incisive fossa, in a slight elevation, the **incisive papilla**. On each side of the raphe, the mucous membrane of the anterior part of the hard palate is thrown into three to four hard transverse ridges, the **transverse palatine folds**, but posteriorly it is comparatively smooth. By careful palpation with a finger or the tongue, the **pterygoid hamulus** [FIG. 298] may be felt immediately posterior to the lingual surface of the third upper molar tooth.

Isthmus of Fauces

The term fauces is often used to indicate the region between the mouth and the pharynx, but it may be defined more precisely as that part of the pharynx which contains the two palatine tonsils between the palatoglossal and palatopharyngeal arches.

The isthmus of the fauces is bounded by the palatoglossal arches and is therefore the communication between the mouth proper and the pharynx [FIG. 113]. The isthmus is best seen in the living body, and should be examined with the mouth wide open and the tongue depressed. Its boundaries are the soft palate, the dorsum of the tongue, and the palatoglossal arches.

Palatoglossal Arch. This is a fold of mucous membrane which covers the palatoglossus muscle. It passes antero-inferiorly from the soft palate to the posterior part of the side of the tongue. The palatopharyngeal arch and the palatine tonsil can be seen also, but they are described with the pharynx to which they belong.

THE PHARYNX

The pharynx is a wide muscular tube, about 12 cm long, which is lined throughout with mucous membrane, and extends from the base of the skull to the level of the body of the sixth cervical vertebra [FIG. 116] where it is continuous with the oesophagus. It lies posterior to the nasal cavities and septum (**nasal part**), the mouth (**oral part**) and the larynx (**laryngeal part**), and it conducts air to and from the larynx and food to the oesophagus from the mouth; the oral part is common to both functions.

It is widest at the base of the skull, posterior to the orifices of the auditory tubes. Thence, it narrows to the level of the palate (**pharyngeal isthmus**) but widens again in the oral and laryngeal parts and then rapidly narrows to the oesophagus.

From the opening of the larynx upwards the walls are not in contact, and thus allow the passage of air via the mouth or nasal cavities to the larynx. Below the opening of the larynx, the anterior and posterior walls are in contact.

Position

Posterior to the pharynx is the pharyngeal venous plexus and a layer of loose areolar tissue which separates it from the prevertebral fascia and allows it to slide freely on that fascia during swallowing. Lateral to the pharynx is the neurovascular bundle of the neck and the styloid process and its muscles. The pharyngeal plexus of nerves ramifies over it, and supplies it with motor and sensory fibres. Anteriorly, the pharynx opens into the nasal cavities, mouth, and larynx, and its muscles are attached to structures in the lateral walls of these apertures.

PHARYNGEAL WALL

The wall of the pharynx consists of five layers. These are: (1) mucous membrane; (2) submucosa;

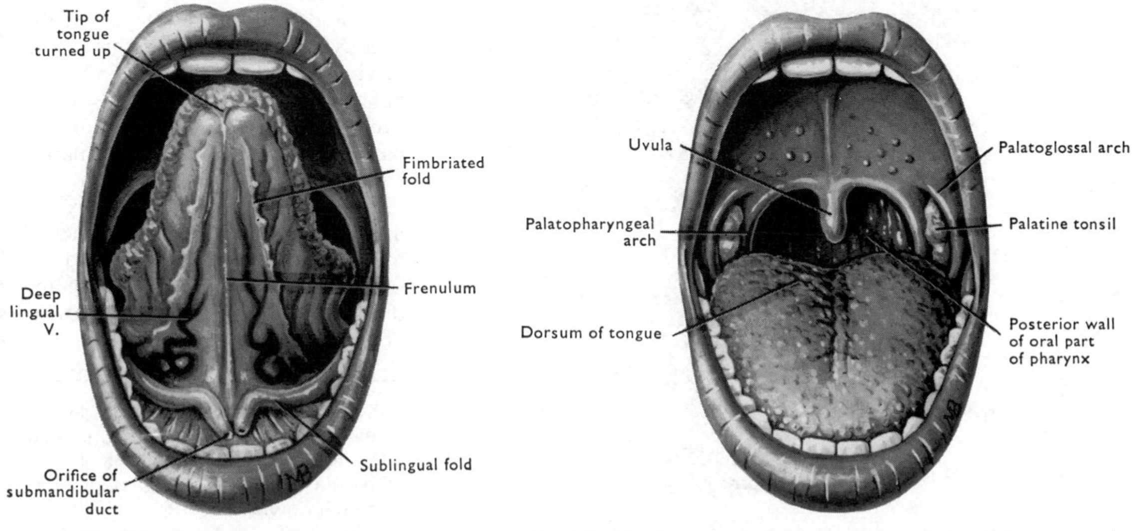

FIG. 112 The inferior surface of the tongue.

FIG. 113 The fauces and its isthmus seen through the widely open mouth.

(3) pharyngobasilar fascia; (4) pharyngeal muscles; (5) buccopharyngeal fascia.

The **buccopharyngeal fascia** covers the external surfaces of buccinator and the pharyngeal muscles. The **pharyngobasilar fascia** lines the internal surfaces of the pharyngeal muscles, attaches the pharynx to the base of the skull, to the auditory tubes and to the lateral margins of the posterior nasal apertures (choanae). It also fills the gap in the pharyngeal wall above the free superior margin of the superior constrictor muscle [FIG. 114].

The muscles of the pharynx consist of the three constrictors, the stylopharyngeus, salpingo-pharyngeus, and palatopharyngeus muscles.

DISSECTION. On the right side, remove the bucco-pharyngeal fascia from the external surfaces of the pharyngeal muscles, moving the knife in the direction of their fibres [FIG. 114]. Note the plexuses of nerves and veins, and remove them.

On the left side, examine the interior of the pharynx [p. 150] and strip off its mucous membrane to expose the pharyngeal muscles from the medial side [FIG. 115].

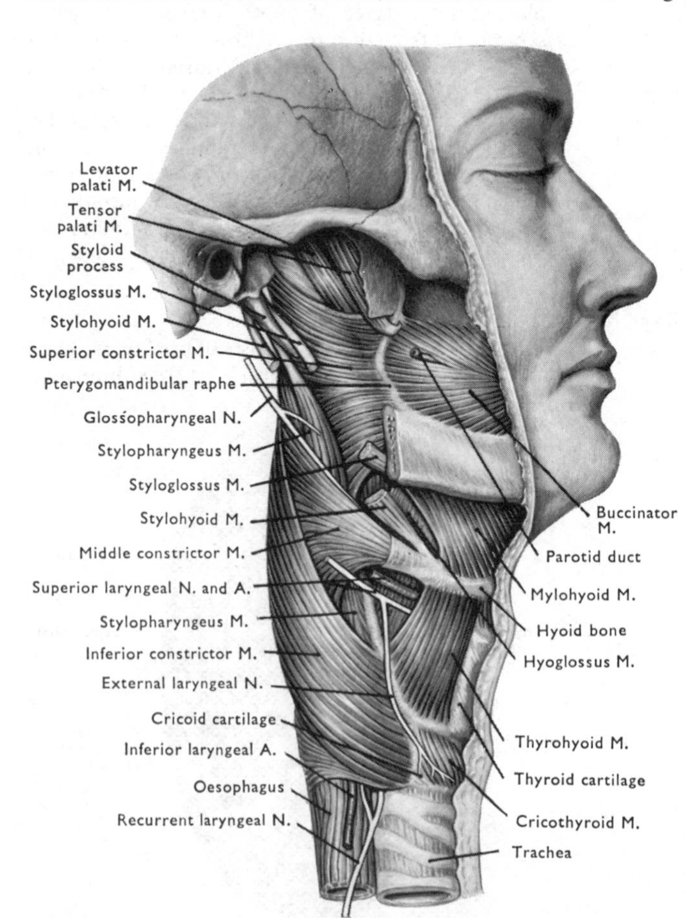

FIG. 114 Lateral view of the constrictors of the pharynx and associated muscles.

Pharyngeal Veins

These veins lie on the posterior wall and borders of the pharynx, and form the **pharyngeal plexus of veins** which receives blood from the pharynx and soft palate. Two or more veins drain from it to each internal jugular vein, and it communicates with the pterygoid plexus and the cavernous sinuses.

CONSTRICTOR MUSCLES

These three muscles form curved sheets which lie in the posterior wall and sides of the pharynx, and overlap each other from below upwards [FIG. 114]. They are inserted into a median fibrous **raphe** which descends in the posterior pharyngeal wall from the **pharyngeal tubercle** on the base of the skull [FIG. 298]. **Nerve supply:** the pharyngeal plexus of nerves, with an additional supply to the inferior constrictor from the external and recurrent laryngeal nerves. **Actions**: p. 120.

Inferior Constrictor

This muscle arises from the side of the cricoid cartilage and the oblique line on the thyroid cartilage. Its fibres sweep posteriorly and medially to the median raphe; the lowest fibres are horizontal, but those above this ascend with increasing obliquity, the highest fibres almost reaching the base of the skull external to the other two constrictors. Inferiorly, it overlaps the beginning of the oesophagus and the recurrent laryngeal nerve and inferior laryngeal artery ascending to the larynx.

Middle Constrictor

This fan-shaped muscle arises from the greater and lesser horns of the hyoid bone and from the lower part of the stylohyoid ligament. From this curved origin, the fibres fan out into the pharyngeal wall, the middle fibres running horizontally. The inferior part passes deep to the inferior constrictor posteriorly, but is separated from it laterally by an interval in which the internal branch of the superior laryngeal nerve and the superior laryngeal artery pierce the thyrohyoid membrane and enter the pharynx [FIG. 114].

DISSECTION. On the right side, the superior constrictor may be brought fully into view by detaching the medial pterygoid muscle from its origin and turning it downwards.

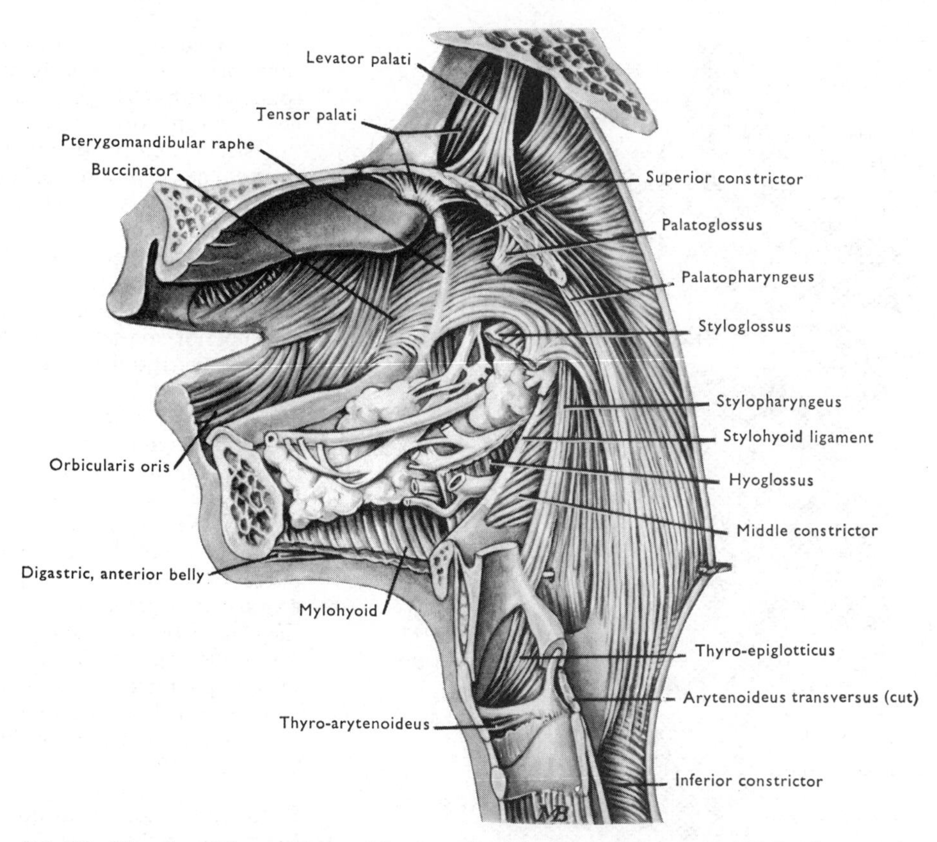

FIG. 115 Dissection of the constrictors of the pharynx and the associated structures which lie adjacent to the mucous membrane lining the mouth, pharynx, and larynx, from the medial side. The tongue has been removed to expose the structures which lie between it and the mylohyoid muscle. See also FIG. 121.

Superior Constrictor

This muscle is in the wall of the nasal and oral parts of the pharynx. It arises from the lower part of the posterior margin of the medial pterygoid lamina and the pterygoid hamulus; from the pterygomandibular raphe; from the mandible near the posterior end of the mylohyoid line; and from the mucous membrane of the mouth and side of the tongue. The fibres curve posteriorly to the median raphe in the posterior wall of the pharynx. The upper fibres ascend to the pharyngeal tubercle [FIG. 12]—the lower fibres descend internal to the middle constrictor with the stylopharyngeus and the glossopharyngeal nerve entering the pharynx between them [FIG. 114].

The free **upper border** of the superior constrictor extends from the medial pterygoid lamina to the pharyngeal tubercle, leaving a gap between it and the skull. This gap is filled by the tensor and levator muscles of the palate with the auditory tube between them [FIG. 126]. The **auditory tube** and the **levator muscle** of the palate enter the pharynx above the superior margin of the superior constrictor together with the ascending palatine artery. The **tensor of the palate** descends external to the upper part of the superior constrictor.

The **pterygomandibular raphe** is formed by the interlacing tendinous fibres of the superior constrictor and buccinator muscles between the pterygoid hamulus and the mandible near the posterior end of the mylohyoid line. The tendinous fibres run horizontally, and so are capable of separating and allowing the raphe to stretch when the mouth is opened.

INTERIOR OF THE PHARYNX

The pharynx is lined throughout with mucous membrane, and the submucosa contains numerous mucous **pharyngeal glands** and nodules of lymph tissue. Aggregations of these **lymph follicles** form the pharyngeal, tubal, and palatine **tonsils**.

Nasal Part of Pharynx [FIGS. 116, 122]

It lies superior to the soft palate, and is continuous inferiorly with the oral part through the narrow **pharyngeal isthmus**, which lies posterior to the palate and is limited laterally by the **palatopharyngeal arches**. These arches are visible internally as a ridge of mucous membrane covering the palato-pharyngeus muscle on each side. An arch begins at

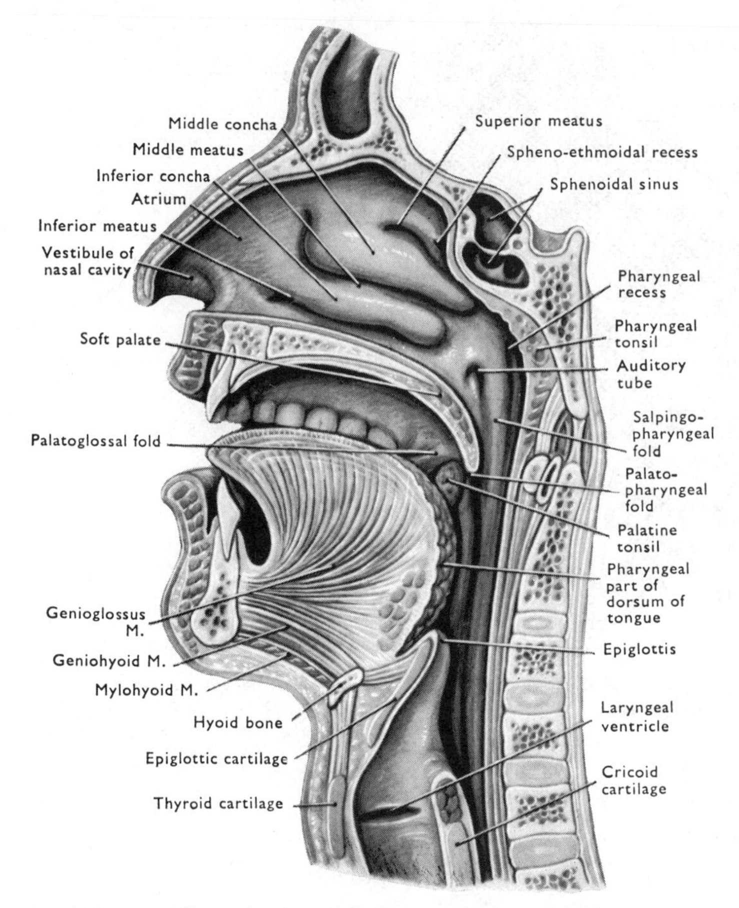

FIG. 116 Paramedian section through the nose, mouth, pharynx, and larynx.

each posterolateral margin of the soft palate, and sweeps postero-inferiorly to fade out on the lateral wall of the pharynx; its superior margin passes almost horizontally backwards around the pharyngeal isthmus.

The nasal cavities open into the nasal part through the oval **choanae** (each approximately 2·5 cm high by 1·5 cm wide). These extend from the base of the skull to the posterior edge of the hard palate, and are separated by the nasal septum.

The **roof** and **posterior wall** of the pharynx form a continuous curved surface of mucous membrane. This covers the inferior surface of the sphenoid, the basilar part of the occipital bone, and the superior part of the longus capitis muscle. Where the roof is continuous with the posterior wall, the **pharyngeal tonsil** in the submucosa bulges the mucous membrane into the cavity of the pharynx. This lymph tissue is often enlarged in children (adenoids) and may be of sufficient size to block the nasal part of the pharynx and give a 'nasal' quality to the voice. A minute, median pit (**pharyngeal bursa**) may be found in its surface.

On each **lateral wall**, the **pharyngeal orifice of the auditory tube** lies at the level of the inferior concha of the nose. The tube leads to the middle ear, and its pharyngeal orifice is bounded superiorly and posteriorly by a firm, rounded **tubal ridge** around which lies the **tubal tonsil**. From the ridge, the slender **salpingopharyngeus muscle** descends into the lateral wall of the pharynx covered by the salpingopharyngeal fold of mucous membrane. Posterior to the tubal ridge, the pharyngeal wall extends laterally, above the upper margin of the superior constrictor, to form the deep **pharyngeal recess**.

Note that the posterior wall and roof of the nasal part of the pharynx can be explored by a finger introduced through the mouth and pharyngeal isthmus.

When the nasal part of the pharynx is illuminated by light reflected from a mirror introduced through the mouth, a view may be obtained in the mirror of the orifices of the auditory tubes, the side walls and roof of the nasal part of the pharynx, and the choanae, through which the posterior ends of the middle and inferior conchae can be seen [FIG. 116].

Structure. The **mucous membrane** lining the pharynx contains a considerable amount of elastic tissue and a number of glands, mainly mucous in type. In the nasal part, the epithelium is of the ciliated columnar type characteristic of the respiratory passages. In the oral and laryngeal parts, there is stratified squamous epithelium of the non-keratinizing type, similar to that in the mouth.

There are many subepithelial collections of **lymph tissue** into which the epithelium tends to pass in the form of narrow clefts or pits (**crypts**). The lymph tissue is similar to that in a lymph node, but differs from it in having only efferent vessels. Many of the lymphoctyes produced in its germinal centres migrate through the covering epithelium into the crypts, and so into the lumen of the pharynx. These collections form the pharyngeal and tubal tonsils in the nasal part and the palatine and lingual tonsils in the oral part of the pharynx. The **lingual tonsil** consists of small, scattered collections which give the pharyngeal part of the tongue its nodular appearance. These masses of lymph tissue form an almost complete ring in the wall of the pharynx where the nasal and oral cavities open into it. This tissue, like that in the small and large intestines, appears to be concerned with protection against ingested and inspired bacteria, etc., and is no doubt involved in the production of antibodies to such invading organisms. The volume of this lymph tissue is normally much greater in children than in adults. Hence simple

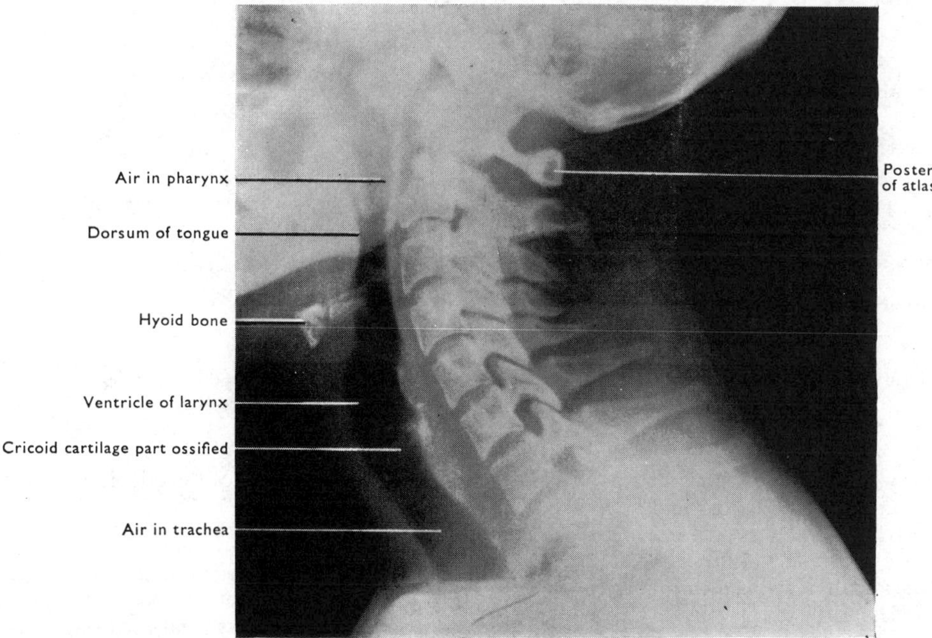

FIG. 117 Lateral radiograph of the neck.

enlargement is not an adequate reason for surgical removal.

Oral Part of Pharynx

This part lies posterior to the palatoglossal arch.

Immediately posterior to the tongue is the **epiglottis**—a leaf-shaped plate of elastic cartilage covered with mucous membrane. The upper part of the epiglottis stands up posterior to the tongue. Pull the epiglottis backwards to expose the **median glosso-epiglottic fold**—a median ridge of mucous membrane between the front of the epiglottis and the back of the tongue. On each side of it is a depression, the **epiglottic vallecula**. The **lateral glosso-epiglottic folds** are ridges of mucous membrane which form the lateral boundaries of the epiglottic valleculae, and extend from the margins of the epiglottis to the side walls of the pharynx at their junctions with the tongue [FIG. 119].

Palatine Tonsils. These masses of lymph tissue lie in the mucous membrane of the lateral walls of the oral part of the pharynx, between the palatoglossal and palatopharyngeal arches. They lie opposite the angle of the mandible between the back of the tongue and the soft palate. In children they are larger than the fossae between the arches, and so bulge into the pharynx, and extend superiorly into the soft palate and anteriorly lateral to the palatoglossal arch [FIG. 116].

The medial surface is covered by mucous membrane which forms about twelve deep **tonsillar crypts**. Superiorly the tonsil is bounded by a mucosal fold under which lies the **supratonsillar fossa**.

The lateral surface is covered by a thin fibrous **capsule** which is attached to the pharyngobasilar fascia by loose areolar tissue and to the sheath of the palatoglossus muscle by a fibrous band which helps to hold the tonsil in position. The superior constrictor is lateral to these structures, and separates them from the arch of the facial artery.

Vessels and Nerves of the Palatine Tonsil. The chief artery is the **tonsillar branch of the facial artery** [FIG. 59] which pierces the superior constrictor and enters the inferior part of its lateral surface. One or more inconstant **veins** descend from the soft palate, lateral to the tonsillar capsule. They pierce the superior constrictor near the artery, and either end in the pharyngeal plexus or unite to form a single vessel which enters the facial vein. These veins may be a source of troublesome bleeding at tonsillectomy, especially when they unite to form a single larger vein. Efferent lymph vessels pierce the superior constrictor and pass to nodes on the carotid sheath including the jugulodigastric at the angle of the mandible, and to posterior submandibular nodes. **Nerve supply**: the glossopharyngeal and lesser palatine nerves.

Laryngeal Part of Pharynx

This part lies posterior to the larynx. It decreases rapidly in width from above downwards, so that the pharyngo-oesophageal junction is a narrow part of the gut tube.

The **posterior** and **lateral walls** are formed by the middle and inferior constrictor muscles, with the palatopharyngeus and stylopharyngeus internally, lined with smooth mucous membrane.

The **anterior wall** is formed by: (1) the inlet of

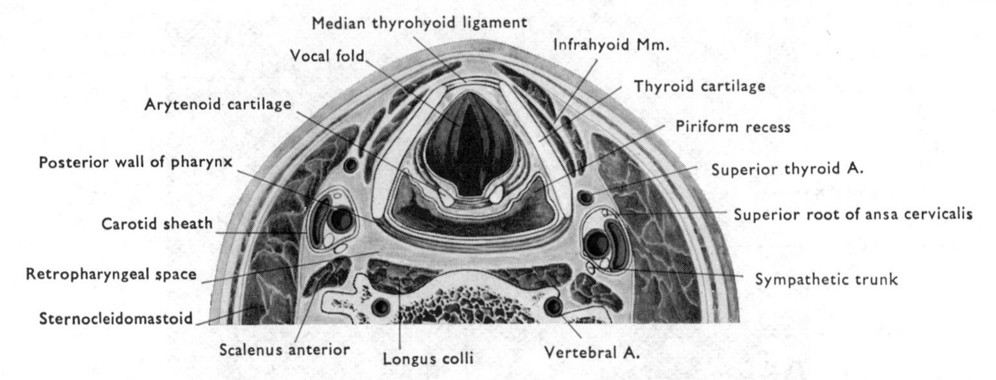

FIG. 118 Transverse section through the anterior part of the neck at the level of the upper part of the thyroid cartilage.

the larynx superiorly; (2) the mucous membrane on the posterior surfaces of the arytenoid and cricoid cartilages, inferiorly, and (3) a piriform recess extending forwards on each side of these, medial to the thyroid cartilage and thyrohyoid membrane. The interarytenoid muscles, the posterior crico-arytenoid muscles, and the attachment of the longitudinal muscle of the oesophagus to the cricoid separate the mucous membrane from these cartilages [FIG. 138].

Inlet of the Larynx. This is a large, almost vertical opening bounded anterosuperiorly by the epiglottis, on each side by the aryepiglottic fold of mucous membrane, and inferiorly by the interarytenoid fold of mucous membrane.

Each **aryepiglottic fold** is a thin, deep fold that extends inferiorly from the margin of the epiglottis to the arytenoid cartilage. It contains the slender aryepiglottic muscle, and, near its inferior end, two small pieces of cartilage which form the **corniculate** and **cuneiform tubercles** in its free edge [Fig. 119].

The **arytenoid cartilages** are a pair of three-sided cartilages placed side by side on the superior border of the lamina of the cricoid cartilage [Fig. 129]. The interarytenoid fold of mucous membrane, passing between them, forms the inferior boundary of the inlet and encloses the muscles which pass between the posterior surfaces of the arytenoid cartilages [Fig. 138].

Piriform Recess. This is a fairly deep gutter which separates the lateral wall of the laryngeal inlet (the aryepiglottic fold) from the posterior part of the lamina of the thyroid cartilage and the thyrohyoid membrane. It is lined with mucous membrane and ends as a blind pocket inferiorly. In this pocket foreign bodies may lodge, and, if sharp, may pierce the mucous membrane.

The **oesophageal orifice** is the narrowest part of the pharynx, and lies opposite the inferior border of the cricoid cartilage.

Soft Palate

The soft palate is a flexible, muscular flap which extends postero-inferiorly from the posterior edge of the hard palate into the pharyngeal cavity. It is also attached to the lateral walls of the pharynx, and has the **uvula** hanging down from the middle of its free posterior border, which is continuous with the palatopharyngeal arch on each side.

The soft palate is a flap valve which, when raised and drawn posteriorly against the posterior pharyngeal wall, shuts off the nasal part of the pharynx. This permits such actions as blowing up a balloon or coughing without air escaping through the nose, and swallowing without regurgitation into the nose. When the soft palate is pulled inferiorly against

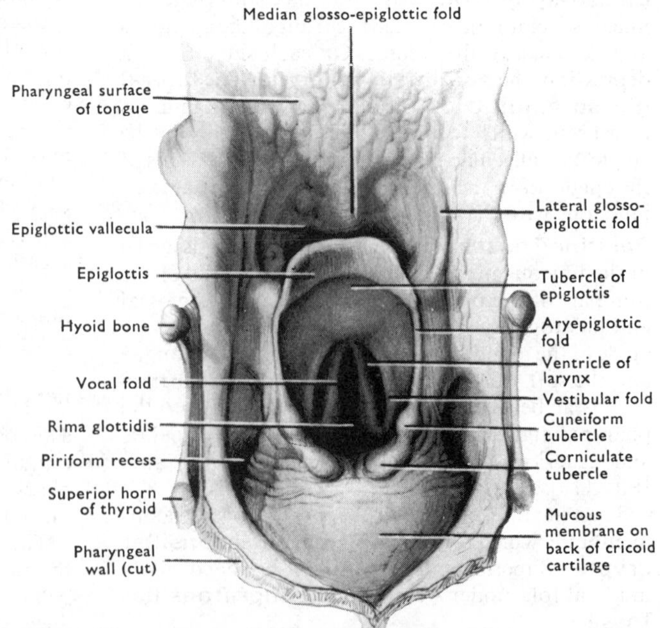

FIG. 119 Anterior wall of the laryngeal part of the pharynx seen from above.

the posterior part of the tongue, it cuts off the mouth from the pharynx, thus permitting respiration to continue during sucking or chewing without danger of inhalation of food or fluid. When tensed, the soft palate assists the tongue in directing food and fluids towards the laryngeal part of the pharynx in swallowing.

Structure. The soft palate is made up of a fold of mucous membrane which encloses parts of five pairs of muscles, of which only the uvular muscles are intrinsic. Each of the remaining pairs of muscles forms a sling, the two muscles meeting in the midline of the palate where they are partly attached to the palatal aponeurosis—an intermediate fibrous sheet formed from the tendons of the tensor palati muscles. The convex superior surface of the soft palate is continuous with the floor of the nasal cavities, and is covered by the same pseudostratified columnar ciliated epithelium, except posteriorly where it is of a stratified squamous type. The **mucous membrane** of the oral surface is much thicker, and contains a considerable layer of tightly packed **mucous glands**. It is covered by the oral type of stratified squamous epithelium, and contains some taste buds, especially in children.

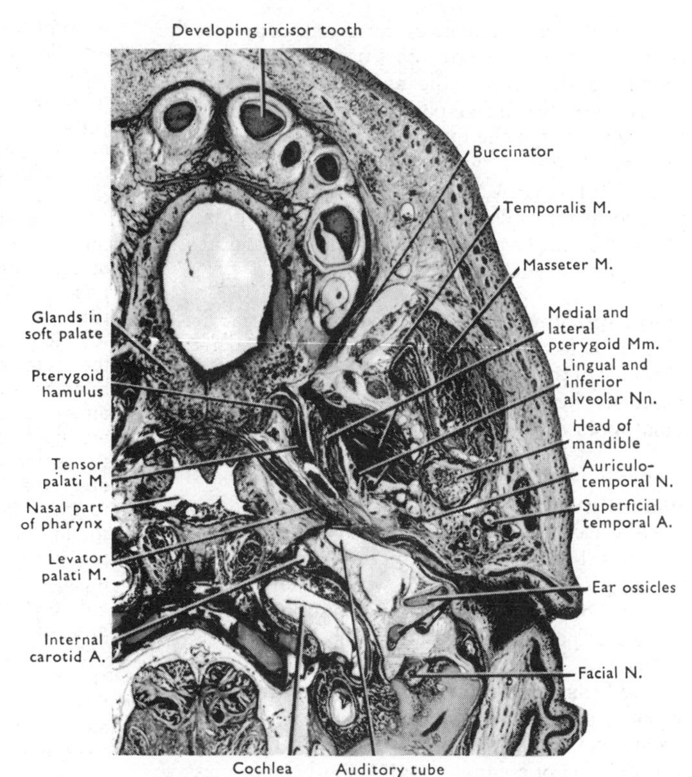

FIG. 120 Horizontal section through part of the head of a human foetus of seven months at the level of the soft palate.

Tensor Palati. This muscle arises from the scaphoid fossa at the base of the medial pterygoid lamina, from the spine of the sphenoid, and from the lateral side of the auditory tube between. It passes downwards, anterior to the auditory tube, and converges on a tendon which hooks round the base of the **pterygoid hamulus** [FIG. 115]. It then spreads out horizontally into the soft palate to meet the opposite tendon in the midline and form the **palatal aponeurosis** [FIG. 154]. This aponeurosis is attached anteriorly to the palatine crests on the inferior surface of the hard palate, and is thick and rigid anteriorly, but rapidly thins posteriorly so that it cannot be identified in the posterior third of the soft palate. The uvular muscles lie on the superior surface of the aponeurosis, and run side by side in the midline from the posterior nasal spine of the palatine bones to the mucous membrane of the uvula. **Action**: tensor palati makes the anterior part of the soft palate rigid. The **musculus uvulae** shortens and tenses the uvula, helping to prevent the soft palate from being everted into the nasal part of the pharynx when it is in contact with the posterior pharyngeal wall, *e.g.*, in coughing.

Levator Palati (Levator Veli Palatini). This muscle arises from the medial side of the auditory tube and the adjacent part of the petrous temporal bone. It descends behind the auditory tube inside the free upper border of the superior constrictor muscle, and curves medially to join the opposite muscle and be partially attached to the superior surface of the palatal aponeurosis. **Action**: the two muscles raise the palate symmetrically [FIG. 115].

Palatoglossus. This is a small counterpart of the levator palati on the inferior surface of the palate. It is attached to the inferior surface of the palatal aponeurosis and meets the opposite muscle in the midline. Thence it converges on the palatoglossal arch, and runs through it to mingle with the muscles of the posterolateral part of the tongue. **Action**: the two muscles, acting together, draw the soft palate inferiorly on to the posterior part of the dorsum of the tongue.

Palatopharyngeus. This muscle arises from the superior surface of the soft palate and the posterior margin of the hard palate. Most of the muscle converges on the palatopharyngeal arch and runs inferiorly in it, on the internal surfaces of the constrictor muscles. Superiorly, it is joined by the salpingopharyngeus, and inferiorly by the stylopharyngeus at the interval between the middle and superior constrictors. It is inserted into the posterior border of the lamina of the thyroid cartilage and fans out into the posterior pharyngeal wall. A few of its fibres, which arise from the hard palate, pass horizontally backwards with the fibres of the superior constrictor and so surround the pharyngeal isthmus.

Action: the main mass of the muscle depresses the palate on to the posterior part of the dorsum of the tongue, and prevents the soft palate from being forced into the nasal part of the pharynx when blowing through the mouth against resistance. The *horizontal fibres*, with those of the superior constrictor, narrow the pharyngeal isthmus and raise a ridge in its wall against which the soft palate is elevated by the levator palati to separate the oral and nasal parts of the pharynx.

Salpingopharyngeus. This slender muscle arises by one or two slips from the inferior border of the cartilage of the auditory tube at its pharyngeal end. It descends in the salpingopharyngeal fold to join palatopharyngeus.

Vessels and Nerves of Soft Palate. The **ascending palatine branch of the facial artery** ascends on the lateral pharyngeal wall to the superior border of the superior constrictor, and hooking over this, descends with the levator to the palate. This main arterial supply is supplemented by the lesser palatine branches of the greater palatine artery.

The **lesser palatine** and **glossopharyngeal nerves** supply the mucous membrane. The tensor palati is supplied by the **mandibular nerve** through the otic ganglion; all the other muscles are supplied from the vago-accessory complex by fibres that reach the **pharyngeal plexus** through the pharyngeal branch of the vagus.

DISSECTION. With careful removal of the mucous membrane and gentle blunt dissection, the main muscles of the soft palate can be shown satisfactorily. Begin by removing the mucous membrane from both surfaces, from the palatoglossal and palatopharyngeal arches, and from the salpingopharyngeal fold, to expose the muscles within them [FIG. 116].

To display the levator and tensor palati, remove the mucous membrane, the submucosa, and pharyngobasilar fascia from the lateral wall of the pharynx anterior and posterior to the opening of the auditory tube. Identify the levator posterior to the auditory tube, and follow it into the palate. The ascending palatine artery may be seen descending beside the levator to the palate. Trace the tensor inferiorly lateral to the medial pterygoid lamina [FIG. 115].

Identify the superior constrictor lateral to the levator palati. Clean its superior border and remove the mucous membrane from the parts of its medial surface which are not covered by the palatal muscles. Dissect out the palatine tonsil and uncover the anterior part of the superior constrictor muscle. This part is difficult to define as some of its fibres sweep inferiorly into the tongue, and it is often partly covered by thin sheets of muscle fibres passing inferiorly from the palate lateral to the tonsil. Anterior to palatoglossus, strip the thick glandular mucous membrane from the inferior surface of the most anterior part of the soft palate, and uncover the pterygoid hamulus with the tendon of tensor palati hooking round it to form the palatal aponeurosis. Identify the pterygomandibular raphe passing from the hamulus to the mandible, and follow the superior constrictor anteriorly to the raphe.

Anterior to the pterygomandibular raphe, strip the mucous membrane from the internal surface of the buccinator and identify its attachments, and the opening of the parotid duct [FIG. 115].

Identify the greater horn of the hyoid bone, and remove the mucous membrane from the medial surface of the middle constrictor, leaving the palatopharyngeus in position on its medial aspect. Follow **palatopharyngeus** to the posterior border of the lamina of the thyroid cartilage. Anterior to palatopharyngeus, identify the **stylopharyngeus** entering the pharynx between the middle and superior constrictor muscles to spread anteroposteriorly in such a manner that its posterior fibres are inserted with those of the palatopharyngeus, while the anterior fibres pass to the lateral aspect of the epiglottis. The intermediate fibres form a thin layer medial to the superior part of the thyrohyoid membrane, and the internal laryngeal nerve may be found entering the pharynx through that membrane below the fibres of the muscle [FIG. 121]. Find the **glossopharyngeal** nerve anterolateral to the stylopharyngeus where it enters the pharynx, and trace the nerve to the tongue.

Finally remove the mucous membrane from the medial surface of the inferior constrictor, the upper part of the oesophagus, and the piriform recess. In the recess identify the medial surface of the thyroid cartilage and the thyrohyoid membrane with the superior laryngeal vessels and the internal branch of the superior laryngeal nerve piercing it.

Swallowing

Now that the walls of the pharynx have been seen from the medial aspect, it is possible to visualize the mechanism of swallowing.

In the first phase, the mouth is closed, the tip of the *tongue is raised* against the hard palate anterior to the bolus of food or fluid, and this is squeezed posteriorly by pressing progressively more posterior parts of the tongue against the palate (intrinsic muscles of the tongue, mylohyoid and styloglossus). As this movement passes backwards, the elevation of the posterior part of the tongue against the tensed anterior part of the soft palate (tensor palati) is achieved by raising the hyoid bone (digastric, stylohyoid). Geniohyoid carries the hyoid bone anteriorly, and this increases the anteroposterior diameter of the oral part of the pharynx to receive the bolus, the middle and inferior constrictor muscles being relaxed. At this stage the superior constrictor muscle and horizontal fibres of palatopharyngeus contract to draw the upper part of the posterior pharyngeal wall against the raised posterior part of the soft palate (levator palati). This effectively shuts off the nasal part of the pharynx from the oral part and prevents passage of food or fluid into the nasal part of the pharynx.

The second phase is very rapid. There is a

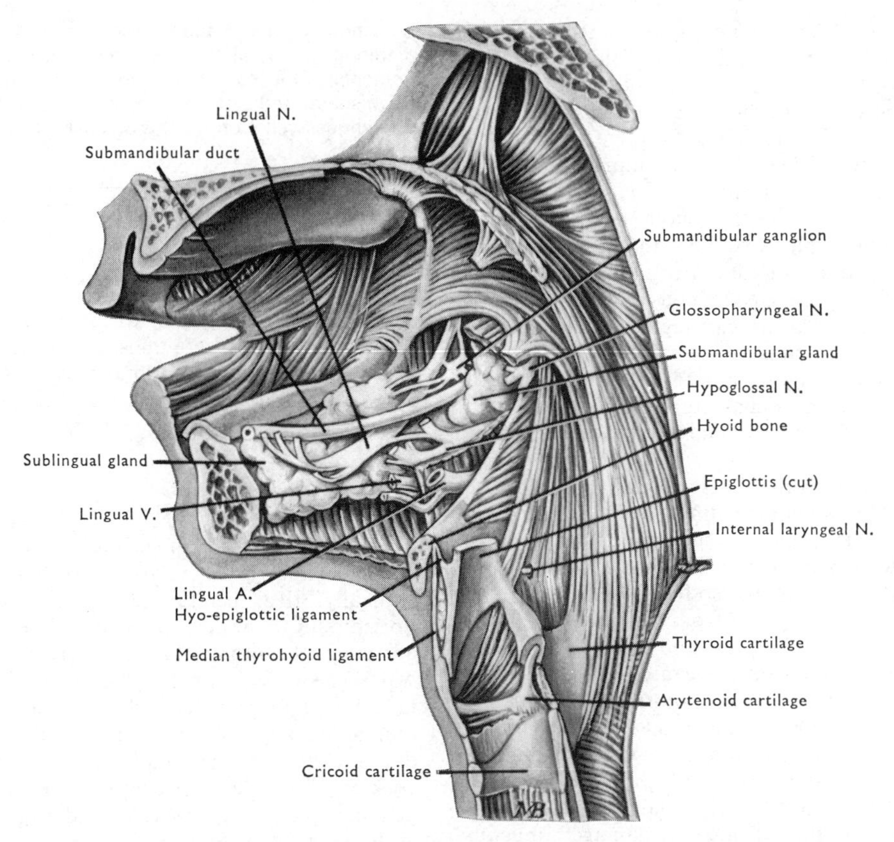

FIG. 121 Dissection of the constrictors of the pharynx and the associated structures which lie adjacent to the mucous membrane lining the mouth, pharynx, and larynx, from the medial side. The tongue has been removed to expose the structures which lie between it and the mylohyoid muscle. See also FIG. 115.

considerable *elevation of the larynx* and of the inferior part of the pharynx, which is attached to it, up to the raised hyoid bone (thyrohyoid, stylopharyngeus, and palatopharyngeus muscles). This brings the thyroid cartilage within the concavity of the hyoid bone and approximates the arytenoid cartilages to the epiglottis, thus *closing the laryngeal orifice*. Contraction of the aryepiglottic muscles [FIG. 138] also helps to draw the epiglottis down on the arytenoid cartilages, but an important factor is the elevation of the base of the **epiglottis** (1) with the thyroid cartilage (to which it is attached) and (2) by the contraction of stylopharyngeus. This pushes the apex of the epiglottis against the bulging posterior surface of the tongue and tips it backwards over the closed laryngeal orifice. The bolus slips from the back of the tongue on to the lingual surface of the epiglottis (which now faces superiorly) and is thrown posteriorly by the elastic epiglottis to be caught and carried inferiorly by the contracting middle and inferior constrictor muscles, aided by the rapid *downward displacement of the larynx and pharynx* (infrahyoid muscles) which follows immediately and reopens the laryngeal orifice.

The importance of the elevation of the hyoid bone and of digastric in this movement is shown by the difficulty of swallowing with the mouth open when digastric cannot act effectively on the hyoid bone.

AUDITORY TUBE

This tube connects the nasal part of the pharynx to the middle ear cavity, and beyond that through the mastoid antrum to the air cells in the mastoid process. The tube is approximately 3·5 cm long. The bony posterolateral 1·5 cm lies between the tympanic and petrous parts of the temporal bone and opens into the middle ear cavity. The cartilaginous anteromedial part lies in the groove between the petrous part of the temporal bone and the posterior border of the greater wing of the sphenoid. Identify the groove and the bony part of the tube on a dried skull [FIG. 12].

Ascertain the direction of the cartilaginous part of the tube by passing a probe into its pharyngeal orifice. At first it runs superiorly and then posterolaterally, and passes for a considerable part of its extent between the tensor and the levator palati muscles [FIGS. 152, 154].

Note that the levator palati forms a rounded prominence inferior to the opening of the auditory tube. Remove the mucous membrane from the mouth of the tube, and note that the superior and medial walls are formed by a folded plate of **cartilage**; the inferolateral part of the tube being completed by dense fibrous tissue joining the edges of the cartilaginous plate. The tubal ridge is formed by the base of the cartilage plate. The lining of the tube is respiratory, pseudostratified ciliated columnar epithelium, which is continuous with the same epithelium in the pharynx and with the epithelium in the middle ear. In the tube it contains some goblet cells, and there are mixed mucous and serous glands in the submucosa of the cartilaginous part near the pharynx. The lumen is narrowest (**isthmus**) where the cartilaginous and bony parts meet, but it gradually increases in diameter from the isthmus to the pharyngeal orifice, which is the widest part of the tube.

The function of this tube is to equalize the pressure in the middle ear with the atmospheric pressure, and so allow free movement of the tympanic membrane which separates the external acoustic meatus from the middle ear cavity. It also forms a route through which infections may pass from the nasal part of the pharynx to the middle ear, and is readily blocked even by mild infections, because the walls of its cartilaginous part lie in apposition. When the auditory tube is blocked, the residual air in the middle ear is absorbed into the blood vessels of its mucous membrane, causing a fall of pressure in it. If associated with infection of the middle ear, there is an outpouring of fluid from the damaged mucous membrane, which, failing to escape along the tube, raises the pressure in the middle ear. In both cases, the free movement of the tympanic membrane is impeded so that hearing is affected. With raised pressure, the normally concave tympanic membrane [p. 142] bulges into the external acoustic meatus—a feature readily confirmed with an auroscope.

DISSECTION. If the otic ganglion has not yet been seen [p. 103], free the opening of the auditory tube from the medial pterygoid lamina and turn it posteriorly. Separate the cartilaginous part of the tube from the base of the skull and tensor palati. This exposes the tensor tympani—a small slip of muscle arising from the petrous temporal bone superomedial to the tube and passing postero-laterally with it. Detach the tensor palati from the base of the skull and turn it inferiorly. Remove the layer of fascia which is exposed and uncover the mandibular nerve with the otic ganglion on its anteromedial aspect. Immediately posterior to the mandibular nerve lies the middle meningeal artery at the foramen spinosum. Identify the branches of the mandibular nerve as far as possible from this aspect, and note how close is its relation to the pharyngeal wall. Confirm this on the base of a macerated skull.

CAROTID CANAL

The carotid canal lies in the petrous part of the temporal bone. It contains the internal carotid artery, the internal carotid plexus of sympathetic nerve fibres, and a plexus of veins. Its position and course can best be seen by inspecting a macerated skull.

Internal Carotid Artery

The part of the internal carotid artery in the canal is approximately 2 cm long. At first it ascends vertically; then bending anteromedially, it runs horizontally to the apex of the petrous temporal bone and enters the foramen lacerum through its posterior wall. The artery turns upwards in the foramen lacerum, and pierces the endocranium of the middle cranial fossa [p. 48]. In the carotid canal, it lies anteromedial to the middle ear, inferior to the cochlea, the greater petrosal nerve, and the trigeminal ganglion [FIG. 154], and superior to the auditory tube.

Internal Carotid Nerve and Plexus [FIG. 155]

The internal carotid nerve is a large branch which ascends from the superior cervical ganglion, and enters the carotid canal. It divides to form the internal carotid plexus around the internal carotid artery, and secondary plexuses extend from it around the branches of the artery.

Branches. The plexus consists predominantly of postganglionic sympathetic fibres which are distributed to the cerebral vessels, to the middle ear (caroticotympanic nerves), to the nose, palate, air sinuses and pharynx through the branches of the pterygopalatine ganglion (to which it sends the deep petrosal nerve), and to the contents of the orbit, the forehead, and anterior scalp through branches which it gives to the 3rd, 4th, 6th, and ophthalmic branch of the 5th cranial nerves, and to the ophthalmic artery.

THE CAVITY OF THE NOSE

SEPTUM OF NOSE

The septum of the nose divides the nasal cavity into two narrow cavities. It is seldom exactly in the median plane, but bulges to one or other side (more frequently to the right). Immediately above the nostril, the septum is slightly concave where it forms the medial wall of the vestibule of the nose, the skin of which carries a number of stiff hairs or vibrissae. The remainder of the septum is covered with mucous membrane the epithelium of which is pseudostratified columnar ciliated epithelium. It is tightly adherent to the underlying periosteum and perichondrium (**mucoperiosteum** and **mucoperichondrium**). The lower, larger area is known as the

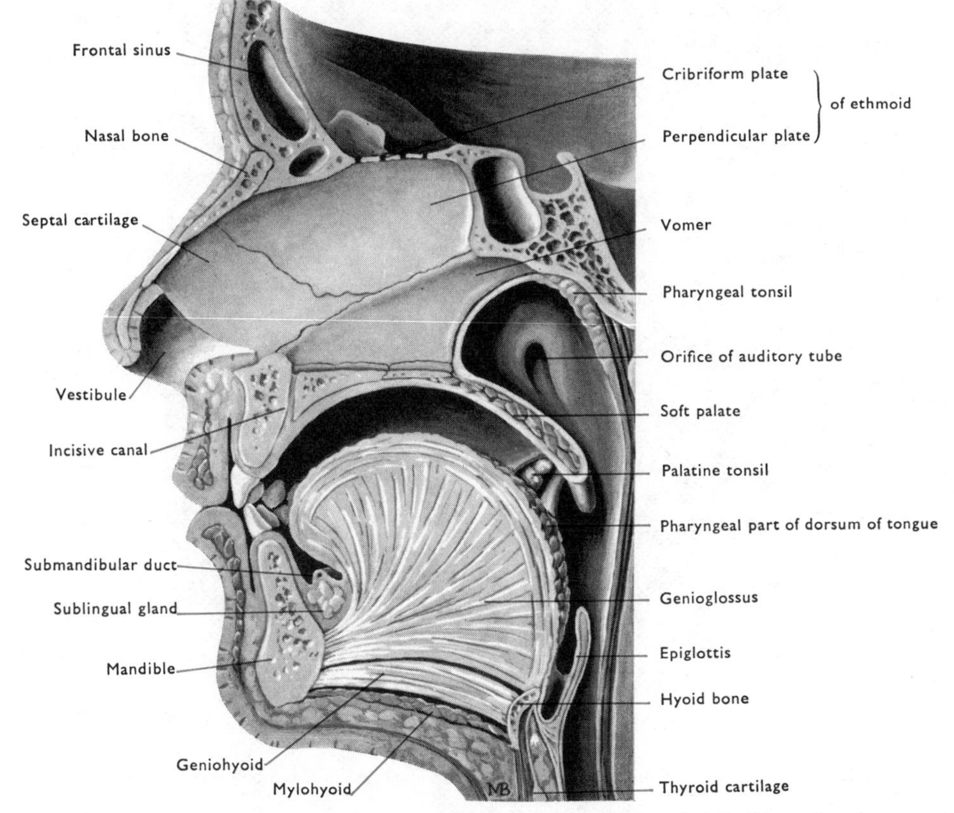

FIG. 122 Sagittal section through the nose, mouth, and pharynx, a little to the left of the median plane.

respiratory region, while the upper third is called the **olfactory region** because its epithelium contains the olfactory nerve cells. The respiratory mucous membrane is thick, spongy, and highly vascular. It contains numerous mucous glands and is capable of swelling to a considerable thickness when the vascular spaces in it are filled with blood. It also contains many arteriovenous anastomoses which increases the flow of blood through it and warm the air passing over it. The olfactory mucous membrane is more delicate, and is yellowish in the fresh state.

Structure. Strip the mucous membrane off the septum, and expose: (1) the vomer; (2) the perpendicular plate of the ethmoid; (3) the septal cartilage; (4) small parts of the maxillary, palatine, nasal, and sphenoid bones. The relative positions of these parts are shown in FIGURE 122.

Note that the anterior angle of the septal cartilage is blunt and rounded, and does not reach the point of the nose which is formed by the greater alar cartilages.

DISSECTION. Remove the septum piecemeal from the mucous membrane on its opposite surface, taking care not to damage the structures in that mucous membrane.

Nerves of Septum (nerves of smell, see p. 126). The **nasopalatine nerve** is a long, slender nerve on the deep surface of the mucous membrane of the septum. It springs from the pterygopalatine ganglion, and enters the nasal cavity through the sphenopalatine foramen with the sphenopalatine branch of the maxillary artery. It runs medially across the roof of the nasal cavity, and then antero-inferiorly, to the floor of the nose, in a groove on the surface of the vomer. It then runs through the incisive canal and median incisive foramen with its fellow and supplies the mucous membrane in the anterior part of the hard palate.

The medial **posterosuperior nasal branches** of the pterygopalatine ganglion, together with twigs from the nerve of the pterygoid canal, supply the posterosuperior parts of the septum, but are too small to be dissected easily.

The medial nasal branches of the **anterior ethmoidal nerve** run on the anterosuperior part of the nasal septum as far as the vestibule.

Arteries of Septum. These are: (1) The sphenopalatine artery, a branch of the maxillary artery; (2) ethmoidal branches of the ophthalmic artery; (3) branches of the superior labial arteries [FIG. 72].

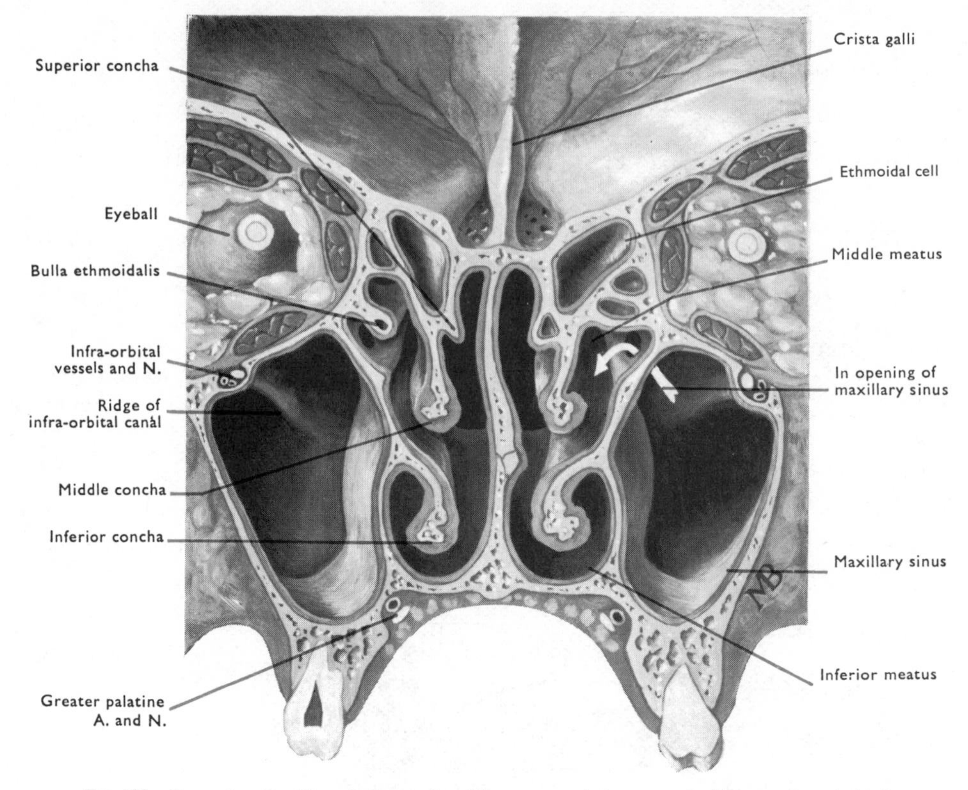

FIG. 123 Coronal section through the nasal cavities, paranasal sinuses, and orbits, seen from behind.

DISSECTION. Disengage the nerves from the mucous membrane of the septum, so that they may later be followed to their origins. Then remove the mucous membrane and open the nasal cavity.

CAVITY OF THE NOSE

Each cavity is approximately 5 cm in height, and 5–7 cm in length. It is narrow transversely, measuring approximately 1·5 cm at the floor, and only 1–2 mm at the roof. The width is further reduced by the conchae, which project into the cavity from the lateral wall.

The oval anterior apertures or nostrils (**nares**) open on the inferior surface of the external nose. The posterior apertures or **choanae** open into the nasal part of the pharynx and face postero-inferiorly.

Roof

The roof, 7–8 cm long, is curved. The middle part is formed by the cribriform plate of the ethmoid, and is nearly horizontal. The anterior and posterior parts are sloping. The anterior is formed by the nasal part of the frontal bone, the nasal bone, and by the junction of lateral and septal cartilages. The posterior part consists of the anterior and inferior surfaces of the body of the sphenoid and the bones in contact with these surfaces.

Floor

The floor is about 5 cm long and 1–1·5 cm in width. It is formed by the palatine process of the maxilla and the horizontal process of the palatine bone. It is concave transversely, and slightly higher anteriorly than posteriorly.

Lateral Wall

This wall is very uneven because of the three projecting **conchae**. The different bones that form the lateral wall of the nose should be studied in a median section of a macerated skull, to which the dissector should constantly refer during the dissection. Adjoining the lateral wall are the air sinuses (air-filled spaces in the bones which communicate with the nasal cavity). The **ethmoidal sinus**, consisting of anterior, middle, and posterior **cells**, lies between the upper part of the nasal cavity and the orbit, while inferior to this is the **maxillary sinus** which lies below the orbit [FIGS. 79, 123].

There are three areas in the lateral wall of the nose.

1. The **vestibule of the nose** [FIGS. 122, 125] lies immediately above the nostril. It is lined with skin from which stout hairs or vibrissae project forming a coarse filter.

2. The **atrium of the middle meatus** [FIG.

124] is above and slightly posterior to the vestibule, and immediately anterior to the middle meatus. Its lateral wall is concave, except close to the nasal bone where a feeble elevation, the agger nasi, represents an additional concha which is present in some mammals.

3. **Nasal conchae and meatuses** [FIGS. 124, 125]. The conchae are three bony plates which project from the lateral wall of the nose and curve inferiorly. The meatuses are the spaces inferior to the conchae. The upper two conchae are processes of the ethmoid, the inferior concha is an independent bone. All are covered with a thick, highly vascular, mucoperiosteum.

The **superior concha** lies in the posterosuperior part of the cavity, and is very short. Its free border begins a little inferior to the middle of the cribriform plate, and passes postero-inferiorly to end immediately anterior to the lower part of the body of the sphenoid. The space posterosuperior to it is known as the **spheno-ethmoidal recess**, and the sphenoidal sinus opens into its posterior part [FIG. 124].

The **middle concha** is much larger and extends from the atrium to the level of the choanae.

The **inferior concha** is slightly longer than the middle concha, and lies about midway between the middle concha and the floor of the nose.

The **superior meatus** is a narrow fissure between the superior and middle conchae. The posterior ethmoidal cells open into its anterosuperior part by one or more orifices. These can be exposed by forcing the margin of the superior concha upwards.

The **middle meatus** is a much longer and deeper passage than the superior meatus. To expose it, tilt the middle concha forcibly upwards. The anterosuperior part of the middle meatus leads into a funnel-shaped opening, the **infundibulum**, which leads to the frontal sinus [FIG. 125].

On the lateral wall of the middle meatus is a deep, curved groove which begins at, or slightly behind, the infundibulum and runs postero-inferiorly. This is the **hiatus semilunaris** [FIG. 125] and the anterior ethmoidal and maxillary sinuses open into it. The upper margin of the hiatus semilunaris is formed by a prominent bulge, the **bulla ethmoidalis**, on which is the opening of the middle ethmoidal cells.

The **opening of the maxillary sinus** [FIG. 123] lies in the posterior part of the hiatus semilunaris, and enters the upper medial wall of the sinus. Occasionally there is a second opening and this leads into the middle meatus superior to the middle of the attachment of the inferior concha. The position of the opening of the frontal sinus favours the flow of material from it to the opening of the maxillary sinus. Consequently infection tends to spread from the frontal to the maxillary sinus.

The **inferior meatus** is the horizontal passage between the inferior concha and the floor of the nose. The nasolacrimal duct opens into the anterior part of the inferior meatus close to the attached border of the inferior concha.

DISSECTION. Remove the anterior part of the inferior concha with scissors, and expose the opening of the nasolacrimal duct. Pass a probe upwards along the duct to confirm its continuity with the lacrimal sac. Separate the medial margin of the aperture from the overlying

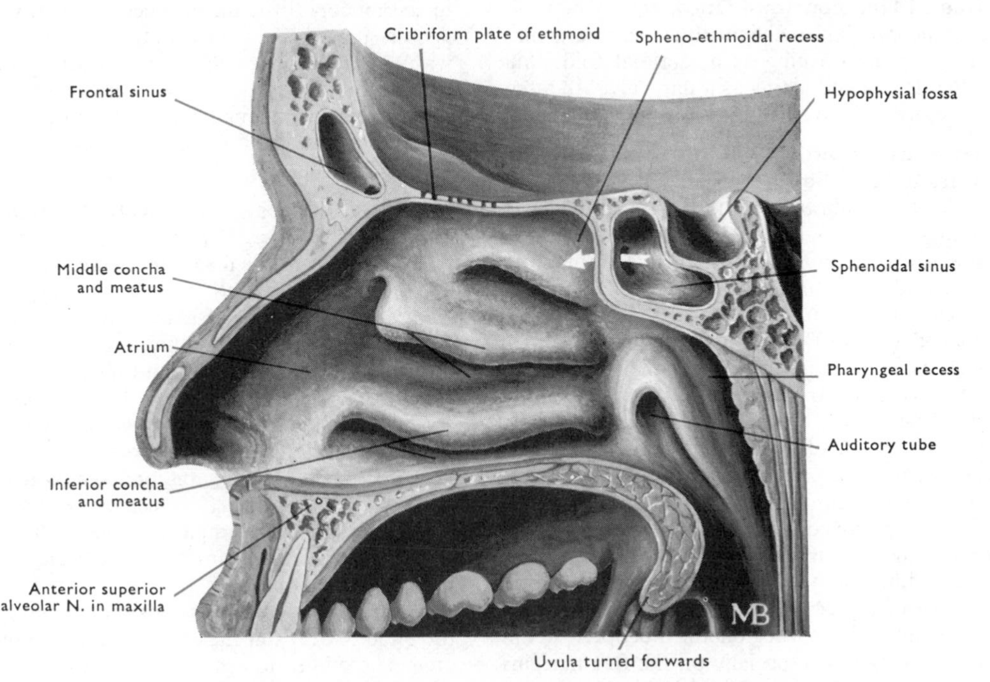

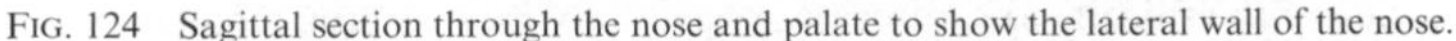
FIG. 124 Sagittal section through the nose and palate to show the lateral wall of the nose.

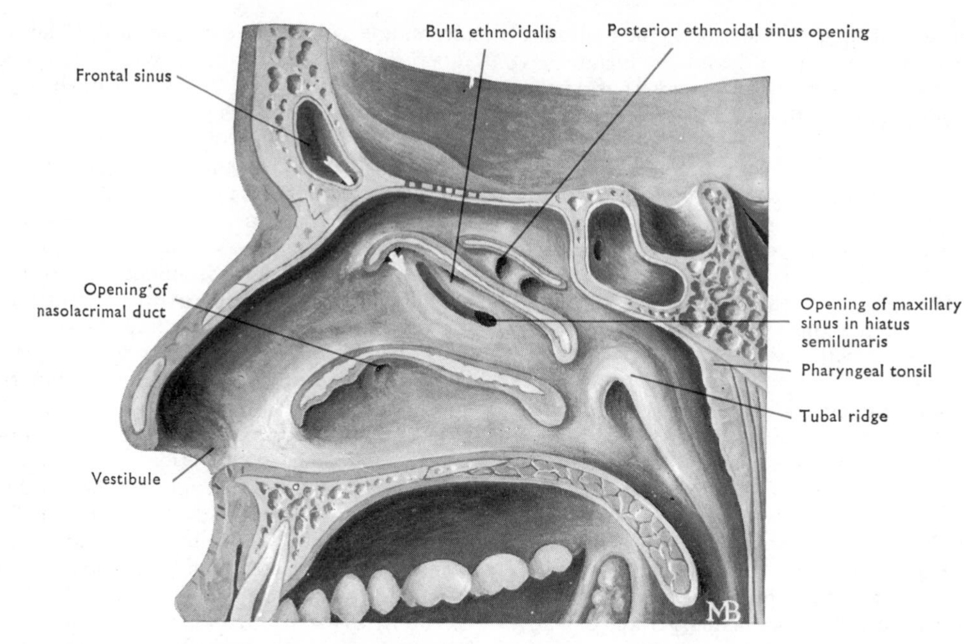

FIG. 125 Sagittal section through the nose and palate. The conchae have been cut away to expose the meatuses and the openings into them. The opening of the infundibulum is unusual.

bone, and pass a probe between the duct and the bone. Break away the thin plate of bone which separates the duct from the nose, and expose the length of the duct and the sac at its superior end by continuing the bony removal to the level of the eye [FIG. 126].

Orifice of Nasolacrimal Duct. It may be wide, patent, and circular, or the mucous membrane may extend over the opening as a lacrimal fold, thus reducing its size and acting as a flap valve. In a few cases the orifice is so small that it is difficult to find.

Mucosa of Lateral Wall of Nasal Cavity. Apart from the vestibule, the lateral wall is covered with mucous membrane which is tightly adherent to the underlying periosteum and forms a **muco-periosteum**—a lining which is found throughout the remainder of the nasal cavity. This muco-periosteum is continuous: (1) through the naso-lacrimal duct, with the ocular conjunctiva; (2) through the various apertures, with the lining of the air sinuses in the frontal, ethmoid, maxilla, and sphenoid bones; (3) through the choanae with the mucous membrane of the nasal part of the pharynx.

The mucoperiosteum on the lateral wall is divisible into the upper, yellowish **olfactory region** in the area of the superior concha, and the remainder which comprises the **respiratory region**. These regions cannot be differentiated by the naked eye because of the absence of any sharp line of demarcation between them. In the respiratory region, the mucoperiosteum is thick and spongy, especially on the free margins and posterior extremities of the middle and inferior conchae. This is due to the presence of rich venous plexuses in the mucoperiosteum, which may even have the character of cavernous tissue, more particularly on the inferior concha. Here the mucoperiosteum may be so swollen by distended venous channels that it impinges on the nasal septum and effectively reduces the cross sectional area of the nasal cavity. This, combined with a rapid blood flow in the mucoperiosteum produced by many arterio-venous anastomoses, ensures that the inspired air is warmed and moistened, and that dust particles are trapped on the mucus which covers the surface from the numerous mucous glands. The mucus and dust particles are wafted towards the anterior aperture of the nose by the cilia.

Nerves and Vessels on Lateral Wall of Nasal Cavity. The nerves of common sensation all arise from branches of the maxillary nerve, except for the **anterior ethmoidal**. This branch of the naso-ciliary nerve in the orbit reaches the anterosuperior part through the anterior ethmoidal foramen and the cribriform plate of the ethmoid. Nerve fibres of the **maxillary nerve** reach the walls of the nasal cavity through branches of the pterygopalatine ganglion and the anterior superior alveolar nerve. All these nerves also convey postganglionic sympathetic fibres from the carotid plexus (via deep petrosal nerve) and postganglionic parasympathetic nerve fibres from the pterygopalatine ganglion to the glands in the mucoperiosteum.

The **olfactory nerves** consist of the processes of the olfactory cells in the epithelium of the olfactory area. These fine, non-myelinated nerve fibres run in shallow grooves and small canals in the bone deep to the mucous membrane, and cannot be dissected. They form twelve to twenty olfactory nerves which pass through the cribriform plate of the ethmoid, and pierce the meninges to enter the olfactory bulb.

DISSECTION. Trace the **nasopalatine nerve** from the nasal septum across the roof of the nasal cavity to the sphenopalatine foramen in the lateral wall. Careful dissection in this region may display one or more of the nasal branches of the pterygopalatine ganglion, and will also display the sphenopalatine artery running with the nasopalatine nerve.

Carefully reflect the mucous membrane on the medial pterygoid lamina anteriorly. Attempt to find the nasal branches of the greater palatine nerve as they pierce the perpendicular plate of the palatine bone.

The **posterior nasal branches of the pterygopalatine ganglion** are minute twigs which pass through the sphenopalatine foramen. They supply the mucous membrane on the posterior part of the septum, the superior and middle conchae, some ethmoidal cells, and the lateral wall of the nasal part of the pharynx [FIG. 126].

Two **nasal branches of the greater palatine nerve** pierce the perpendicular plate of the palatine bone. They supply the mucous membrane on the posterior parts of the conchae and meatuses.

The **anterior ethmoidal nerve** [FIG. 126] descends in a groove on the deep surface of the nasal bone. Its branches supply the anterosuperior parts of the septum, roof, and lateral wall.

The **sphenopalatine branch of the maxillary artery** is the main supply to the mucous membrane. It enters through the sphenopalatine foramen and is distributed with the various nerves. The anterior and posterior **ethmoidal arteries** supplement this supply in the anterosuperior region; the anterior reaching as far inferiorly as the anterior end of the inferior concha.

PTERYGOPALATINE GANGLION, MAXILLARY ARTERY, AND MAXILLARY NERVE

The pterygopalatine ganglion lies in the pterygopalatine fossa, lateral to the sphenopalatine foramen and the perpendicular plate of the palatine bone.

DISSECTION. The mucoperiosteum has already been stripped from the perpendicular plate of the palatine bone, and the greater palatine nerve can be seen shining through this very thin plate of bone, as it descends on the lateral side of the bone to reach the palate, with the **descending palatine artery** from the maxillary artery.

Break through the perpendicular plate of the palatine bone and expose part of the greater palatine nerve; then open up the whole length of the canal by levering off the remainder of the lamina lying medial to the nerve. Superiorly, the **greater palatine nerve** joins the pterygopalatine ganglion at the level of the sphenopalatine foramen, through which the nasopalatine nerve passes to the ganglion. Inferiorly, where the canal reaches the hard palate, cut out a narrow, transverse strip of the hard palate to open into the palatine foramen, through which the greater palatine nerve reaches the hard palate. Remove the fibrous sheath covering the greater palatine nerve and expose the **lesser palatine nerves** which run with it in the upper part of their course, but leave inferiorly to pass through separate bony canals. As far as possible open these canals, and follow the lesser palatine nerves to their termination in the soft palate. One of the lesser palatine nerves is more lateral in position, and hence difficult to follow from the medial side: it may be absent.

Turn to the inferior surface of the palate, and follow the greater palatine nerve and artery in the hard palate.

Greater Palatine Nerve

This is the largest branch of the pterygopalatine ganglion. It descends vertically through the greater palatine canal and foramen with the **descending palatine artery** from the maxillary artery, and enters the inferior surface of the hard palate at its posterolateral corner. It runs forwards in a groove on the inferior surface of the bony palate close to its lateral margin. At the incisive fossa, it communicates with the terminal branches of the nasopalatine nerve. It supplies the gum and the mucous membrane of the hard palate, including the palatine mucous glands which indent the inferior surface of the bone.

Branches. (1) It sends two **posterior nasal** branches through the perpendicular plate of the palatine bone to the mucous membrane of the nose (above). (2) The **lesser palatine nerves** descend through the lesser palatine canals. The more medial of these emerges immediately posterior to the palatine crest, and enters the soft palate to supply its mucous membrane and glands. The more lateral nerve, when present, supplies the mucous membrane of the soft palate near the palatine tonsil.

DISSECTION. Remove the three nasal conchae, and, beginning just posterior to the infundibulum, strip away the thin medial walls of the ethmoidal air cells, noting their continuity with the nasal mucous membrane through the apertures already described. Remove the mucous membrane lining the individual cells and the bony walls between them, and thus expose the medial surface of the orbital lamina of the ethmoid.

Break away the medial wall of the maxillary sinus from the nasolacrimal duct, anteriorly, to the greater palatine canal, posteriorly, and examine the interior of the sinus [FIG. 103]. Remove the orbital process of the palatine bone and as much of the posterior part of the roof of the maxillary sinus as may be necessary to expose the maxillary nerve in the pterygopalatine fossa p. 104]. This also exposes the anterior surface of the pterygopalatine ganglion and the terminal parts of the

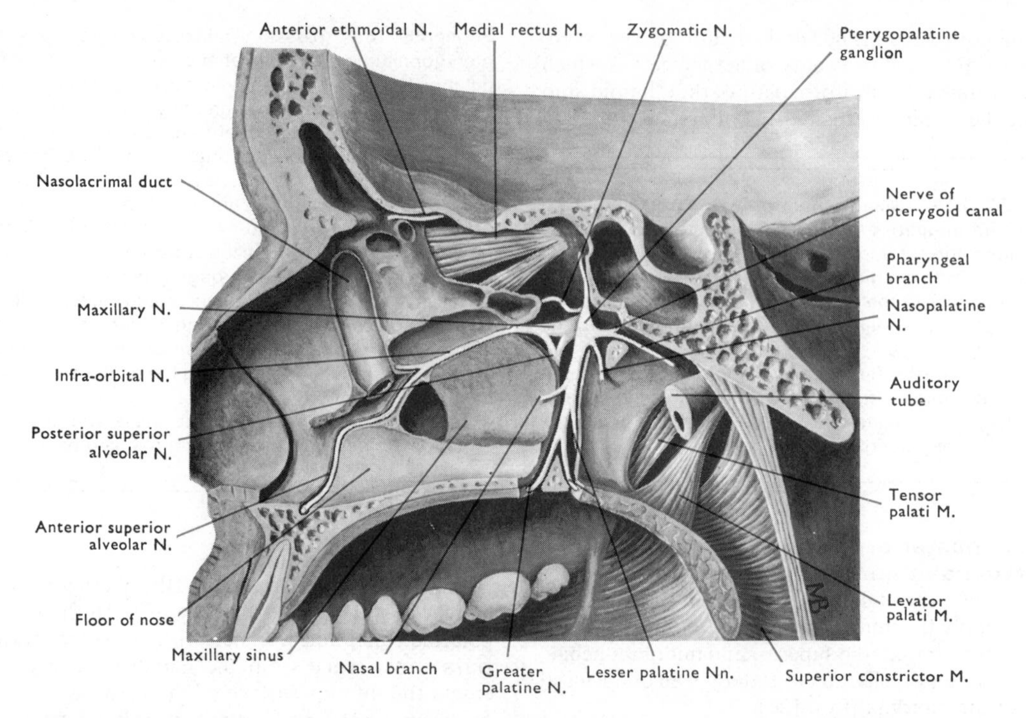

FIG. 126 The same specimen as FIG. 125. The mucous membrane and a large part of the bone of the lateral wall of the nose has been removed to expose the maxillary sinus and break through the ethmoidal sinuses into the orbit (cf. FIG. 123). The mucous membrane has also been stripped from the pharynx and lateral wall of the mouth, and the maxillary, infra-orbital, anterior superior alveolar, and palatine nerves exposed by removal of the bony wall of their canals. The pterygopalatine fossa and ganglion are also exposed.

maxillary artery. Chip away the sphenoid bone medial to the ganglion, taking care to preserve the pharyngeal branch of the ganglion and the nerve of the pterygoid canal which enters the posterior surface of the ganglion.

Follow the infra-orbital nerve [p. 104] anteriorly by chipping away the floor of the infra-orbital groove and canal. Find the **anterior superior alveolar branch** of the nerve, and trace this through its sinuous, bony canal [FIG. 126], below the opening of the nasolacrimal duct, into the anterior part of the floor of the nose, superior to the incisor teeth. Lift the nerve gently out of its canal, and note its branches to the upper teeth, gums, and mucous membrane of the maxillary sinus. Where necessary, remove the accompanying branches of the maxillary artery in order to get a clear view of the nerves.

Pterygopalatine Ganglion

This ganglion is a small, triangular body which lies in the superior part of the pterygopalatine fossa near the sphenopalatine foramen. It is surrounded by the terminal branches of the maxillary artery.

Roots. It is suspended from the inferior aspect of the maxillary nerve by two stout **ganglionic branches**. The sensory trigeminal fibres in these branches pass directly through the ganglion into its branches. Sympathetic and parasympathetic nerve fibres enter the ganglion in the **nerve of the pterygoid canal** which is formed by the greater petrosal (facial nerve, preganglionic parasympathetic) and deep petrosal (internal carotid plexus) nerves [FIG. 155]. Only the preganglionic parasympathetic fibres synapse in the ganglion, and its **postganglionic parasympathetic fibres** supply the lacrimal gland and glands in the nose, palate, and pharynx through the branches of the ganglion which also transmit sensory (maxillary nerve) and sympathetic fibres.

Branches. (1) Palatine [p. 127]. (2) The **orbital branches** are two to three filaments which enter the orbit through the inferior orbital fissure to supply the orbital periosteum (sensory) and the lacrimal gland (secretory). (3) The **nasopalatine** and **posterior nasal nerves** pass through the sphenopalatine foramen to the mucous membrane of the nose [p. 123]. (4) The **pharyngeal branch** passes posteriorly through the palatovaginal canal to the mucous membrane of the sphenoidal air sinus and the roof of the pharynx.

Termination of Maxillary Artery

The third part of the maxillary artery enters the pterygopalatine fossa through the pterygomaxillary fissure. It breaks up into branches which accompany all the nerves in the fossa (infra-orbital, posterior superior alveolar, greater palatine, nasopalatine, pharyngeal, and nerve of the pterygoid canal) and receive the same names.

Maxillary Sinus [FIGS. 123, 126]

This is the largest of the paranasal air sinuses. It occupies the whole of the body of the maxilla, and has the shape of an irregular three-sided pyramid. The **apex** extends into the zygomatic process of the maxilla, and the **base** is the lower part of the lateral wall of the nose. The sides are the orbital, anterior, and infratemporal surfaces of the maxilla. The sinus lies superior to the molar and premolar teeth. The lowest part of this sinus is opposite the second premolar and first molar teeth, and is approximately 1 cm below the level of the floor of the nose.

Nasal Opening. The sinus opens into the middle meatus of the nasal cavity through an aperture in the superior part of its base. This makes it impossible for fluid in the sinus to escape into the nose until the sinus is nearly filled when the head is in the erect position.

The infra-orbital groove and canal run forwards in the bone of the roof of the sinus, and, as the canal bends inferiorly towards the infra-orbital foramen, it produces a marked ridge in the angle between the orbital and anterior surfaces of the sinus. The posterior superior alveolar nerve and vessels run in the lower part of the infratemporal and anterior walls of the sinus; the anterior superior alveolar nerve and vessels are in the orbital and anterior surfaces. The **mucous membrane** of the sinus is supplied by branches of these nerves and by branches of the greater palatine nerve. It is lined with ciliated columnar epithelium, and the cilia waft the mucus on its surface towards the opening into the nose. In some situations, the bone which separates the nerves from the mucous membrane of the sinus may be absent, and this, combined with the fact that the alveolar nerves supply the teeth and the mucous lining of the sinus, may be responsible for the sensation of toothache which frequently accompanies inflammation in the sinus.

Sphenoidal Air Sinuses [FIGS. 45, 116, 124]

Examine these sinuses on the two halves of the bisected skull [FIG. 124]. As there are two, and the septum between them is not median, it may be necessary to break down the septum to open the sinus which is not exposed by the cut through the skull. Find the openings into the sinuses in their anterior walls and pass a probe into both to determine which has been opened. Open the other.

These sinuses occupy a variable amount of the body of the sphenoid bone. They may extend into the wings and pterygoid processes and even into the basilar part of the occipital bone. Each opens by a small, round hole into the spheno-ethmoidal recess [FIG. 124].

Position. Each sinus is related to the nasal cavity anteriorly, and to the nasal part of the pharynx inferiorly. Posteriorly a thick layer of bone usually separates it from the cranial cavity, the basilar artery, and the pons. Above lie the hypophysis and intercavernous sinuses with the cavernous sinuses further laterally.

THE LARYNX

STRUCTURE AND POSITION

The larynx is the upper, expanded part of the windpipe. It extends from the trachea, at the level of the sixth cervical vertebra, to the pharynx, and is specially modified for the production of the voice.

It lies anterior to and parallel with the laryngeal part of the pharynx, and opens into its anterior wall by a long, almost vertical orifice which stretches from the apex of the epiglottis down to the level of the middle of the thyroid cartilage [FIG. 116]. The margin of the orifice (composed of the mucous membrane covering the free edge of the epiglottis, the aryepiglottic folds, and the arytenoid cartilages [FIG. 119]) projects backwards into the cavity of the pharynx, which extends anteriorly to form a deep gutter on each side of it, the **piriform recess**, medial to the laminae of the thyroid cartilage and thyrohyoid membrane.

The walls of the larynx are supported by the cricoid, thyroid, epiglottic, and arytenoid cartilages [FIG. 129], and the rigidity of the lower part of each aryepiglottic fold is increased by two small nodules, the **corniculate** and **cuneiform cartilages**.

Cricoid Cartilage

Inferiorly, the larynx is surrounded by the cricoid cartilage. It is shaped like a signet ring, horizontal inferiorly, with the narrow part of the ring (**arch**) anteriorly [FIG. 127]. Behind this it deepens progressively [FIG. 132] so that its upper margin passes deep to the thyroid cartilage and forms a vertical **lamina** posteriorly [FIG. 129], between the free posterior margins of the laminae of the thyroid cartilage. The superior surface of the lamina of the cricoid is surmounted by the two arytenoid cartilages, one on each side [FIG. 129] at the inferior margin of the laryngeal orifice. Here the mucous membrane of the posterior wall of the larynx becomes continuous with that of the anterior wall of the pharynx, which covers the posterior surfaces of the arytenoid cartilages, the lamina of the cricoid cartilage, and extends into the oesophagus where that tube is attached to the posterior surface of that lamina.

The arch of the cricoid cartilage is attached anteriorly to the thyroid cartilage by the **crico-thyroid ligament**. On each side the gap between

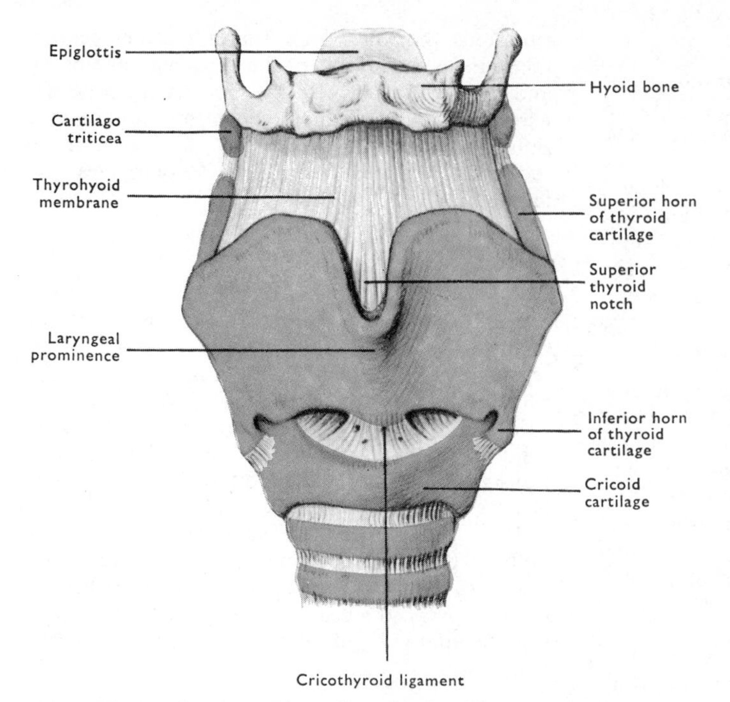

FIG. 127 Anterior aspect of the cartilages (blue) and ligaments of the larynx.

these cartilages is filled by the cricothyroid muscle [FIG. 137], and the inferior horn of the thyroid cartilage descends across it to articulate with the lateral surface of the cricoid cartilage [FIG. 128]. Inferiorly, the cricoid is attached to the first tracheal ring by the cricotracheal ligament.

DISSECTION. Turn the sternothyroid muscle upwards and define its attachment to the thyroid cartilage. Identify the attachments of the inferior constrictor to the thyroid and cricoid and to the fascial arch crossing the cricothyroid muscle. Expose this muscle fully by a horizontal cut through the fibrous arch, reflecting the divided parts of the inferior constrictor. Trace the inferior horn of the thyroid cartilage to its articulation with the cricoid cartilage. Expose the cricothyroid ligament to the anterior border of the cricothyroid muscle. Behind this it passes deep to the muscle as the conus elasticus.

Thyroid Cartilage

This is the largest of the laryngeal cartilages consisting of two roughly quadrilateral laminae of hyaline cartilage. These are fused anteriorly in their inferior two-thirds, but are separated above by the deep **superior thyroid notch** [FIG. 127] where they extend furthest anteriorly to form the laryngeal prominence. The angle at which the laminae meet varies (90–120 degrees). It is acutest in the male, thus making it more prominent anteriorly and deeper anteroposteriorly than in the female.

Each thyroid lamina ends posteriorly in a vertical posterior margin. These margins are far apart, and each is separated from the laryngeal orifice by the corresponding **piriform recess** of the pharynx. Each posterior margin extends superiorly and inferiorly to form the slender horns (**cornua**) of the thyroid cartilage [FIG. 128]. The **superior horns** are attached to the tips of the corresponding greater horns of the hyoid bone by the thyrohyoid ligaments; the **inferior horns** articulate with the cricoid cartilage.

The lateral surfaces of the thyroid cartilage are relatively flat, but where they thicken to form the posterior margins, there is on each a raised, oblique line which extends from the superior to the inferior tubercle [FIG. 128]. The inferior constrictor of the pharynx, the pretracheal fascia (sheath of the thyroid gland) and the sternothyroid and thyrohyoid muscles are attached to the oblique line.

Thyrohyoid Membrane and Ligaments. The ligaments each contain a small cartilage nodule (**triticeal cartilage**) and are the thickened posterior margins of the thyrohyoid membrane. The membrane connects the sinuous superior margin of the thyroid cartilage to the upper margin of the hyoid bone internally. The membrane ascends within the concavity of the hyoid bone, is partly lined by the mucous membrane of the piriform recess, and is pierced by the internal branch of the superior laryngeal nerve and the superior laryngeal vessels. Anteriorly, the membrane is thickened to form the **median thyrohyoid ligament**. This is separated from the posterior surface of the body of the hyoid bone by a bursa which lessens the friction between them when the upper border of the thyroid cartilage is drawn superiorly behind the hyoid bone in swallowing.

DISSECTION. Cut through the thyrohyoid muscle to expose the thyrohyoid membrane and the vessels and nerve piercing it. Define the attachments of the membrane.

Cricothyroid Joint. The inferior horns of the thyroid cartilage each articulate with the postero-inferior part of the lateral surface of the cricoid cartilage by a synovial joint [FIG. 128]. These two joints allow the cricoid cartilage to rock around a horizontal axis passing through both of them. This swings the superior margin of the lamina of the cricoid and the attached arytenoid cartilages towards or away from the anterior part of the thyroid cartilage. This slackens or tightens the elastic vocal ligament [FIG. 132] which unites each arytenoid cartilage to the thyroid cartilage.

Epiglottis

The rigidity of the epiglottis is provided by a thin, curved lamina of elastic fibrocartilage. It forms the upper part of the anterior wall and the superior margin of the laryngeal orifice. It is posterior to the base of the tongue, the hyoid bone, and the median thyrohyoid ligament, and tapers inferiorly to be attached to the posterior surface of the thyroid cartilage, in the midline, by the strong **thyro-epiglottic ligament**. The whole of the posterior surface and the upper part of the anterior surface are covered with mucous membrane which is stratified squamous in the upper posterior and anterior parts, but columnar ciliated elsewhere. Numerous mucous and serous **glands** lie in pits and perforations of the cartilage which is convex anteriorly in its superior part, and posteriorly in the lower part which bulges into the larynx as the **epiglottic tubercle**.

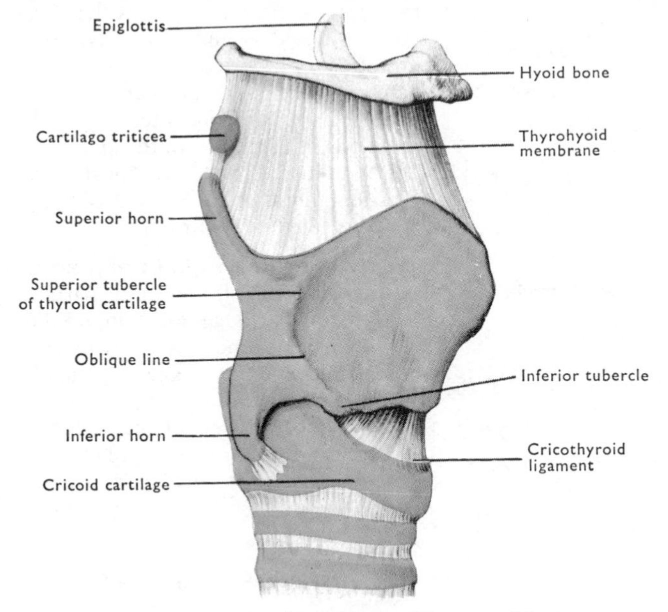

FIG. 128 Profile view of the cartilages (blue) and ligaments of the larynx.

Ligaments. In addition to the thyro-epiglottic ligament, the anterior surface of the epiglottis is attached to the upper surface of the hyoid bone by a loose, fibro-elastic **hyo-epiglottic ligament**. This is separated from the thyro-epiglottic and median thyrohyoid ligaments by fat which is displaced when the thyroid cartilage is drawn up inside the hyoid bone. The epiglottis is attached to the tongue by the median and lateral **glosso-epiglottic folds** of mucous membrane [FIGS. 119, 140] and to the arytenoid cartilages by the **aryepi-glottic folds**.

DISSECTION. On the sectioned surface of the larynx, identify the epiglottis, the thyro-epiglottic and hyo-epiglottic ligaments, and note their relation to the thyroid cartilage, the hyoid bone, and the thyrohyoid ligament [FIG. 121].

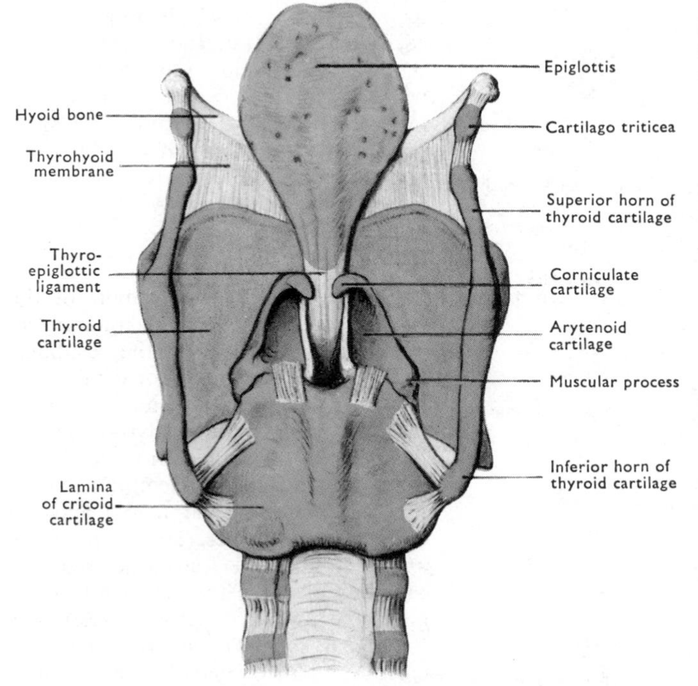

FIG. 129 Posterior aspect of the cartilages (blue) and ligaments of the larynx.

Arytenoid Cartilages

These are a pair of three-sided pyramids [FIG. 129]. The base of each forms a synovial joint with the superior border of the lamina of the cricoid cartilage [FIGS. 129, 132]. The cartilage projects laterally to form the **muscular process** (to which the crico-arytenoid muscles are attached) and anteriorly to form the **vocal process** (to which the vocal ligament is attached). The apex of each cartilage extends upwards, curving posteromedially to support the corniculate cartilage. The transverse arytenoid muscle is attached to the posterior surface of each arytenoid cartilage [FIG. 138] while the anterolateral surfaces have the thyro-arytenoid and vocalis muscles attached to them [FIGS. 130, 139].

Crico-arytenoid Joints. These synovial joints allow the arytenoid cartilages to glide transversely on the lamina of the cricoid cartilage, so that they are able to move closer together or further apart. Rotation of each arytenoid cartilage around its vertical axis swings the vocal process laterally and medially, thus separating or approximating the vocal ligaments. The arytenoid cartilages are prevented from slipping anteriorly by the strong posterior capsule of the joint [FIG. 129], and thus they are carried posteriorly with the lamina of the cricoid cartilage when it is tilted backwards.

Structure of Laryngeal Cartilages. The thyroid, cricoid, and basal parts of the arytenoid cartilages are composed of hyaline cartilage, and tend to ossify even in early adult life; in old age they may be completely transformed into bone. The apex and vocal process of the arytenoid cartilage and the other cartilages are formed of elastic fibrocartilage and do not ossify.

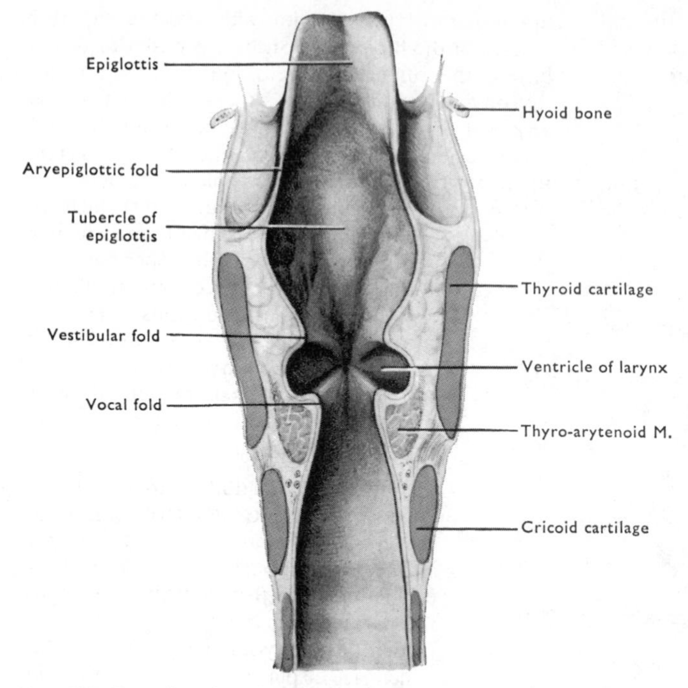

FIG. 130 Coronal section through the larynx to show its compartments. Cartilage blue.

INTERIOR OF LARYNX

The cavity of the larynx is nearly divided into superior and inferior parts by two anteroposterior **vocal folds** of mucous membrane, one of which projects from each lateral wall [FIG. 130]. Above each of these is a subsidiary **vestibular fold** which is separated from the corresponding vocal fold by a narrow, horizontal groove (the **ventricle of the larynx**). The two pairs of folds narrow the middle part of the laryngeal cavity.

Vestibule of Larynx

The superior part, or vestibule of the larynx, extends from the pharyngeal opening of the larynx to the vestibular folds. It has a long **anterior wall** which consists of the mucous membrane covering the epiglottis and thyro-epiglottic ligament, but a short **posterior wall** formed by the mucous membrane covering the apex of the arytenoid and the corniculate cartilages. The lateral walls are the aryepiglottic folds which separate the vestibule from the piriform recesses and slope inwards towards the vestibular folds [FIG. 130]. The aryepiglottic folds enclose in their margins the slender aryepiglottic and thyro-epiglottic

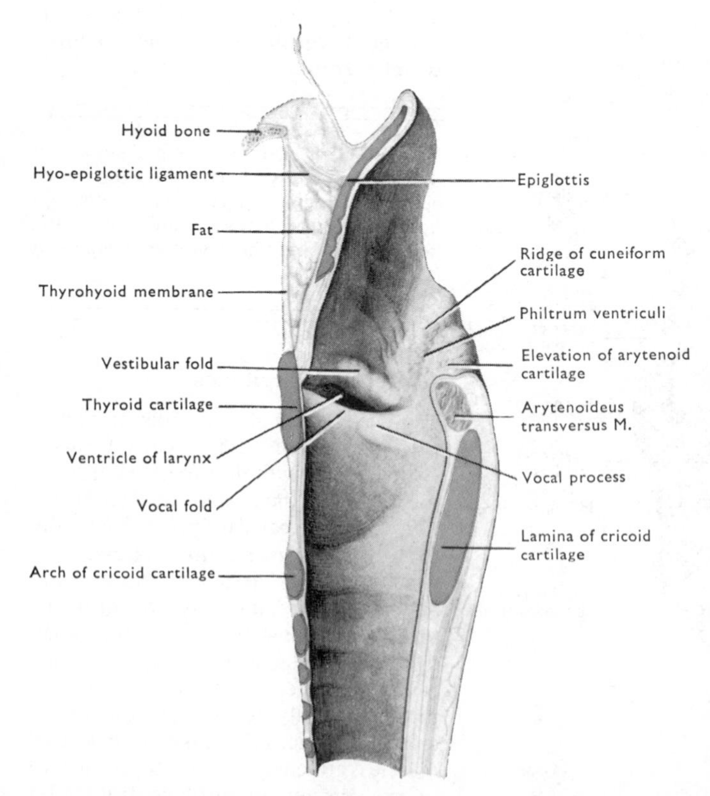

FIG. 131 Median section through the larynx to show the side-wall of its right half. Cartilage blue.

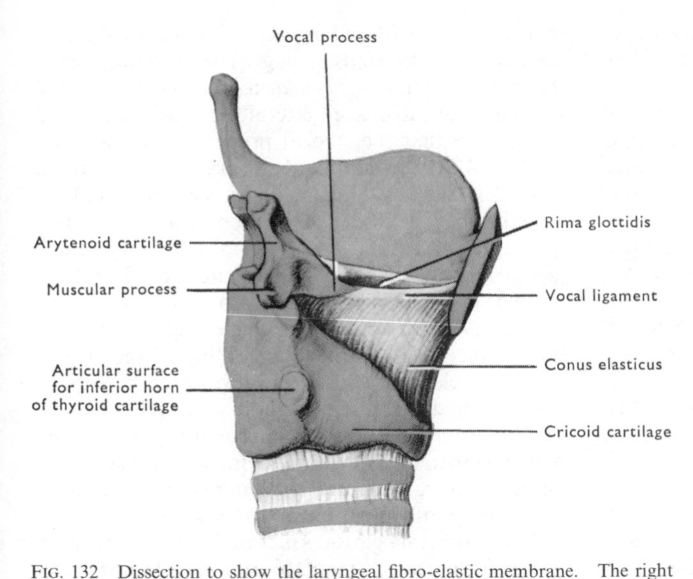

FIG. 132 Dissection to show the laryngeal fibro-elastic membrane. The right lamina of the thyroid cartilage has been removed. Cartilage blue.

muscles, and the corniculate and cuneiform cartilages [FIGS. 119, 139].

Vestibular Folds

These are soft, flaccid folds of mucous membrane which extend between the thyroid and arytenoid cartilages, and contain: (1) numerous mucous glands; (2) a feeble band of fibro-elastic tissue; (3) a few muscle fibres. They lie further apart than the vocal folds, and play little or no part in the production of the voice which is unimpaired by their destruction. The space between the two vestibular folds is the **rima vestibuli**.

Ventricle and Saccule of Larynx

The ventricle of the larynx is the narrow groove between the vestibular and vocal folds, and it partly undermines the vestibular fold. If a blunt seeker is passed along the roof of the ventricle, it enters the saccule of the larynx. This is a narrow, blind diverticulum which passes posterosuperiorly between the vestibular fold and the thyroid cartilage, and may reach the upper border of the cartilage.

DISSECTION. Cut the vestibular fold away from the upper part of the arytenoid cartilage, and strip it carefully forwards from the wall of the larynx, avoiding injury to the underlying muscle (thyro-epiglotticus). Separate the fold from the thyroid cartilage, and open the saccule.

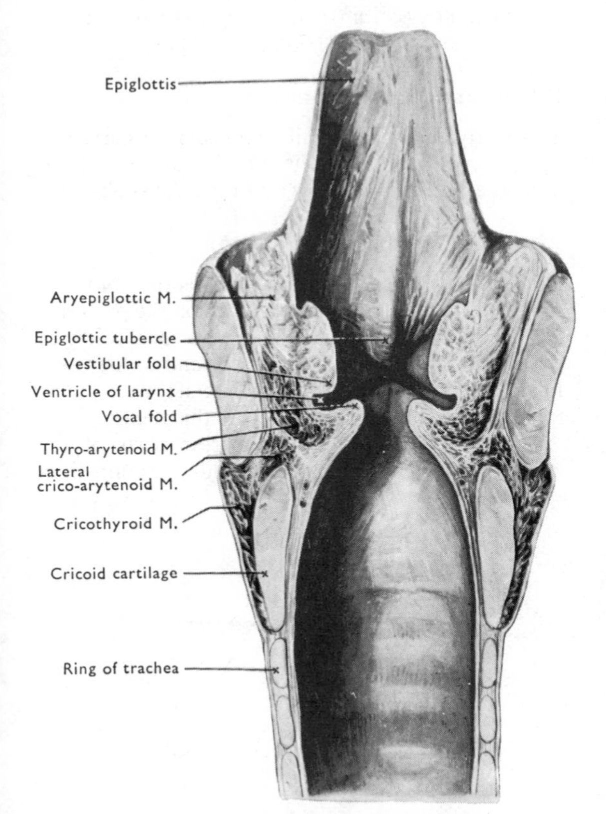

FIG. 133 Coronal section of the larynx to show the position of the muscles.

Vocal Folds

In coronal section [FIG. 130] each vocal fold is wedge-shaped. The apex projects medially into the laryngeal cavity. The base lies against the lamina of the thyroid cartilage. The inferior surface slopes inferolaterally to the superior border of the cricoid cartilage. The superior surface forms the inferior wall of the ventricle. Each vocal fold consists of the conus elasticus, the vocal ligament, and muscle fibres, all covered with mucous membrane.

Conus Elasticus and Vocal Ligament. The inferomedial surface of the vocal fold consists of a thin layer of fibro-elastic tissue (conus elasticus) separated from the laryngeal cavity by mucous membrane only. The conus elasticus is attached inferiorly to the upper border of the cricoid cartilage, and slopes superomedially to end in the free edge of the vocal fold as a thickened elastic band, the vocal ligament. Anteriorly, the conus elasticus is attached to the deep surface of the cricothyroid ligament and the posterior surface of the thyroid cartilage; posteriorly, it extends to the vocal process and medial side of the arytenoid cartilage. Thus the vocal ligament [FIG. 132] is attached from the posterior surface of the thyroid cartilage (close to the midline and the opposite vocal ligament) to the vocal process of the arytenoid cartilage. The mucous membrane covering the vocal ligament is tightly bound to it, and is of the stratified squamous type, unlike the ciliated columnar epithelium found in most of the remainder of the larynx. This change in the epithelium on the free surface of the vocal fold gives it a whitish appearance in life, and allows it to withstand the stresses applied to this vibrating margin.

The muscle fibres in the vocal fold lie between the conus elasticus and the lamina of the thyroid cartilage. They arise from the thyroid cartilage and vocal ligament anteriorly. Most of them pass horizontally backwards to the arytenoid cartilage (**thyro-arytenoideus**). The most medial fibres are from the vocal ligament (**vocalis muscle**). The most lateral fibres sweep superiorly to the epiglottis (**thyro-epiglotticus**; FIG. 139).

DISSECTION. Strip the mucous membrane from the inferior surface of the vocal fold and expose the conus elasticus and vocal ligament. Separate the conus from the superior border of the cricoid cartilage and turn it superiorly, dissecting it away from the muscle which lies lateral to it [FIG. 133] but leave the vocal ligament intact. Strip the mucous membrane from the superior surface of the vocal fold, thus exposing the upper surface of the thyro-arytenoid muscle and its continuity with the thyro-epiglotticus which was exposed when the vestibular fold was removed.

Rima Glottidis

This is the narrow anteroposterior fissure separating the free margins of the vocal folds and the vocal processes of the arytenoid cartilages. It is the narrowest part of the laryngeal cavity, and lies behind the thyroid cartilage. The shape of the rima glottidis is continually changing [FIG. 134] due to the movements of the arytenoid cartilages on the cricoid, and of the cricoid and arytenoid cartilages together relative to the thyroid cartilage. The posterior part of the rima is progressively *widened* when the arytenoid cartilages are displaced laterally on the cricoid and rotated so that their vocal processes turn laterally [FIG. 134]. If the lamina of the cricoid is then tilted forwards, the vocal ligaments are slackened and the opening can be widened further to allow the free passage of air in forced respiration. The rima glottidis is *narrowed or closed* if the arytenoid cartilages are drawn together and rotated so that their vocal processes are in apposition. The additional tightening of the vocal ligaments by tilting of the cricoid lamina backwards effectively prevents the passage of air. This is the position taken up as a preliminary to the explosive discharge of air in coughing or sneezing, the tension in the vocal folds being released suddenly after the intrathoracic pressure has been raised by contraction of the expiratory muscles.

When the rima glottidis is closed, attempts to draw air into the thorax tend to press the sloping vocal folds more firmly together and prevent inspiration—a situation which arises in laryngeal spasm.

Delicate variations (1) in the tension and length of the vocal cords, (2) in the width of the rima, and (3) in the intensity of expiratory effort are together responsible for producing changes in the pitch of the voice. The lower range of pitch in the male voice is due to the greater length of the vocal folds (approximately 2·5 cm) as compared with those in the female (approximately 1·7 cm).

Infraglottic Part of Larynx

Superiorly, the smooth wall of this cavity is narrowed by the vocal folds, but is circular at the cricoid cartilage, and continuous with the trachea inferiorly.

Mucous Membrane of Larynx

Over most of the larynx the mucous membrane is loosely adherent to the walls except on the posterior surface of the epiglottis and over the vocal folds. In the latter position it is so tightly bound to the vocal ligament that inflammatory swelling of the vestibular mucosa is unable to spread across this region.

On the vocal folds, the superior parts of the epiglottis, and the aryepiglottic folds the epithelium

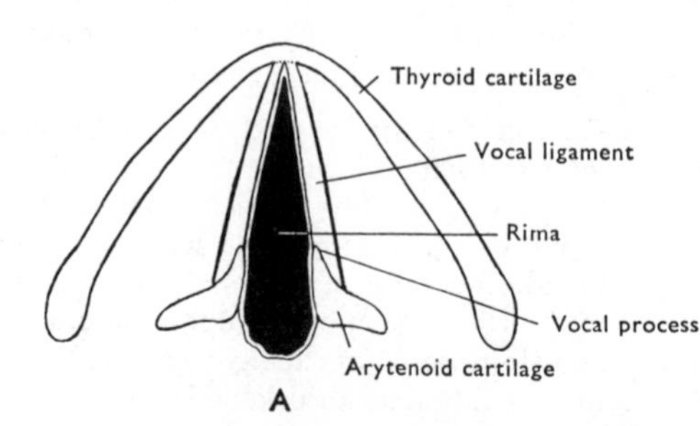

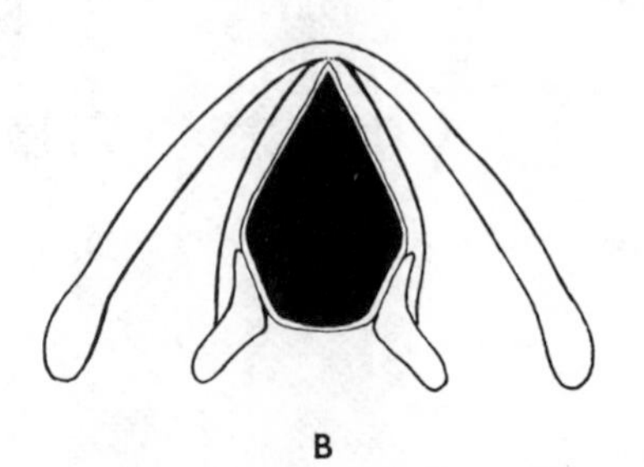

FIG. 134 Two diagrams to show the movements of the arytenoid cartilages by which the rima glottidis is opened and closed. A. Position during quiet breathing. B. Position during forced respiration.

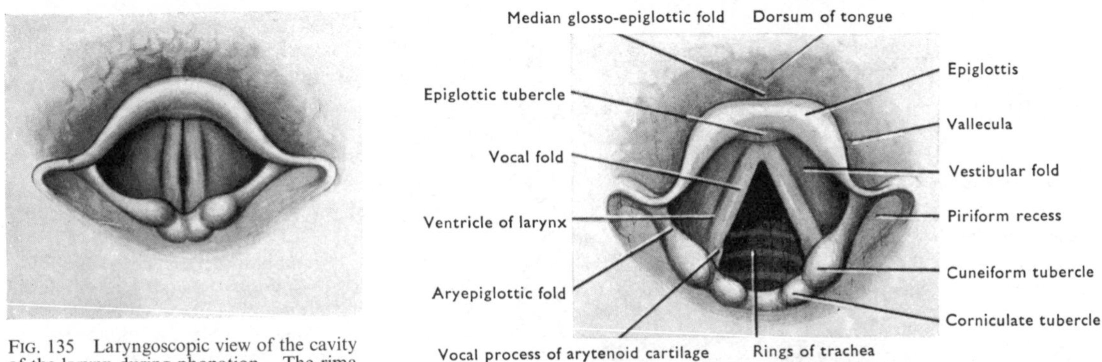

FIG. 135 Laryngoscopic view of the cavity of the larynx during phonation. The rima glottidis is closed by approximation of the vocal folds.

FIG. 136 Laryngoscopic view of the cavity of the larynx during moderate respiration. The rima glottidis is widely open.

is stratified squamous in type, but elsewhere is columnar ciliated. Taste buds are found on the lateral parts of the posterior surface of the epiglottis, on the aryepiglottic folds, and on the arytenoid cartilages.

MUSCLES OF LARYNX

Intrinsic Muscles

These small muscles move the parts of the larynx on each other, and are particularly concerned with alterations in the length and tension of the vocal folds in the production of the voice, and in changing the size of the rima glottidis to facilitate or prevent the passage of air to and from the lungs.

Cricothyroid. This muscle passes posterosuperiorly from the anterolateral part of the cricoid cartilage to the inferior horn of the thyroid cartilage and the inferior margin and lower part of the deep surface of its lamina. **Nerve supply**: external branch of the superior laryngeal nerve. **Action**: it draws the arch of the cricoid posterosuperiorly, rotating the entire cartilage around the cricothyroid joints, so that the lamina is tilted posteriorly. This elongates and tightens the vocal ligaments, thus raising the pitch of the voice.

DISSECTION. Identify the recurrent laryngeal nerve entering the larynx deep to the inferior constrictor muscle. Strip the mucous membrane from the posterior surfaces of the arytenoid and cricoid cartilages. Note the attachment of the longitudinal oesophageal muscle fibres by a tendon to the median part of the cricoid lamina, and identify [FIG. 138] the posterior crico-arytenoid, transverse arytenoid, and oblique arytenoid muscles. The last two have been divided in the midline, but the oblique arytenoid can be followed into continuity with the aryepiglottic muscle by stripping the mucous membrane from the margin of the aryepiglottic fold.

Posterior Crico-arytenoid. This muscle arises from the posterior surface of the lamina of the cricoid, and converges on the laterally directed muscular process of the arytenoid cartilage. **Nerve supply**: *this and all the other intrinsic muscles of the larynx (except cricothyroid) are supplied by the recurrent laryngeal nerve.* **Action**: the upper, more horizontal fibres rotate the arytenoid so that its vocal process swings laterally, opening the rima glottidis; the lower, more vertical fibres pull the arytenoid laterally, further increasing the size of the rima. It is the sole abductor of the folds.

Transverse and Oblique Arytenoid Muscles. They cross between the arytenoid cartilages and draw them together, closing the rima glottidis. The continuity of the oblique and aryepiglottic muscles ensures that the arytenoid cartilages are drawn

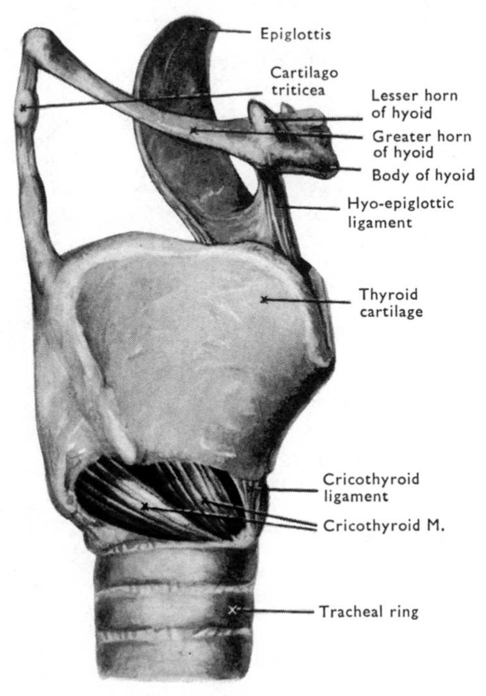

FIG. 137 Right cricothyroid muscle.

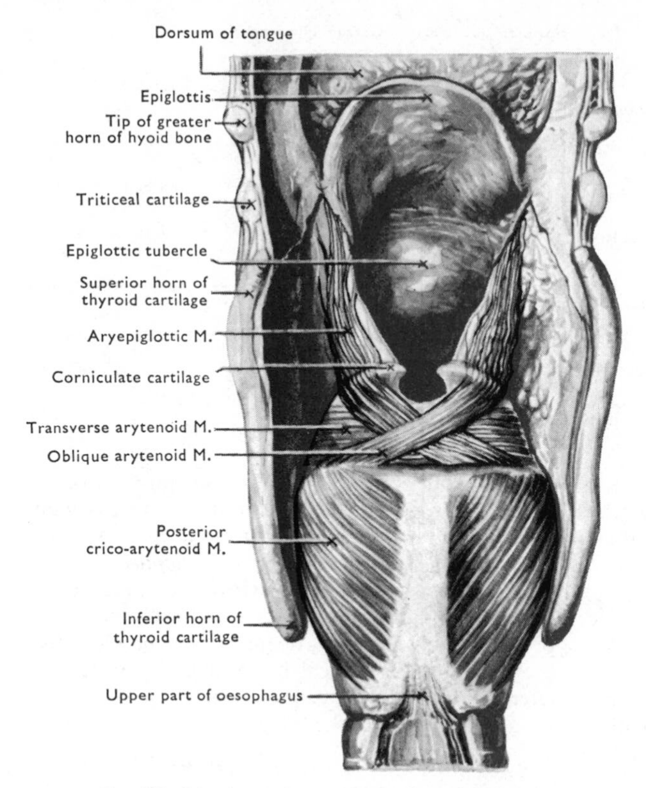

FIG. 138 Muscles on the posterior surface of the larynx.

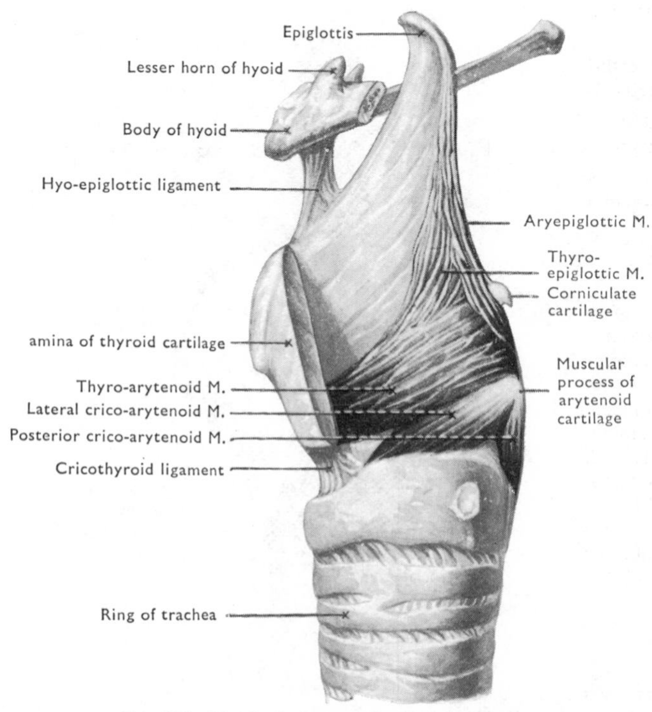

FIG. 139 Muscles in the lateral wall of the larynx.

together, closing the rima, while the epiglottis is pulled down over the opening into the larynx, during the passage of food or fluid through the pharynx.

DISSECTION. Remove the cricothyroid muscle and, on one side, the lower part of the lamina and inferior horn of the thyroid cartilage, thus opening the cricothyroid joint. Take care not to injure the continuation of the recurrent laryngeal nerve (the **inferior laryngeal nerve**) which is deep to the posterior part of the thyroid cartilage. The thyro-arytenoid muscle in the vocal fold is now exposed [FIG. 139] with the lateral crico-arytenoid muscle inferior to it.

Lateral Crico-arytenoid Muscle. It passes from the superior border of the posterior part of the arch of the cricoid to the muscular process of the arytenoid muscle. **Action**: it pulls the muscular process anteriorly, rotating the arytenoid so that its vocal process swings medially and closes the rima glottidis.

Thyro-arytenoid. This muscle arises from the posterior surface of the thyroid cartilage close to the midline, and is attached to the anterolateral surface of the arytenoid cartilage. Some of the deeper fibres arise from the vocal ligament and pass to the vocal process of the arytenoid, the **vocalis muscle**. The upper, lateral fibres sweep superiorly to the epiglottis, the **thyro-epiglottic muscle**. **Action**: the main mass pulls the arytenoid anteriorly, slackening the vocal ligament. The vocalis tends to tighten the anterior part of the ligament while slackening the posterior part—a position taken up by the vocal cords in whispering. The thyro-epiglottic and aryepiglottic muscles are minor elements in closing the laryngeal orifice [pp. 120–1].

Extrinsic Muscles [pp. 36, 70, 119]

In swallowing, the **thyrohyoid**, acting from a fixed hyoid bone, raises the larynx, assisted by **stylopharyngeus** and **palatopharyngeus**. This pushes the epiglottis against the bulging posterior surface of the tongue which tips it backwards into contact with the elevated arytenoid cartilages (assisted by **aryepiglotticus** and **thyro-epiglotticus**) and closes the laryngeal orifice. **Sternothyroid** returns the larynx to its original position, and opens the laryngeal orifice.

NERVES AND VESSELS OF LARYNX

Superior Laryngeal Nerve. This branch of the vagus divides into internal and external [p. 70] branches. The internal branch (sensory and autonomic) pierces the thyrohyoid membrane with the superior laryngeal artery, and breaks up into several branches in the wall of the piriform recess, which it supplies. Some branches pass in the aryepiglottic fold to the base of the tongue and epiglottis; others descend to supply the mucous membrane of the internal surface of the larynx as far as the vocal folds, and the mucous membrane covering the posterior surfaces of the arytenoid and cricoid cartilages. One branch descends deep to the thyroid cartilage to join a similar branch from the recurrent laryngeal nerve.

Recurrent Laryngeal Nerve. It ascends in the groove between the trachea and oesophagus, and enters the larynx by passing deep to the lower border of the inferior constrictor. It supplies all the intrinsic muscles of the larynx, except cricothyroid, and the mucous membrane below the rima glottidis. It communicates with the internal branch of the superior laryngeal nerve. It is accompanied by the inferior laryngeal artery, a branch of the inferior thyroid artery, and runs posterior to the cricothyroid joint.

THE TONGUE

The tongue is a mobile organ which bulges upwards from the floor of the mouth, and its posterior part forms the anterior wall of the oral part of the pharynx. It is covered by stratified squamous epithelium, and consists of a mass of striated muscle interspersed with a little fat and numerous glands, especially in the posterior part.

It is separated from the teeth by a deep **alveololingual sulcus** which is filled in by the palatoglossal fold posterior to the last molar tooth. The sulcus partly undermines the lateral margins of the tongue, and extends beneath its free anterior third. In the depths of the sulcus, the mucous membrane passes from the root of the tongue across the floor of the mouth on to the internal aspect of the mandible, and becomes continuous superiorly with that on the gum. Internal to the sulcus, the **root of the tongue** contains the muscles which connect the tongue to the hyoid bone and mandible, and transmits the nerves and vessels which supply it.

DORSUM OF TONGUE

The dorsum of the tongue extends from the tip to the anterior surface of the epiglottis. It is separated into palatine and pharyngeal parts by a V-shaped **sulcus terminalis**, the apex of which points posteriorly and is marked by a pit, the **foramen caecum** [FIG. 140]. A shallow median groove extends from the tip of the tongue to the foramen caecum.

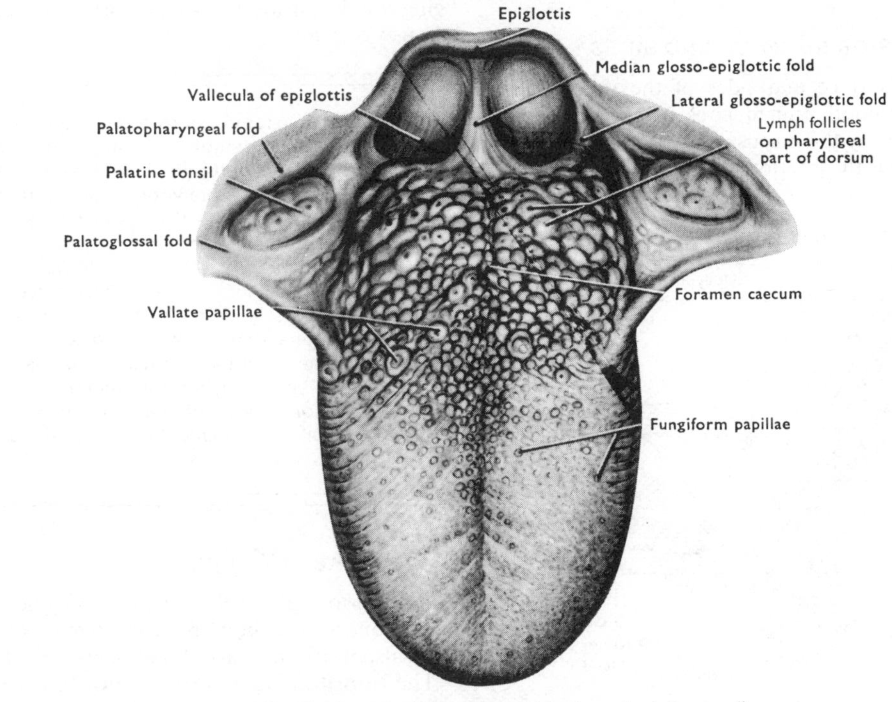

FIG. 140 The dorsum of the tongue, epiglottis, and palatine tonsils.

The thick mucous membrane of the palatine part is roughened by the presence of papillae. In the pharyngeal part it is smooth, thinner, and finely nodular in appearance, due to the presence of small **lymph follicles** in the submucosa. Each of these has a small, central epithelial pit. Posteriorly the lingual mucous membrane is continuous with that on the anterior surface of the epiglottis over the median and lateral glosso-epiglottic folds and the valleculae of the epiglottis between them.

Lingual Papillae

The largest are seven to twelve **vallate papillae**, which lie immediately anterior to the sulcus terminalis. Each has the shape of a short cylinder sunk into the surface of the tongue, with a deep trench around it. The opposing walls of the trench are studded with taste buds.

Fungiform papillae, smaller and more numerous, are the bright red spots seen principally on the tip and margins of the living tongue, but scattered over the remainder of the dorsum also. Each is attached by a narrow base, and expands into a rounded knob-like free extremity. Most of them carry taste buds.

Filiform papillae are very numerous, minute, pointed projections which cover all of the palatine part of the dorsum and the margins of the tongue. They are arranged in rows which are more or less parallel to the sulcus terminalis posteriorly, but become more transverse anteriorly. Their apices are cornified and may be broken up into thread-like processes.

INFERIOR SURFACE AND SIDES

The inferior surface and sides of the tongue are covered with smooth, thin mucous membrane. In the midline anteriorly the mucosa is raised into a sharp fold which joins the inferior surface of the tongue to the floor of the mouth (**frenulum linguae**, FIG. 112). On each side of the frenulum, the deep lingual vein may be seen through the mucous membrane in the living subject, and lateral to this, the fringed **fimbriated fold** of mucous membrane [FIG. 112] sweeps inferolaterally. On each side of the frenulum on the floor of the mouth, is the opening of the submandibular duct on the **sublingual papilla**. Passing posterolaterally from this, in the floor of the mouth, is the rounded **sublingual fold** which is produced by the sublingual gland and the submandibular duct, and on which open a number of the ductules of the sublingual gland.

On the sides of the tongue, anterior to the lingual attachment of the palatoglossal arch, are five short, vertical folds of mucous membrane (**folia linguae**). These carry taste buds and are much better developed in some animals, *e.g.*, the hare and rabbit.

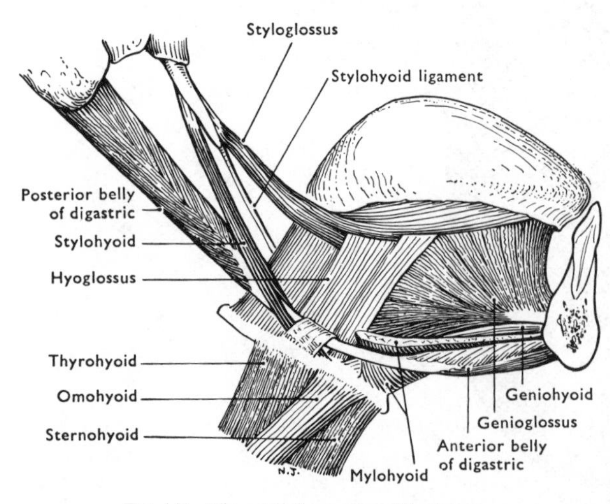

FIG. 141 The extrinsic muscles of the tongue.

MUSCLES OF TONGUE

The tongue is divided into halves by a median fibrous septum, and the muscles of each half consist of an extrinsic and an intrinsic group:

EXTRINSIC	INTRINSIC
Genioglossus [p. 108].	Superior longitudinal.
Hyoglossus [p. 108].	Inferior longitudinal.
Styloglossus [p. 108].	Vertical.
Palatoglossus [p. 119].	Transverse.

The extrinsic muscles take origin from parts outside the tongue, and can therefore move the tongue as well as alter its shape. The intrinsic muscles, being wholly inside the tongue, can only produce changes of shape.

DISSECTION. On the cut surface of the tongue, identify genioglossus and geniohyoid, and confirm their attachments and position [FIG. 116]. On the right side, separate the buccinator, pterygomandibular raphe, and superior constrictor from their attachments to the mandible, and turn the remainder of the body of the mandible downwards to expose the lateral surface of the tongue. Avoid injury to the lingual nerve and the palatoglossus muscle. Remove the remainder of the mucous membrane from the lateral surface of the tongue, and follow the various extrinsic muscles into its substance. When removing the mucous membrane from the inferior surface of the tongue near the tip, identify a small, oval glandular mass, the **anterior lingual gland**.

Movements of Tongue

The posterior part of the tongue is attached to the hyoid bone. Hence the muscles which move the hyoid bone also move this part of the tongue [p. 109].

The **hyoglossus** and **genioglossus muscles** enter the tongue from below, the former vertically

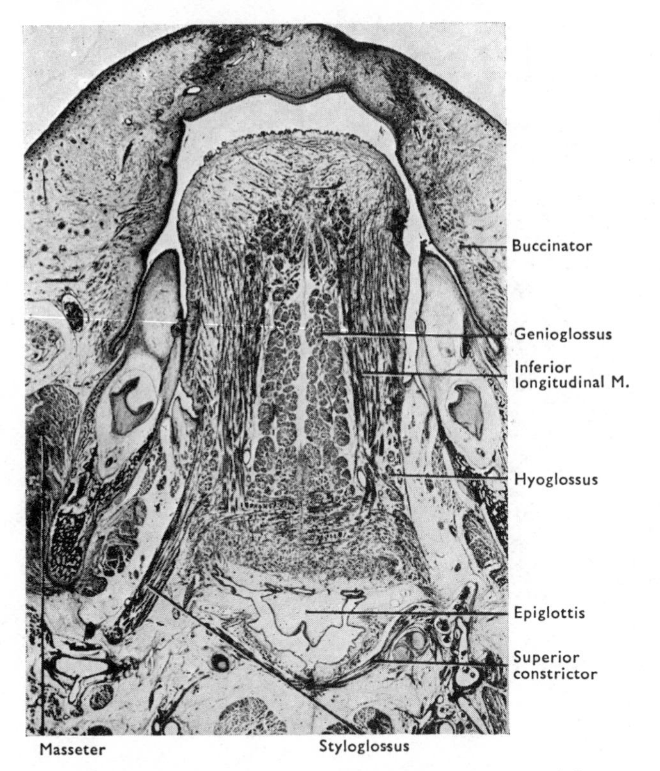

FIG. 142 Horizontal section through the tongue of a seven-month human foetus.

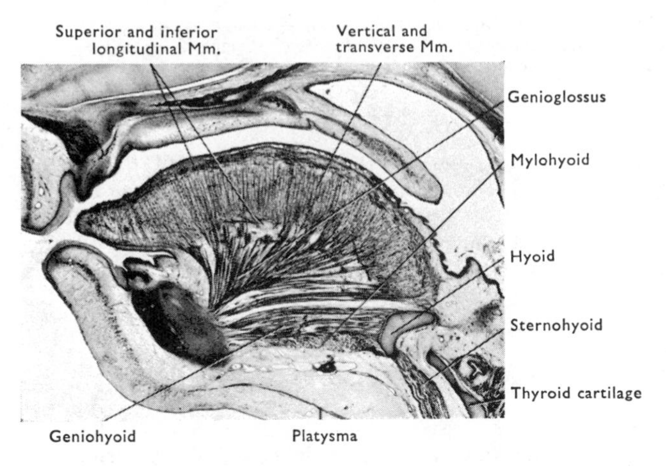

FIG. 143 Paramedian section through the tongue of a fourteen-week human foetus.

along its side, the latter in a paramedian position. Thus they depress the lateral and median parts of the tongue respectively. The genioglossus is fan-shaped [FIG. 143]. Hence its posterior fibres pull the tongue forwards and help to protrude it (as does geniohyoid) while the anterior fibres depress and retract the tip. The **palatoglossus** and **styloglossus** enter the lateral part of the tongue from above. The palatoglossus passes almost transversely and is continuous with the intrinsic transverse fibres; the styloglossus runs anteriorly along the lateral margin. Both muscles elevate the posterior part of the tongue, the styloglossus also retracts it, and the palatoglossus draws the palate down on to the tongue. Their combined action helps to close off the mouth from the oral part of the pharynx.

The **superior longitudinal muscle** forms a layer on the dorsum of the tongue; it curls the tip upwards and rolls it posteriorly. The **inferior longitudinal muscles** lie in the lower part of the tongue, one on each side lateral to genioglossus [FIG. 142]. They curl the tip of the tongue inferiorly, and act with the superior muscle to retract and widen the tongue.

The **transverse muscle fibres** lie inferior to the superior longitudinal muscle, and run from the septum to the margins between the vertical fibres of genioglossus, hyoglossus, and the vertical muscle. They narrow the tongue and increase its height.

The **vertical muscle fibres** run inferolaterally from the dorsum. They flatten the dorsum, increase the transverse diameter, and tend to roll up the margins. Acting with the transverse muscles, they increase the length of the tongue and assist with protrusion.

The actions given above represent only a few of the possible movements; many others are produced by complex combinations of the muscles. The tongue is bilaterally symmetrical, and unilateral action of any muscle or group of muscles will cause the tongue to deviate from the midline. Thus in *protrusion of the tongue* with one side paralysed, the tip deviates towards the paralysed side because that side fails to act, and, lagging behind, swings the active side towards it.

Septum of Tongue [FIG. 110]

The septum is best seen in a transverse section through the tongue. It is a median fibrous partition, which is strongest posteriorly where it is attached to the hyoid bone. It does not reach the mucous membrane of the dorsum, but is separated from it by the superior longitudinal muscle.

Glands of Tongue

Small serous and mucous glands lie between the muscle fibres deep to the mucous membrane of the pharyngeal surface, tip, and margins. Small serous glands lie near the vallate papillae and open into their trenches. The **anterior lingual gland** consists of mucous and serous alveoli, and it lies on the inferior surface of the tongue near to its tip.

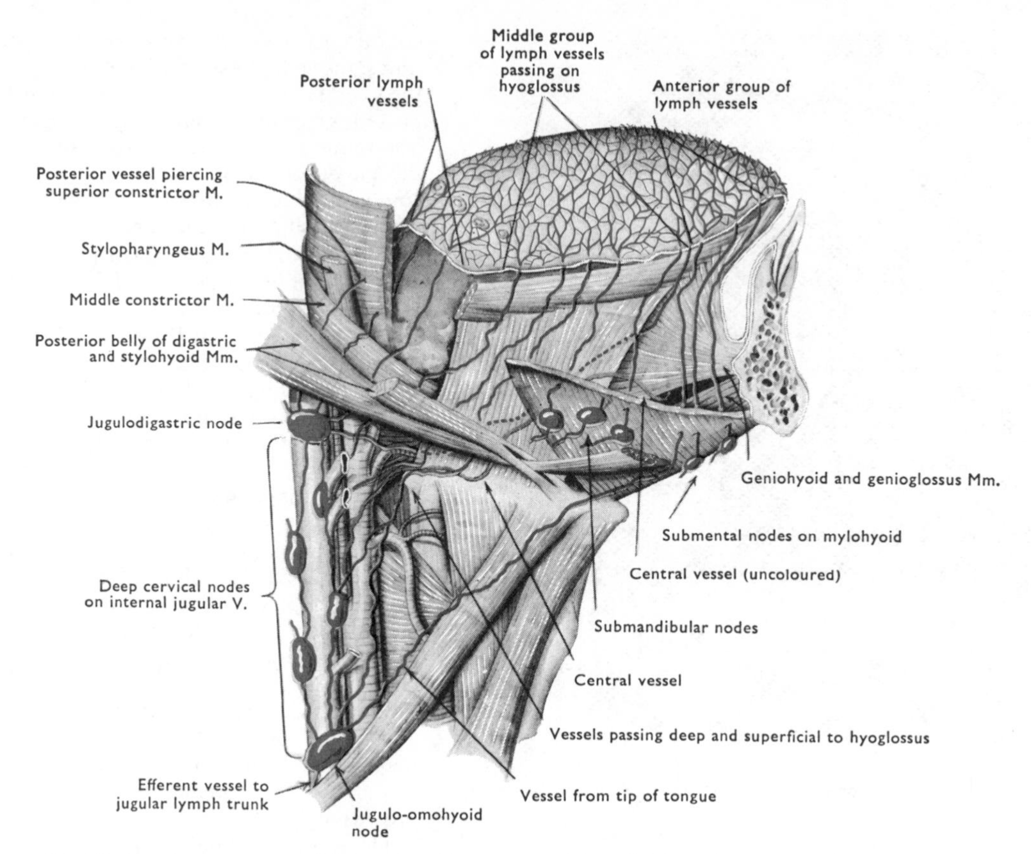

FIG. 144 Diagram of the course of the lymph vessels of tongue. It is doubtful whether any lingual lymph vessels reach the submental nodes which mostly drain the floor of the mouth and symphysis region of the jaw. Lymph vessels may cross the median plane to reach the opposite deep cervical lymph nodes.

NERVES OF TONGUE

The mucous membrane and muscles of the tongue have entirely separate nerve supplies. The supply to the mucous membrane is complex and consists of: (1) the **lingual** (including taste fibres from the chorda tympani) to the anterior two-thirds; (2) the **glossopharyngeal** (including taste fibres) to the posterior third; (3) twigs of the internal branch of the **superior laryngeal** nerve to a small area adjacent to the epiglottis. The glossopharyngeal nerve supplies the vallate papillae. The nerves also carry parasympathetic secretomotor fibres to the glands buried in the substance of the tongue. The supply to the muscles of the tongue is from the **hypoglossal nerve** which innervates all the intrinsic and extrinsic muscles except palatoglossus.

This difference in innervation arises because the mucous membrane is derived from the floor of the embryonic pharynx, while the muscle originates from occipital somites adjacent to those (cervical) which produce the infrahyoid muscles. This series of muscles, from tongue to sternum, is innervated by the hypoglossal and first three cervical nerves (the nerves of these somites), entirely separate from the nerves of the pharynx, even though these muscles and the muscles of the pharynx act together.

VESSELS OF TONGUE

The chief **arteries** are branches of the lingual [p. 111]; the deep artery of the tongue to the anterior part; the dorsales linguae arteries to the posterior part. Trace the deep artery to the tip, where it forms an anastomotic loop with its fellow. This is the only significant anastomosis across the midline of the tongue. The **veins** are described on page 111.

Lymph Vessels of Tongue [FIG. 144]

These vessels cannot be dissected, but they are important because of the common involvement of the tongue in cancer and its spread by these channels.

There is a rich lymph capillary plexus in the mucous membrane which has been described as draining by a number of separate routes. There is no doubt, however, that the great majority of the lymph vessels from the mucous membrane of the tongue drain along the route of the vessels which supply it. They therefore pass posteriorly, either deep or superficial to hyoglossus, and running superior to the hyoid bone reach the upper deep cervical nodes in the carotid triangle close to the angle of the mandible, especially the jugulodigastric node. Drainage either to the submental or to the submandibular nodes, or direct drainage to the lower deep cervical nodes (jugulo-omohyoid, FIG. 144) is rare as a primary event in cancer of the tongue. Nevertheless they may be implicated because adjacent superficial lymph plexuses communicate freely, and thus blockage of the drainage route from one area can lead to drainage through an abnormal pathway with the involvement of unusual groups of lymph nodes.

The same principle of drainage along the blood vessels applies in the case of the submental and submandibular nodes: the former receive lymph from the territory supplied by the submental artery (anterior part of floor of mouth) while the latter drain lymph from the territory of the facial artery, particularly the lips and cheek and the posterior part of the floor of the mouth.

There is no separation of the superficial lymph plexus on the right and left sides of the tongue, and the paramedian areas may drain equally to both sides.

THE ORGANS OF HEARING AND EQUILIBRATION

The 'ear' consists of the auditory apparatus and the organs concerned with balance which record both rotary movements of the head and the direction of the gravitational field acting on it. It is readily divisible into three parts; the external ear, the middle ear, and the internal ear.

The **external ear** consists of the auricle and the external acoustic meatus. The auricle collects the sound waves, and the external acoustic meatus transmits them medially to the tympanic membrane, which separates the external ear from the middle ear. The **middle ear** is a narrow chamber whose air-filled **tympanic cavity** lies between the tympanic membrane and the lateral wall of the internal ear, and is maintained at atmospheric pressure by communicating with the nasal part of the pharynx through the auditory tube. Three small auditory ossicles lie in the middle ear, and stretch across it from the tympanic membrane to the lateral wall of the internal ear to transmit the vibrations of the tympanic membrane to the internal ear. The **internal ear** consists of a complex system of communicating cavities (the **bony labyrinth**) situated in the densest part of the petrous temporal bone. Suspended in it is a similarly shaped, but narrower complex of membranous tubes and sacs (the **membranous labyrinth**) on which the sensory nerve fibres of the vestibulocochlear nerve end.

THE EXTERNAL EAR

The auricle has been examined already [pp. 11, 13].

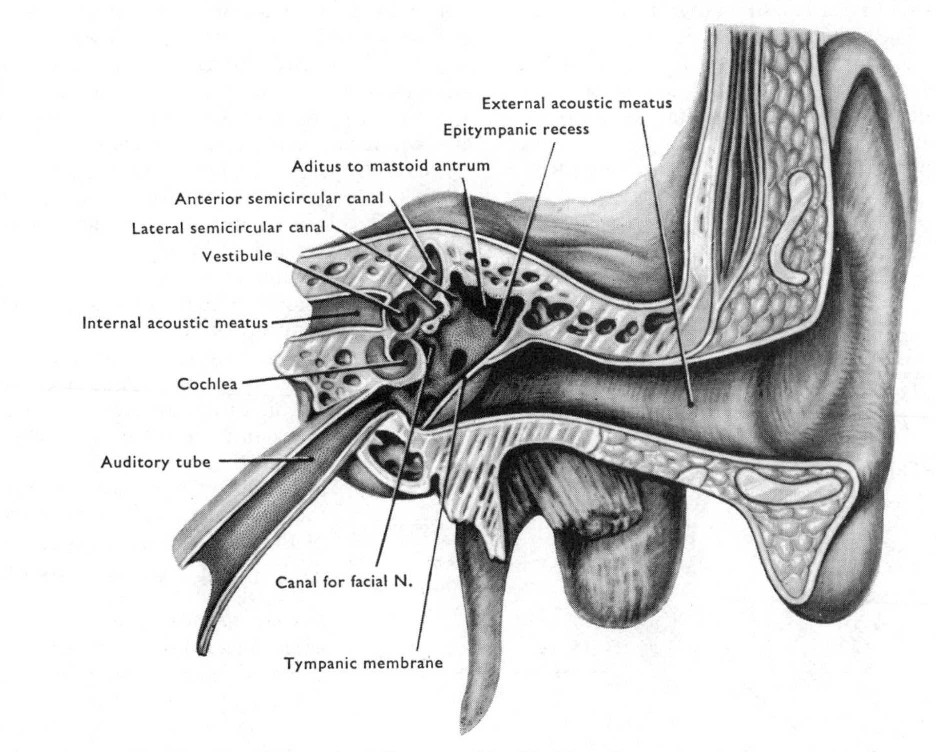

FIG. 145 The parts of the ear (semi-diagrammatic). The blue colour represents the mucous membrane lining the middle ear and auditory tube.

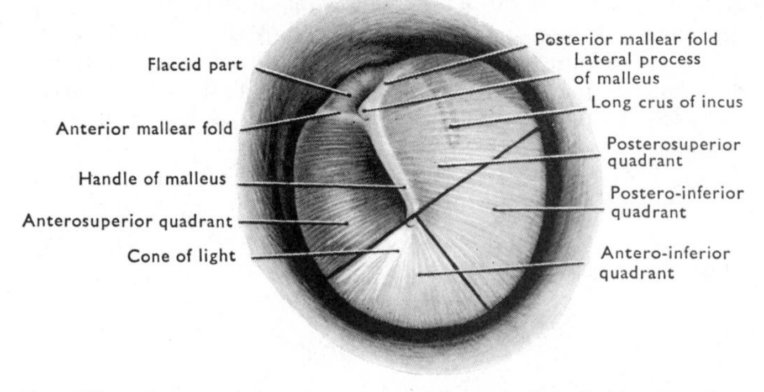

FIG. 146 Left tympanic membrane as seen from the lateral side. The four arbitrary quadrants are indicated by solid lines and by the handle of the malleus.

DISSECTION. Cut away the tragus of the auricle to expose the orifice of the external meatus. Remove the anterior wall of the cartilaginous part of the meatus with knife or scissors. Pass a probe into the bony part of the meatus to determine its length, and using the probe as a guide, cut away the anterior wall of the bony part of the meatus (tympanic bone) taking care not to injure the tympanic membrane.

EXTERNAL ACOUSTIC MEATUS

The external meatus runs medially with a slight anterior inclination. It is almost exactly in line with the internal acoustic meatus, and their shadows are superimposed in a true lateral radiograph of the skull [FIG. 149]. The external meatus is approximately 24 mm long, of which two-thirds is the bony part, but the tympanic membrane lies obliquely so that the anterior wall and floor are longer than the posterior wall and roof. The diameter is not uniform; the narrowest point, the **isthmus**, is about 5 mm from the tympanic membrane. The vertical diameter is greatest at the lateral end, whilst the anteroposterior diameter is greatest at the medial end, and the meatus is slightly curved with an upward convexity. All these points should be remembered when attempting to remove foreign bodies from the meatus, and in addition the fact that the bony part of the meatus is absent *in the newborn* and the tympanic membrane is considerably nearer the surface and much more oblique than in the adult.

The skin of the cartilaginous portion contains many **ceruminous glands** and hairs. The hairs are directly laterally and prevent the entry of small objects. In the bony part of the canal the skin is thin and tightly adherent to the periosteum. It has no hairs, few if any glands, and is continued, as a very delicate layer, over the lateral surface of the tympanic membrane.

Tympanic Membrane

This is an elliptical disc which is stretched across the medial end of the external acoustic meatus, and forms the greater part of the lateral wall of the middle ear cavity. It slopes obliquely downwards, forwards, and medially, and its lateral surface is deeply concave. The medial surface is correspondingly convex, and the point of maximum convexity is called the **umbo**, which is reached by the lower end of a bar of bone, the **handle of the malleus**, attached to the membrane. The malleus is the most lateral of the three auditory ossicles, and its handle can be seen in the tympanic membrane. It extends superiorly towards the upper edge of the membrane, near which the lateral process of the malleus bulges the membrane into the meatus. The small part of the membrane superior to the lateral process is less tense than the remainder and forms the **flaccid part**.

The tympanic membrane, composed of a taut fibrous layer with the handle of the malleus attached to its internal aspect, is covered externally by the thin skin of the meatus and internally by the mucous membrane of the middle ear. The thickened margin of the fibrous part is inserted into a distinct tympanic groove in a ring-like ridge of the tympanic bone at the medial end of the meatus. Above the lateral process of the malleus, the ring and groove are replaced by a shallow depression, the **tympanic notch**. Here the thickened edge of the membrane leaves the bone and passes to the lateral process of the malleus, thus

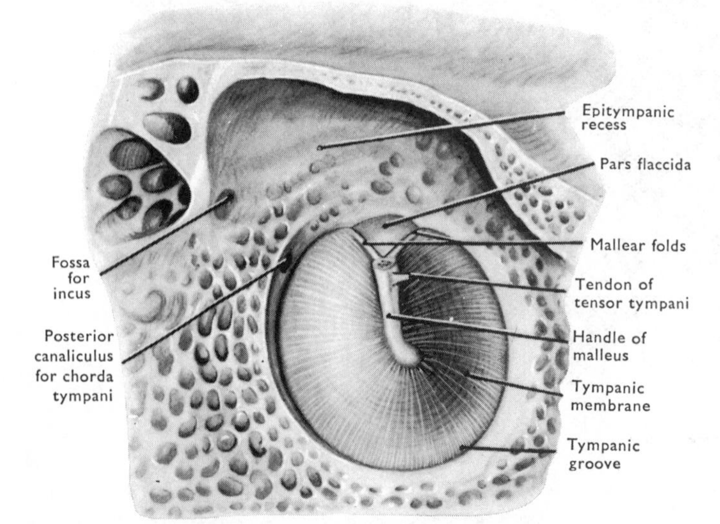

FIG. 147 View of the left tympanic membrane and epitympanic recess from the medial side. The head and neck of the malleus have been removed.

forming the anterior and posterior **mallear folds** [FIG. 147]. Between these folds and the margin of the tympanic notch lies the flaccid part of the membrane, which differs from the rest in not having a well-defined fibrous layer.

When the living tympanic membrane is examined, its surface appears highly polished, and a cone of light extends antero-inferiorly from the tip of the handle of the malleus. The mallear folds can be seen as two striae outlining the flaccid part, and the **long process of the incus** (the intermediate auditory ossicle) can be seen dimly through the membrane, parallel and posterior to the handle of the malleus [FIG. 146].

DISSECTION. Strip the dura mater and endocranium from the floor of the middle cranial fossa as far anteriorly as the mandibular nerve. The greater petrosal nerve will be found emerging from the anterior surface of the petrous temporal bone and running anteromedially inferior to the trigeminal ganglion.

The cavity of the middle ear is separated from the middle cranial fossa by a thin layer of bone (**tegmen tympani**) and lies parallel and slightly lateral to the greater petrosal nerve. Force a small aperture in the tegmen with the point of a rigid knife or seeker, and slipping a strong seeker through the aperture under the tegmen, lever it up and break it away, first in an anteromedial and then in a posterolateral direction towards the junction of the transverse and sigmoid sinuses. Keep the seeker close to the tegmen to avoid damage to the contents of the middle ear. Clean away the broken edges of the tegmen and expose the cavity of the middle ear. Avoid injury to the parts of the auditory ossicles which extend superiorly towards the tegmen.

Pass a blunt seeker into the anteromedial part of the cavity, and slide it anteromedially till it appears through the pharyngeal opening of the auditory tube. The cavity of the middle ear is a deep trench which extends inferomedially. Its posterior (mastoid) wall, immediately behind the auditory ossicles, has an opening (**aditus**) in its upper part. This links the middle ear with the mastoid antrum, and so with the mastoid air cells.

Identify the rounded head of the malleus articulating with the incus, the short crus of which passes posteriorly towards the superior surface of the posterior wall of the cavity. Inferior to the head of the malleus note the sloping tympanic membrane and the handle of the malleus lying in it. A slender, tough strand passes from the malleus to the medial wall of the middle ear; this is the tendon of **tensor tympani** which turns through an angle of nearly 90 degrees at the medial wall to become continuous with the muscle belly. The muscle runs in a semicanal superomedial to the auditory tube. Break through the thin wall of this canal and expose the muscle. Immediately superior to the attachment of the tendon to the malleus, the chorda tympani nerve [FIG. 151], which traverses the tympanic membrane and forms a narrow ridge on it, runs over the medial surface of the malleus. Look for the long crus of the incus which passes inferiorly to the stapes, which is not visible at this stage.

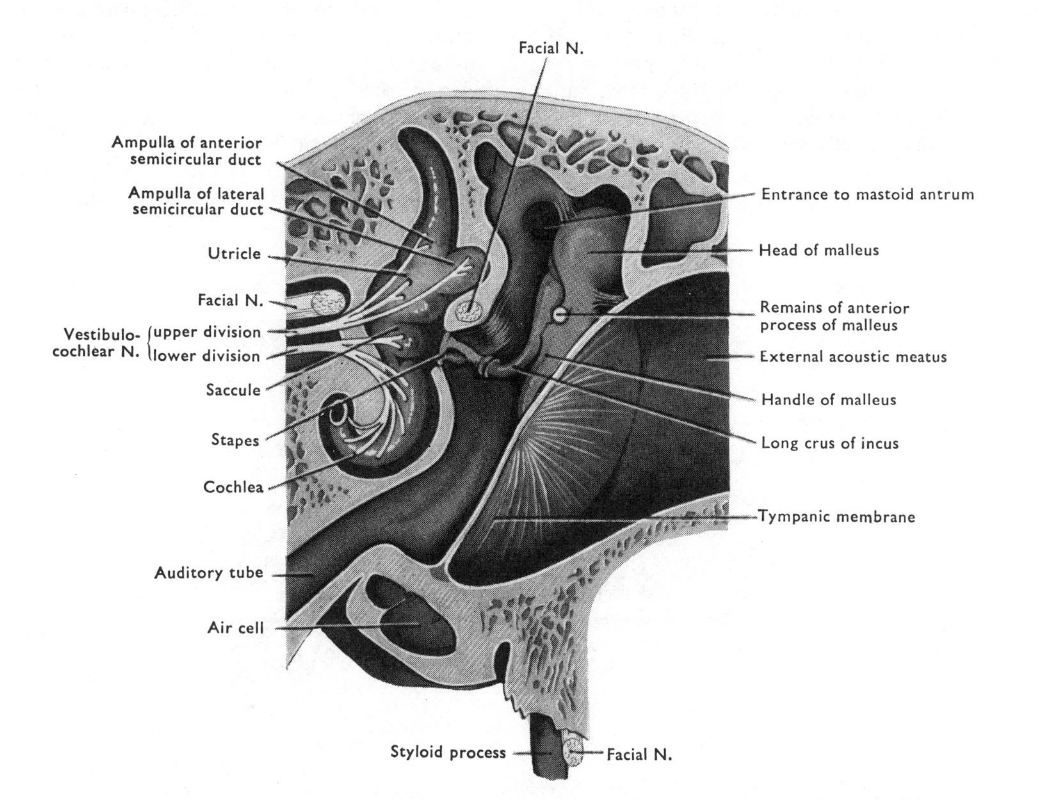

FIG. 148 Coronal section through the left tympanic cavity viewed from the front (semi-diagrammatic).

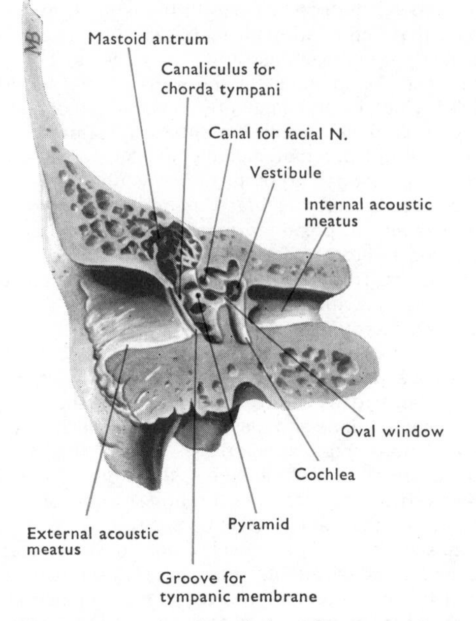

FIG. 149 Coronal section through the right temporal bone seen from in front.

THE MIDDLE EAR, MASTOID ANTRUM, AND AUDITORY TUBE

MIDDLE EAR

The narrow middle ear cavity together with the tympanic membrane is known as the tympanum. It is lined with mucous membrane, and is filled with air. It communicates, anteriorly, with the nasal part of the pharynx through the auditory tube [FIG. 154], and, posteriorly, through the mastoid antrum [FIG. 150] with the mastoid air cells, which are small, air-filled cavities in the mastoid process.

The middle ear also contains: (1) the auditory ossicles; malleus, incus, and stapes; (2) the tendons of stapedius and tensor tympani muscles; (3) the chorda tympani nerve and the tympanic plexus of nerves.

The vertical and anteroposterior diameters are each approximately 15 mm; the width varies from 6 mm above to 4 mm below, and is even less centrally where the medial and lateral walls bulge into the cavity.

The part of the cavity superior to the tympanic membrane is called the **epitympanic recess** [FIGS. 145, 151].

The **roof** or **tegmental wall** is the thin **tegmen tympani**, which separates the cavity from the middle cranial fossa. Chronic inflammatory conditions of the middle ear may spread through the tegmen to the meninges of the brain.

The **floor** or **jugular wall** is narrow. It is formed by a thin bony lamina which separates the cavity from the jugular fossa and its contained superior bulb of the jugular vein. Extension of an inflammatory condition of the middle ear through this bone may involve the vein, and may lead to thrombosis (clotting) of the blood in the vein and to the spread of infection via the blood stream.

The **posterior** or **mastoid wall** [FIG. 149] has a series of openings in it: (1) in its upper part is the opening (**aditus**) which leads from the epitympanic recess into the mastoid antrum; (2) inferiorly, close to the medial wall, is a small aperture on the apex of a hollow, conical projection called the **pyramid** [FIG. 151]. This contains the **stapedius muscle** with its tendon emerging through the aperture on the summit, to pass to the stapes; (3) lateral to the pyramid, the opening through which the **chorda tympani nerve** enters the middle ear from the facial nerve posteriorly.

The **anterior** or **carotid wall** is narrow, because the medial and lateral walls converge anteriorly. It

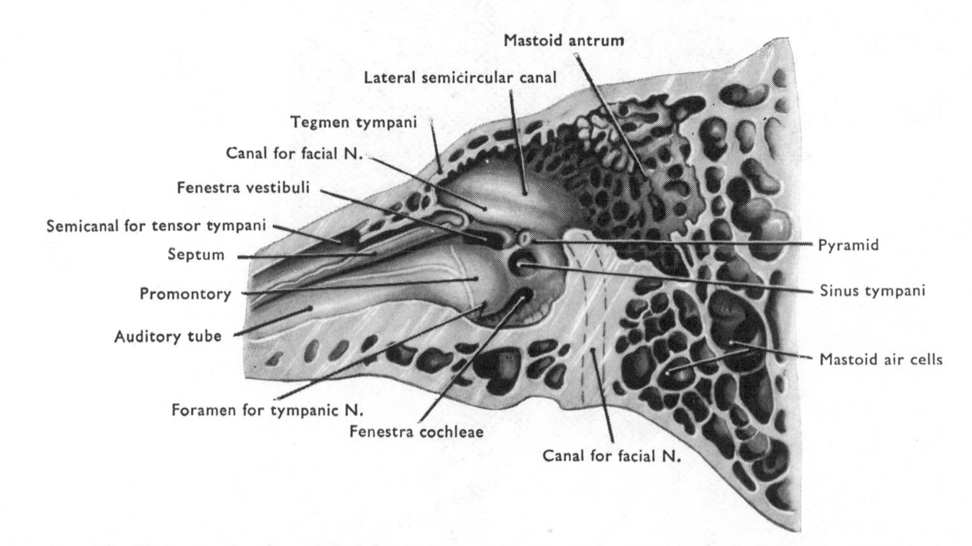

FIG. 150 Vertical section through the left middle ear and auditory tube to expose the medial or labyrinthine wall of the middle ear. The auditory ossicles have been removed.

consists of three parts: (1) superiorly, the opening of the semicanal for tensor tympani; (2) the orifice of the auditory tube is intermediate; (3) inferiorly, a lamina of bone which separates the cavity from the carotid canal [FIG. 154]. The bony septum between the **semicanals** for the tensor tympani and the auditory tube is continued posteriorly on the medial wall of the middle ear as a shelf (**processus cochleariformis**). The posterior end of this forms a pulley around which the tendon of tensor tympani turns laterally through 90 degrees to run to the malleus.

The **medial** or **labyrinthine wall** separates the middle ear from the internal ear, and is entirely bony except for two small apertures [FIG. 150]: (1) the **fenestra vestibuli** (oval window) which is filled by the base of the stapes [FIG. 148] mounted in an elastic ring; (2) the **fenestra cochleae** (round window) which is closed by the delicate **secondary tympanic membrane**. These two apertures lead into the cavity of the bony labyrinth, and movements of the stapes are transmitted in the form of pressure waves to the fluid (perilymph) in that cavity. Since the fluid is incompressible, movements of the stapes are allowed by corresponding movements of the secondary tympanic membrane, which thus prevents damping of the stapedial oscillations. FIGURE 150 shows the principal features, the cochlea of the internal ear forming the promontory anteriorly, while the facial nerve and the lateral semicircular canal, enclosed in the bone, produce parallel bulges (superiorly) which extend posteriorly into the aditus. The **sinus tympani** is a depression between the two windows.

Mucous Membrane of Middle Ear Cavity. This is a thin delicate lining, but the cavity is so narrow that it may be blocked when the mucous membrane swells during inflammation. The mucous membrane lines the walls of the cavity and covers the auditory ossicles. It is continuous with the linings of the auditory tube, mastoid antrum, and mastoid air cells, and its epithelium is a simple cuboidal layer.

Auditory Ossicles [FIG. 151]

These little bones extend in a chain from the tympanic membrane to the fenestra vestibuli. The malleus and stapes are attached to these two parts respectively with the incus slung between them.

Malleus. The **head** is the rounded superior part that lies in the epitympanic recess; the slender **manubrium** or handle is attached to the tympanic membrane.

Incus. The **body** of the incus lies in the epitympanic recess and articulates with the head of the malleus. It sends its **long crus** inferiorly to articulate with the head of the stapes, and a **short crus**, posteriorly, to be attached to the floor of the aditus by a ligament.

Stapes. It is shaped like a stirrup, and its foot piece or **base** is held in the fenestra vestibuli by an **anular** elastic **ligament**.

When the tympanic membrane and the handle of the malleus move, the head imparts a rotatory movement to the incus around the axis of its short crus, the long crus swinging towards then away from the fenestra vestibuli and imparting a similar movement to the stapes.

Tympanic Muscles

Stapedius. It lies within the pyramid and the canal which curves inferiorly from it. Its tendon enters the middle ear through the summit of the pyramid, and is inserted into the posterior surface of the neck of the stapes. **Nerve supply**: the facial nerve. **Action**: it damps the movement of the stapes, and its paralysis is associated with excessive acuteness of hearing (hyperacusia) owing to the uninhibited movement of the stapes.

Tensor Tympani. It arises from the superior surface of the cartilaginous part of the auditory tube and from the adjacent parts of the greater wing of the sphenoid and petrous temporal bone. It passes posterolaterally in the semicanal superior to the bony

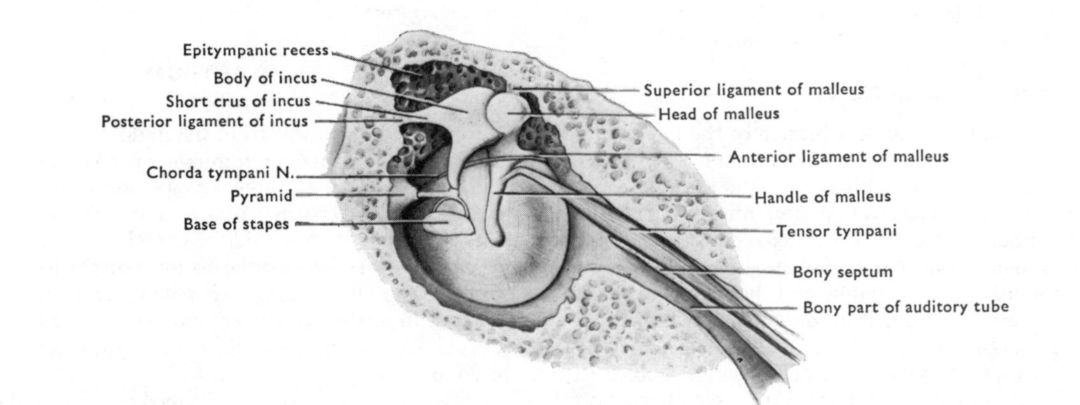

FIG. 151 Left tympanic membrane and auditory ossicles seen from the medial side. Note the auditory tube, and superior to it, tensor tympani and its semicanal.

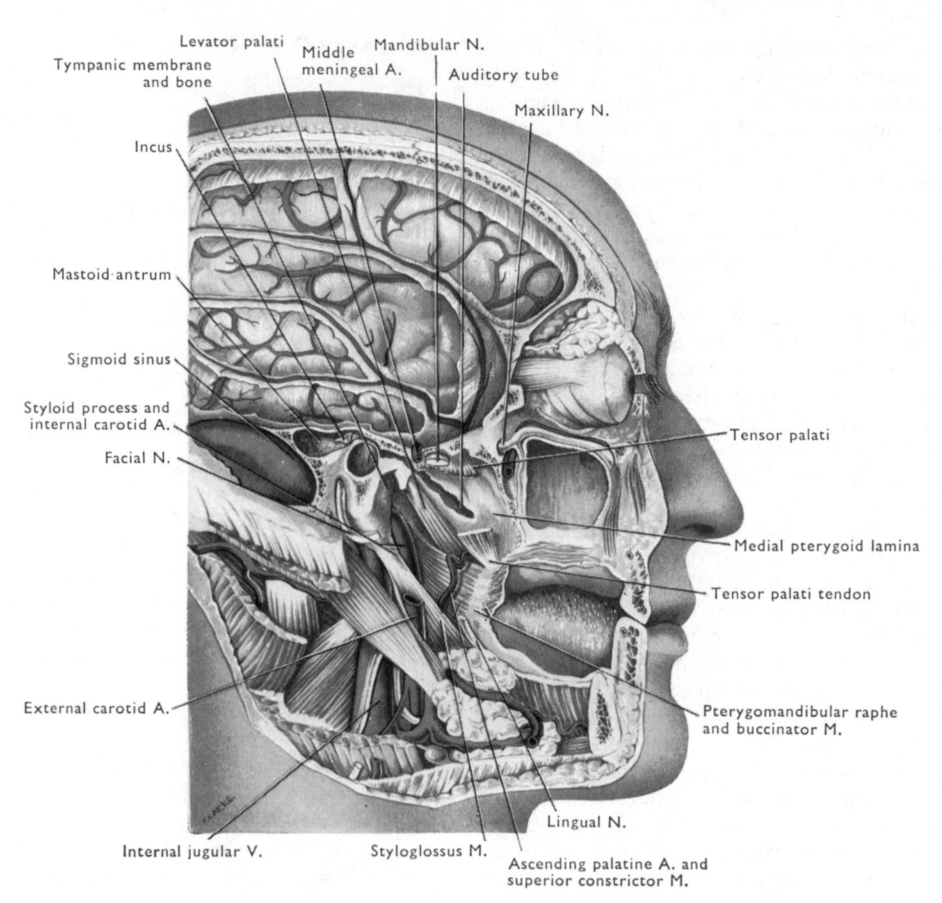

FIG. 152 Dissection of the auditory tube and mastoid antrum from the lateral side.

part of the auditory tube, and turns at 90 degrees round the processus cochleariformis in the middle ear cavity, to run laterally to the superior part of the handle of the malleus. **Nerve supply**: the mandibular nerve, through fibres which traverse the otic ganglion. **Action**: it tenses the tympanic membrane, increases its medial convexity, and restricts its freedom of movement. Thus it prevents wide excursions of the ear ossicles and potential damage to the inner ear when exposed to loud sounds.

Chorda Tympani Nerve

The chorda tympani is a branch of the facial nerve passing to the lingual nerve. It contains taste fibres to the anterior two-thirds of the tongue, and preganglionic parasympathetic fibres to the submandibular ganglion. The latter synapse with cells in the ganglion which innervate the submandibular and sublingual salivary glands and the glands buried in the anterior two-thirds of the tongue [p. 110].

The chorda tympani arises from the facial nerve in the temporal bone, a short distance above the stylomastoid foramen [FIG. 155]. It ascends in a bony canaliculus to the posterior wall of the middle ear, and passes forwards across the medial aspect of the upper part of the tympanic membrane and the neck of the malleus, outside the mucous membrane. At the superior part of the anterior wall of the middle ear, it enters another canaliculus through which it descends to the medial end of the squamotympanic fissure (petrotympanic part) on the base of the skull. Thence it runs antero-inferiorly, medial to the spine of the sphenoid, and joins the lingual nerve a short distance inferior to the skull.

MASTOID ANTRUM

[FIG. 150]

This air-filled extension from the middle ear cavity lies posterior to the epitympanic recess, and is continuous with it through an opening called the **aditus ad antrum**, immediately above the posterior wall of the middle ear. In the adult, the antrum is 1·5 cm medial to the **suprameatal triangle** (the small triangular area on the surface of the skull immediately posterosuperior to the bony external acoustic meatus) but is more superficial in the child [FIG. 4].

The walls of the antrum are formed by parts of the temporal bone. The tegmen tympani forms the **roof**. The **floor** and **posterior wall** are the mastoid

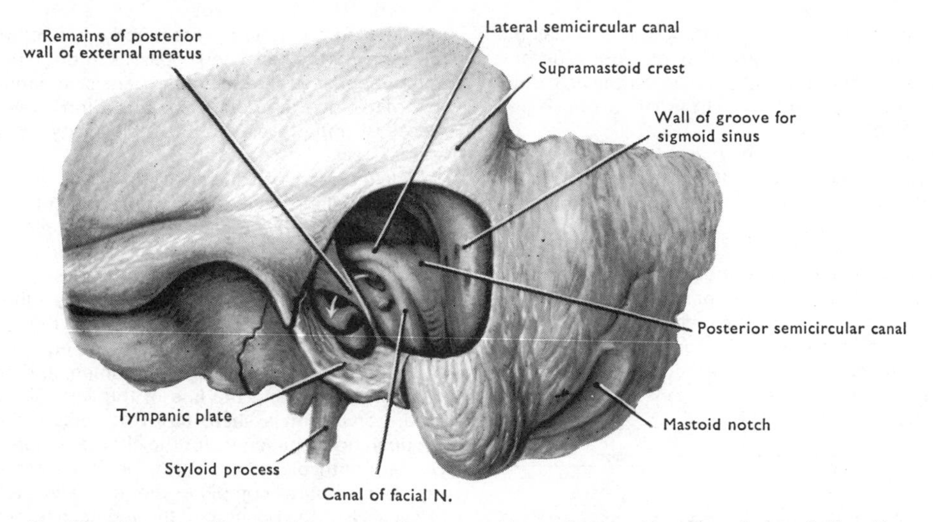

FIG. 153 Dissection of the mastoid antrum and petromastoid part of temporal bone from the lateral side. The arrow passes from the mastoid antrum into the tympanic cavity through the aditus.

process, and these have the openings of the mastoid air cells in them. The sigmoid venous sinus lies close to the posterior wall and may only be separated from it by a thin plate of bone [FIG. 152], while the jugular bulb may have the same close relation to the floor. Infections of the antrum may involve both veins in septic thrombosis, and this may sometimes spread along their tributaries to form an abscess in the adjacent part of the brain (cerebellum). The **medial wall** is the petrous temporal bone, and the ridge produced by the underlying lateral semicircular canal extends posteriorly into it from the aditus [FIG. 153].

The ridge on the medial wall of the middle ear, which is produced by the **facial nerve** in its bony canal, does not reach the antrum, but turns inferiorly (at the aditus) into the wall of bone which separates the middle ear cavity from the mastoid air cells and the inferior part of the antrum. In this bone the facial nerve descends vertically to emerge at the stylo-mastoid foramen [FIG. 150].

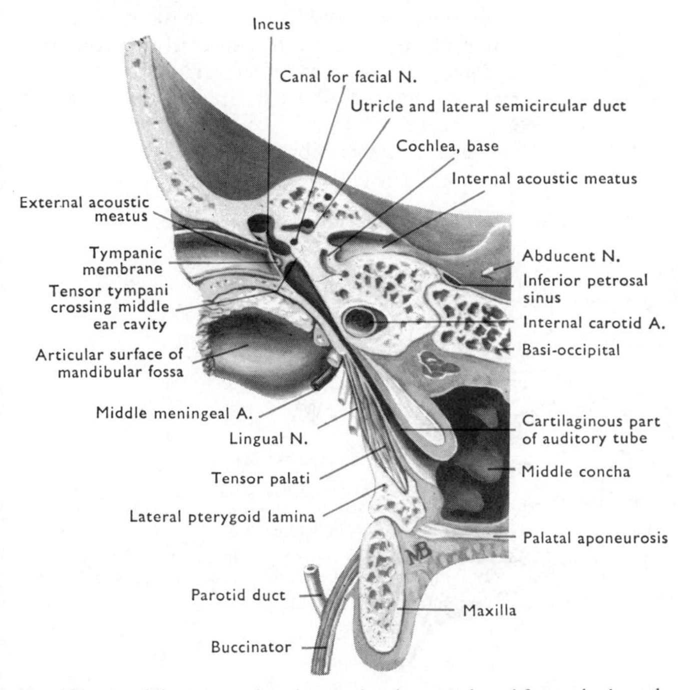

FIG. 154 An oblique coronal section passing downwards and forwards along the line of the auditory tube, viewed from behind. The mandible and muscles of mastication have been removed leaving the branches of the mandibular nerve on the lateral aspect of tensor palati.

AUDITORY TUBE

This pharyngotympanic tube consists of bony and cartilaginous parts. The cartilaginous part, approximately 2·5 cm long, is described on page 121.

The **bony part**, approximately 1·5 cm long, is widest at its junction with the middle ear cavity, and narrowest where it joins the cartilaginous part on the base of the skull posteromedial to the spine of the sphenoid. The bony part lies between the petrous and tympanic parts of the temporal bone, below the semicanal for the tensor tympani, and lateral to the internal carotid artery in the carotid canal [FIG. 154].

DISSECTION. Remove all the soft parts, including periosteum, from the lateral surface of the mastoid temporal bone, and identify the suprameatal triangle and the supramastoid crest. Begin by exposing the mastoid air cells and the mastoid antrum, without injuring the external meatus or entering the sigmoid sinus. With a fine chisel, remove the cortical bone from the supra-meatal triangle, and extend the bony removal anteromedially, parallel with the posterior wall of the external acoustic meatus, until the mastoid antrum is opened. Remove the spongy bone from the mastoid area until the compact bone posterior and medial to it is exposed. The extent of the **mastoid air cells** and antrum is variable.

Identify the structures which cause projections of the bony walls of the aditus and antrum; particularly the canal for the facial nerve, the sigmoid sinus, and the lateral semicircular canal [FIG. 153].

Cut away the posterior and superior walls of the external acoustic meatus up to the level of the roof of the mastoid antrum. Remove the tympanic membrane (try to identify the chorda tympani nerve at its posterosuperior margin) and the handle of the malleus, to obtain a clear view of the long crus of the incus and the position of the stapes and the stapedius tendon. Review the medial wall of the middle ear [FIGS. 148, 150].

INTRAPETROUS PARTS OF FACIAL AND VESTIBULOCOCHLEAR NERVES

The facial and vestibulocochlear nerves have already been traced to the internal acoustic meatus [p. 50]. These nerves should now be followed into the petrous temporal bone; the vestibulocochlear ends in the internal ear, medial to the middle ear; the facial nerve follows an angulated course through dense bone to emerge at the stylomastoid foramen on the inferior surface of the temporal bone.

DISSECTION. Identify the **arcuate eminence** on the anterior surface of the petrous temporal bone; it marks the position of the **anterior semicircular canal**. On the posterior surface of the petrous temporal bone, identify the internal acoustic meatus and the nerves entering it. Place a chisel horizontally across the upper part of the internal acoustic meatus, and, with a sharp tap from a mallet, attempt to break off the superior part of the petrous temporal bone. Frequently it fractures along the canal for the facial nerve, passes through parts of the anterior and lateral semicircular canals, and removes the roof of the internal acoustic meatus. Chip away any extra pieces of bone to expose the facial nerve as far as the aditus.

Note the sharp bend (**geniculum**) on the facial nerve, and the swelling on it at this point (genicular ganglion) where it gives off the greater petrosal nerve, and turns posteriorly into the medial wall of the middle ear. Follow the facial nerve posteriorly above the fenestra vestibuli till it turns inferiorly in the medial wall of the aditus.

Turn to the inferior surface and identify the facial nerve at the stylomastoid foramen. Place the edge of a chisel across the lateral margin of the foramen, and attempt to split the bone along the line of the vertical part of the facial nerve with a sharp tap. If it splits satisfactorily, the posterior canaliculus for the chorda tympani and the whole length of the canal for the facial nerve will be exposed; if not, complete the exposure with bone forceps. Note the position of the vertical part of the facial nerve relative to the middle ear cavity and the mastoid air cells and antrum. Finally break away one wall of the pyramid and expose stapedius.

Facial Nerve [FIG. 155]

This nerve runs laterally through the internal acoustic meatus anterosuperior to the vestibulocochlear nerve. Here it is joined by the **nervus intermedius**. At the lateral extremity of the meatus, it enters its canal in the bone, and continues laterally for a short distance, above the vestibule of the internal ear, to reach the **genicular ganglion**. At the ganglion, it gives off the **greater petrosal nerve** [p. 49] and sends a branch to the lesser petrosal nerve.

The nerve turns abruptly backwards at the

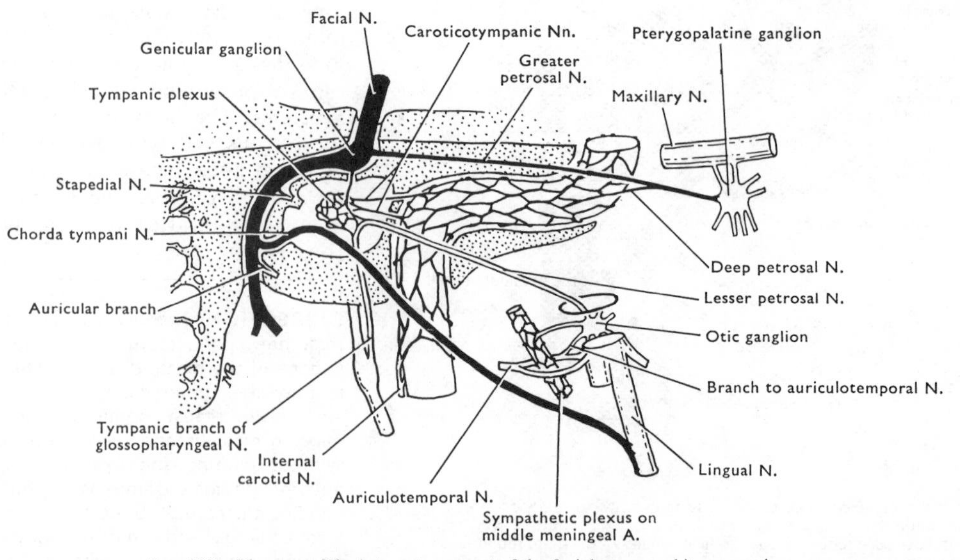

FIG. 155 Diagram of the intrapetrous part of the facial nerve and its connexions.

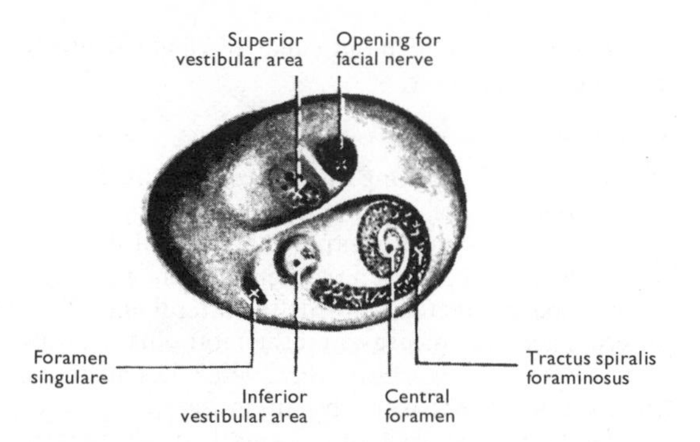

FIG. 156 Fundus of left internal acoustic meatus divided into upper and lower areas by transverse crest.

ganglion, and running posteriorly in the bone of the upper part of the medial wall of the middle ear cavity, swings inferiorly at the aditus, to run vertically through the bone between the middle ear cavity, anteriorly, and the mastoid antrum and air cells, posteriorly. In the vertical part of its course it gives off: (1) the nerve to stapedius; (2) the chorda tympani; (3) twigs to the auricular branch of the vagus.

Clinically it is possible to differentiate *injuries to the facial nerve* at, or distal to, the stylomastoid foramen, which show only paralysis of the facial muscles, from those at the brain or in the internal acoustic meatus, which also produce loss of taste on the anterior two-thirds of the corresponding half of the tongue (chorda tympani) and hyperacusia (nerve to stapedius). Clearly, an injury in the internal acoustic meatus is also likely to involve the vestibulocochlear nerve.

Vestibulocochlear Nerve

This nerve is postero-inferior to the facial nerve in the internal acoustic meatus. At the lateral end of the meatus, the vestibulocochlear nerve splits into cochlear and vestibular parts. Each of these divides and passes through foramina in the lateral end of the meatus to supply the various parts of the membranous labyrinth of the internal ear [FIG. 148].

THE INTERNAL EAR

The internal ear consists of a series of communicating bony spaces (the **bony labyrinth**) in the petrous temporal bone, and a series of membranous tubes and sacs, the **membranous labyrinth**, contained within it. The cochlear portion is concerned with hearing, the remainder with equilibration, for it records the direction of gravity acting on the head, and rotational movements of the head. The membranous labyrinth is filled with a clear fluid (**endolymph**) and is separated from the surrounding bone by a considerable space which contains a similar fluid (**perilymph**) and some delicate connective tissue.

DISSECTION. Free the facial nerve from its canal as far as the aditus, but retain its continuity with the greater petrosal nerve. Using the chisel in the same way as before, cut more bone from the superior surface of the petrous temporal, till the level of the middle of the internal acoustic meatus is reached. The bone should be cracked by sharp taps from a mallet so as to avoid driving the chisel into the middle ear cavity and the ear ossicles. As each flake of bone is removed, examine the holes in the bone produced by the semicircular canals, and try to see the semicircular ducts of the membranous labyrinth lying free within them. Note also the branches of the vestibulocochlear nerve entering the bone at the lateral end of the meatus. As the level of the middle of the meatus is reached, the vestibule and cochlea will be broken into. The vestibule is the cavity lying immediately lateral to the internal meatus, and separating it from the medial wall of the middle ear; the cochlea lies anterolateral to the lateral end of the meatus.

The above procedure shows the main features of the bony labyrinth, and it is desirable at each stage to establish the continuity of its various parts by passing a fine wire through each foramen that is exposed. If time permits, a better dissection may be obtained by decalcifying an intact temporal bone in dilute acid for several weeks, and then dissecting it carefully with a sharp knife.

BONY LABYRINTH

The bony labyrinth is divided into three parts: (1) a small chamber called the **vestibule**; (2) a coiled tube, the **cochlea**, anterior to the vestibule; (3) three **semicircular canals** posterior to the vestibule. The cochlea and canals communicate freely with the vestibule, and are lined throughout by a delicate endosteum.

Vestibule [FIGS. 149, 157, 158]

This is a small, ovoid, bony chamber about 5 mm in length, placed between the lateral end of the internal acoustic meatus and the medial wall of the middle ear. The three semicircular canals open into its posterior part. The **fenestra vestibuli**, closed by the base of the stapes and the endosteal lining, is on the lateral wall. When these are removed, the

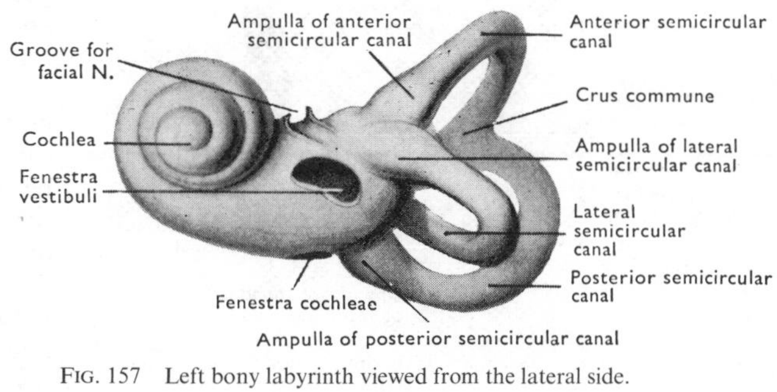

FIG. 157 Left bony labyrinth viewed from the lateral side.

vestibule communicates with the middle ear cavity. On the antero-inferior wall of the vestibule is the opening of the **scala vestibuli**, one of the two perilymph tubes which pass into the coiled cochlea. In the posterior part of the medial wall is the mouth of a small canal (**aqueduct of the vestibule**) which passes to the posterior surface of the petrous temporal bone, where it opens between the bone and the dura mater.

Semicircular Canals

Each of the three semicircular canals (anterior, posterior, and lateral) with a swelling (**ampulla**) at one end, forms considerably more than half a circle. They lie in planes such that each forms one of three adjacent faces of an obliquely set cube. The anterior and posterior canals lie in approximately vertical planes, and their adjoining ends meet in a common limb (**crus commune** [FIG. 158]) which runs along the edge of the cube to the vestibule. Thus the three canals have five points of entry to the vestibule.

The **anterior canal** is at right angles to the long axis of the petrous temporal bone, and produces the **arcuate eminence** on the anterior surface of that bone. The **posterior canal** is postero-inferior to the anterior canal, and lies almost parallel to the posterior surface of the petrous temporal bone, deep to the slit for the aqueduct of the vestibule. The **lateral canal** is nearly horizontal, but slopes a little anterosuperiorly in the angle between the attached ends of the other two. Its most lateral part makes a bulge in the medial wall of the mastoid antrum and its aditus.

The **ampullae** lie at the anterior ends of the anterior and lateral canals, while that on the posterior canal lies at its posterior opening into the vestibule.

The semicircular canals of one inner ear are mirror images of those in the other ear. Thus, the two lateral canals lie in the same plane, and the anterior canal of one side is parallel with the posterior canal of the other. In any parallel pair, the ampullae are so arranged that endolymph in the semicircular ducts (flowing in the same direction in both as a result of rotation of the head in the plane of that pair) moves towards the ampulla in one canal and away from the ampulla in the other.

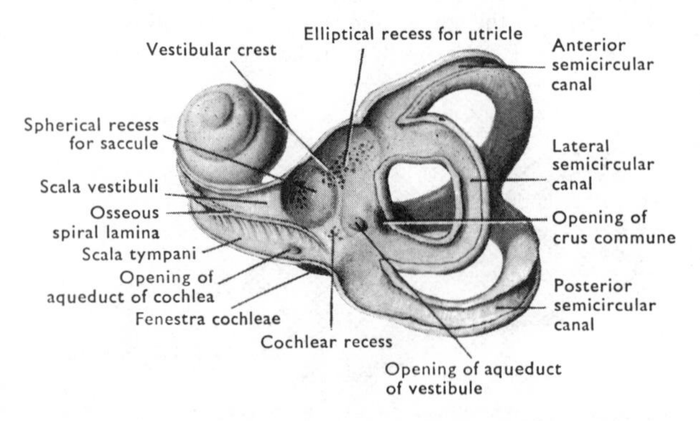

FIG. 158 Interior of the left bony labyrinth viewed from the lateral side.

Cochlea

The cochlea is a tapering tube coiled spirally, for two and one-half turns, around a central pillar, the modiolus. It has the appearance of a spiral shell laid on its side. The cochlea lies anterior to the vestibule, with its base directed towards the lateral end of the internal acoustic meatus, and the modiolus pointing anterolaterally towards its apex, which lies medial to the canal for the tensor tympani muscle.

The bony cochlear tube rapidly diminishes in diameter as it is traced towards its apex; the large first turn producing the bulge of the **promontory** on the medial wall of the middle ear cavity [FIGS. 150, 157].

The **modiolus** is thick at its base where it abuts on the internal meatus, but it rapidly tapers towards the apex, forming the internal wall of the cochlear tube. Winding spirally round the modiolus, like the thread of a screw, is a thin, narrow plate of bone (the **spiral lamina**) which partially divides the cochlear tube into two canals.

The modiolus is traversed by a number of minute, longitudinal canals which turn outwards towards the spiral lamina and enter the **spiral canal** of the modiolus. This canal runs in the base of the spiral lamina and lodges the sensory, **spiral ganglion** of the cochlea. The cells of the ganglion send their peripheral processes through minute canals in the spiral lamina to the sensory organ of the cochlea, and their central processes run in the canals of the modiolus, to emerge from its base in the internal acoustic meatus as the cochlear part of the vestibulocochlear nerve.

The **duct of the cochlea** is a spiral, tubular extension of the membranous labyrinth, which passes into the bony cochlear tube. It occupies the central part of the cochlear tube, lying between the spiral lamina and the lateral wall; the remainder of the space on each side of it being filled by the **scala tympani** and **scala vestibuli**. These two perilymph tubes run parallel to and on either side of the cochlear duct, and are continuous at the apex of the cochlea (**helicotrema**) around the blind end of the cochlear duct.

The scala vestibuli is continuous with the perilymph cavity of the vestibule at the base of the cochlea, while the scala tympani ends on the **secondary tympanic membrane** which fills the round window (fenestra cochleae). Thus pressure waves generated in the vestibular perilymph by movements of the stapes are readily transferred through the scala vestibuli and the scala tympani because of the corresponding oscillations of the secondary tympanic membrane. Thus the cochlear duct and its sensory organ lying between the scalae is directly involved.

The scala tympani is also continuous with the aqueduct of the cochlea which passes through the **cochlear canaliculus** [FIG. 158] to appear

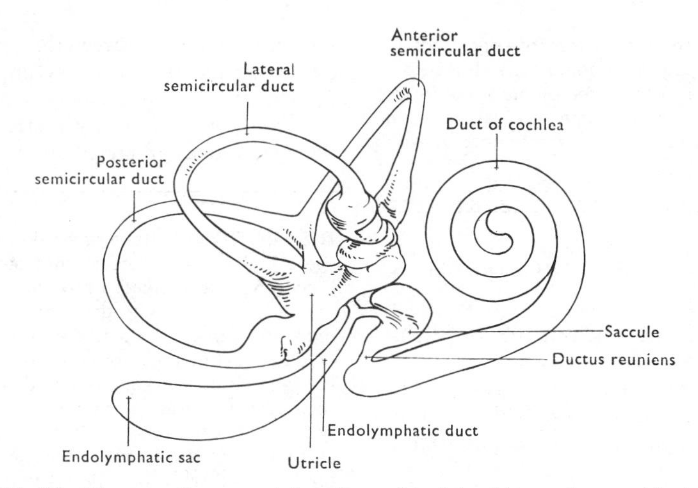

FIG. 159 Diagrammatic representation of a model of the right membranous labyrinth.

immediately superior to the jugular foramen, in close association with the glossopharyngeal nerve. This duct is said to be continuous with the subarachnoid space, but it probably ends blindly on the dura mater.

MEMBRANOUS LABYRINTH

The chief parts of the membranous labyrinth are: (1) the duct of the cochlea; (2) the utricle and saccule, two small membranous sacs which lie in the vestibule, and are joined by a narrow tube through the root of the **endolymphatic duct**. The saccule is continuous with the base of the cochlear duct through the **ductus reuniens**, and the utricle has the membranous semicircular ducts arising from it. (3) Three membranous tubes, the semicircular ducts, which lie in the semicircular canals [FIGS. 148, 159].

The **semicircular ducts** are considerably narrower than the canals, and lie against the peripheral bony walls of the canals. They open into the utricle and have ampullae in the same position as the ampullae of the canals. The **ampullae** of the ducts contain a fold of their lining (crista ampullaris) which extends transversely across the ampulla. It is surmounted by hair cells embedded in a gelatinous **cupula**, and is the sensory organ of the duct. It records movements of the endolymph in the ampulla which result from rotation of the head in the plane of the duct, and which may be simulated by convection currents produced by introducing hot or cold water into the external acoustic meatus. It is innervated by the vestibular part of the vestibulocochlear nerve, which ends on the hair cells of the epithelium of the crista.

The **utricle** occupies a depression in the superior wall of the vestibule. The smaller **saccule** lies antero-inferior to it, and the endolymphatic duct (which unites them) passes through the bony aqueduct of the vestibule, to end as a dilated **endolymphatic sac** external to the dura mater. The utricle and saccule each have a small area of thickened epithelium (**macula**) in their wall, with hair cells on which lie a number of concretions of calcium salts (**statoconia**). These maculae record the direction of the gravitational field relative to the head, the macula of the utricle lying in a horizontal plane, that of the saccule in a vertical plane. Both are innervated by the vestibular part of the vestibulocochlear nerve.

THE EYEBALL

The eyeball lies in the anterior part of the orbit, enclosed in a fascial sheath which separates it from the orbital muscles and fat.

The eyeball is about 2·5 cm in diameter, but is not perfectly spherical, for the anterior, clear part (the cornea) has a smaller radius of curvature than the rest of the globe, and protrudes from the anterior surface of the eyeball.

It is rarely possible to make a satisfactory dissection of the eyeball of a cadaver, as good results can only be achieved either with the fresh eye, or with a fresh one which has been hardened for a few days in formaldehyde. The student is therefore advised to procure a number of fresh eyes from pig, sheep, or ox, and store them in formaldehyde solution, except for one which is best laid aside for a day or two to demonstrate the vitreous body.

DISSECTION. Clean the loose tissue off the surface of the eyeball. Pick up the conjunctiva and fascial sheath close to the corneal margin, and cut through these layers around the margin of the cornea. Proceed to strip all these soft parts from the surface of the white part of the eye (the sclera) working steadily backwards towards the entry of the optic nerve. The venae vorticosae will be found piercing the sclera a little posterior to the equator (in the coronal plane), and the posterior ciliary arteries and ciliary nerves will be seen entering the sclera around the attachment of the optic nerve.

To obtain a general idea of the parts which form the

eyeball, make sections through hardened specimens in different planes: (1) through the equator; (2) a horizontal section through two specimens, removing the vitreous body [FIG. 160] from one. Place the sections in formaldehyde solution and keep for reference during dissection.

GENERAL STRUCTURE OF THE EYEBALL

The eyeball consists of three concentric coats, which enclose a cavity filled with three of the four light-refracting media.

The coats are: (1) An external fibrous coat, the posterior five-sixths of which is the white, opaque **sclera**; while the transparent, anterior one-sixth is the more highly curved **cornea**. (2) A middle coat which is vascular and muscular. It consists of the **iris** anteriorly, the **choroid** posteriorly, and the thickest part, the **ciliary body**, which is intermediate in position. (3) The internal coat, the **retina**, contains the light-sensitive elements (rods and cones) and gives rise to the nerve fibres of the optic nerve.

The **refracting media** are: (1) the cornea; (2) the aqueous humour, a watery medium which lies between the cornea and the lens, and fills the **anterior chamber** (between the cornea and the iris) and the **posterior chamber** (between the iris and the lens) of the eye. These two chambers are directly continuous through the aperture in the middle of the iris, the **pupil**. (3) The lens separates the posterior chamber from (4) the vitreous body, which occupies the largest part of the cavity of the eyeball, the **vitreous chamber**. Because the greatest change in refractive index is between the air and the cornea, its curved surface plays a very important part in the optical system, and changes in its shape, from whatever cause, markedly interfere with the ability of the eye to focus an image on the retina.

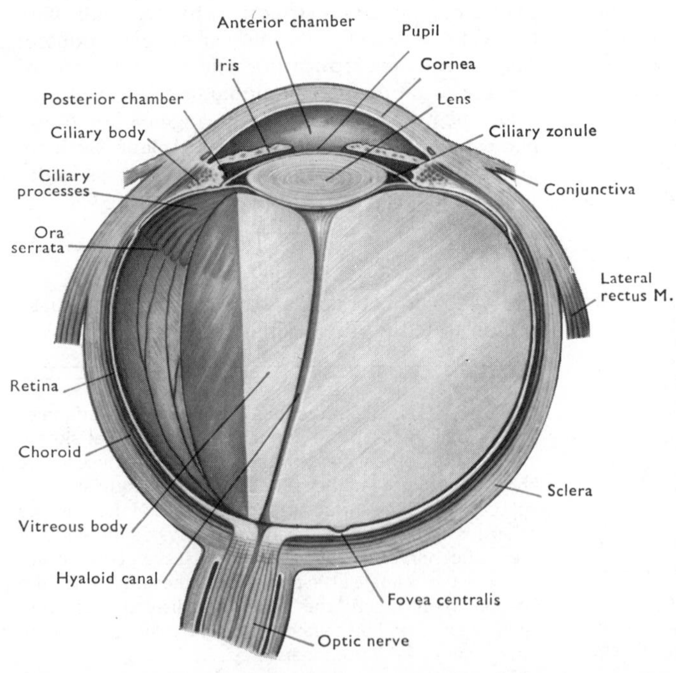

FIG. 160 Diagrammatic section of the eyeball.

DISSECTION. With a sharp knife, make an incision through the sclera at the equator, stopping as soon as the black choroid appears. Introduce one blade of a blunt ended pair of scissors through the cut and extend the incision through the sclera around the equator, taking care to separate the choroid from its deep surface as you proceed. Raise the anterior and posterior parts of the divided sclera from the choroid, and by careful, blunt dissection turn these parts anteriorly and posteriorly, stripping them from the choroid. Some resistance will be encountered with the anterior part close to the cornea, due to the attachment of the ciliary muscle to the deep surface of the sclera. As this attachment is broken down by continued blunt dissection, the aqueous humour will escape. The posterior part can finally be removed by dividing the optic nerve fibres where they enter the deep surface of the sclera.

The eyeball, denuded of its fibrous coat, should be placed in water for further investigation.

FIBROUS COAT

Sclera [FIG. 160]

The sclera is a dense, collagenous coat, which covers the posterior five-sixths of the eyeball, and is loosely attached to the choroid by some delicate, pigmented areolar tissue, but firmly adherent to the ciliary body.

The point where the optic nerve pierces the sclera is approximately 3 mm to the nasal side of the posterior pole. The bundles of optic nerve fibres, with the blood vessels to the retina among them, pass through a number of holes in the sclera. Here the sheath of the optic nerve, consisting of all three meninges of the brain, fuses with the sclera; the subdural and subarachnoid spaces ending at this point.

The sclera is also pierced by numerous blood vessels and nerves, which supply the fibrous and vascular coats. The long and short **posterior ciliary arteries** and the **ciliary nerves** pierce it around the optic nerve. Four to five **venae vorticosae** emerge a short distance posterior to the equator, more or less equally spaced around the eyeball. The **anterior ciliary arteries** pierce it near the corneal margin [FIG. 163].

The sclera is directly continuous with the cornea, and close to their junction, a minute canal in the internal part of the sclera (**sinus venosus sclerae**) encircles the

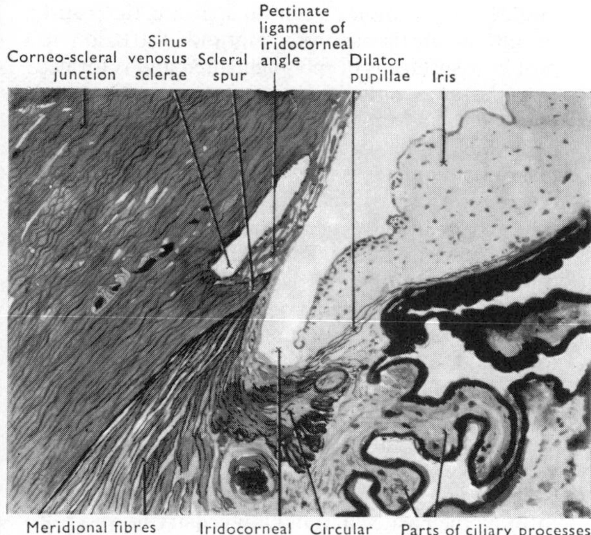

FIG. 161 Section of iridocorneal angle.

margin of the cornea, and drains the aqueous humour [FIG. 167].

Cornea

The cornea is the transparent one-sixth of the fibrous coat of the eyeball, and it is separated from the iris by the aqueous humour in the anterior chamber.

The anterior surface of the cornea is covered with a layer of epithelium which is firmly bound to it, and is continuous with the conjunctiva at the margin of the cornea. On the posterior surface of the cornea is the **posterior limiting lamina**; an elastic glassy layer which becomes wrinkled when the tension in the cornea is released, and may be torn away in shreds from the proper corneal tissue.

Pectinate Ligament of Iris. At the margin of the cornea, the posterior limiting lamina breaks up into bundles of fibres. Some of these pass posteriorly into the choroid and sclera; others arch medially into the iris (pectinate ligament of the iris) crossing the angle between the cornea and the iris in the form of a number of separate bundles, with the minute **spaces of the iridocorneal angle** between them. These spaces seem to form a communication between the anterior chamber of the eyeball and the sinus venosus sclerae, through which aqueous humour can filter away into the venous system. Blockage of this system is associated with a serious condition in which the intra-ocular tension is greatly increased (glaucoma) [FIG. 161].

MIDDLE COAT

Choroid

The choroid is the largest part of the middle coat, and lies between the sclera and the retina. It is thickest posteriorly, where it is pierced by the optic nerve, and becomes thinner as it approaches the ciliary body. It is connected with the sclera by some delicate, pigmented areolar tissue, and also by the blood vessels and nerves which pass between them. The deep surface of the choroid is moulded on the retina and is in contact with a layer of deeply pigmented cells which belong to the retina, but which normally adhere to the choroid, since a potential space (the remnant of the cavity in the optic outgrowth of the embryo) exists between this pigmented layer of the retina and the retina proper.

The choroid consists mainly of blood vessels which are arranged in two layers: (1) a deep, close-meshed capillary net; (2) a more superficial venous layer from which the venae vorticosae arise. The short posterior ciliary arteries run between these layers.

The eyeball, from which the fibrous coat has been removed, should be gently brushed under water with a camel hair brush, to remove the pigment and expose the curved tributaries of the venae vorticosae, which appear as white lines.

Ciliary Body

This region of the middle coat is divisible into (1) an external part, the ciliary muscle, and (2) an internal part consisting of a number of radial ridges, the ciliary processes [FIG. 162].

Ciliary Muscle. It is composed of involuntary muscle fibres which are arranged in two groups, radiating and circular, which are only visible in microscopic preparations [FIG. 161].

The **radiating fibres** arise from the deep surface of the sclera close to the cornea, and radiate posteriorly into the ciliary processes. The **circular fibres** are arranged in two or three bundles that lie on the deep surface of the radiating fibres, and form a muscular ring close to the peripheral margin of the iris. **Nerve supply**: through the short ciliary nerves by postganglionic parasympathetic nerve fibres which originate in the ciliary ganglion; the preganglionic fibres reach the ganglion through the oculomotor nerve. **Action**: see page 157.

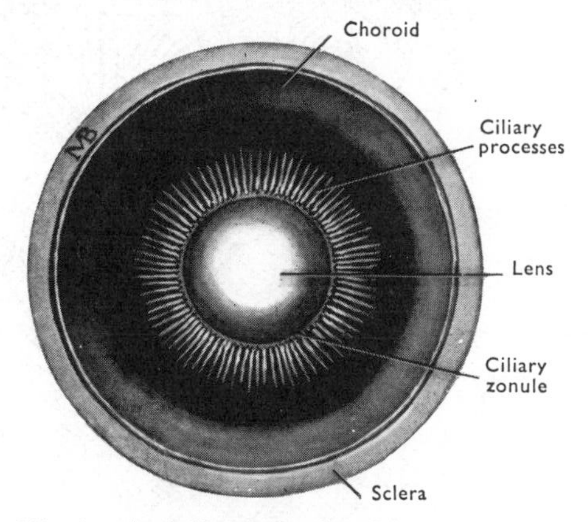

FIG. 162 Anterior half of the interior of the eyeball viewed from behind after removal of the vitreous.

DISSECTION. Expose the ciliary processes by making a coronal section through an eyeball anterior to the equator, and remove the vitreous body from the anterior segment. Then wash out the pigment from the anterior part of the middle coat to expose the ciliary processes more fully.

Alternatively, remove the cornea by cutting round the corneoscleral junction, and examine the exposed iris. Make a number of radial cuts into the anterior part of the sclera, and fold each segment of sclera outwards, stripping it from the ciliary body. Remove the iris.

Ciliary Processes. These are approximately seventy radial folds which extend from the anterior margin of the choroid across the posterior surface of the ciliary body, towards the margin of the lens. They end behind the peripheral margin of the iris, posterior to the posterior chamber of the eyeball. The deep surfaces of the processes are applied to the **vitreous membrane**, the outer condensed part of the vitreous body, and also to the peripheral part of the ciliary zonule [p. 156] to which they are attached.

Iris

The iris (the coloured part of the eye) lies anterior to the lens and the posterior chamber. It is separated from the cornea by the anterior chamber, and is bathed by aqueous humour on both surfaces. Its circumference is continuous with the ciliary body, and is connected to the cornea by the pectinate ligament.

The iris varies greatly in colour. It is circular in outline and surrounds a central aperture, the **pupil**. Its anterior surface shows a faint radial striation: its posterior surface is deeply pigmented and is formed from the same epithelial layer as the retina. The pupil is circular in Man, but may be slit-like in other mammals, either in the horizontal or the vertical plane. Its diameter is constantly varying to control the amount of light reaching the retina, and this is achieved by the presence of two sets of involuntary muscle fibres in the iris. The pupil is decreased in size by the circular fibres of the **sphincter pupillae** (which lie in the pupillary margin of the iris) and dilated by the radial myoepithelial cells, which pass towards the periphery of the iris—the **dilator pupillae**. **Nerve supply**: postganglionic parasympathetic fibres from the ciliary ganglion pass to the sphincter via the short ciliary nerves. The dilator is supplied by postganglionic sympathetic fibres which originate in the carotid plexus, and reach it through the nasociliary and long ciliary nerves.

Ciliary Nerves [Fig. 163]

The **short ciliary nerves** arise from the ciliary ganglion and pierce the sclera around the optic nerve as twelve or more fine filaments. On the internal aspect of the sclera, they run anteriorly and break up into fine plexiform branches to the ciliary muscle, iris, and cornea [Fig. 163]. The short ciliary nerves contain: (1) **postganglionic parasympathetic fibres** from the cells of the ciliary ganglion to the ciliary muscle and the sphincter of the pupil; (2) **postganglionic sympathetic fibres** from the internal carotid plexus to the vessels of the eyeball; (3) **sensory fibres** from the nasociliary nerve. Groups (2) and (3) have no connexion with the cells of the ciliary ganglion.

The **long ciliary nerves** transmit sensory and sympathetic nerve fibres from the nasociliary nerve through the sclera, close to the optic nerve. The sympathetic fibres come from the internal carotid plexus, and some of them supply the dilator pupillae.

Ciliary Arteries [Figs. 163, 164]

The **short posterior ciliary arteries** are branches of the ophthalmic artery which pierce the sclera close to the optic nerve and end in the choroid.

The **long posterior ciliary arteries** are two branches of the ophthalmic artery which pierce the sclera some distance from the optic nerve. They run anteriorly, between sclera and choroid, on opposite sides of the eyeball, and branching in the ciliary region, anastomose with the anterior ciliary arteries to form the **greater arterial circle** at the periphery of the iris [Fig. 164]. This circle supplies

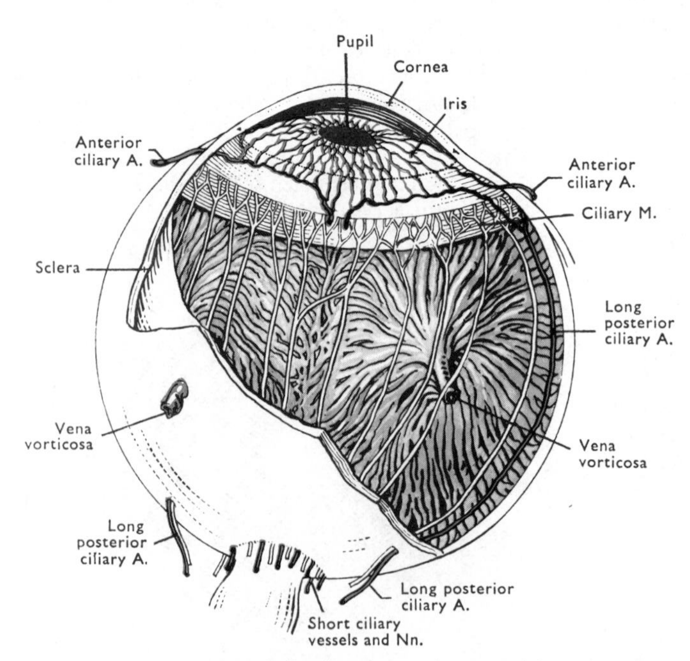

Fig. 163 Dissection of the eyeball to show the vascular coat and the arrangement of ciliary nerves and vessels.

the iris, the ciliary muscle, and the ciliary processes.

The **anterior ciliary arteries** are minute twigs which arise from the arteries to the rectus muscles. They pierce the sclera close to the cornea, and take part in the formation of the greater arterial circle.

The **lesser arterial circle** is a small anastomotic ring in the iris at the external border of the sphincter pupillae.

Venae Vorticosae [FIG. 163]

A large vein arises from each venous vortex in the choroid, and pierces the sclera immediately posterior to the equator. Four to five of these veins drain into the ophthalmic veins.

DISSECTION. Remove the vitreous body and retina from the posterior segment of the specimen used to show the ciliary processes. Using a stream of water, raise the choroid from the sclera and expose the venae vorticosae entering the sclera. Cut these veins, and continue to strip the choroid from the sclera until the short ciliary nerves are exposed.

In the specimen from which the sclera and cornea have been removed, strip off the iris, ciliary processes, and choroid, piecemeal, under water. This exposes the external surface of the retina.

Retina

The retina consists of: (1) a thin, outer **pigmented layer** which is adherent to the choroid and is removed with it; (2) the thicker, internal, **nervous layer**. The nervous layer is only attached to the vitreous and pigmented layers where the optic nerve pierces the sclera. Hence it is readily detached from

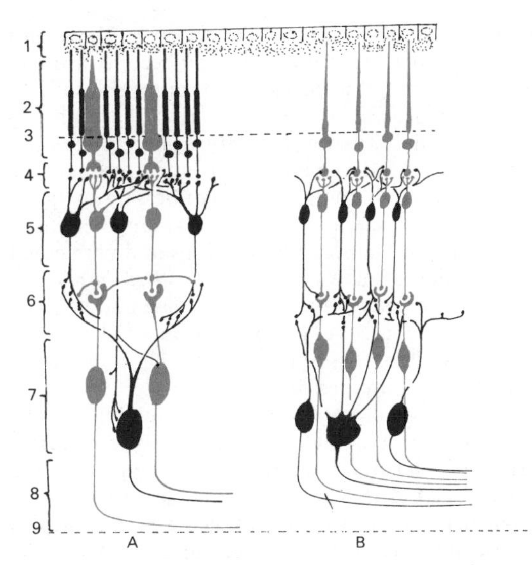

FIG. 165 Diagram of the arrangement of cells in the peripheral (A) and central (B) parts of the retina. Only the rods, cones, bipolar, and ganglion cells are shown. The cones and specific cone connexions are shown in red. In B the tightly packed cones have been separated for diagrammatic purposes. (After Polyak.)

the pigmented layer even during life. Such a detachment of the retina makes it impossible to focus an image on the nervous layer, and may interfere with its nutrition (see below).

The pigmented layer absorbs the light which has passed through the nervous layer so that it is not reflected back as scattered rays which would interfere with the sharpness of the image on the retina. In many nocturnal animals the choroid forms a reflecting layer (tapetum) to ensure the passage of low intensities of light twice through the retina, and so increase its sensitivity.

The optic part of the retina ends, anteriorly, at a wavy margin, the **ora serrata**, close to the posterior margin of the ciliary body. The two layers are continued anterior to this as a very thin layer (consisting of two single layers of cells) over the ciliary processes (**pars ciliaris retinae**) and the posterior surface of the iris (**pars iridica retinae**).

The transparent optic part of the retina soon becomes opaque after death. It consists of three layers of cells, the outermost are the light-sensitive **rods** and **cones**, and the innermost are the nerve (**ganglion**) cells that give rise to the nerve fibres which pass to the brain in the optic nerve. These nerve fibres course over the internal surface of the retina and converge on the end of the optic nerve. Opposite this they build up into a circular elevation, which is slightly hollowed out in the centre, the **optic disc**. This is situated approximately 3 mm to the nasal side of the anteroposterior axis of the

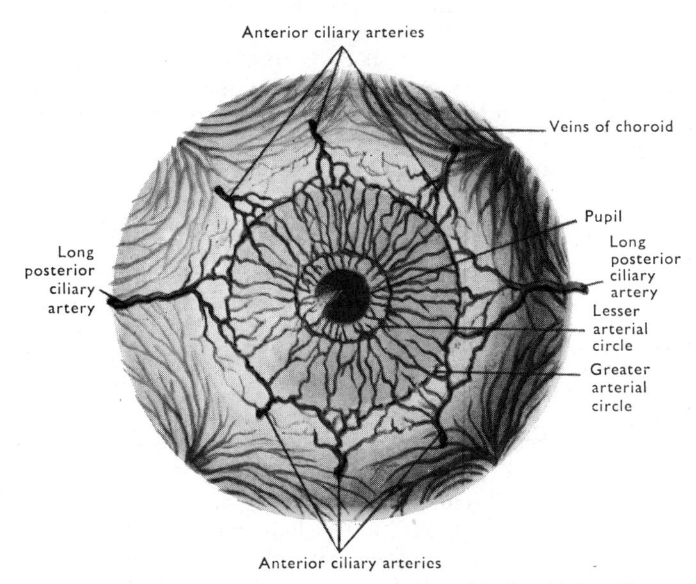

FIG. 164 Blood vessels of the iris and the anterior part of the choroid viewed from in front.

eyeball, and is the site where the nerve fibres turn posteriorly into the optic nerve and become white in colour, owing to the development of myelin sheaths on their surfaces. At the optic disc there are no sensory retinal elements, but only the nerve fibres and the retinal blood vessels which enter through the optic nerve. At this point the nerve fibres tie down the retina to the sclera, though elsewhere there is a potential space between the nervous and pigmented layers of the retina. *No blood vessels enter the nervous layer except through the optic disc.* The middle layer of the retina consists of **bipolar cells** which connect the rods and cones to the ganglion cells, and certain other important types of cells.

In the centre of the human retina, and lying in the visual axis, is a small yellowish spot, the **macula lutea**, which has a slight depression at its centre, the **fovea centralis**. At the fovea, the inner two layers of the nervous layer of the retina are swept aside to expose the cones on the internal surface. This allows light to fall directly on these elements without passing through the other layers, and so improves the quality of the image on the cones in this tiny area. At this point also the resolving power of the retina is at its maximum, for here the sensory elements (cones) are slender and very tightly packed. The nerve fibres and blood vessels which pass to and from the optic disc skirt the macula.

Retinal Arteries and Veins. These are the only vessels in the retina, and the only supply to its inner layers. Thus blindness results if the central artery is blocked. The outer retinal layers receive at least some of their nutrition from the choroidal vessels by diffusion—an important point in retinal detachment.

The central artery of the retina reaches it from the ophthalmic artery by piercing the sheaths of the optic nerve and entering its substance. Here it runs with the corresponding central vein, and appears in the eyeball at the optic disc, dividing into superior and inferior branches. Each of these divides into a large temporal and a smaller nasal arteriole [FIG. 166], the branches of which run on the internal surface of the retina to the ora serrata. They neither anastomose with each other nor with any of the other arteries of the eyeball. Hence blockage of a branch leads to death of the part of the retina it supplies, and blindness in the corresponding part of the visual field of that eye.

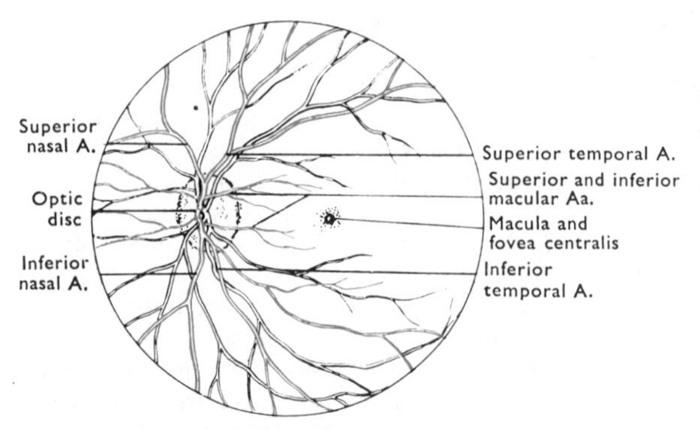

FIG. 166 Retinal blood vessels as seen through the ophthalmoscope.

The **retinal veins** converge on the optic disc, and enter the optic nerve as two small trunks which soon unite to form the **central vein of the retina**. The latter vein drains into the superior ophthalmic vein, but it must pierce the meningeal sheaths of the optic nerve and the contained sleeve of subarachnoid space to reach it. Increased pressure in the subarachnoid space tends to compress the vein and cause distension of its tributaries in the retina; a feature which can be seen with the ophthalmoscope.

The retina is the one situation where blood vessels can be examined in the living, and signs of vascular pathology observed at first hand. The student should take every opportunity of using an ophthalmoscope and seeing the normal retina and blood vessels of his fellows.

DISSECTION. Take the eye which was put aside to lose its freshness and divide the coats around the equator. Strip the coats gently from the underlying vitreous and lens. It is advisable to place the vitreous and lens in a strong solution of picrocarmine for a few minutes, and then wash well in water. This stains the vitreous membrane, the capsule of the lens, and the ciliary zonule red, and makes their connexions readily visible.

Vitreous Body

This is a soft, transparent, jelly-like body which fills the posterior four-fifths of the eyeball. It abuts on the retina, the ciliary processes, and the lens which forms a deep concavity on the anterior surface of the vitreous, the **hyaloid fossa**.

The vitreous body is condensed superficially to form a transparent envelope, the **vitreous membrane**. A minute canal runs through it from the optic disc to the posterior surface of the lens, and represents the remains of a branch of the central artery of the retina (**hyaloid artery**) which supplied the developing lens in the foetus, but which subsequently disappeared. This **hyaloid canal** cannot be seen in an ordinary dissection, but may be visualized in the living with a slit lamp. The minute remnant of the hyaloid artery attaches the vitreous body to the optic disc, though elsewhere it is entirely free from the retina.

Ciliary Zonule [FIG. 167]

This is a thickened part of the vitreous membrane which is fitted to the posterior surfaces of the ciliary processes. At the medial margin of the ciliary processes it splits into two layers. The posterior layer is an extremely thin membrane which lines the hyaloid fossa. The anterior layer is thicker

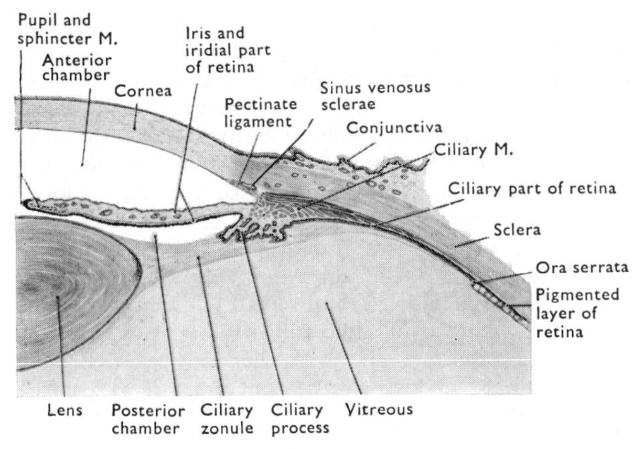

FIG. 167 Meridional section through the ciliary region of human eyeball. (After Schaffer.)

(**suspensory ligament of the lens**) and is attached principally to the anterior surface of the capsule of the lens a short distance from its margin, but it is also attached to the margin of the lens and the posterior surface close to the margin.

The suspensory ligament holds the lens firmly in the hyaloid fossa, and maintains a degree of tension on the lens when the eye is at rest, thus drawing it out radially, flattening it, and maintaining it at a long focal length. When the ciliary muscle contracts, it pulls the ciliary processes and zonule anteromedially, relaxes the suspensory ligament, and allows the elastic lens to round up and thus shorten its focal length. This is the mechanism of focusing on near objects (accommodation) which is intimately linked with convergence of the eyes and with constriction of the pupil in the *accommodation-convergence reflex*.

The **zonular spaces** are enclosed in the suspensory ligament at its multiple attachments to the lens, and form a multilocular lymph space around the margin of the lens.

DISSECTION. Remove the lens by cutting through the suspensory ligament with scissors.

Lens

This is a biconvex, transparent, elastic solid which lies between the iris and the vitreous body, enclosed in a glassy, elastic capsule. It contains no blood vessels.

The **anterior surface** is less highly curved than the posterior surface. Its central part is opposite the pupil, and lateral to this the margin of the iris is in contact with the lens, though further laterally the lens and iris are separated by aqueous humour in the posterior chamber. The **posterior surface** fits into the hyaloid fossa, and the equator is a blunt margin that forms one boundary of the zonular spaces.

DISSECTION. Incise the anterior surface of the lens with a sharp knife, and apply a little pressure to the margin between finger and thumb. This will extrude the body of the lens and allow the capsule to be studied separately.

Compress the body of the lens between finger and thumb, and note that the central part (nucleus) is firmer than the remainder. The lens is formed of a number of concentric laminae, a fact which may easily be proved after the lens has been hardened in alcohol.

The lens becomes tougher and less elastic with age, and this accounts for the decreasing ability to focus on near objects. It also becomes less transparent, and several varieties of opacities may form in it (*cataract*).

THE CONTENTS OF THE VERTEBRAL CANAL

DISSECTION. The lower lumbar and sacral parts of the vertebral canal may have been opened already, but it is desirable to open the remainder in order to see the entire length of the spinal medulla and the course taken by the spinal nerves to their exits.

Remove all the remaining muscles of the back to expose the laminae and spines of the vertebrae and the dorsum of the sacrum. Follow the dorsal rami of the spinal nerves towards the intervertebral foramina, noting their course relative to the vertebrae and especially the articular processes. Remove the posterior wall of the vertebral canal by sawing through the lateral parts of the laminae and dividing the ligamenta flava in a coronal plane. Do not saw completely through the laminae, but complete the division with a sharp tap on a chisel or with bone forceps to avoid cutting into the spinal medulla and its coverings (meninges). It is often possible to cut the cervical laminae entirely with bone forceps by starting at the upper end where the right half of the skull has been removed.

When a length of spines, laminae, and ligamenta flava has been removed, test the elasticity of these ligaments by stretching the specimen. Try to avoid removing with the laminae the extensive venous plexus which surrounds the spinal dura mater—it is very obvious when filled with blood.

EPIDURAL SPACE

This is the interval which separates the periosteum in the vertebral canal from the spinal dura mater. The space is filled with loose areolar tissue, semi-liquid fat, a network of veins, and the small arteries which supply the structures in the vertebral canal. The veins correspond to the venous sinuses of the dura mater in the skull, and are continuous with them through the foramen magnum.

Spinal Arteries

These minute arteries arise from the ascending cervical and vertebral arteries, from the posterior branches of the posterior intercostal and lumbar arteries, and from the lateral sacral arteries. They enter through the intervertebral foramina, and supply the spinal medulla, its nerve roots and meninges, and the surrounding bone and ligaments.

Internal Vertebral Venous Plexus [FIG. 168]

This plexus extends throughout the length of the vertebral canal, and may be divided into four subordinate longitudinal channels: two anterior and two posterior.

The posterior plexuses lie on the deep surfaces of the laminae and ligamenta flava; the anterior plexuses are on the posterior surfaces of the vertebral bodies near the edges of the posterior longitudinal ligament. They are united by dense transverse plexuses opposite the laminae, and receive the veins from the bones and contents of the vertebral canal, the **basivertebral veins** from the back of each vertebral body being especially large. The plexus communicates through the intervertebral foramina with the body wall veins—the vertebral, posterior intercostal, lumbar, and lateral sacral veins. Since *the plexus contains no valves* and is capable of considerable distention, it may drain into any of these veins or receive blood from them, transmitting it from areas of high venous pressure to areas of lower venous pressure, *e.g.*, from tributaries of the inferior vena cava (lateral sacral and lumbar veins) to tributaries of the superior vena cava (posterior intercostal and vertebral veins). Thus it is implicated in the spread of some pelvic tumours (*e.g.*, of the prostate) to the vertebral bodies and even to the skull without particles of the tumours having to pass through the lungs.

MENINGES OF THE SPINAL MEDULLA

These consist of exactly the same layers as the cranial meninges, and extend down the vertebral canal as separate tubular sheaths to the level of the second sacral vertebra. Here the three meninges fuse to form a median filament which continues to the coccyx.

DISSECTION. Remove the veins and fat from the external surface of the **dura mater**, and follow at least one of the prolongations which it gives over a spinal nerve into an intervertebral foramen by cutting away the articular processes. Note the large veins which accompany the dural sheath of the nerve through the intervertebral foramen.

Make a small incision in the dura mater in the midline. This allows the **arachnoid** to fall away from it. Introduce the point of a pair of scissors into the incision, and split the dura mater longitudinally, taking care not to damage the transparent arachnoid. If the arachnoid is not damaged, the subarachnoid space may be inflated by injecting water or air beneath it with a syringe and needle.

Split the arachnoid longitudinally, and pulling its edges apart, note the trabeculae attaching it to the pia mater on the surface of the spinal medulla. Below the first lumbar vertebra, the spinal medulla is replaced by the lower lumbar and sacral spinal nerve roots (the **cauda equina**) descending in the subarachnoid space to their points of exit.

Gently pull the spinal medulla to one side to expose the flange-like extension of the pia mater from its lateral aspect—the **ligamentum denticulatum**. This sends pointed extensions (teeth) through the arachnoid to the dura between the points of emergence of the spinal nerves. It is more readily exposed if the dorsal rootlets of one or two spinal nerves are cut away from the spinal medulla and reflected laterally.

Fold the cut edges of the dura mater and arachnoid laterally to expose the emergence of one or more spinal nerves through them. Note that each spinal nerve makes two separate openings in the dura and arachnoid—one for the dorsal root, and one for the ventral. Follow the dorsal root to the

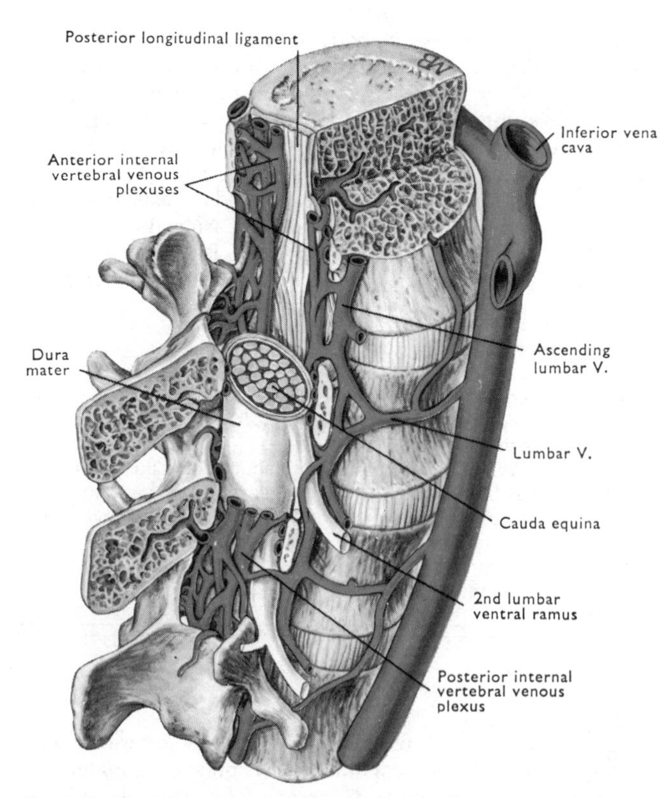

FIG. 168 Dissection of the upper four lumbar vertebrae to show the internal vertebral venous plexuses and their communications with the inferior vena cava.

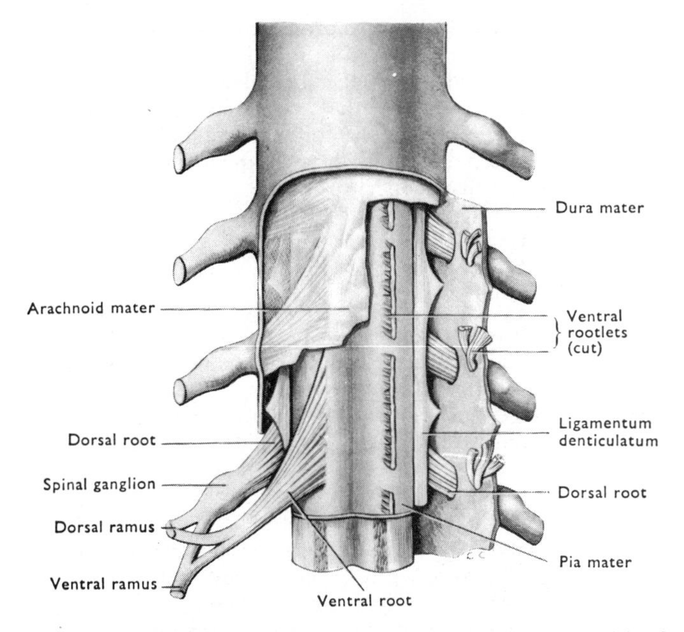

FIG. 169 Membranes (meninges) of the spinal medulla and mode of origin of spinal nerves. The contribution of each root to both rami is shown diagrammatically in the lowest spinal nerve on the left.

spinal medulla, and note how the rootlets into which it branches enter the spinal medulla. Follow one or more ventral roots in a similar manner, dividing one or more teeth of the ligamentum denticulatum so that the spinal medulla may be rotated to expose the exit of the rootlets which make up the ventral root.

Spinal Dura Mater [FIG. 169]

This tough, fibrous sleeve extends from the margin of the foramen magnum (where it is fused with the endocranium) to the second sacral vertebra, where it closes on to the arachnoid and the central filum terminale which attaches it to the coccyx. The dural sleeve is separated from the walls of the vertebral canal by the contents of the epidural space (see above), but is attached (1) to the posterior longitudinal ligament on the bodies of the second and third cervical vertebrae, and is lightly adherent to it elsewhere; (2) to the intervertebral foramina by the extensions around the roots of the spinal nerves. These attachments hold the dural sheath close to the vertebral bodies so that the spinal medulla is as near as possible to the axis of movement between the vertebrae, and hence is not greatly affected by these movements.

The dura mater forms a separate sheath for each dorsal and ventral root. The adjacent surfaces of these sheaths fuse at the proximal surface of the spinal ganglion, but the dura on the posterior surface of the dorsal root remains as a separate layer until it fuses with the distal part of the ganglion and becomes continuous with the **epineurium** of the nerve [FIG. 170].

Spinal Arachnoid [FIG. 21]

This thin, transparent membrane lines the dural sac throughout, but is separated from it by the bursa-like **subdural space** except where the two meninges fuse with the spinal nerves and with the filum terminale, and also where the teeth of the ligamenta denticulata (see below) pass through the arachnoid to the dura mater. The arachnoid mater only extends a short distance along the spinal nerve roots, and fuses with the proximal part of the spinal ganglia.

Deep to the arachnoid lies a relatively wide **subarachnoid space** containing the spinal **cerebrospinal fluid**. Only a few connecting strands pass from the deep surface of the arachnoid to the pia mater investing the spinal medulla, and these are mainly on the posterior aspect of the spinal medulla. The space is also crossed by the ligamenta denticulata. There is evidence that the spinal cerebrospinal fluid can be absorbed into the lymphatic and venous systems, mainly along the spinal nerves, but the amounts are probably not large, and it appears not to be replaced when isolated from the subarachnoid space surrounding the brain by blockage of the sub-arachnoid space at a higher level. In addition to its

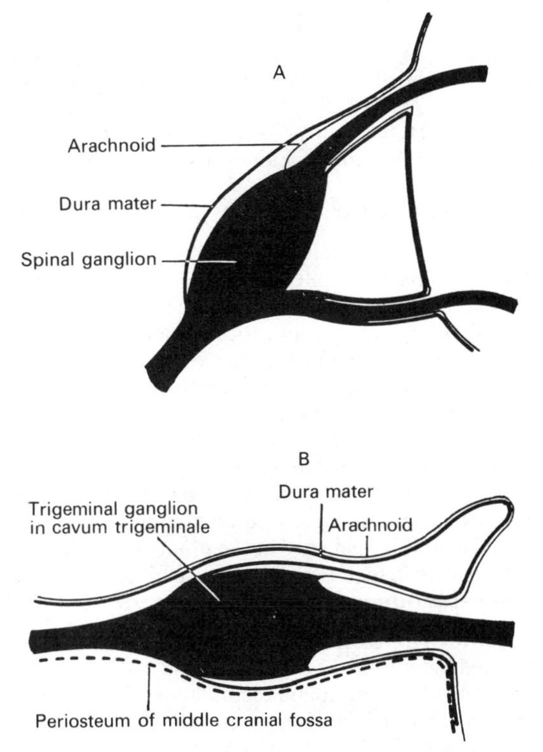

FIG. 170 Diagram to indicate the meningeal relation of A, a spinal nerve, and B, the trigeminal ganglion in the cavum trigeminale.

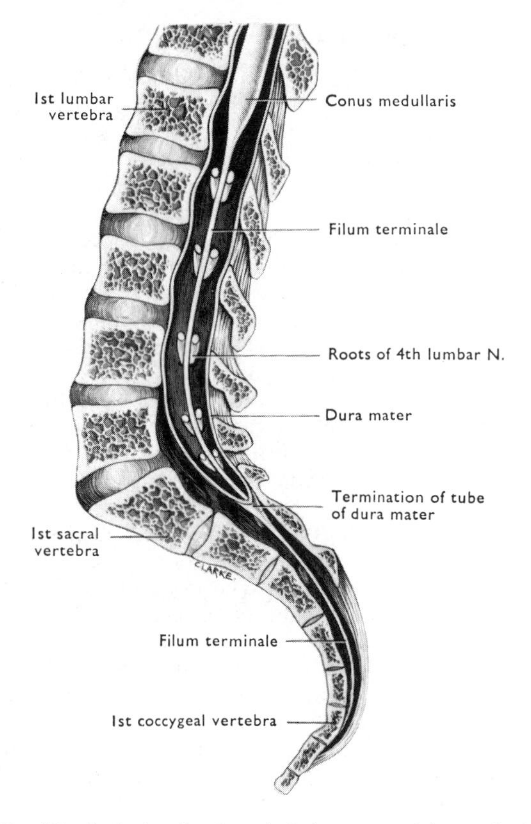

FIG. 171 Sagittal section through the lower part of the vertebral canal.

metabolic function in the nervous system, the cerebrospinal fluid may well transport substances from one part of the central nervous system to another. It certainly gives some mechanical protection by acting as a kind of water bath.

Spinal Pia Mater

This is a firm vascular membrane which adheres closely to the spinal medulla, is continued over the filum terminale at the lower end of the spinal medulla, and ensheathes every rootlet of each spinal nerve as it crosses the subarachnoid space. In this way it becomes continuous with the **perineurium** of the spinal nerves. It also extends for a short distance over each vessel entering the spinal medulla, and so carries a short sleeve of subarachnoid space around the vessel—the **perivascular space**. The pia mater extends into the anterior median fissure of the spinal medulla, and is thickened at the mouth of the fissure to form a median, longitudinal, glistening band—the **linea splendens**.

Ligamenta Denticulata [FIG. 169]. These toothed ligaments appear as a thin ridge of pia mater on each side of the spinal medulla midway between the ventral and dorsal nerve root attachments. They extend from the foramen magnum to the lower limit of the spinal medulla at the first lumbar vertebra. Twenty-one pointed projections extend from its free edge to be attached to the dura mater through the arachnoid, each in the interval between two adjacent nerves piercing the dura mater. The highest projection is at the foramen magnum; the lowest between the twelfth thoracic and first lumbar nerves. These ligaments suspend the spinal medulla in the subarachnoid space from these points, thus permitting some movement of the dura mater without a corresponding movement of the spinal medulla.

Filum Terminale. This delicate filament extends from the tapering end of the spinal medulla (**conus medullaris**) to the coccyx. It lies among the descending roots of the lumbar and sacral nerves (cauda equina) but is readily differentiated from them by its silvery appearance and attachments [FIG. 171].

It is composed chiefly of pia mater surrounding a continuation of the central canal of the spinal medulla [p. 163] and some nervous elements which can be traced in it for nearly half its length. At the level of the second sacral vertebra, the sheaths of arachnoid and dura mater enclosing the cauda equina close down on and fuse with the filum. Together they traverse the sacral canal and fuse with the periosteum on the back of the coccyx or the last piece of the sacrum.

THE SPINAL MEDULLA

The spinal medulla begins at the foramen magnum as the inferior continuation of the medulla oblongata of the brain. It ends opposite the intervertebral disc between the first and second lumbar vertebra in the adult and approximately one vertebra lower in the new-born. The point of termination is variable—it rises when the vertebral column is flexed and descends when it is extended.

The average length is 45 cm. It is nearly cylindrical in shape though its girth increases considerably in the regions giving rise to the large nerves of the limbs. These enlargements are known as the **cervical and lumbar swellings**. They arise because of the increased number of nerve cells within the spinal medulla at these regions. The cervical swelling extends from the foramen magnum to the first thoracic vertebra (maximum transverse diameter is 14 mm at the sixth cervical vertebra); the lumbar from the tenth to twelfth thoracic vertebrae (maximum transverse diameter, 12 mm at twelfth thoracic vertebra) and then rapidly tapers to a point, forming the **conus medullaris**.

SPINAL NERVES

Thirty-one pairs of nerves are attached to the spinal medulla. Of these eight are cervical, twelve are thoracic, five are lumbar, five are sacral, and one is coccygeal. The cervical nerves leave the vertebral canal above the corresponding vertebrae, with the exception of the eighth which emerges between the seventh cervical and first thoracic vertebrae. The remainder emerge below the corresponding vertebrae.

Spinal Nerve Roots [Figs. 169, 170]

Each spinal nerve is attached to the spinal medulla by a dorsal and a ventral root. The **dorsal root** is larger (except in the first cervical which frequently has no dorsal root) and has an oval swelling, the **spinal ganglion**, where it joins the ventral root. The dorsal root is composed of afferent (sensory) nerve fibres which are processes of the cells in the spinal ganglion, the **ventral root** of efferent (motor) nerve fibres which arise from cells in the spinal medulla. The two roots unite at the spinal ganglion to form the mixed spinal nerve, which contains both types of fibres.

Each of the two roots is made up of a number of separate rootlets which diverge as they approach the spinal medulla. The rootlets of the dorsal roots enter the dorsolateral aspect of the spinal medulla consecutively in a continuous straight line at the bottom of a slight furrow (**posterolateral sulcus**). The rootlets of the ventral roots emerge from the ventrolateral aspect of the spinal medulla over a broader strip.

The part of the spinal medulla to which any one pair of nerves is attached is called a **spinal segment**.

The *size of the roots* varies directly with the amount of tissue each supplies. Thus the roots of nerves contributing to the limb plexuses are large, especially those which innervate the lower limb. The thoracic and upper cervical roots are small except the first thoracic which innervates part of the upper limb.

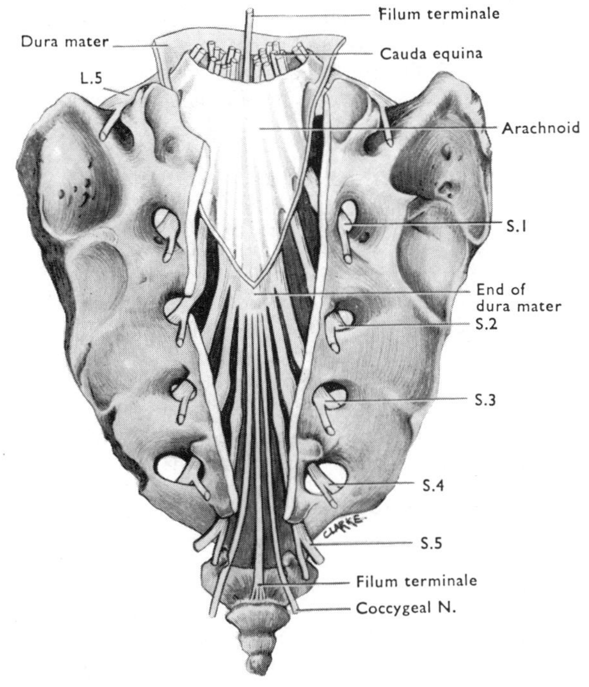

Fig. 172 Posterior view of the meninges and spinal nerves in the sacral canal.

The roots are very variable in *length* and in the direction they take in the vertebral canal, because the spinal medulla is shorter than the vertebral canal. The relative shortening of the spinal medulla occurs mainly during intra-uterine development, and causes its lower end to rise progressively in the vertebral canal. By birth the lower end of the spinal medulla has moved from the coccyx to the third lumbar vertebra, and in the adult lies between the first and second. This shortening causes a progressive elongation of those roots which have a greater distance to run to reach the appropriate intervertebral foramen. The upper part of the spinal medulla remaining fixed, has short (1·5–2 cm) horizontal roots, while the lower roots are progressively elongated (up to 27 cm—S. 5) and become increasingly oblique. The lumbar, sacral, and coccygeal nerves descend vertically as a leash of dorsal and ventral roots, the **cauda equina**, in the subarachnoid space inferior to the end of the spinal medulla.

Lumbar Puncture. The importance of the high level of termination of the spinal medulla, and the presence only of a leash of roots in the subarachnoid space inferior to it is: (1) that injuries to the vertebral column below the second lumbar vertebra can damage only spinal nerve roots, though the roots of many nerves may be damaged by an injury at one level; (2) that samples of cerebrospinal fluid can be obtained for diagnostic purposes without injury to the spinal medulla by introducing a hollow needle into the subarachnoid space between the laminae of the lower lumbar vertebrae. This is lumbar puncture.

Lengths and Levels of Spinal Segments. The relative shortening of the spinal medulla does not affect all segments equally. Strangely the degree of shortening is directly proportional to the amount of tissue supplied by each spinal segment. Thus the lumbosacral segments are the shortest and the thoracic the longest.

The shortening results in the individual spinal segments being displaced from their corresponding vertebral levels—a fact of great clinical importance in determining the level of vertebral injury from signs of injury to a particular spinal segment.

Spinal Segments	Vertebral Levels (spines)
Cervical 1–8.	Foramen magnum—6th Cervical.
Thoracic 1–6.	6th Cervical—4th Thoracic.
Thoracic 7–12.	4th Thoracic—9th Thoracic.
Lumbar and Sacral.	10th Thoracic—1st Lumbar.

Exit of Spinal Nerves. Each spinal nerve emerges through an intervertebral foramen, except the first cervical which lies above the posterior arch of the atlas, the second between this and the vertebral arch of the axis, and the fifth sacral and coccygeal nerves which emerge through the lower end of the sacral canal. The other sacral nerves additionally have a separate sacral foramen for each ramus.

Meningeal Branch. This slender twig, partly from the spinal nerve and partly from the sympathetic trunk, runs back through the intervertebral foramen. It is sensory to the walls and contents of the vertebral canal, and supplies its blood vessels. It requires careful dissection to demonstrate this branch.

Spinal Ganglia. These are collections of nerve cell bodies each of which gives rise to a sensory fibre in the peripheral nerve and a nerve fibre in the dorsal root on which the ganglion lies. *There are no synapses in these ganglia.* The ganglia lie in the intervertebral foramina, except the sacral and coccygeal which lie in the sacral canal, and the first two cervical which are in the corresponding position above and below the atlas, behind the articular facets.

Each spinal nerve, formed by the union of a dorsal and a ventral root at the spinal ganglion, begins at the distal end of the ganglion, and divides almost immediately into a **dorsal** and a **ventral ramus**.

DISSECTION. Cut across the spinal nerves in the intervertebral foramina leaving as long a piece of each nerve attached to the dura as possible. Pull on the dura and divide its attachments to the posterior longitudinal ligament, so as to remove the spinal medulla and meninges in one piece. Split the dura mater along the median plane anteriorly to expose the arachnoid and pia mater with the ligamenta denticulata. Find the rootlets of the **spinal part of the accessory nerve** posterior to the ligamentum denticulatum in the upper cervical region. Follow them to the nerve, and trace it cranially in front of the dorsal roots [FIG. 47].

Remove the linea splendens from the ventral surface of the anterior median fissure of the spinal medulla to expose the anterior spinal artery. Follow the artery cranially and caudally, and try to expose the arteries entering it on some of the ventral roots. With a hand lens, it is possible to see small branches of the artery passing on to the pial surface of the spinal medulla, and the perforating branches passing into the anterior median sulcus when the artery is lifted gently away from the spinal medulla.

ARTERIES OF THE SPINAL MEDULLA

Many thin-walled arteries reach the spinal medulla on the roots of the spinal nerves. They arise from the vertebral, ascending cervical, posterior intercostal, lumbar, and lateral sacral arteries, and are smaller and more numerous on the dorsal than on the ventral roots. They enter three longitudinal vessels on the surface of the spinal medulla—one anterior and two posterior spinal arteries.

The anterior spinal artery is formed on the anterior surface of the medulla oblongata of the brain by a branch from each intracranial vertebral artery [FIG. 187]. It descends throughout the length of the spinal medulla on the mouth of the anteromedian fissure, dorsal to the linea splendens, and receives three to ten tributaries from vessels on the ventral roots. The largest of these supplies the lumbar swelling, and usually enters from the left, frequently on the tenth thoracic ventral root.

Branches. 1. Many small perforating vessels enter the anteromedian fissure. They branch to supply the central parts of the spinal medulla, including most of the grey matter (nerve cell bodies) and the deeper parts of the surrounding white matter (nerve fibres).

2. Small vessels pass backwards on the surface of the spinal medulla to anastomose with the **posterior spinal arteries** which course longitudinally beside the entering dorsal rootlets. These vessels send branches on to the posterior surface of the spinal medulla to complete an irregular, circular anastomosis from which branches pass radially into the white matter. *All the branches which enter the spinal medulla are end-arteries. This pattern of a surface anastomosis with perforating end-arteries is present throughout the central nervous system.*

On the conus medullaris, each posterior spinal artery sweeps ventrally to join the anterior spinal artery, which continues as a slender vessel on the filum terminale.

The posterior spinal arteries are formed at their cranial ends as branches of the posterior inferior cerebellar or vertebral arteries on the medulla oblongata. They are fed by numerous small arteries on the dorsal roots, and are much smaller than the anterior spinal artery which rarely measures 1 mm in diameter, and that only where it receives a major tributary.

VEINS OF THE SPINAL MEDULLA

The veins, though small, numerous, and very tortuous, are larger than the arteries. They anastomose freely on the surface of the spinal medulla, and form six more or less perfect longitudinal channels—one anteromedian, one posteromedian, and one close to the attachment of each set of dorsal and ventral rootlets. They are continuous above with the veins of the medulla oblongata, and drain laterally along the spinal nerve roots to the internal vertebral venous plexus.

STRUCTURE OF THE SPINAL MEDULLA

In transverse section, the spinal medulla is seen to consist of an H-shaped core of **grey matter**, consisting principally of nerve cell bodies and neuroglia (the cellular connective tissue of the central nervous system), and an external layer of **white matter**—nerve fibres. The horizontal bar of the H is the **grey commissure** (crossing). Ventral to this

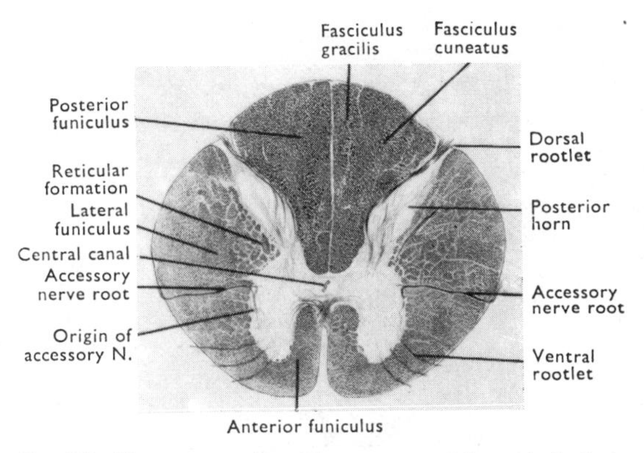

FIG. 173 Transverse section of the upper part of the cervical spinal medulla.

nerve fibres cross the midline forming the **white commissure**. The two commissures are the only neural continuity between the two halves of the spinal medulla, which is otherwise divided into right and left halves by the **anteromedian fissure** extending inwards to the white commissure, and the **postero-median** (glial) **septum** extending from the posteromedian sulcus to the grey commissure [FIG. 174].

The grey commissure is traversed longitudinally by a narrow tunnel, the **central canal**. This is lined by ependyma, and is continuous superiorly with the fourth ventricle of the brain and inferiorly with a slight expansion in the conus medullaris, the **terminal ventricle**. It ends in the filum terminale. The canal contains cerebrospinal fluid, but is often absent in aged individuals who show only a clump of ependymal cells.

The anterior and posterior limbs of the H-shaped grey matter are the **anterior** and **posterior horns**. The **dorsal roots** on each side enter the **posterolateral sulcus** near the tip of the posterior horn. Here each fibre of the root divides into an ascending and a descending branch which run longitudinally in the spinal medulla giving off numerous collaterals (branches) into the grey matter. The **ventral roots** arise from large **motor cells** in the anterior horn. In the cervical and lumbosacral swellings, the motor cells innervating the muscles of the limbs lie in lateral extensions of these horns, those which supply the proximal muscles of the limb lying furthest anteriorly. Since the horns extend throughout the length of the spinal medulla, they are also known as anterior and posterior **grey columns**.

The amount of grey matter at any level is proportional to the amount of tissue supplied by the spinal nerves arising there. Thus the horns are large in the swellings (and are mainly responsible for them) and small in the thoracic and upper cervical regions.

Throughout the spinal medulla, the tips of the posterior horns, adjacent to the entry of the posterior roots, are composed of the translucent **substantia gelatinosa** [FIG. 174]. Between the first thoracic and second or third lumbar segments there is a spike-like extension of the lateral aspect of the grey matter—the **lateral grey column** or **horn**. This contains the nerve cells which give rise to the preganglionic fibres of the sympathetic system—the fibres that form the white rami communicantes of the thoracic and upper lumbar nerves. A similar spike in the sacral region (S. 2–4) gives rise to the preganglionic pelvic parasympathetic nerve fibres. The **thoracic nucleus** also lies in the thoracic and upper lumbar segments at the medial side of the base of the posterior grey column [FIG. 174]. It receives

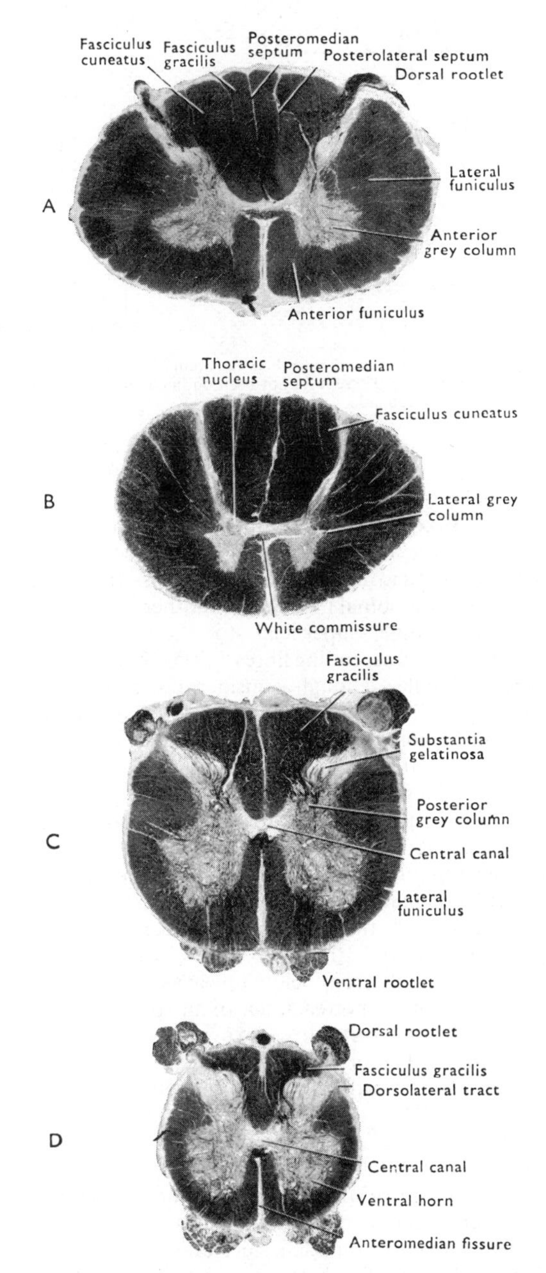

FIG. 174 Transverse sections through the cervical (A), thoracic (B), lumbar (C), and sacral (D) regions of the spinal medulla.

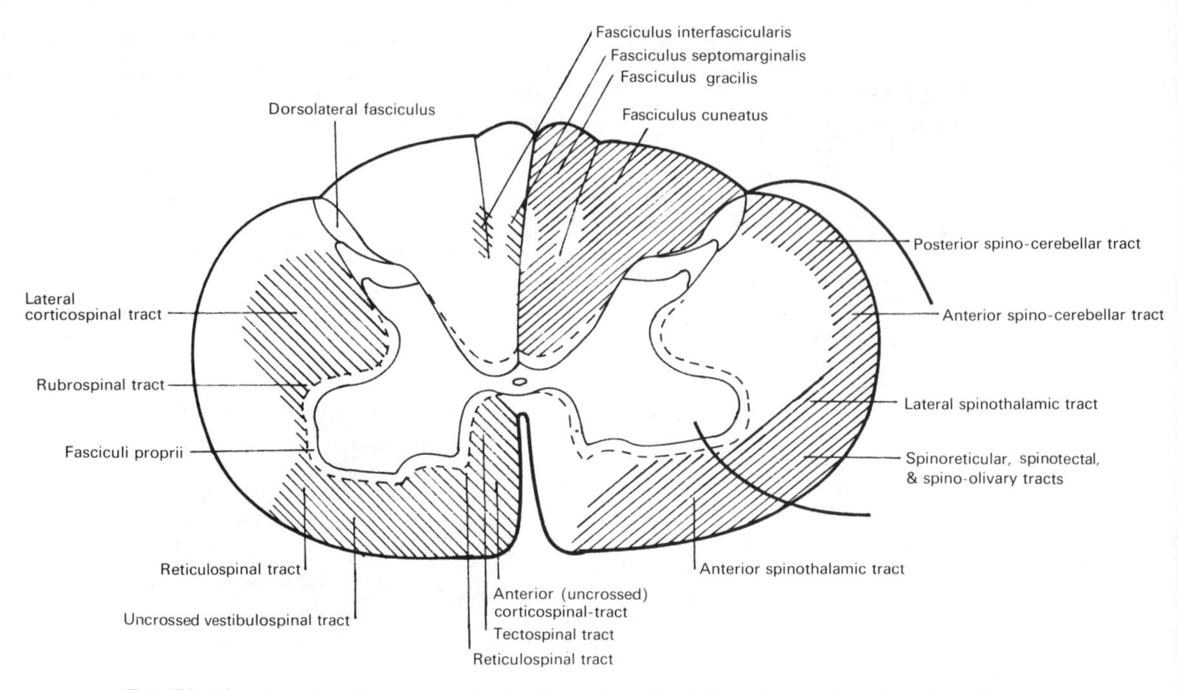

FIG. 175 Diagram to show the main ascending (right) and descending (left) pathways in the white matter of the spinal medulla. These have not been outlined individually because of the mixture of different pathways in the anterior and lateral funiculi.

fibres from the fasciculus gracilis (see below) and sends its axons to the cerebellum through the **posterior spinocerebellar tract**.

The **white matter** is composed of longitudinally running nerve fibres and forms a surface coating on the grey matter. The deepest fibres pass from one segment of the spinal medulla to another (**fasciculi proprii**), while superficial to these are long descending and ascending fibres linking the brain and spinal medulla. Though considerably mixed, the ascending fibres tend to lie superficial to the descending fibres in the lateral and anterior parts of the white matter [FIG. 175], and fibres with similar origins and terminations are gathered into groups in the white matter to form **tracts** or **fasciculi**. From each segment of the spinal medulla fibres are added to the ascending fasciculi, and fibres in the descending fasciculi end in the grey matter of each segment. For these reasons, both groups of fibres increase in size superiorly. Hence the volume of white matter increases steadily from below upwards.

The white matter on each side of the spinal medulla is divided into **posterior**, **lateral**, and **anterior funiculi** by the dorsal and ventral roots piercing it. The **posterior funiculi** (posterior white columns), between the dorsal roots of the two sides, consist mainly of ascending processes of fibres which enter the spinal medulla in the dorsal roots. The *thicker fibres in the dorsal roots* (predominantly concerned with tactile and proprioceptive sensations) enter the posterior column on each side, and many ascend there to the medulla oblongata giving collaterals for reflex purposes deep into the posterior horn and even to the motor cells of the anterior horn for monosynaptic reflexes. New fibres added at each level increase the width of the posterior column, the fibres entering through the highest roots lying most lateral. In the upper half of the spinal medulla, each posterior column is subdivided by a **posterolateral** (glial) **septum** [FIG. 174] into a slender, medial **fasciculus gracilis** (which carries such fibres from the lower half of the body) and a broader, lateral **fasciculus cuneatus** transmitting fibres from the upper half of the body. The position of this septum is marked on the surface by the **postero-intermediate sulcus**. *The finer fibres in the dorsal roots* (principally concerned with pain and temperature sensations) enter further laterally, and dividing, ascend and descend for a short distance only. Together they form the **dorsolateral tract** overlying the lateral part of the substantia gelatinosa. They give many branches into the grey matter of the tip of the posterior horn concerned with spinal reflexes. Their impulses are transmitted through cells deeper in the horn which are subject to the influence of the collaterals from the posterior columns whose impulses reach this area earlier because of the greater diameter and speed of conduction of their fibres in the peripheral nerves. Thus proprioceptive (etc.) impulses are able to modify the spinal reflexes initiated by painful and thermal impulses—a feature common to many descending tracts from the brain. Impulses from the dorsolateral tract are also transmitted to the brain through these cells and the **spinothalamic tract** fibres [p. 185; FIG. 175] which arise from them. Because of the overlapping territories of the dorsolateral tract branches and convergence in the posterior horn cells, individual fibres in the spinothalamic tract probably transmit impulses from several sensory endings, superficial and deep—a feature which could explain the inability of individuals to localize pain accurately when the specific pathways of the posterior columns are destroyed, and the tendency to refer pain impulses arising in one branch of a nerve to the territory of another branch of that or an adjacent nerve.

THE JOINTS OF THE NECK

DISSECTION. These joints have already been partly dissected when the right half of the head was removed. This opened the right atlanto-occipital joint and cut through the right alar ligament and the longitudinal fibres of the cruciate ligament [Fig. 179].

Remove any remnants of muscle from the cervical articular processes and from the laminae and spines which have been removed in a piece, in order to expose the ligaments uniting the cervical vertebrae.

The joints between the second to fifth cervical vertebrae are similar to those in the other parts of the vertebral column, and allow flexion and extension together with lateral flexion, but little rotation. The joints between the first and second cervical vertebrae are designed to allow rotation, while those between the first cervical vertebra (atlas) and the skull are so arranged that they permit nodding movements of the skull on the vertebral column.

TYPICAL CERVICAL JOINTS

These are the joints between the lower six cervical vertebrae. The **bodies** of these vertebrae are firmly bound to each other by a flexible fibrous disc (**anulus fibrosus**) of moderate thickness. This contains a central gelatinous structure (**nucleus pulposus**) and allows a moderate degree of movement. The intervertebral discs are strengthened anteriorly and posteriorly by the anterior and posterior **longitudinal ligaments**, which are attached principally to the discs and the adjacent parts of the bodies. The discs are the major factor in producing the cervical curvature, and they do not cover the entire surfaces of the vertebral bodies, but are replaced laterally by small **synovial joints** situated where the margins of the inferior vertebra overlap the vertebral body above [Fig. 176].

The **vertebral arches** are united by the synovial joints between the articular processes and by a number of ligaments. The cervical articular processes lie in an oblique coronal plane, which passes anterosuperiorly, and the processes on the two sides are parallel to each other. The capsules of these joints are lax, and permit a considerable range of movement.

Ligaments of Vertebral Arches

Ligamenta Flava [Fig. 177]. These wide, flat bands of yellow elastic tissue pass between adjacent laminae. They help to maintain the position of the vertebral column, and restore it to that position after they have been stretched by flexion.

Interspinous Ligaments. These are weak in the cervical region. They pass between adjacent spines, and are directly continuous with the supraspinous ligaments and the ligamentum nuchae [p. 27].

JOINTS OF ATLAS, AXIS, AND OCCIPITAL BONE

The atlas consists of a ring of bone with a lateral mass on each side. The lateral masses articulate superiorly with the occipital condyles and inferiorly with the superior articular facets of the axis. Each has a long,

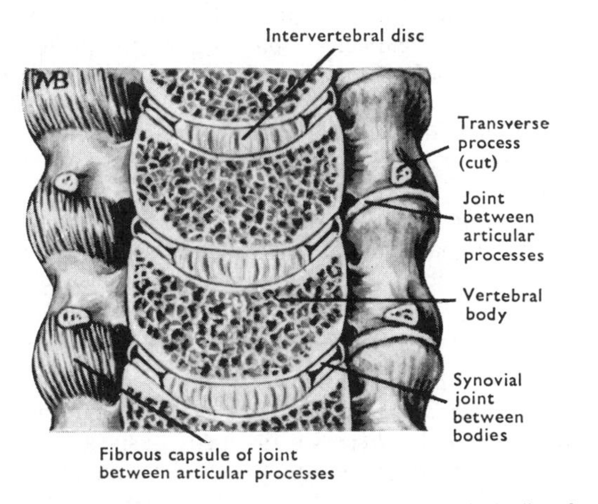

Fig. 176 Coronal section through the joints between the bodies of the cervical vertebrae.

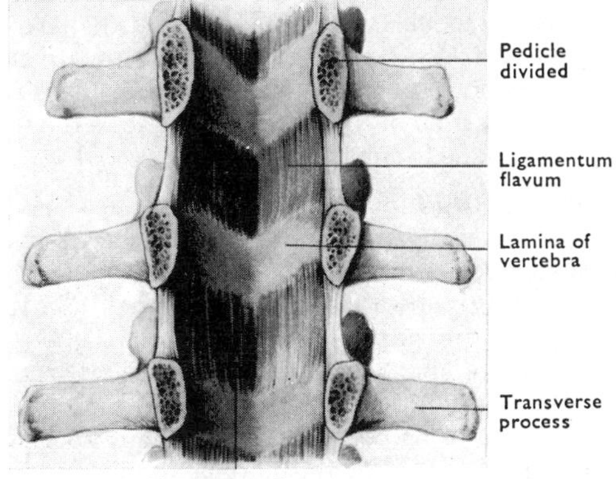

Fig. 177 Ligamenta flava seen from the front after removal of the bodies of the vertebrae by saw cuts through the pedicles. Lumbar region.

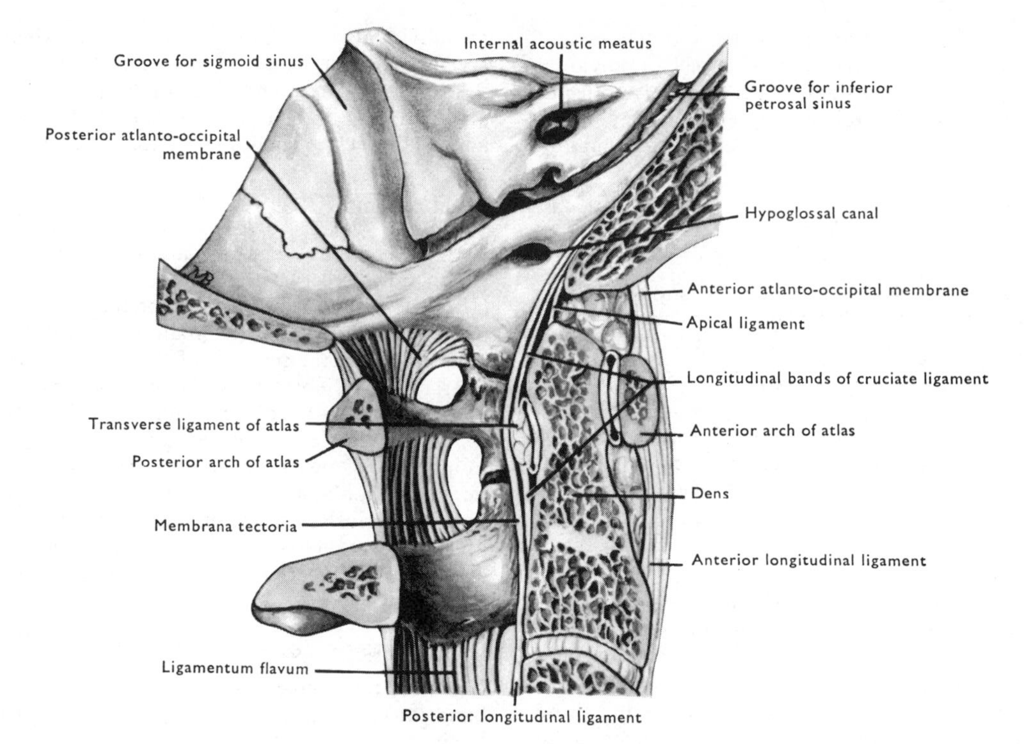

FIG. 178 Median section through the foramen magnum and first two cervical vertebrae. The brain, spinal medulla, and meninges have been removed.

stout transverse process projecting laterally from it. The occipital bone and the axis are firmly bound to each other by strong ligaments, so that the atlas is held, like a washer, between them. The articular surfaces which the atlas has with the skull and the axis are of different configuration. Those with the skull permit flexion and extension, while those with the axis permit rotation.

Ligaments of Joints of Atlas

Anterior Longitudinal Ligament. This narrows superiorly to be attached to the anterior tubercle of the atlas. Above this it continues as a narrow band to the base of the skull, strengthening the median part of the **anterior atlanto-occipital membrane**. This membrane passes from the superior margin of the anterior arch of the atlas to the base of the skull anterior to the foramen magnum [FIG. 178].

Ligamentum Flavum. This is a delicate structure between the atlas and axis and also between the posterior arch of the atlas and the occipital bone where it is known as the **posterior atlanto-occipital membrane**. It passes from the part of the posterior arch of the atlas between the grooves for the vertebral arteries, to the margin of the foramen magnum posterior to the atlanto-occipital joints. Each of its lateral margins arches over the corresponding vertebral artery to reach the posterior surface of the lateral mass of the atlas. These margins may be ossified.

DISSECTION. Remove any remaining parts of the posterior arch of the atlas and of the laminae of the axis to expose the parts of the membrana tectoria [FIG. 178] not damaged in the removal of the right half of the head. Cut through the membrana tectoria and reflect it to expose the undamaged parts of the cruciate ligaments and the accessory atlanto-axial ligaments.

Membrana Tectoria. This broad ligamentous sheet is the superior continuation of the posterior longitudinal ligament. It passes from the posterior surface of the body of the axis to the cranial surface of the occipital bone [FIGS. 178, 179]. It holds the axis to the skull, and covers the posterior surface of the dens and its ligaments and the anterior margin of the foramen magnum.

The **accessory atlanto-axial ligaments** [FIG. 179] pass from the posterior surface of the body of the axis to the corresponding lateral mass of the atlas. These strong ligaments help to limit the rotation of the atlas on the axis.

Cruciate Ligament. This ligament is formed by the transverse ligament of the atlas (which passes between the tubercles on the medial aspects of the lateral masses of the atlas) and the superior and inferior longitudinal bands which pass from the transverse ligament to the cranial surface of the occipital bone and the body of the axis respectively. The **transverse ligament of the atlas** curves over the posterior surface of the dens, and unites the medial aspects of the lateral masses of the atlas. It forms a strap which holds the **dens** firmly against the posterior surface of the anterior arch of the atlas. There is a synovial joint between the dens and the ligament and another between the dens and the anterior arch of the atlas. Thus the atlas is capable of rotating round the dens, but cannot be displaced anteroposteriorly on it.

DISSECTION. If the division of the skull was kept to the right of the midline, the apical ligament of the dens [FIG. 178] should still be present with the left half. It lies anterior to the superior band of the cruciate ligament and may be exposed by removing this.

Identify the divided right alar ligament, and expose the left alar ligament passing from the side of the apex of the dens to the medial aspect of the occipital condyle.

Apical Ligament of the Dens. This weak, cord-like ligament stretches from the apex of the dens to the cranial surface of the occipital bone immediately above the margin of the foramen magnum. It develops around a part of the notochord (see below), but does not play a significant part in strengthening the joints.

Alar Ligaments [FIG. 179]. These very powerful ligaments arise from the sloping sides of the summit of the dens. They pass laterally and slightly upwards to the medial sides of the occipital condyles. They hold the skull to the axis and they tighten when the atlas, carrying the skull, rotates round the dens. They are thus the main factor in limiting rotation at the atlanto-axial joints. The alar ligaments are attached to the skull on the axis of the nodding movements which the skull makes on the atlas, and so do not hinder these movements.

Atlanto-occipital Joint

The kidney-shaped occipital condyles lie on the anterolateral aspects of the foramen magnum, and are shaped as two segments of an obtuse, inverted cone. They fit into the superior articular facets of the atlas which are also kidney-shaped. The joints allow flexion, extension, and slight side to side rocking of the head, but no rotation. The stability of these joints does not depend on the loose articular capsule, but rather on the alar ligaments, the membrana tectoria, and the longitudinal bands of the cruciate ligament, all of which bind the skull to the axis.

The Atlanto-axial Joints

These joints are formed between the large, nearly circular, slightly curved facets on the adjacent surfaces of the atlas and axis. These facets slope downwards and outwards and form segments of a flat cone which permits the atlas, carrying the skull, to rotate round the dens of the axis. These joints are stabilized partly by the articulation of the anterior arch of the atlas and its transverse ligament with the dens, and partly by the ligaments which bind the axis and the skull together so that the atlas is held firmly between them.

In the atlanto-occipital and atlanto-axial joints there are no articular facets corresponding to the articular facets of the other cervical vertebrae. The articular surfaces of these two joints correspond in position to the small synovial joints at the lateral edges of the vertebral bodies. Thus the emerging spinal nerves (first and second cervical) pass posterior to the joint capsules and not anterior to them as in the case of the articular facets between the vertebral arches of the remaining cervical vertebrae.

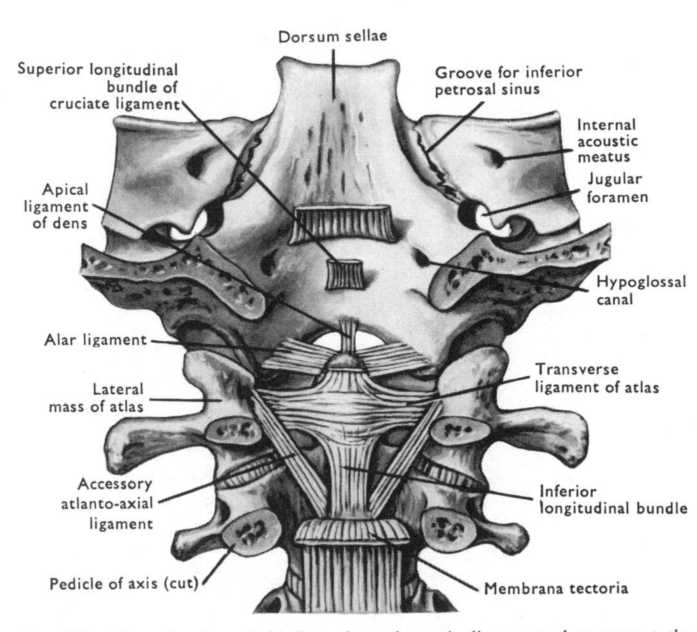

FIG. 179 Dissection from behind to show the main ligaments that connect the occipital bone, the atlas, and the axis.

The Notochord

This cellular rod, formed from the cephalic margin of the primitive streak in the early embryo, is the primitive axial supporting structure of the embryo. It extends from the buccopharyngeal membrane [p. 243] to the cloacal membrane. The greater part of the notochord is surrounded by the mesoderm which forms the vertebral bodies and intervertebral discs. The formation of the cartilaginous vertebral bodies breaks the notochord into separate pieces, one in each developing **intervertebral disc**. Here they play a considerable part in the formation of the **nucleus pulposus**.

The cranial end of the notochord is attached to the ectoderm and entoderm of the buccopharyngeal membrane. When the stomodaeal part of the hypophysis is drawn upwards [p. 244] the ectodermal attachment of the notochord is carried with it into the mesoderm which later forms the posterior part of the body of the sphenoid, while an entodermal attachment may persist in the roof of the pharynx. Caudal to this, the notochord is first enclosed in the mesoderm which forms the basilar part of the occipital bone, then lies between that mesoderm and the roof of the pharynx, re-entering the mesoderm caudal to this to leave it at the foramen magnum and pass to the rudiment of the dens. This last part is believed to form the **apical ligament of the dens**. The relation of the notochord to the base of the skull is important because certain tumours (chordomata) are believed to arise from remnants of it, and are found in the sites which it occupied between the hypophysial fossa and the foramen magnum.

THE BRAIN

INTRODUCTION

The central nervous system contains two types of cells: (1) the **nerve cells** proper; (2) the connective tissue of the central nervous system (**neuroglia**); including the **ependymal** lining of the brain cavities. In addition to these two types, there are the cells of the meninges which cover: (a) the central nervous system; (b) the blood vessels which enter it, (c) the nerves which arise from it.

The **neuroglia** consists of three types of cells. Two are derived from ectoderm in common with most of the nervous system—the astrocytes and the oligodendroglia. The third type forms the microglia—a phagocytic type of cell derived from the mesoderm as are the blood vessels.

Astrocytes are star-shaped cells with processes radiating from them. They pervade the central nervous system, forming a surface layer throughout it. At least one of the numerous processes passes to form an end-foot on an adjacent capillary, so that together they ensheathe the capillaries of the central nervous system. Because of this and because their biochemistry alters with that of the nerve cells adjacent to them, it is believed that they play an important part in nerve cell metabolism and the transfer of substances from the blood to the nerve cells. Two main types are present with many intermediate forms. 1. The **protoplasmic astrocytes** whose processes branch and rebranch to form a dense bush, and which are found in areas where nerve cells predominate (grey matter). 2. The **fibrillary astrocytes** are found predominantly among the bundles of nerve fibres (white matter). Their long, thin processes branch infrequently and appear smooth in stained preparations. Astrocytes are numerous in old injuries of the nervous system. They seem to react as fibrocytes do in other parts of body to form a scar, but there is no good evidence that they undergo mitotic division. However, their rapid accumulation at the site of an injury, and the numerous processes which they form there have been suggested as the reason for the failure of central nervous system axons to grow across wounds.

Oligodendroglia are small cells with few, short processes. They are found surrounding nerve cell bodies and on their dendrites, but are particularly numerous among nerve fibres before and during the development of their fatty myelin sheaths. Many functions have been ascribed to these cells, but only the formation of central system myelin sheaths can be ascribed to them with any certainty. Each oligodendrocyte seems to form a segment of myelin for each of several nerve fibres.

Microglia. These small cells are often rod-shaped. They are difficult to identify in the normal nervous system, only becoming readily visible where there is tissue damage and particles to be phagocytosed. In nerve fibre injury, they ingest particles of degenerating myelin and develop a foam-like cytoplasm. It may be that the microglia are present in the normal nervous system in some unrecognized form, but their rapid appearance after injury is reminiscent of the appearance of macrophages elsewhere in similar circumstances.

Ependyma. This is the epithelium which lines the cavities of the brain and covers the vascular pia mater which invaginates the ventricles to form the choroid plexuses. In the embryo, the ependyma is the germinal epithelium of the nervous system and extends from the lumen of the neural tube to its external surface. This situation only persists where the wall of the tube remains thin; elsewhere in the adult the ependyma is a single layer of low ciliated columnar or cubical epithelium.

Nerve cells are very variable in size and shape, but all of them have cytoplasmic processes, some of which extend for considerable distances from the cell body. These processes are of two types, axons and dendrites. The **dendrites** are the processes which receive stimuli from the axons of other nerve cells that end on them. They vary greatly from cell to cell, but are commonly branched in a complicated fashion and restricted to the vicinity of the cell body. The **axons** are usually thinner than the dendrites, and transmit impulses [Vol. 1, p. 3] from the cell body either to other nerve cells, or to peripheral tissues through the peripheral nerves. They also transmit substances in both directions in their cytoplasm. Thus substances produced in the cell bodies, which have many of the characteristics of secretory cells, can be passed along the axon to the tissue which it innervates (*e.g.*, substantia nigra to corpus striatum) or held in storage for future release (*e.g.*, posterior lobe of the hypophysis). The axon can also convey materials absorbed at its distal end to the cell body—a mechanism which may control the activity of the cell in relation to that of the tissue which it innervates. Some axons are of considerable length (up to 1 m) and all are capable of extensive branching so that the impulses they carry can be widely disseminated, but they may also pass to strictly circumscribed regions of the central nervous system.

Most of the axons in the central nervous system are covered with a fatty **myelin sheath**. This is in the form of a series of discontinuous segments separated by **nodes** (where the myelin is absent) and is white in colour, in contrast to the grey colour of the fresh cell bodies and dendrites which have no covering of myelin. The axon together with its myelin sheath is known as a **nerve fibre**, but the same term is applied to the thinnest axons which do not possess a myelin sheath, but are nevertheless enclosed in the same type of cells.

Throughout most of the central nervous system, the axons and cell bodies are segregated from each other. The axons form bundles (**tracts**), while the cell bodies with their associated dendrites form clusters or **nuclei**. Because of their difference in colour in the fresh state, the tracts form the **white matter** of the central nervous system, while the nuclei make up the **grey matter**.

In the spinal medulla and in most of the brain stem (which are continuous with each other through the foramen magnum) the grey matter lies internally, and is covered by a superficial layer of white matter, which consists of longitudinally running nerve fibres.

In the cerebellum, the cerebral hemispheres, and certain other situations, there is an additional layer of nerve cells on the external surface. This rind or **cortex** of grey matter greatly increases the number of nerve cells and hence the complexity of their interactions.

Functionally the central nervous system is concerned with the receipt and integration of sensory information from all parts of the body, and with producing responses which are appropriate to the sum of the sensory information reaching it at any moment, in the light of past experience. This extremely complicated activity is achieved by the passage of impulses through the interconnected networks of cells in the central nervous system. Each cellular unit is linked into the system by junctions (**synapses**) at which the surfaces of the nerve cells come together and a mechanism exists for the transfer of activity from one cell to another. Whether the activity passes from cell to cell or dies out at the synapse depends on a number of complex factors, but it is the discontinuity at the synapses and the ability to facilitate or inhibit the passage of impulses across them in different circumstances which allows for the extreme variability of response; a variability which is also dependent on the total number of cells and synapses involved.

It is not possible to give more than the most rudimentary analysis of this complex system, but a knowledge of the arrangement of its parts and of the major tracts by which they are interrelated is an essential prerequisite to an understanding of its functions. In the following pages the gross anatomy of the central nervous system is dealt with, and some of the details of its finer structure are indicated briefly.

The **brain** is that part of the central nervous system which lies within the cranial cavity. It is surrounded by the same three membranes (**meninges**) as the spinal medulla (dura mater, arachnoid and pia mater) and they are continuous with their spinal counterparts through the foramen magnum. In the skull the dura mater is fused with the periosteum lining the cranial cavity (endocranium) so that, when the brain is removed, the arachnoid separates from the dura at the subdural space [p. 37] and the pia and arachnoid come away with the brain.

The main **blood vessels of the brain** and their principal branches lie in the subarachnoid space between the arachnoid and pia mater, and the smaller vessels ramify on the pia mater before sinking into the substance of the brain.

The first stage in the dissection of the brain is the examination of the arachnoid and the pia mater (leptomeninges) and the blood vessels which lie between them. The student must have some knowledge of the main parts of the brain before proceeding, and the following introduction, together with the information obtained when the brain was removed from the skull [p. 40] is intended to meet this need.

The student should avoid damage to the meninges during this preliminary examination, and will be greatly helped by having a brain from which the meninges and blood vessels have been removed [Fig. 180].

The fresh brain is extremely soft and unsuitable for dissection, and the specimens used are hardened in formalin so that they retain their shape. Avoid drying

Parts of Brain

			Cavity
Forebrain	Cerebrum		Right and left lateral ventricles
	Diencephalon	Thalamus Hypothalamus	Third ventricle
Midbrain		Brain Stem	Cerebral aqueduct
Hindbrain	Pons Medulla oblongata	Brain Stem	Fourth ventricle and central canal
	Cerebellum		

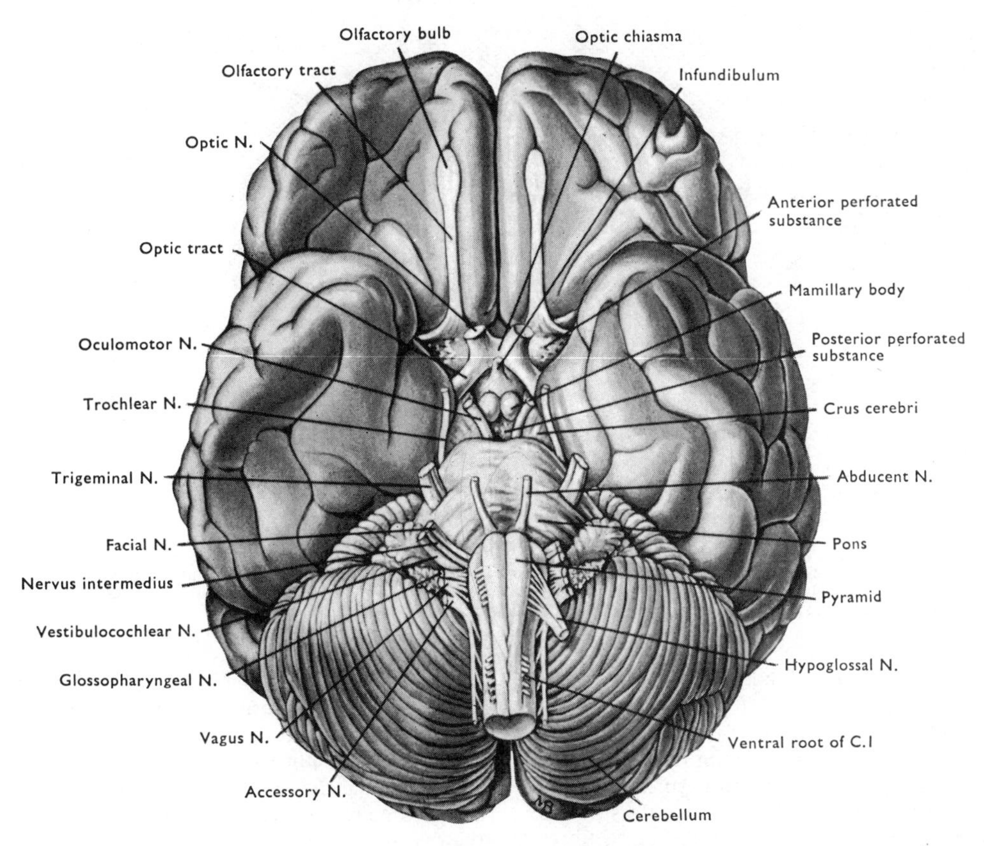

FIG. 180 The base of the brain and the cranial nerves.

of the brain during dissection by storing it in a plastic bag with a preservative which will inhibit the growth of moulds.

With the assistance of FIGURES 180 and 227, identify the major parts of the brain; the details of each part will be dealt with later when they are studied individually. A skull should be available to confirm the relations of the parts of the brain to the skull.

The **cerebrum** is the largest part. It is composed of the two cerebral hemispheres which are partially separated by the falx cerebri lying in the longitudinal fissure. The hemispheres cover the other parts of the brain so that these can only be seen on the inferior surface [FIG. 180]. The surface of each hemisphere is increased in area by extensive folding. It thus presents a number of grooves (**sulci**) between which are blunt ridges (**gyri**). These sulci and gyri cannot be seen clearly until the meninges are removed, but it should be noted that the arrangement of these is never quite alike in any two brains, and even the right and left hemispheres of one brain show marked differences.

Each hemisphere has three surfaces; (1) the convex **superolateral surface**; (2) the medial surface, partly in contact with the falx cerebri; (3) the **inferior surface**, which consists of orbital and tentorial parts. The **orbital part** lies on the roof of the orbit and of the nasal cavity (floor of anterior cranial fossa), and is separated from the tentorial part by a deep horizontal fissure (stem of the lateral sulcus) into which the lesser wing of the sphenoid fits. Posteriorly, the **tentorial part** rests on the tentorium and ends in the occipital pole, which fits into the fossa on the occipital bone in the angle between the grooves for the superior sagittal and transverse sinuses. Anteriorly, it lies on the floor of the middle cranial fossa, and extends forwards below the lesser wing of the sphenoid into the anterior extremity of the fossa (temporal pole).

The **diencephalon** is almost entirely hidden from view by the cerebral hemispheres. It consists of a dorsal **thalamus** and a ventral **hypothalamus**, and only the floor of the latter is visible on the base of the brain. Here it appears as the small area bounded by the optic chiasma and optic tracts, anteriorly and anterolaterally, and the crura cerebri posterolaterally [FIG. 180]. It lies superior to the sella turcica of the sphenoid bone, and forms the roof of the **interpeduncular fossa**. This is a deep median hollow on the inferior surface of the brain, which lies between the temporal lobes of the brain laterally, the optic chiasma anteriorly, and the pons posteriorly. If the meninges are intact, the interpeduncular fossa is hidden by a layer of arachnoid which spans it, and forms its floor. If the arachnoid is torn, the tuber cinereum and mamillary bodies can be seen on the

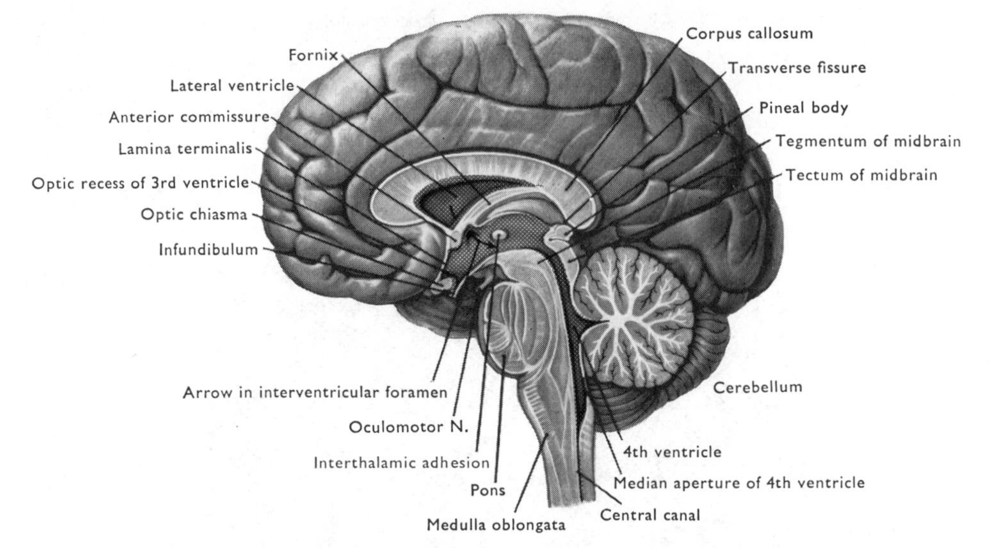

FIG. 181 Medial surface of the right half of a bisected brain. The septum pellucidum has been removed to expose the lateral ventricle. The arrow passes through the interventicular foramen from the lateral to the third ventricle, and lies in the hypothalamic sulcus. Ependymal lining of ventricles, blue.

inferior surface of the hypothalamus [FIG. 180] in the fossa.

The **midbrain** is the narrow neck which passes through the tentorial notch and joins the forebrain to the hindbrain. Only the anterior surface of the midbrain can be seen at this stage. This consists of the **crura cerebri** (two broad bundles of nerve fibres, one of which issues from each cerebral hemisphere, while both enter the pons inferiorly) with the posterior part of the interpeduncular fossa between them. The posterior surface (**tectum**) of the midbrain consists of four small swellings (the **colliculi**) but is deeply buried between the cerebellum and the cerebral hemispheres, and no attempt should be made to expose it at this juncture as this may lead to the midbrain being torn across. Between the tectum and the crura cerebri lie the **tegmentum** and the **substantia nigra** of the midbrain [FIG. 213] only a small part of which can be seen in the interpeduncular fossa between the crura cerebri. In each half of the midbrain, the crus cerebri, substantia nigra, and tegmentum together constitute a **cerebral peduncle**. The oculomotor (third cranial) nerves emerge from the anterior surface of the midbrain, medial to the crura cerebri.

The **cerebellum** is the second largest mass of nervous tissue. It lies in the posterior cranial fossa, inferior to the tentorium cerebelli, and overlaps the posterior surfaces of the midbrain, pons, and medulla oblongata. It is easily recognized by the large number of closely set transverse fissures which cross its surface. The major parts of the cerebellum may be identified with the help of FIGURES 203 and 204, but it is sufficient, at this stage, to note that it consists of two hemispheres, and a median portion (the **vermis**) which, on the inferior surface, is deeply set between the hemispheres in a median groove, the **vallecula cerebelli**. The vallecula is partly filled by the falx cerebelli, a blunt dural fold which lies vertically between the fossae on the occipital bone for the cerebellar hemispheres.

The **pons** is the white, bulging bridge which arches across the anterior aspect of the hindbrain between the two halves of the cerebellum. On each side, the pons narrows posteriorly to form the rounded **middle cerebellar peduncle** which extends posteriorly into the cerebullum. The dividing mark between the pons and the peduncle is the thick **trigeminal** (fifth cranial) **nerve**, the only nerve which issues through the substance of the pons [FIG. 180].

The **medulla oblongata** is the conical, white body which extends inferiorly from the pons to join the spinal medulla at the foramen magnum. A **median fissure** divides the ventral surface into right and left halves, and a longitudinal ridge, on each side of the fissure, is known as the **pyramid** because it tapers to a point inferiorly. Posterolateral to the pyramid is an oval swelling, the **olive** [FIGS. 192–194].

The sixth to twelfth **cranial nerves** are attached to the medulla oblongata. The sixth and twelfth arise ventrally, between the pyramid and the olive; the remaining nerves are attached to the lateral surface, posterior to the olive. The seventh and eighth emerge at the lower border of the pons, and, inferior to this, the ninth, tenth, and eleventh form a linear series of rootlets, the most inferior of which arise at the level of the fifth cervical segment of the spinal medulla.

The **brain stem**. This term is usually applied to the midbrain, pons, and medulla oblongata, and is used in this sense here, though it is sometimes considered to include the diencephalon.

THE MEMBRANES OF THE BRAIN (MENINGES)

DURA MATER

This has been seen and described during the dissection of the head [pp. 37–51] but its parts should be reviewed in relation to the brain.

ARACHNOID

[p. 37]

This exceedingly thin, almost transparent membrane lines the internal surface of the dura mater, and has exactly the same shape as the dural sac except where its arachnoid granulations (*q.v.*) pierce the dura mater. The arachnoid is separated from the dura mater by a bursa-like, capillary space (the **subdural space**) containing a film of fluid. This forms a sliding plane where movement is possible between the dura mater and the brain enclosed in the arachnoid and pia mater, except where the arachnoid and dura are fused, *i.e.*, where both are pierced by structures entering or leaving the brain (*e.g.*, nerves and blood vessels), where the arachnoid granulations pierce the dura mater, and where the ligamenta denticulata are attached to the dura mater [p. 160].

SUBARACHNOID SPACE

This space, between the arachnoid and pia mater is filled with **cerebrospinal fluid** which enters it from the cavities (ventricles) of the brain, and acts as a mobile buffer to distribute and equalize pressures within the skull.

Filaments or **trabeculae** traverse the subarachnoid space from the arachnoid to the pia mater, and

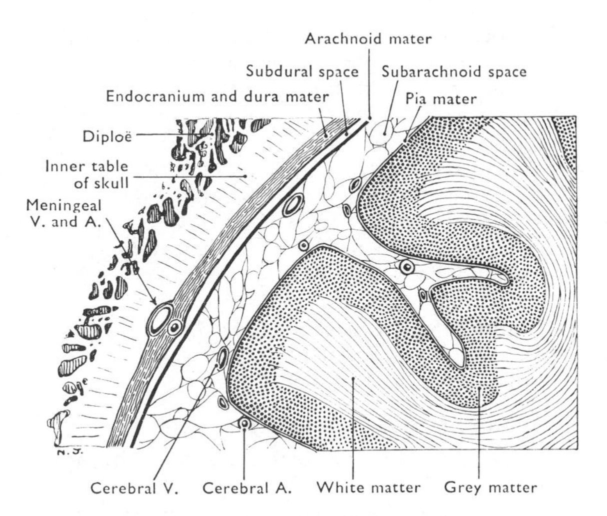

FIG. 182 Diagrammatic section to show the relation of the meninges to the skull and brain. Note that the meningeal arteries are in the endocranium, while the cerebral blood vessels lie in the subarachnoid space.

are numerous in some situations, *e.g.*, on the surfaces of the cerebral hemispheres. Here the dense trabeculae form a kind of fluid-filled sponge which may help to protect the surface of the brain from damage against the skull and the dural folds (the falx and tentorium) when it moves within the dura mater. They also bind the pia and arachnoid tightly together in these situations. Elsewhere, and especially where the pia mater closely, investing the brain, is widely separated from the arachnoid lining the dura mater,

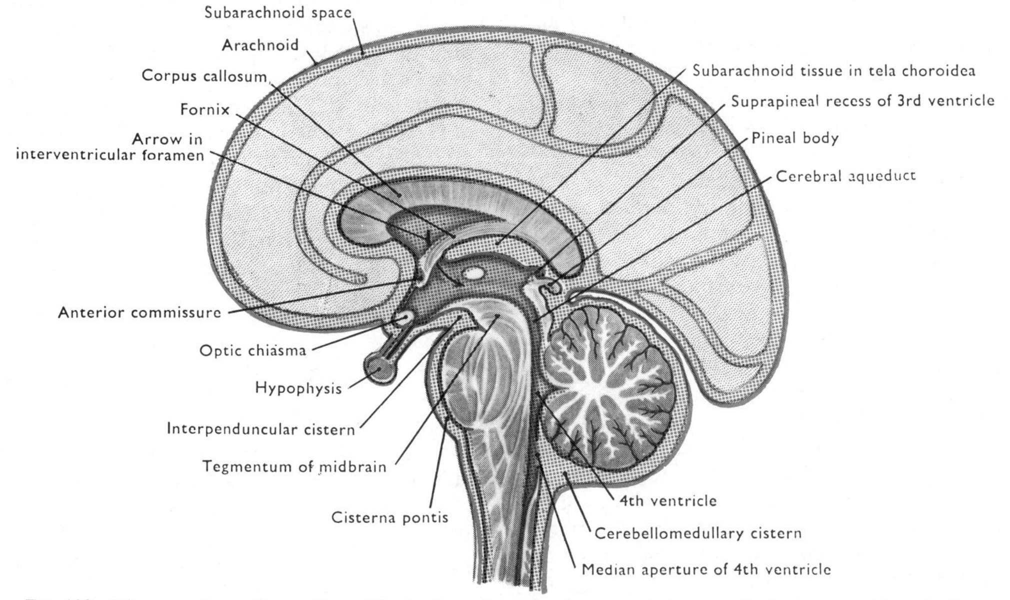

FIG. 183 Diagram of a median section of the brain to show membranes and cisterns. Red, pia mater; blue stipple, subarachnoid space and ventricles; blue, arachnoid; pink, surface view of shallow subarachnoid space. The lines of blue stipple running over the pink area indicate the places where the subarachnoid space is deeper around the branches of the anterior and posterior cerebral arteries.

the mesh is less dense and the cerebrospinal fluid can flow more freely. The larger arteries and veins of the brain lie in the subarachnoid space, and entering or leaving the brain traverse it. In some situations (*e.g.*, the stem of the lateral sulcus) the space forms a tight sleeve around the artery, so that its pulsations force the cerebrospinal fluid along the sleeve. Where the vessels on the external surface of the pia mater send branches into the substance of the brain, a sleeve of pia mater and subarachnoid space (**perivascular space**) surrounds each branch for a short distance until the pia mater fuses with the wall of the vessel.

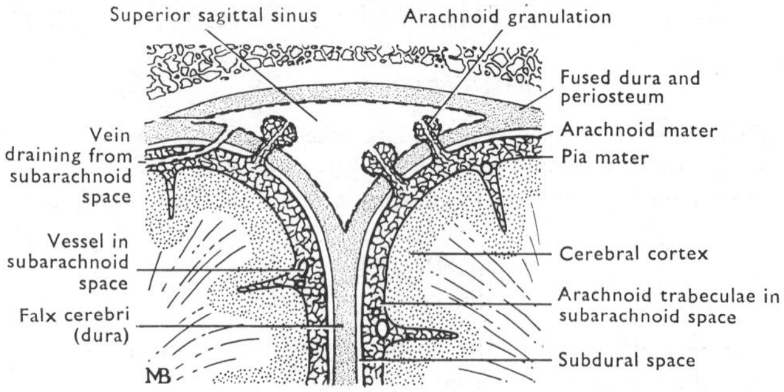

FIG. 184 Diagrammatic transverse section through the superior sagittal sinus and surrounding structures. Note the arrangement of the arachnoid granulations.

Subarachnoid Cisterns [Fig. 183]

These pools of cerebrospinal fluid are found where the brain, closely covered with pia mater, lies some distance from the arachnoid lining the dura mater. Cisterns are principally found around the brain stem and cerebellum, on the base of the brain, around the free margin of the tentorium cerebelli, and in association with the major blood vessels.

The **cerebellomedullary cistern** lies in the angle between the cerebellum, medulla oblongata, and the occipital bone. It is directly continuous, inferiorly, with the posterior part of the spinal subarachnoid space [FIG. 183] and is accessible to a needle introduced anterosuperiorly through the posterior atlanto-occipital membrane, between the posterior arch of the atlas and the posterior margin of the foramen magnum (*cisternal puncture*; *cf.*, lumbar puncture, page 161).

The **cisterna pontis** lies anterior to the pons and medulla oblongata, and contains the vertebral and basilar arteries. It is continuous: (1) posteriorly, around the medulla oblongata with the cerebellomedullary cistern; (2) inferiorly, with the spinal subarachnoid space; (3) superiorly, with the interpeduncular cistern.

The **interpeduncular cistern** fills the interpeduncular fossa. At the superior border of the pons, the arachnoid turns anteriorly above the sella turcica and is stretched between the temporal lobes of the hemispheres. Here it forms the floor of the interpeduncular cistern. This cistern contains the circulus arteriosus [FIG. 186] and is continuous, laterally, with the subarachnoid spaces surrounding the middle and posterior cerebral arteries, and anteriorly (around the anterior cerbral arteries) with the cisterna chiasmatis. It seems certain that the pulsations of these arteries help to force cerebrospinal fluid from the cistern on to the surfaces of the hemispheres. That which passes along the middle cerebral artery finds its way to the cistern of the lateral fossa of the cerebrum, while, along the posterior cerebral arteries, it runs around the midbrain, along the margin of the tentorium cerebelli, the cisterna ambiens.

The various cisterns communicate freely, but there is *no communication with the subdural space*, and *the only communications between the cavities of the brain and the subrarachnoid space are through three small openings in the roof of the fourth ventricle in the hindbrain.* Blockage of these apertures by scar tissue following injury or infection interferes with the flow of cerebrospinal fluid, as will scar tissue that forms wherever there is a dense mesh of trabeculae uniting the pia and arachnoid.

Arachnoid Villi and Granulations

Arachnoid villi are minute protrusions of the arachnoid through apertures in the dura mater into the venous sinuses of the dura mater, especially the superior sagittal sinus and its lateral lacunae. They contain subarachnoid tissue and cerebrospinal fluid, and a number of capillary tubules which pass through the middle of each villus and become continuous with the cavity of the venous sinus at the apex of the villus. The tubules and the spaces of the mesh of arachnoid trabeculae in the villus are both continuous with the subarachnoid space at the base of the villus. When the pressure of the cerebrospinal fluid exceeds that in the venous sinus, the spaces fill with cerebrospinal fluid, the villi bulge into the sinus, and the central tubules become patent, allowing the escape of cerebrospinal fluid into the venous sinus. If the venous pressure exceeds that of the cerebrospinal fluid, the villi collapse, effectively closing the central tubules and preventing the reflux of blood into the subarachnoid space. Hence they are one-way valves for the escape of cerebrospinal fluid into the venous system.

With increasing age, the size of the villi increases till they form arachnoid granulations which often indent the overlying bone of the skull [p. 38].

DISSECTION. If the arachnoid is complete, divide it along the anterior aspect of the medulla oblongata and pons to expose the vertebral and basilar arteries and their branches [FIG. 187]. Note the arachnoid trabeculae. Cut posteriorly through the arachnoid into the cerebellomedullary cistern, and identify the posterior inferior cerebellar branch of the vertebral artery. This branch winds posteriorly on the medulla oblongata into the cistern, on its way to the cerebellum.

Identify the apertures in the roof of the cavity of the hindbrain (fourth ventricle) by the tuft of finely granular material (**choroid plexus**) which protrudes through each [FIG. 192]. The **median aperture** can be seen in the depths of the cerebellomedullary cistern by gently depressing the medulla away from the cerebellum [FIG. 208]. The **lateral apertures** face anteriorly, and lie immediately posterior to the corresponding glossopharyngeal nerve. Pick up the tuft of choroid plexus protruding through the lateral aperture and pull it gently towards the cerebellum: the aperture will be seen immediately anterior to the choroid plexus at the end of a sleeve-like extension of the thin roof of the fourth ventricle [FIGS. 192, 208].

Extend the median incision in the arachnoid into the interpeduncular fossa which lodges the **interpeduncular cistern**. Note the blood vessels in the cistern, and the fact that its floor is perforated posteriorly by the oculomotor nerves and anteriorly by the internal carotid arteries and the infundibulum (stalk of the hypophysis).

PIA MATER OF THE BRAIN

The pia mater is adherent to the surface of the brain and follows the irregularities of its surface closely. The pia is thick on the medulla oblongata and spinal medulla, but is more delicate elsewhere.

The blood vessels of the brain run in the subarachnoid space on the surface of the pia mater. They break up into branches on the pia mater and anastomose there before passing into the substance of the brain as end-arteries. Thus if a small piece of pia mater is stripped from the surface of the brain, its deep surface is covered with a number of minute processes which are the small blood vessels entering the brain substance. If this manœuvre is carried out on a living animal, the piece of brain lying under the excised pia mater will degenerate from lack of blood supply, because of the lack of anastomoses with vessels in the adjacent brain tissue.

In certain situations the walls of the cavities of the brain (ventricles) are thin, and consist of a layer of the lining epithelium (**ependyma**) only. In these regions the pia mater on the external surface is invaginated into the cavities as a series of vascular tufts which carry the ependyma before them, and thus form the **choroid plexuses** of the ventricles. The pial element of this complex is known as the **tela choroidea**. The choroid plexuses are the main source of cerebrospinal fluid within the ventricles, whence it escapes into the subarachnoid space through the apertures in the roof of the fourth ventricle. The cerebrospinal fluid circulates slowly in the subarachnoid space, assisted partly by the pulsations of the arteries therein and partly by the fact that it is removed from the space through the arachnoid granulations and villi, and is passed into lymph vessels and veins associated with the extradural parts of the spinal and cranial nerves. It is probable that the cerebrospinal fluid formed by the choroid plexuses is supplemented from the ependyma which lines the remainder of the cavities of the brain as well as covering the choroid plexuses, and by fluid produced in the subarachnoid space itself.

THE BLOOD VESSELS OF THE BRAIN

All the intracranial vessels, with the exception of the venous sinuses [pp. 38, 44, 51] which are enclosed in the thick dura mater, have much thinner walls than extracranial vessels of comparable size; for they are enclosed in and supported by the surrounding cranium. The walls of all but the largest veins are so thin that they are virtually invisible unless filled with blood, and thus some of them are difficult to identify. The veins are capable of considerable distention, and if the venous drainage is partially blocked by pressure on the internal jugular veins, the pooling of blood in the intracranial veins produces a marked rise in intracranial pressure, and this may be measured by means of a manometer attached to a hollow needle introduced into the subarachnoid space by lumbar or cisternal puncture. This feature may be used to confirm the continuity of the cranial and lumbar subarachnoid spaces in clinical investigations (Queckenstedt's test).

VEINS OF THE CEREBRAL HEMISPHERES

Most of these veins lie on the surfaces of the hemisphere in the subarachnoid space. They drain into the venous sinuses of the dura mater by crossing the subarachnoid and subdural spaces, and piercing the arachnoid and dura.

Veins of Superolateral Surface

The **superior cerebral veins** [FIG. 185] converge on the superior sagittal sinus, the anterior and posterior veins entering the sinus obliquely from in front and behind, and all passing inferior to the lateral lacunae of the sinus. Most of the **inferior cerebral veins** converge on the superficial middle cerebral vein, but those of the occipital lobe run inferiorly into the transverse sinus.

The **superficial middle cerebral vein** runs anteriorly between the lips of the posterior ramus of

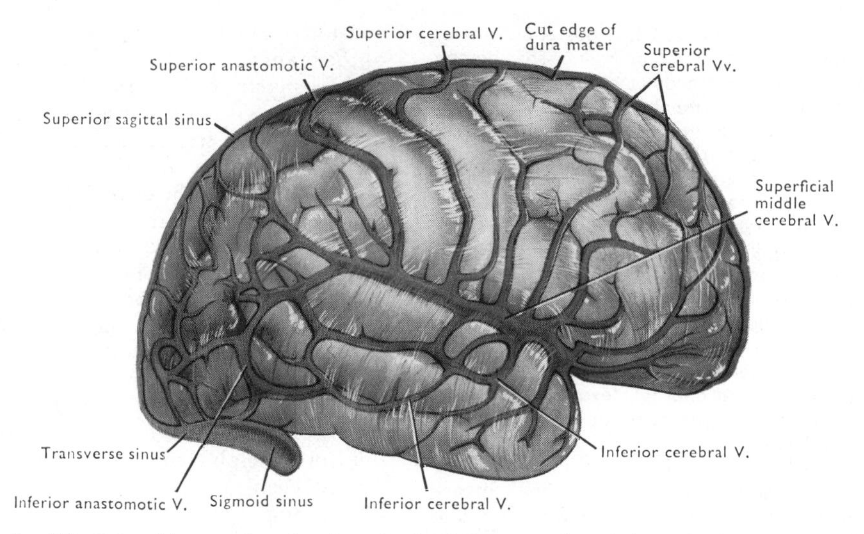

FIG. 185 Veins of the superolateral surface of the hemisphere, seen through the arachnoid mater.

the lateral sulcus [FIG. 227] and curving medially into the stem of that sulcus, ends in the cavernous sinus. Its cut end will be found piercing the arachnoid near the medial end of the stem of the lateral sulcus. Posteriorly, the superficial middle cerebral vein is frequently connected to the superior sagittal sinus by a wide **superior anastomotic vein**, and commonly to the transverse sinus by the **inferior anastomotic vein** [FIG. 185]. These anastomotic veins may become important if the cavernous sinus is blocked, but the superficial veins also connect with veins in the interior and on the base of the hemisphere by small veins which pierce the hemisphere.

Veins of Inferior Surface of Hemisphere

These drain in a number of directions: (1) anteriorly, to the anterior and superficial middle cerebral veins; (2) posteriorly, to the basal vein (*q.v.*); and (3) directly to the superior petrosal, straight, and tranverse sinuses.

Veins of Medial Surface of Hemisphere

These veins will be seen later when the hemispheres are separated, but parts of this system can be seen now, so a brief description is given.

The main vein on the medial aspect is the **great cerebral vein**. It emerges from beneath the posterior end (splenium) of the corpus callosum (the mass of white matter which passes between the two hemispheres [FIGURE 181]) and curves superiorly on the splenium to join the inferior sagittal sinus and form the straight sinus [FIG. 37]. If the occipital lobes of the hemispheres are gently separated, the cut end of the great cerebral vein will be seen close to the splenium.

The great cerebral vein is joined by a number of symmetrical tributaries from the midbrain and cerebellum inferior to it, and by the basal veins which curve round the sides of the midbrain to reach it from the inferior surface of the brain.

Each **basal vein** is formed deep in the medial part of the stem of the lateral sulcus by the confluence of: (1) the **anterior cerebral vein**, which runs with the corresponding artery and enters the basal vein from in front; (2) the **deep middle cerebral vein**, which lies in the depths of the lateral sulcus on the insula; (3) the **striate vein** or veins which descend through the substance of the brain and emerge close to the formation of the basal vein through the anterior perforated substance [FIG. 180].

DISSECTION. Divide the arachnoid over the stem of the lateral sulcus and depress the temporal pole so as to expose the beginning of the middle cerebral artery, the deep middle cerebral vein, the striate veins, and the origin of the basal vein. If the latter is filled with blood it will be seen passing posteriorly close to the optic tract [FIG. 180] and should be traced as far as possible towards the great cerebral vein.

The veins of the brain stem and cerebellum mainly drain in a posterior direction to reach the basal and great cerebral veins, and the adjacent venous sinuses. The veins of the medulla oblongata communicate with those of the spinal medulla.

ARTERIES OF THE BRAIN

Two internal carotid and two vertebral arteries carry the total blood supply of the brain. The vertebral arteries enter the skull through the foramen magnum. Each internal carotid traverses the skull in the carotid canal and the superior part of the foramen lacerum. It then takes a sinuous course through the

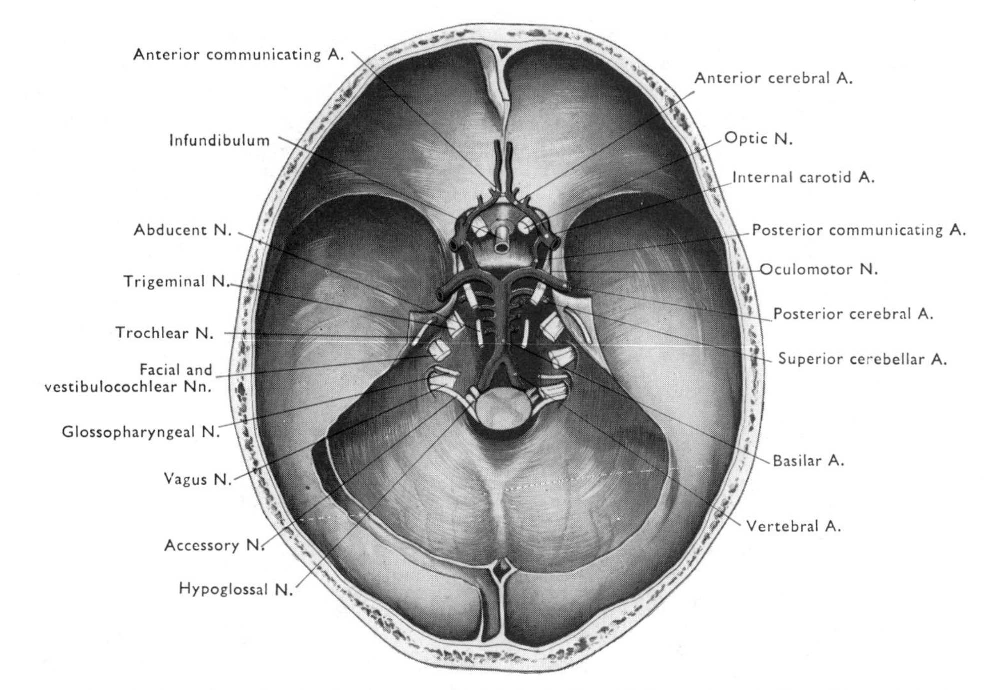

FIG. 186 The floor of the cranial cavity after removal of the brain, but with the arteries at the base of the brain *in situ*.

cavernous sinus, pierces the dural roof of the sinus, and ends immediately lateral to the optic chiasma, inferior to the anterior perforated substance [FIG. 180], by dividing into its terminal branches.

Vertebral Arteries

Each vertebral artery enters the subarachnoid space in the upper part of the vertebral canal by piercing the lateral aspect of the dura mater and arachnoid. It ascends anterosuperiorly through the foramen magnum, and curves round the ventrolateral aspect of the medulla oblongata, close to the rootlets of the hypoglossal nerve, to unite with its fellow at the lower border of the pons and form the median **basilar artery**. The two vertebral arteries are often of very different calibre.

Intracranial Branches. (1) The **posterior spinal artery** is the first intracranial branch. It passes inferiorly on the spinal medulla among the dorsal rootlets of the spinal nerves [p. 162]. It commonly arises from the posterior inferior cerebellar artery. (2) The **posterior inferior cerebellar artery** is the largest branch. It arises from the vertebral artery soon after it pierces the meninges, and pursues a tortuous course posteriorly on the side of the medulla oblongata, among the rootlets of the hypoglossal, vagus, and glossopharyngeal nerves. It supplies branches to this part of the medulla oblongata, occasionally as far cranially as the inferior border of the pons. Reaching the posterior surface of the medulla oblongata, between the thin roof of its cavity (the fourth ventricle) and the cerebellum [p. 193], it supplies the choroid plexus of that ventricle, and turns downwards on the surface of the cerebellum, supplying it. It varies inversely in size with the anterior inferior cerebellar artery, but is usually much larger than that vessel. (3) The **anterior spinal artery** arises near the inferior border of the pons, and runs inferomedially to join its fellow on the ventral surface of the medulla oblongata. It supplies the median part of the medulla oblongata, and continues inferiorly throughout the length of the spinal medulla [p. 162].

Basilar Artery

This artery is formed by the junction of the two vertebral arteries at the inferior border of the pons. It ends at the superior border of the pons by dividing into the two posterior cerebral arteries. It lies in the median groove of the pons in the cisterna pontis, on the basilar part of the occipital bone and the dorsum sellae of the sphenoid [FIG. 186].

Branches [FIG. 187]. (1) The **anterior inferior cerebellar artery** arises at the lower border of the pons, and runs laterally along it superficial to and supplying the sixth, seventh, and eighth cranial nerves. It then loops over the flocculus [FIG. 192] and supplies the antero-inferior part of the cerebellar surface. (2) The **artery of the labyrinth** arises beside or from the anterior inferior cerebellar artery. It accompanies the vestibulocochlear nerve to the internal ear, and supplies the adjacent part of the facial nerve. (3) Numerous, slender **pontine branches** pierce the pons, some in its medial part, others further laterally. (4) The large **superior cerebellar artery** arises close to the superior border of the pons. It winds posteriorly along the superior border of the pons and the middle cerebellar peduncle, supplying both and the part of the midbrain which lies immediately superior to them. It also sends many branches on to the superior surface of the cerebellum, inferior to the tentorium cerebelli. (5) The two, large **posterior cerebral arteries** diverge at the superior border of the pons. Each sends a number of fine **central branches** into the ventral surface of the midbrain (interpeduncular fossa), and then curves posteriorly on the lateral surface of the upper part of the midbrain, on to the inferomedial surface of the corresponding hemisphere. Here it passes towards the occipital pole, giving branches as shown in FIGURES 188 and 216. The posterior cerebral and superior cerebellar arteries lie on the superior and inferior parts of the midbrain respectively, and hence are above and below the tentorium; the third and fourth cranial nerves, which arise from the midbrain, pass forwards between them.

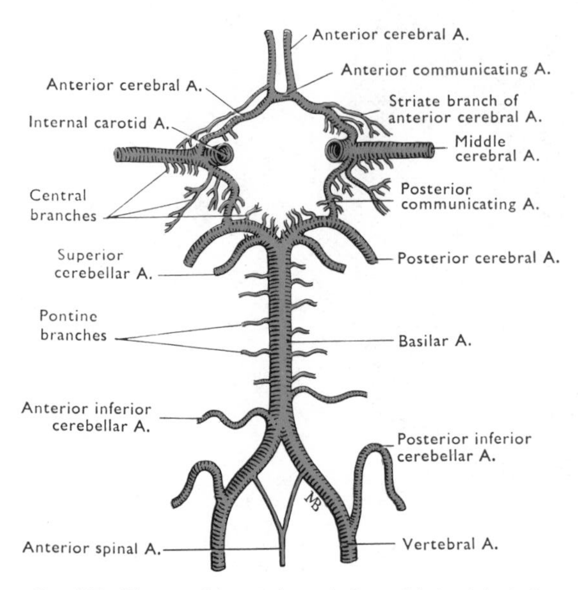

FIG. 187 Diagram of the arteries on the base of the brain including the circulus arteriosus.

Branches of the Posterior Cerebral Arteries. (1) Small **central branches** pierce the ventral surface of the midbrain in the interpeduncular fossa, thus forming the **posterior perforated substance**. Similiar branches pierce the lateral surface of the midbrain and caudal diencephalon. (2) **Posterior choroid branches** arise on the lateral surface of the midbrain, and pass forwards to supply the greater part of the choroid plexus in the cavity of

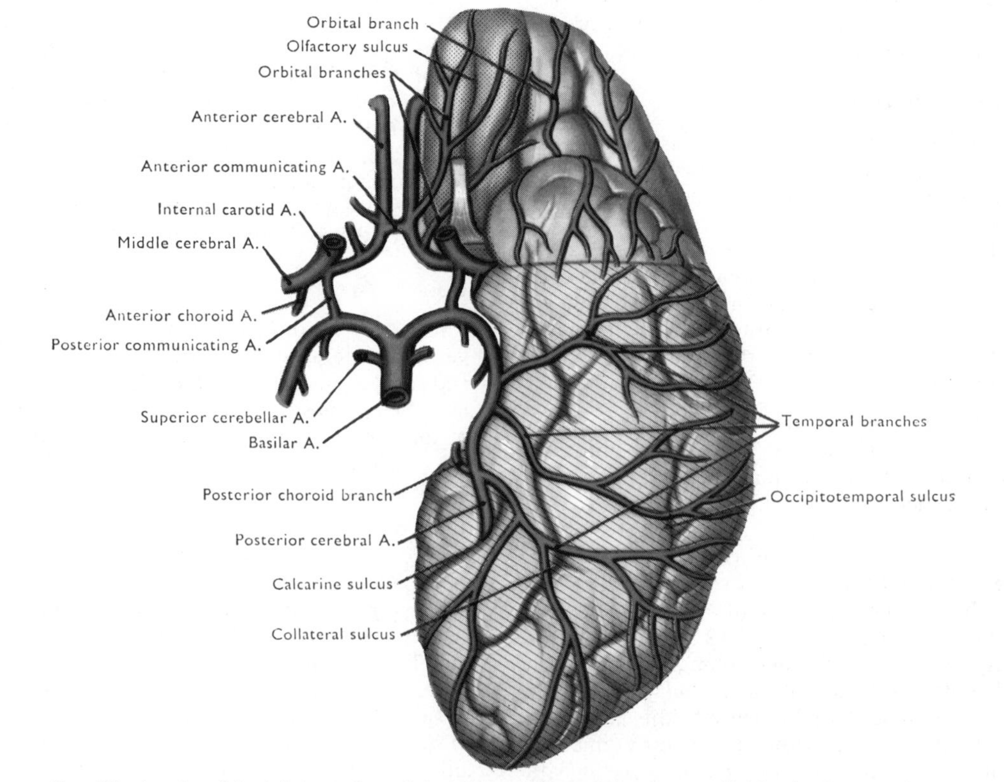

FIG. 188 Arteries of the inferior surface of the left hemisphere. The areas supplied by the three arteries are indicated by: red stipple, anterior cerebral artery; red cross-hatching, posterior cerebral artery; no colour, middle cerebral artery.

the hemisphere (lateral ventricle) and of the diencephalon (third ventricle). (3) Branches **to the cerebral cortex**, most of which will be seen when the medial aspect of the hemisphere is exposed.

On the anterior surface of the midbrain, each posterior cerebral artery receives a slender **posterior communicating** branch from the corresponding internal carotid artery. This forms part of the arterial circle in the interpeduncular fossa, and sends branches to the hypothalamus and hypophysis.

Internal Carotid Arteries

The cut end of each internal carotid artery can be seen immediately lateral to the optic chiasma. It is passing into a shallow pit (the vallecula) immediately inferior to the anterior perforated substance [FIG. 180]. Here the artery gives off two small branches, and divides into the middle and anterior cerebral arteries. The small branches are:

(1) The **posterior communicating artery** passes posteriorly, across the crus cerebri to join the posterior cerebral artery. It gives off minute branches to the crus, optic tract, hypophysis, and hypothalamus. It is usually a slender vessel, but it may be large on one or both sides, and form the major route for blood passing to the posterior cerebral artery. In such cases, occlusion of the internal carotid may lead to damage in the territory of the posterior cerebral artery. (2) The **anterior choroid artery** arises superior to (1), and passes posterolaterally, close to the optic tract. It gives branches into the crus cerebri, and turns laterally, superior to the medial aspect of the temporal lobe (uncus, FIGS. 229, 244) to enter the choroid plexus of the inferior horn of the lateral ventricle which lies in the temporal lobe.

Anterior Cerebral Artery. Only a small part of this artery can be seen before the medial aspect of the hemisphere is exposed. This part runs horizontally in an anteromedial direction above the optic chiasma [FIGS. 188, 216]. It then bends sharply upwards into the longitudinal fissure, and is connected to its fellow of the opposite side by the short **anterior communicating artery** [FIG. 187] which lies anterosuperior to the optic chiasma. Beyond this it runs on the medial surface of the hemisphere [FIG. 216].

Branches. (1) Several slender branches pierce the brain anterior to the optic chiasma and enter the anterior hypothalamus. (2) Branches pass to the optic chiasma and the optic nerve. (3) One or more larger recurrent branches arise in the region of the anterior communicating artery and run back to the

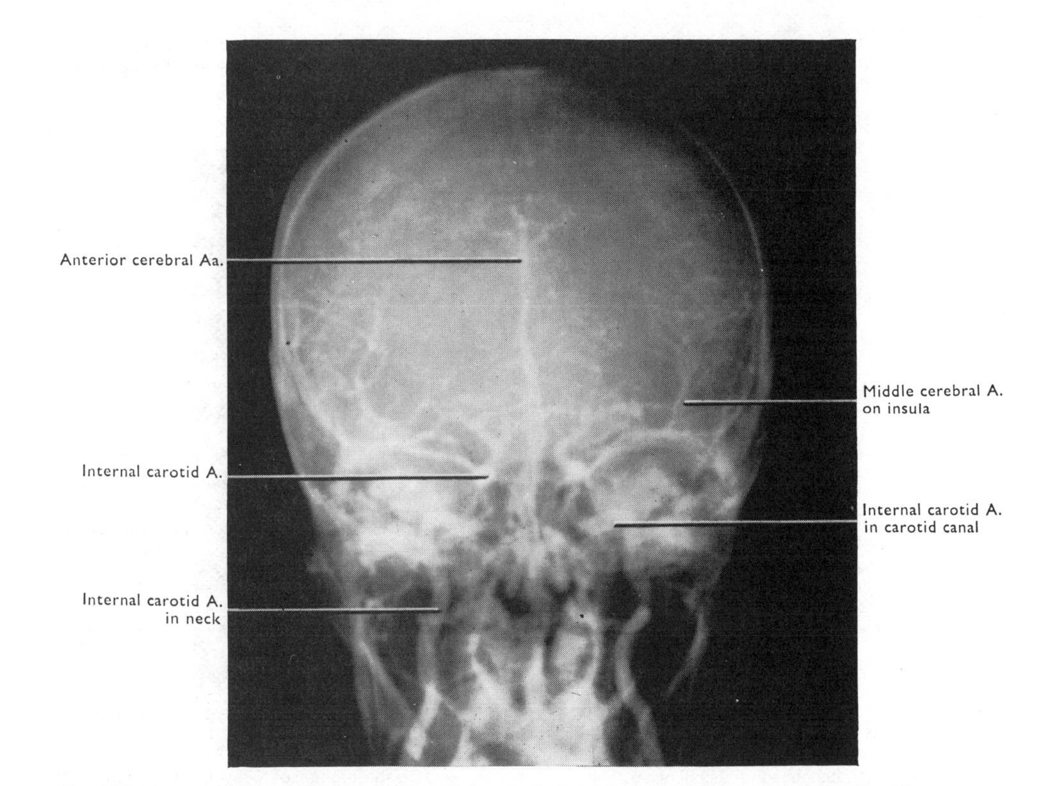

FIG. 189 Anteroposterior radiograph of the skull showing both internal carotid arteries injected with an X-ray opaque medium. This has been carried by the circulating blood into the middle and anterior cerebral arteries.

medial part of the anterior perforated substance. Here they send a number of perforating branches into the brain substance. (4) The cortical branches will be seen later.

Middle Cerebral Artery. This large branch lies in line with the internal carotid artery, and thus particulate matter passing through that artery enters the middle cerebral artery more frequently than the anterior cerebral. The middle cerebral artery runs laterally in the stem of the lateral sulcus, and breaks up into a number of branches on the **insula**—a part of the cerebral cortex buried in the depths of the posterior ramus of the lateral sulcus [FIG. 217] which separates the temporal lobe from the frontal and parietal lobes [FIG. 227]. These branches emerge on to the superolateral surface of the hemisphere through the lateral sulcus, and supply most of that surface [FIG. 190] and also the adjacent parts of the orbital and tentorial surfaces, including the temporal

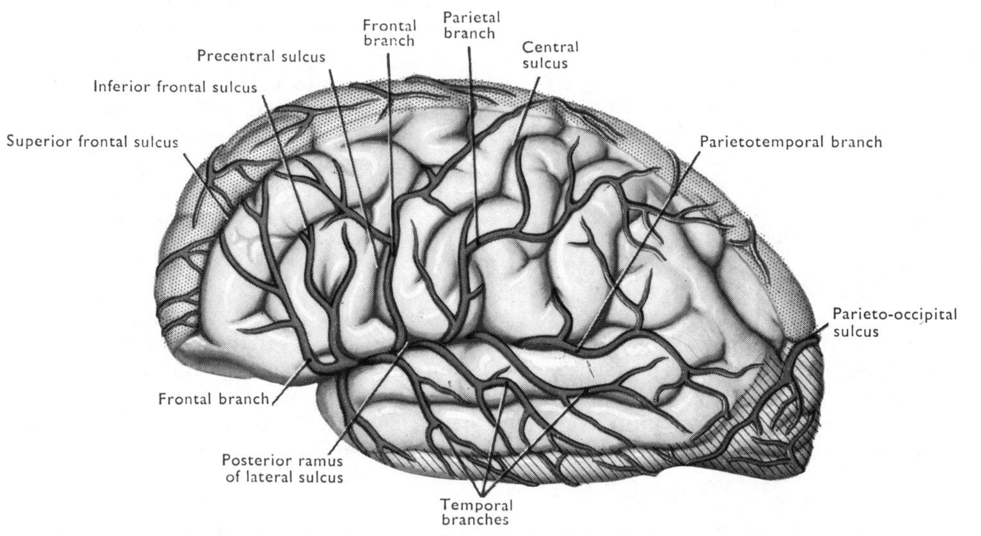

FIG. 190 Arteries of the superolateral surface of the left cerebral hemisphere. The areas supplied by the three arteries are indicated by: red stipple, anterior cerebral artery; no colour, middle cerebral artery; red cross-hatching, posterior cerebral artery.

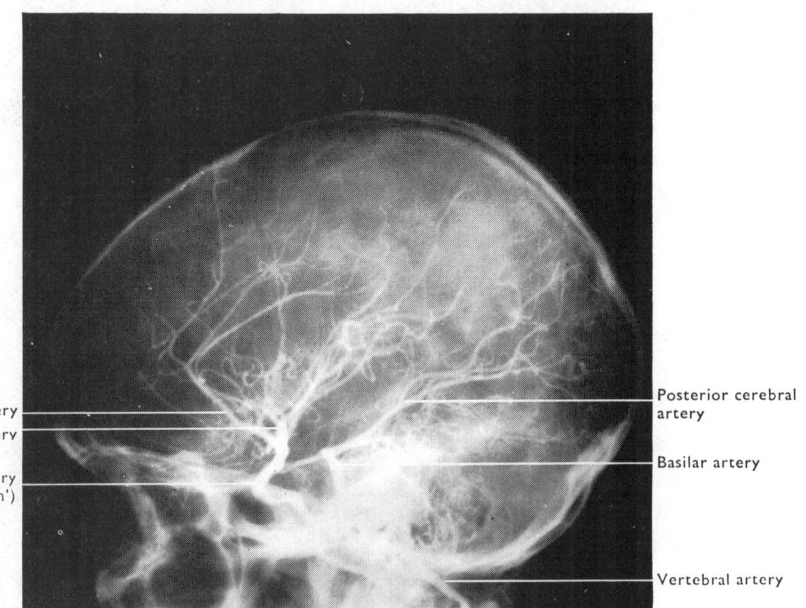

FIG. 191 Lateral radiograph of the skull of a patient taken immediately after the injection of X-ray opaque material into one internal carotid and one vertebral artery. The distribution of these two arteries is outlined. Note the posterior communicating artery apparently joining the superior end of the basilar artery to the internal carotid artery.

pole [FIG. 188]. This area includes most of the 'motor' and 'sensory' areas and all of the 'auditory' area of the cerebral cortex.

The **central branches** are numerous small **striate arteries** that pass superiorly through the anterior perforated substance (principally its lateral part) to supply the deep nuclei of the hemisphere, chiefly the corpus striatum (*q.v.*).

Circulus Arteriosus [FIG. 187]

This arterial circle extends from the superior border of the pons to the longitudinal fissure, and it lies principally in the interpeduncular fossa. It is composed of the posterior cerebral, posterior communicating, anterior cerebral, and anterior communicating arteries. It forms a route through which blood entering by either internal carotid artery or the basilar artery may be distributed to any part of the cerebral hemispheres. Because of variations in the calibre of the vessels which form the circulus, it is important, before tying off a carotid artery for any reason, to determine the nature of the circle by carotid or vertebral angiography [FIG. 191]. combined with compression of one or more of the feeding vessels to determine the efficacy of the parts of the circle.

Two types of branches arise from the circle and its major branches; these are arbitrarily divided into cortical and central, and they differ from each other in the extent of their anastomoses. The **central branches** are very numerous and slender; they tend to arise in groups and immediately pierce the surface of the brain to supply its internal parts. The largest collections of these pass through the anterior and posterior perforated substances. They do not anastomose to a significant extent within the brain substance. The **cortical branches** ramify over the surface of the cortex and anastomose fairly freely on the pia mater. They give rise to numerous small branches which enter the cortex at right angles and, like the central branches, do not anastomose in it. It follows that blockage of an artery on the pia mater may produce little if any damage to the brain, but damage to branches entering the substance of the brain leads to the destruction of brain tissue.

The arteries of the brain are liberally supplied with **sympathetic nerves** which run on to them from the carotid and vertebral plexuses. They are extremely sensitive to injury and readily react by passing into prolonged spasm. This of itself may be sufficient to cause damage to brain tissue since even the least sensitive of neurons cannot withstand absolute loss of blood supply for a period exceeding seven minutes.

DISSECTION. Remove the blood vessels and the remains of the arachnoid and pia mater from the base of the brain, cutting the cerebral arteries as they leave the interpeduncular cistern, but leaving their distal parts intact. Do not attempt to remove the pia mater from the brain stem as this will result in removal of the cranial nerves. Note the points of entry of the central branches of the posterior cerebral arteries at the posterior perforated substance, and of the central branches of the middle cerebral arteries at the anterior perforated substance.

THE BASE OF THE BRAIN

Interpeduncular Fossa

This is the rhomboidal space bounded by the pons, the crura cerebri, and the optic tracts and chiasma. The crura cerebri [FIG. 180] emerge from the cerebral hemispheres, and are crossed by the optic tracts, which are applied to their lateral surfaces. Inferiorly, the crura descend on the anterior surface of the midbrain and converge into the pons.

Structures in Interpeduncular Fossa. These are: (1) The **oculomotor nerves**, each of which emerges immediately dorsomedial to the corresponding crus. (2) The **posterior perforated substance** is a layer of grey matter in the angle between the crura cerebri. It is pierced by the central branches of the posterior cerebral arteries which arise close to the origin of these vessels. (3) The **mamillary bodies** are a pair of small, white, spherical bodies which protrude, side by side, from the ventral surface of the hypothalamus immediately anterior to the posterior perforated substance. (4) The **tuber cinereum** is a slightly raised area of grey matter between the mamillary bodies and the optic chiasma. The **infundibulum** arises from the tuber cinereum. It is the narrow stalk which connects the hypothalamus to the hypophysis (pituitary gland) but it was cut when the brain was removed.

Anterior Perforated Substance [FIGS. 180, 228]

This small area of grey matter forms the roof of the vallecula of the cerebrum, and is pierced by the central branches of the middle and anterior cerebral arteries. It continues laterally as the roof of the stem of the lateral sulcus. Anteriorly, it is bounded by the diverging striae of the olfactory tract; medially, by the optic chiasma and tract and by the diagonal band; posteriorly, by the uncus [FIG. 228]. The anterior perforated substance is directly continuous, superiorly, with the corpus striatum in the interior of the hemisphere.

Lamina Terminalis

This thin, grey membrane extends superiorly from the optic chiasma, and forms the anterior wall of the third ventricle, the median cavity which separates the two halves of the diencephalon. The lamina terminalis may be seen by bending the optic chiasma gently downwards, when the continuity of the lamina terminalis with the anterior perforated substance may be confirmed on both sides [FIG. 228].

SUPERFICIAL ATTACHMENTS OF THE CRANIAL NERVES

Twelve pairs of cranial nerves are attached to the brain, but of these, the second is not a true nerve as it is developed from an outgrowth of the full thickness of the brain tube, and not by the outgrowth of axons either from cells situated within the central nervous system or within ganglia closely associated with it. The nerve fibres which it contains are, therefore, more akin to a tract than to a peripheral nerve. The twelve pairs of nerves are:

1. Olfactory.
2. Optic.
3. Oculomotor.
4. Trochlear.
5. Trigeminal.
6. Abducent.
7. Facial.
8. Vestibulocochlear.
9. Glossopharyngeal.
10. Vagus.
11. Accessory.
12. Hypoglossal.

A minute bundle of nerve fibres attached to the cerebrum posterior to the striae of the olfactory tract has been described as a thirteenth pair (the nervi terminales). They accompany the corresponding olfactory tract anteriorly, and are distributed with it to the nose, but their function is unknown.

Each cranial nerve enters or leaves the brain surface at its **superficial attachment**, and the fibres which it contains either arise from (efferent or motor fibres) or terminate on (afferent or sensory fibres) nuclei within the brain. These constitute respectively the **deep origins** and the **deep terminations** of the cranial nerves.

The first two pairs are attached to the forebrain; the third and fourth to the midbrain; the fifth to the pons; the remainder to the medulla oblongata, and, in the case of the eleventh, to the cervical spinal medulla also.

The cranial nerves fall naturally into three groups according to the position of their superficial attachment to the brain, which may either be ventral, or lateral, or dorsal [p. 250].

Cranial Nerves with Ventral Attachments

Olfactory Nerves. Approximately twenty of these on each side pass through the cribriform plate of the ethmoid, and end in the olfactory bulb.They are unusual in that they arise in the olfactory cells of the

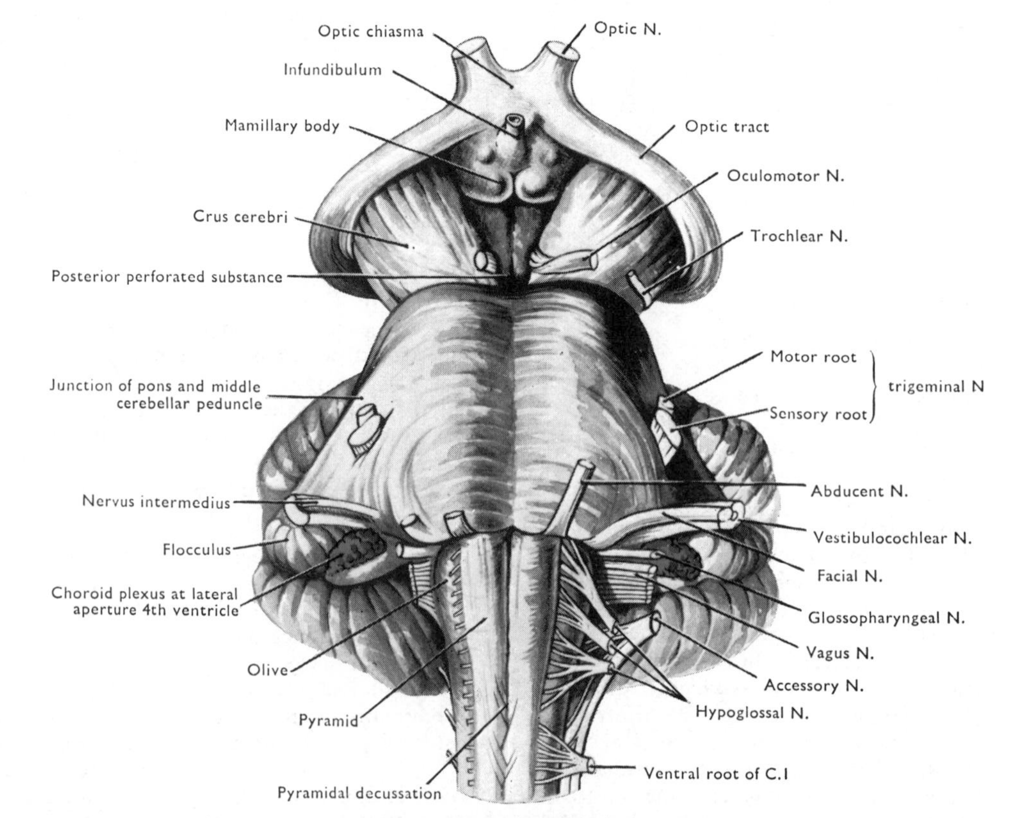

FIG. 192 Anterior surface of the brain stem.

nasal mucous membrane, and consist of bundles of minute, non-myelinated nerve fibres which are so delicate that no trace of them will be seen on the olfactory bulb [FIG. 180].

Optic Nerve. This is a thick, cylindrical nerve. It is composed of myelinated nerve fibres which arise in the retina, and it joins the anterolateral angle of the optic chiasma.

Oculomotor Nerve. It arises as a compact bundle of rootlets from the groove on the medial aspect of the cerebral peduncle in the posterior part of the interpeduncular fossa.

Abducent Nerve. It emerges at the inferior border of the pons immediately lateral to the pyramid [FIG. 192].

Hypoglossal Nerve. It is formed by a row of rootlets which arise from the anterior aspect of the medulla oblongata in the groove between the pyramid and the olive [FIG. 192]. They are directly in line with the ventral rootlets of the first cervical nerve.

Cranial Nerves with Lateral Attachments

Trigeminal Nerve. This is the largest of the cranial nerves. It is attached to the junction of the pons with the middle cerebellar peduncle, and consists of two roots: a larger, posterolateral, **sensory root**, consisting of loosely packed nerve bundles, and a smaller, anterosuperior, **motor root** which is compact and applied to the sensory root.

Facial and Vestibulocochlear Nerves. These two nerves, with the small **nervus intermedius** between, emerge on the inferior border of the pons, posterior to the olive, and in the same vertical line as the other laterally attached cranial nerves.

The facial nerve is motor, while the nervus intermedius [FIG. 192] carries its sensory and parasympathetic fibres. It lies anterior to the larger vestibulocochlear nerve which splits at the medulla oblongata into a cochlear part, which passes posterior to the inferior cerebellar peduncle [FIG. 194], and a vestibular part which passes anterior to it.

Glosspharyngeal, Vagus, and Accessory Nerves. These nerves arise as a vertical series of rootlets from a groove posterior to the olive in the medulla oblongata, and from the lateral aspect of the spinal medulla as far inferiorly as the fifth cervical spinal segment. The rootlets which form the glossopharyngeal nerves can only be differentiated from the others because they pierce the dura mater separately at the jugular foramen. A part of the accessory nerve may still be attached to the brain, and, if a portion of the spinal medulla is present, its spinal and cranial roots may be seen; the spinal more widely spaced than the cranial [FIG. 192].

Cranial Nerve with Dorsal Attachment

Trochlear Nerve. This slender nerve decussates with its fellow immediately deep to the dorsal surface of the brain stem, just inferior to the midbrain, and emerges from this surface, lateral to the median plane. It will be seen later when the anterior lobe of the cerebellum is removed, but it may be found winding round the lateral aspect of the midbrain towards the interpeduncular region, immediately superior to the pons [FIGS. 41, 207].

DISSECTION. When the attachments of the nerves have been identified, remove the small blood vessels and pia mater from the medulla oblongata, leaving the nerve roots in position as far as possible.

THE HINDBRAIN

THE MEDULLA OBLONGATA

This conical part of the brain extends from the pons to the spinal medulla, which it joins at the foramen magnum. It is approximately 2·5 cm long, and was once known as the bulb, a term which is occasionally used in some clinical conditions, *e.g.*, bulbar paralysis.

The anterior surface of the medulla oblongata lies against the basilar part of the occipital bone, while the posterior surface is lodged in a groove on the anterior surface of the cerebellum, the vallecula cerebelli. The inferior half is tunnelled by the narrow central canal, which ascends through it from the spinal medulla, and opens out into the fourth ventricle on the posterior surface of the superior half.

The medulla oblongata is partially divided by an **anterior median fissure** which is continuous with that of the spinal medulla, but is interrupted in the lower medulla oblongata, where bundles of fibres from the pyramids cross the midline through it, interdigitating with each other (**decussation of the pyramids**, FIG. 193).

The **posterior median sulcus** of the spinal medulla extends upwards to the inferior angle of the fourth ventricle, separating the upper parts of the fasiculi gracilis and, at their superior extremities, the tubercles of the nuclei gracilis [FIGS. 208, 211, and p. 185].

SURFACE FEATURES

Pyramid

On the anterior surface note the two pyramids separated by the anterior median fissure, except at the decussation. The pyramids are bundles of nerve fibres which originate from cells in the cerebral cortex, principally in the precentral gyrus [FIG. 217], and descend through the hemispheres, the crura

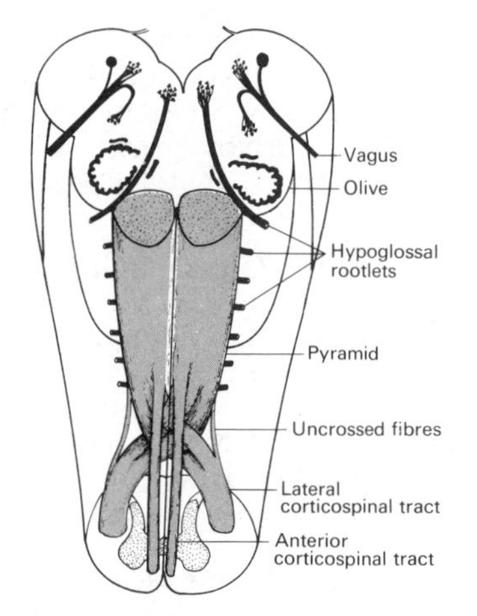

FIG. 193 Diagram of the medulla oblongata and spinal medulla to show the decussation of the pyramids (corticospinal tracts).

cerebri, and the pons to form the pyramid in the medulla oblongata. Most of the fibres cross in the decussation to form the **lateral corticospinal tract** in the deeper parts of the lateral funiculus of the spinal medulla [FIG. 174]; the remainder descend (without crossing) either in the anterior funiculus, the **anterior corticospinal tract,** or anterior to the lateral corticospinal tract. Because the fibres pass through the pyramid, the tracts they form are commonly called the **pyramidal tracts**. Each lateral corticospinal tract extends throughout the length of the spinal medulla, and contributes fibres to it at every level. Such fibres end mainly in the base of the posterior horn of the grey matter, though a number pass to the most posterior motor cells in the swellings of the spinal medulla. These are the motor cells which supply the distal parts of the limbs. The anterior corticospinal tract reaches the thoracic region of the spinal medulla, its fibres crossing at every level superior to this, to end in the grey matter of the opposite side. The size of the pyramidal decussation, and hence the size of the anterior corticospinal tracts, varies considerably, but the termination of the fibres appears to be the same irrespective of the number that cross in the decussation.

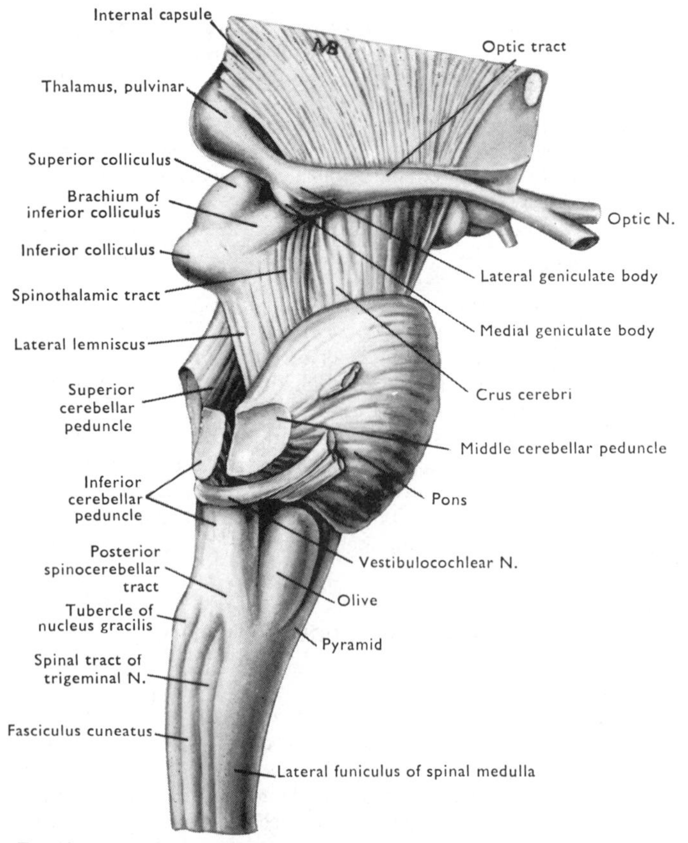

FIG. 194 Lateral view of the brain stem. See also FIGS. 192 and 211.

Olive

Posterolateral to the pyramid, and separated from it by the rootlets of the hypoglossal nerve, is the oval elevation of the olive [FIG. 194]. It is produced by the olivary nucleus, the main part of which has the shape of a crumpled bag of grey matter with its open mouth facing medially. Its efferent fibres emerge from this mouth and pass to the opposite half of the cerebellum via the inferior cerebellar peduncle [FIG. 197]. The surface of the olive is often crossed by a number of bundles of nerve fibres passing posteriorly from the region of the anteromedian fissure towards the inferior cerebellar peduncle. These are the **anterior external arcuate fibres**, and they arise in the **arcuate nucleus** which lies on the pyramid. This nucleus and the fibres represent an inferior extension of the pontine nuclei and the pontocerebellar fibres which arise from them. Similar fibres arise in the arcuate nucleus [FIG. 198], pass posteriorly through the midline of the medulla oblongata, and run across the floor of the fourth ventricle to the cerebellum forming the **striae medullares** of the fourth ventricle, which will be seen later [FIG. 211].

Inferior Cerebellar Peduncle

Posterolateral to the olive, and separated from it by a shallow groove in which the seventh to eleventh cranial nerves are attached, is a thick bundle of fibres on the posterolateral margin of the medulla oblongata. This inferior cerebellar peduncle begins near the middle of the medulla oblongata, and passes upwards and slightly laterally to the inferior border of the pons. Here it is covered by the narrow grey ridge composed of the dorsal and ventral cochlear nuclei, and then passes medial to the middle cerebellar peduncle, curving posteriorly into the cerebellum. This peduncle forms a route of communication from the medulla oblongata and the spinal medulla to the cerebellum.

Spinal Tract of the Trigeminal Nerve [Fig. 194]

Immediately anterior to the inferior cerebellar peduncle is a poorly defined longitudinal ridge, through which the rootlets of the ninth and tenth cranial nerves emerge. This descending or spinal tract of the trigeminal nerve is a bundle of the sensory fibres of that nerve, which descends through the pons and medulla oblongata into the superior part of the spinal medulla. Here it is continuous with the **dorsolateral tract** [Fig. 174]; a similar bundle formed from the dorsal rootlets of the spinal nerves, which also consists principally of fibres conveying impulses from pain and thermal sensory endings. The ridge formed by the spinal tract is often poorly defined in its intermediate part, because a bundle of fibres from the lateral funiculus of the spinal medulla (**posterior spinocerebellar tract**) crosses it obliquely to enter the inferior cerebellar peduncle. The spinal tract carries sensory fibres from the trigeminal area into the reflex zone of the neck muscles, so that the head may be moved rapidly away from a painful stimulus applied to the trigeminal area.

Gracile and Cuneate Fasciculi [Fig. 175]

These fasciculi lie on the posterior surface of the inferior half of the medulla oblongata. The fasciculus gracilis, one on each side of the posterior median sulcus, ends superiorly in the **gracile tubercle**. The fasciculus cuneatus, immediately lateral to each fasciculus gracilis, ends at a slightly higher level in a poorly defined eminence, the **cuneate tubercle** [Figs. 208, 211].

These fasciculi are composed of nerve fibres which originate in the cells of the spinal ganglia of the same side. They ascend without interruption to the nuclei gracilis and cuneatus which produce the corresponding tubercles at their superior extremities. The fibres in the fasciculus gracilis originate in the inferior half of the body, those in the fasciculus cuneatus arise in the superior half. Both groups of fibres give off collaterals to the grey matter throughout their length.

The **gracile** and **cuneate nuclei** lie at the lower end of the inferior cerebellar peduncle, and their cells, which receive impulses from the fibres of the fasciculi through synapses, give rise to: (1) the **internal arcuate fibres** which sweep anteriorly from the nuclei through the substance of the medulla oblongata, and, crossing the midline anterior to the central canal as the great sensory **decussation of the lemnisci** [Fig. 196], ascend as the **medial lemniscus** [Fig. 271] to the thalamus. In the medulla oblongata, the medial lemnisci are in the form of flat bands adjacent to the midline and dorsal to the corresponding pyramid [Fig. 197] but they diverge laterally in the pons and midbrain, reaching the lateral surface of the midbrain. This pathway from the spinal ganglia to the thalamus, conveys impulses which originate from sensory endings in tendons, and joints, from pressure and touch endings in the skin, and is concerned with the conscious appreciation of these senses, for the thalamus relays the impulses to the cerebral cortex. (2) **Posterior external arcuate fibres** arise from the accessory cuneate nucleus [Fig. 196] on the lateral aspect of the main cuneate nucleus, and pass superiorly into the inferior cerebellar peduncle. The accessory cuneate nucleus receives collaterals (branches) from the fibres in the fasciculus cuneatus, and transmits ascending impulses to the cerebellum from the upper half of the body on the same side [Fig. 196].

Spinocerebellar Tracts [Figs. 194, 209]

There are two other pathways which connect the spinal medulla with the cerebellum. The **posterior spinocerebellar tract** arises from the cells of the thoracic nucleus [Fig. 174]; present from the first thoracic to the second lumbar segments of the spinal medulla. It receives impulses through branches of the fibres in the fasciculus gracilis. The axons of the tract ascend in the most posterolateral part of the lateral funiculus of the same side, and entering the medulla oblongata anterior to the spinal tract of the trigeminal nerve, pass posteriorly over its lateral surface to enter the inferior cerebellar peduncle [Figs. 194, 196]. This tract transmits impulses from the inferior half of the body to the cerebellum.

The **anterior spinocerebellar tract** ascends in the lateral funiculus immediately anterior to the posterior spinocerebellar tract. Its exact origin in the spinal medulla is unknown, but it passes through the medulla oblongata and pons, anterior to the spinal tract of the trigeminal nerve, and turns posteriorly into the cerebellum over the superior cerebellar peduncle [Figs. 194, 209] above the level of entry of the trigeminal nerve.

Spinothalamic Tract [Figs. 175, 212, 271]

This tract arises throughout the length of the spinal medulla from cells in the posterior horn which receive impulses indirectly from the fibres of the dorsolateral fasciculus. Most of the fibres run obliquely upwards, crossing in the white commissure [Fig. 174] to ascend in a position anterior to the anterior spinocerebellar tract. They pass through the medulla oblongata posterior to the olivary nucleus, giving many fibres to the reticular formation there (**spinoreticular**

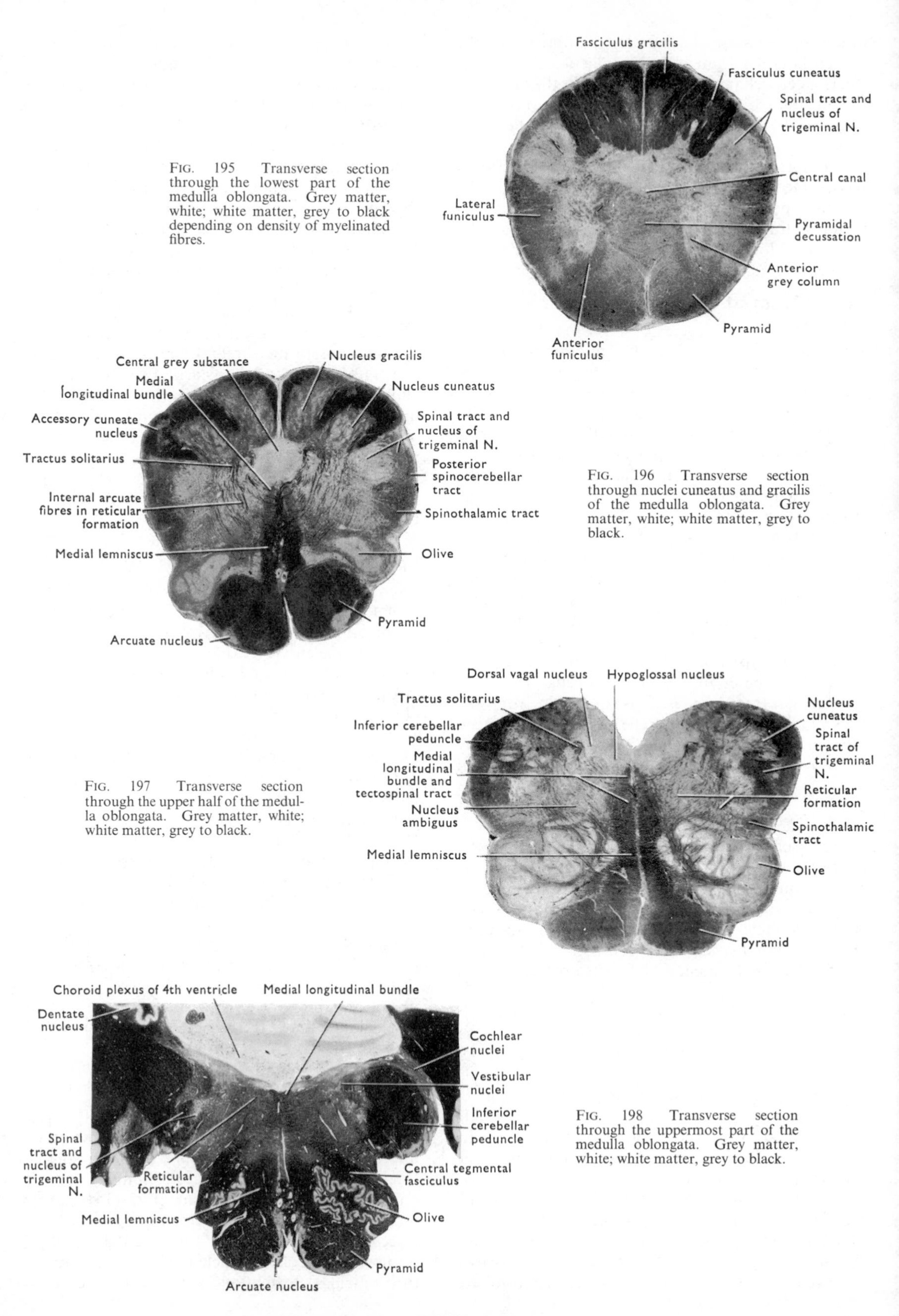

FIG. 195 Transverse section through the lowest part of the medulla oblongata. Grey matter, white; white matter, grey to black depending on density of myelinated fibres.

FIG. 196 Transverse section through nuclei cuneatus and gracilis of the medulla oblongata. Grey matter, white; white matter, grey to black.

FIG. 197 Transverse section through the upper half of the medulla oblongata. Grey matter, white; white matter, grey to black.

FIG. 198 Transverse section through the uppermost part of the medulla oblongata. Grey matter, white; white matter, grey to black.

tract). The remainder join the medial lemniscus in the pons, and run with it to the thalamus. The tract conveys impulses which mainly evoke the conscious sensations of pain, temperature, and touch. The fibres in the spinothalamic tract and medial lemniscus are topographically arranged [FIG. 199] so that impulses from different parts of the body are carried in specific parts of the tracts. They are also the main tracts concerned with the conscious appreciation of sensations other than those concerned with taste, smell, sight, hearing, and balance, and they carry impulses predominantly from the opposite side of the body. Fibres which carry similar information to the cerebellum do so predominantly from the same side of the body but *do not evoke conscious sensations*.

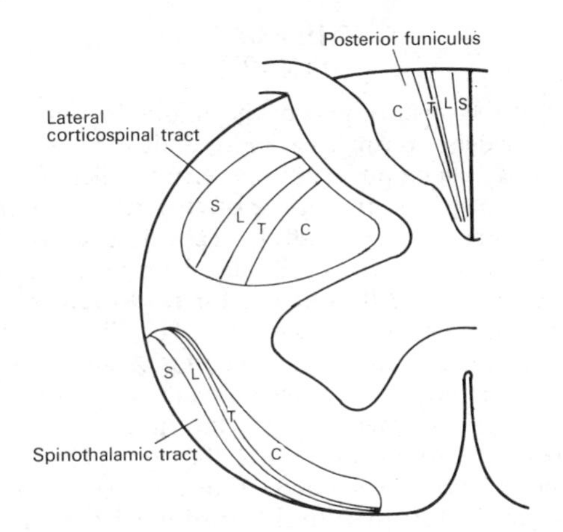

FIG. 199 Diagrammatic transverse section through half of the cervical spinal medulla to show the relative positions in the posterior funiculi and spinothalamic tracts of fibres with originate at cervical (C), thoracic (T), lumbar (L), and sacral (S) levels, and of fibres in the lateral corticospinal tract which end at the corresponding levels.

Grey and White Matter of Medulla Oblongata

The above tracts form a large part of the white matter of the surface of the medulla oblongata. In the interior there is a scattered mixture of grey and white matter (the **reticular formation**), certain deeply placed tracts, and a number of well-defined nuclei. In the inferior part, a compact sleeve of grey matter surrounds the narrow central canal. Half way along the medulla oblongata, the central canal opens into the inferior angle of the fourth ventricle (a diamond-shaped cavity which lies on the posterior surfaces of the medulla oblongata and pons) and its surrounding grey matter becomes continuous with the grey matter in the floor of the fourth ventricle. Most of the nuclei of the cranial nerves lie in this grey matter.

For the following sections it is advisable to have available microscopic sections of representative regions of the brain stem which have been stained to show the myelin of the nerve fibres. If this is not possible, FIGURES 195–198, 200–202, 212–215, are photographs of such sections designed to show the main features. They should be used in conjunction with FIGURES 209, 261, 271–272, which are diagrams based on these sections.

CRANIAL NERVES WITH NUCLEI IN THE MEDULLA OBLONGATA [p. 250]

Trigeminal Nerve. The fibres of the **spinal tract** of the trigeminal nerve extend throughout the length of the medulla oblongata, and terminate at all levels in the **spinal nucleus** which lies on its medial side [FIGS. 195–198]. The fibres which end in the nucleus caudal to the fourth ventricle are principally concerned with the sensations of pain and temperature from the trigeminal area, but the spinal tract and nucleus also also receive fibres of general sensation from the other cranial nerves entering the medulla oblongata through it; notably the ninth and tenth. The spinal nucleus gives rise to nerve fibres which cross the midline obliquely to join the spinothalamic tract at a higher level [FIG. 271].

Facial Nerve. The motor nuclei of the facial nerve lie in the inferior part of the pons [FIG. 272]. Its visceral sensory fibres (taste from the anterior two-thirds of the tongue) enter the **tractus solitarius** [FIGS. 197, 271] and terminate on the nucleus of that tract in the medulla oblongata.

Vestibulocochlear Nerve. Sensory fibres from the cochlea have already been seen passing into the **cochlear nuclei** which overlie the superior part of the inferior cerebellar peduncle in the medulla oblongata. The vestibular fibres pass anterior to the same peduncle and fan out into the **vestibular nuclei** in the floor of the fourth ventricle. These nuclei underlie the **vestibular area** [FIG. 211] which is partly in the pons also [FIGS. 198, 272].

Glossopharyngeal, Vagus, and Cranial Part of Accessory. The preganglionic parasympathetic fibres in these nerves arise in the cells of the **dorsal nucleus of the vagus** [FIG. 197] the superior extremity of which is known as the **inferior salivatory nucleus** and sends fibres into the glossopharyngeal nerve for distribution to the otic ganglion. Efferent nerve fibres to pharyngeal and laryngeal muscles arise in the **nucleus ambiguus** [FIGS. 197, 272] and pass into all three nerves. Visceral sensory nerve fibres, including taste fibres in the ninth and tenth nerves, enter the **tractus solitarius** and end in its nucleus, which surrounds the tractus solitarius. General sensory fibres enter the spinal tract of the trigeminal (see above) and in this respect correspond with the arrangement of dorsal spinal roots, all of which send fibres into the dorsolateral tract for termination in the posterior horn of the spinal medulla.

Hypoglossal Nerve. This pure motor nerve arises from a single, paramedian nucleus [FIG. 197] which extends throughout the greater part of the medulla oblongata, and gives rise to the long row of hypoglossal rootlets [FIGS. 192, 272].

THE PONS

[FIG. 192]

This is the superior part of the hindbrain, and has a bulky ridge covering its anterior and lateral aspects. This ridge is composed of transverse bundles of nerve fibres which arise from the **pontine nuclei** buried in its substance [FIG. 201]. These fibres cross the median plane, and on each side converge posteriorly to form the **middle cerebellar peduncle** where the trigeminal nerve pierces it. The fibres in the peduncle pass to the cerebellar cortex. Anterior to the trigeminal nerves, the ridge forms the **ventral part of the pons**, which receives the crura cerebri at its superior margin, and emits the pyramids at its inferior margin. The fibres of the crura cerebri split up into the **longitudinal bundles of the pons**, and most of them end in the pontine nuclei (**corticopontine fibres**); the greater part of the remainder emerges as the pyramids (**corticospinal fibres**) and a few cross to the region of the motor cranial nerve nuclei of the opposite half of the pons (**corticonuclear fibres**). Anteriorly, the ventral part of the pons is grooved by the **basilar sulcus** (which lodges the basilar artery) and it lies adjacent to the basi-occiput and dorsum sellae of the skull.

The **dorsal part of the pons** separates the ventral part from the floor of the superior half of the fourth ventricle [FIGS. 201, 205], and cannot be seen at present. It is continuous superiorly with the tegmentum of the midbrain and, inferiorly, with the medulla oblongata. Structures such as the medial lemniscus and spinothalamic tract, which are continued into the pons from the medulla oblongata, pass into the dorsal part, with the exception of (1) the pyramid (which extends inferiorly from the ventral part) and (2) the inferior cerebellar peduncle (which passes dorsally into the cerebellum). In addition the dorsal part contains the superior extension of the **reticular formation** of the medulla oblongata and several discrete nuclei, including the nuclei of certain cranial nerves.

The Reticular Formation

This consists of numerous interneurones arranged in ill-defined groups, intermingled with fine bundles of nerve fibres which run mainly in a longitudinal direction. Many of these fibres arise in the reticular formation and unite its parts in the medulla, pons, and midbrain [p. 201]. They also descend in the ventral part of the white matter of the spinal medulla (**reticulospinal tracts**) from the pons and medulla oblongata, and ascend into the diencephalon—thalamus and hypothalamus. The reticular formation receives collaterals of ascending pathways from the spinal medulla (*e.g.*, spinothalamic tracts) and afferent information from the cranial nerves (*e.g.*, nucleus of tractus solitarius and the vestibular nuclei). It has extensive reciprocal connexions with the cerebellum, which thereby receives spinothalamic impulses, and with many other parts of the central nervous system (see below).

In many vertebrates with less well developed nervous systems than Man, the reticular formation produces the main descending pathways from the brain to the spinal medulla, and it is certainly the main route through which the cerebellum acts on the spinal medulla.

In the fishes, the hindbrain lies at the same level as the branchial (gill) arches and the heart, and its nerves innervate them. Because of this, the interneurones in this part of the brain (mainly the reticular formation) are responsible for integrating sensory information from these structures, and for controlling their activity. Thus one of the functions of the reticular formation is to act as a control mechanism for respiratory and cardiovascular activity—a function which it continues to exercise after the heart descends into the thorax and the gills are replaced by lungs. This is achieved through the reticulospinal fibres which pass to nerve cells innervating respiratory muscles, and to those giving rise to preganglionic autonomic nerve fibres in the spinal and cranial nerves. In addition, the reticular formation is situated where ascending information from the spinal medulla comes into association with that entering the central nervous system in the cranial nerves. Thus it can form an integrated picture of the changing sensory experience of the entire animal—a feature which seems to have given it control over adjustments in the muscles of the body and in the viscera (autonomic system) in the light of such experience. Such a complex, already having a mechanism for modifying activity in the spinal medulla (reticulospinal tracts), becomes the apparatus through which most of the more recently developed parts of the central nervous system exercise their effects by modifying the activity of the reticular formation—a feature which requires the development of reciprocal connexions between that formation and these parts. Such connexions continue to form an important element in the functioning nervous system despite the formation of long ascending and descending pathways which bypass the reticular formation and link the spinal medulla directly to higher levels of the brain—*e.g.*, the spinothalamic and corticospinal pathways. Because of the sensory functions of the reticular formation (see above), its ascending fibres not only transmit information about activity in the formation, but also an integrated sensory picture less specific than that conveyed by the long ascending pathways. No doubt this is the basis of the ascending activating impulses through which the reticular formation produces arousal of the cerebral cortex through the thalamus.

The reticular formation is also responsible for complex body movements such as swimming (fishes) and walking (amphibia). In higher forms it is concerned with complex adjustments of muscle tone in the trunk and proximal parts of the limbs to achieve such actions as the righting reflexes, at least

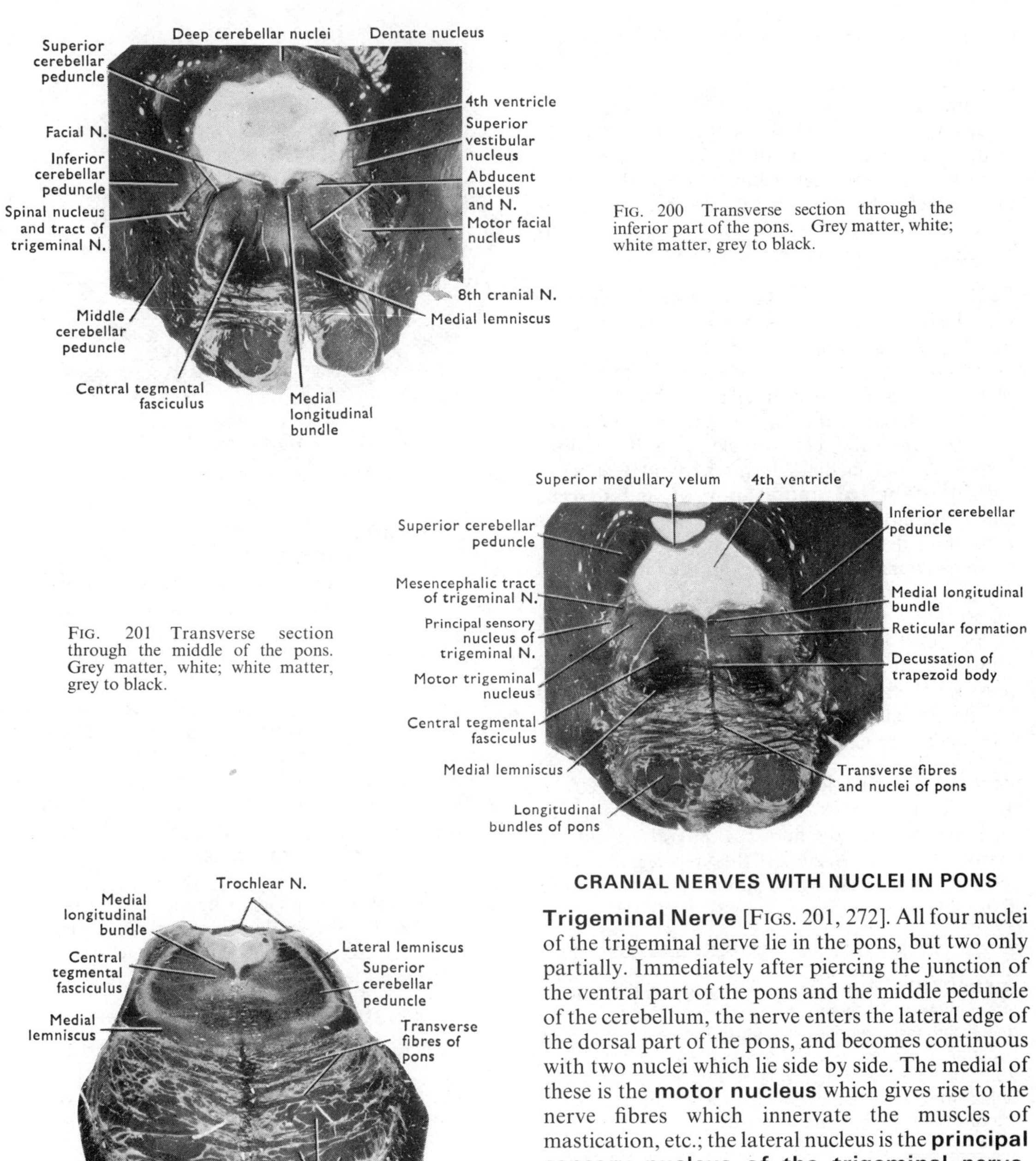

FIG. 200 Transverse section through the inferior part of the pons. Grey matter, white; white matter, grey to black.

FIG. 201 Transverse section through the middle of the pons. Grey matter, white; white matter, grey to black.

FIG. 202 Transverse section through the upper part of the pons. Grey matter, white; white matter, grey to black.

partly through changes in the tension of the **muscle spindles** reflexly producing muscle contractions. It may well be concerned with more complex phasic changes in muscle tone induced by the activity of the cerebellum, corpus striatum, and cerebral cortex, all of which send pathways to it.

CRANIAL NERVES WITH NUCLEI IN PONS

Trigeminal Nerve [FIGS. 201, 272]. All four nuclei of the trigeminal nerve lie in the pons, but two only partially. Immediately after piercing the junction of the ventral part of the pons and the middle peduncle of the cerebellum, the nerve enters the lateral edge of the dorsal part of the pons, and becomes continuous with two nuclei which lie side by side. The medial of these is the **motor nucleus** which gives rise to the nerve fibres which innervate the muscles of mastication, etc.; the lateral nucleus is the **principal sensory nucleus of the trigeminal nerve**, which probably receives fibres concerned with tactile information from the trigeminal area. The **spinal tract and nucleus** extend inferiorly from the point of entry of the trigeminal nerve, and are found at the lateral margin of the dorsal part of the pons in its inferior half. A small bundle of fibres passes dorsally from the trigeminal nerve, between the motor and principal sensory nuclei, to the lateral margin of the floor of the fourth ventricle, where it turns superiorly to traverse the midbrain. This **mesencephalic tract of the trigeminal nerve** [FIG. 271] contains the proprioceptive sensory fibres from that nerve. It differs from all other peripheral sensory neurons in that its cells of origin lie scattered along the tract and are not situated in the trigeminal ganglion.

Abducent Nerve. This small nucleus lies immediately anterior to the floor of the fourth ventricle, in the most inferior part of the pons, close to the midline. It raises a small protuberance on the floor of the ventricle which is called the **facial colliculus** [FIG. 211] because the facial nerve is closely associated with this nucleus [FIGS. 200, 272].

Facial Nerve. The **motor nuclei** of the facial nerve lie in the most inferior part of the pons. (1) The nucleus which supplies the muscles of facial expression, etc., lies immediately anteromedial to the spinal nucleus of the trigeminal nerve [FIG. 200]. The axons which arise in the nucleus pass posteromedially towards the abducent nucleus, and running superiorly on its medial side, hook round its superior margin (**genu of facial nerve**) and descend anterolaterally to emerge at the inferior border of the pons, immediately lateral to their nucleus of origin. (2) In its course through the pons the facial nerve is joined by preganglionic parasympathetic fibres from the **superior salivatory nucleus**, which lies in the inferior part of the pons, superior to the inferior salivatory nucleus. These parasympathetic fibres pass peripherally through the greater petrosal and chorda tympani nerves to the pterygopalatine and submandibular ganglia respectively.

Vestibulocochlear Nerve. The superior part of the vestibular area extends into the inferior part of the pons, between the floor and lateral wall of the fourth ventricle [FIG. 200]. In this area lie parts of the vestibular nuclei on which fibres of the vestibular division terminate.

DISSECTION. Remove the meninges from the superior and posterior surfaces of the cerebellum, but avoid depressing the cerebellum too strongly to expose all of its superior surface. With the assistance of FIGURES 203–208 identify its main parts, including the fissura prima, horizontal fissure, the superior and inferior vermis, and the hemispheres and flocculus.

THE CEREBELLUM

The cerebellum consists of a median part, the **vermis**, which is only clearly separated from the lateral parts, **hemispheres**, on the antero-inferior surface. The cerebellum is wrapped around the posterior surface of the brain stem, which is lodged in a wide groove on its anterior surface. Postero-inferiorly it is notched by the falx cerebelli.

The **superior vermis** is the median ridge on the superior surface. It lies inferior to the line of junction of the tentorium cerebelli with the falx cerebri, containing the straight sinus.

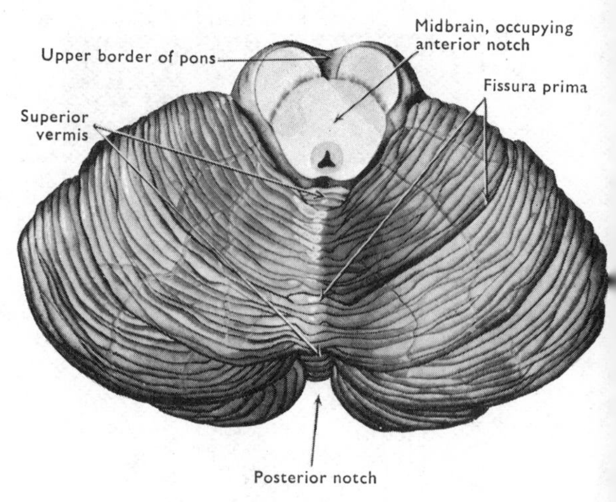

FIG. 203 Superior surface of the cerebellum.

The **inferior vermis** bulges into a deep median hollow on the antero-inferior surface of the cerebellum (the **vallecula of the cerebellum**). Here the vermis is separated from the hemispheres by deep furrows, but each part is continuous with a part of the hemispheres—a feature which is more obvious in the superior vermis. Thus the nodule is continuous with the flocculus, forming the **flocculonodular lobe**; the uvula is continuous with the tonsil, and so on. The cerebellum is divided by a number of transversely placed fissures of considerable depth. Between these the surface has many small subsidiary folds which greatly increase its surface area.

The complexity of the cerebellum was no doubt responsible for the number of highly fanciful names which have been applied to its various parts. It is inadvisable for the student to burden his memory with all of these terms, and only those which have a functional or descriptive value are used here.

FISSURES OF THE CEREBELLUM

The general arrangement of the cerebellum can most easily be understood if it is appreciated that it develops from a ridge of tissue which lies transversely in the thin roof of the fourth ventricle opposite the pontine part of the brain stem. This ridge expands posteriorly to an enormous degree but does not increase its area of attachment to the brain stem to a corresponding extent. Thus the superior and inferior margins of the original ridge remain close together in the roof of the fourth ventricle, while the intermediate part expands superiorly over the posterior surface of the midbrain, and inferiorly over the medulla oblongata, hiding both from view posteriorly. Because of this method of growth, the cerebellum becomes greatly folded, and the roof of the fourth ventricle is drawn posteriorly into the centre of the cerebellum in a tent-like recess [FIG. 205]. As a result the parts of the thin roof of the fourth ventricle superior and inferior to the attachment of the cerebellum sweep anterosuperiorly

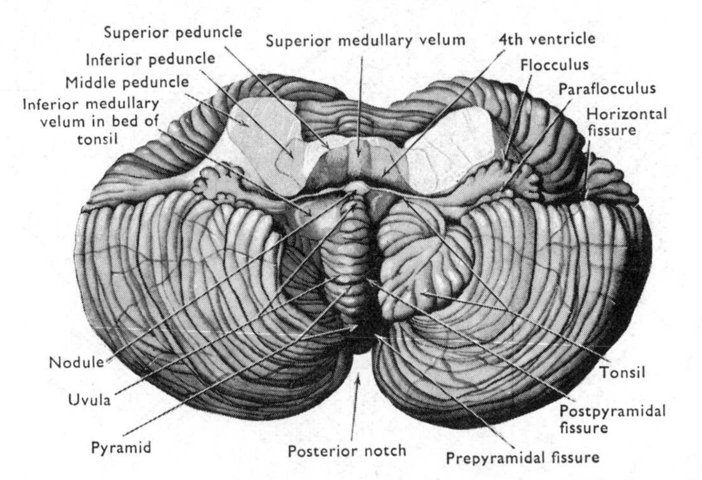

FIG. 204 Antero-inferior surface of the cerebellum. The right tonsil has been removed so as to display more fully the inferior medullary velum.

and antero-inferiorly from the recess to form the **superior and inferior medullary vela** [FIGS. 205, 208; p. 251]. Of the great number of fissures formed in the cerebellum, three are of particular importance:

1. **Fissura Prima** [FIGS. 203, 205]. This deep fissure cuts across the superior surface of the cerebellum and separates the anterior and posterior lobes.

2. **Horizontal Fissure** [FIG. 204]. This fissure extends from one middle cerebellar peduncle to the other, and lies close to the junction of the superior and inferior surfaces of the cerebellum.

3. **Posterolateral Fissure.** This fissure lies on the antero-inferior surface of the cerebellum, and separates the nodule and flocculus from the remainder of the cerebellum. This **flocculonodular lobe** has connexions with the vestibular apparatus and is therefore primarily concerned with equilibration. Hence disease of this part of the cerebellum manifests itself in disturbances of balance, while disease of the remainder is primarily associated with disturbances of muscle tone and coordination.

CEREBELLAR CORTEX

The surface of the cerebellum is a thin layer of grey matter which overlies the deeper white matter. This cerebellar cortex is uniform in structure throughout, and consists of three layers: (1) The deepest layer of small granule cells on which most of the incoming fibres end. (2) The middle layer, a single row of large cell bodies of the **Purkinje cells** which send their axons deeply into the white matter, and their branching dendrites superficially into the third layer

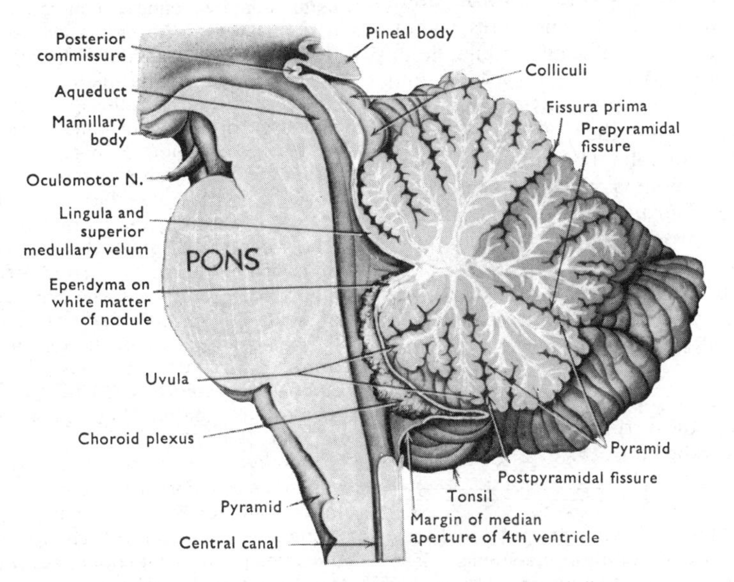

FIG. 205 Median section through the brain stem and cerebellum.

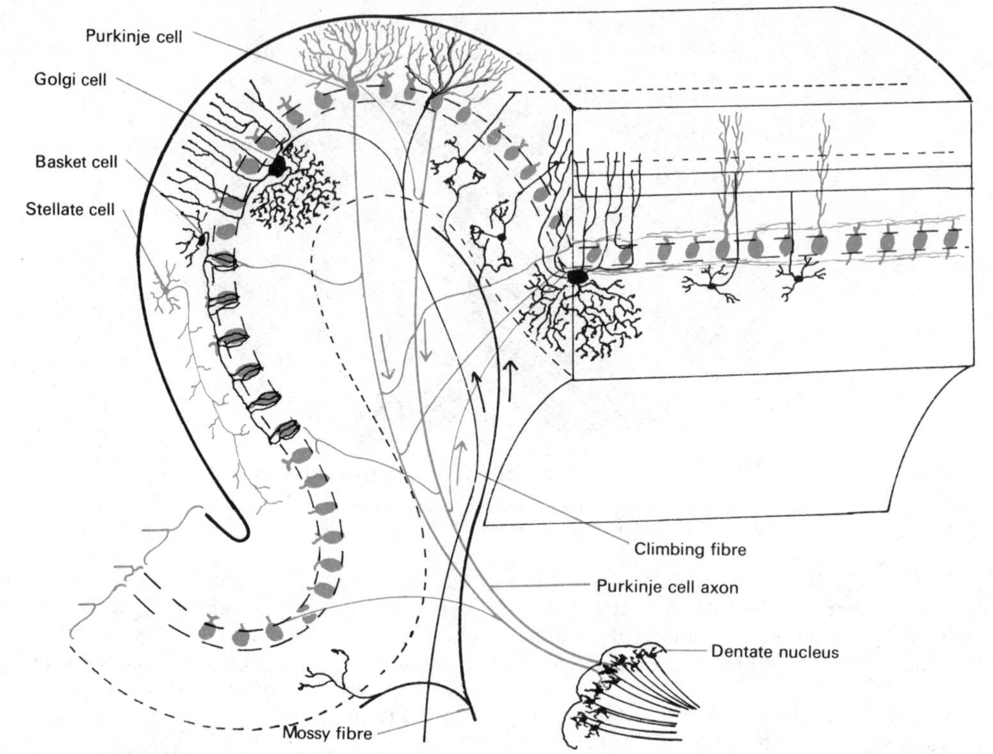

FIG. 206 A diagram of the main types of nerve cells and their processes in the cerebellar cortex. A single folium is shown cut longitudinally (right) and transversely (left). The Golgi and stellate (basket) cells have inhibitory functions, while the mossy and climbing fibres are excitatory. Both mossy and climbing fibres send collaterals to the cerebellar nuclei. Purkinje and stellate cells are shown red.

where they spread out in a plane at right angles to the fissures and to the ridges between them (the **folia**). (3) A surface layer into which the axons of the granule cells pass, and running parallel to the folia, synapse with the dendrites of the Purkinje cells, along which the olivocerebellar **climbing fibres** pass. The structure of the cortex is uniform throughout, but the distribution of the afferent fibres is not. Vestibular fibres pass to the flocculonodular lobe; spinocerebellar fibres pass to the vermis and paravermal regions of the anterior and adjacent parts of the posterior lobes, and to the pyramid and uvula [FIG. 205] of the posterior lobe. The entire cerebellum receives fibres from the pons via the middle cerebellar peduncle, and also from the olive via the inferior peduncle. The fibres from the pons relay impulses which arise in the cerebral cortex and reach the pons through the crura cerebri. The very large size of the hemispheres of the cerebellum in Man is related to the increased size of the cerebral cortex. In other mammals where the cerebral cortex is less well developed, there is a corresponding reduction in size of the hemispheres, the vermis forming a greater proportion of the cerebellum.

DISSECTION. Split the cerebellum into two equal parts by a median sagittal section through the vermis. Make this incision carefully, separating the two parts as the incision is deepened. Approximately half way between the superior and inferior margins of the cerebellum, a narrow slit will be opened into; this is the **cerebellar recess of the fourth ventricle**. Superior and inferior to this recess, progress with care to avoid cutting through the superior and inferior medullary vela [FIG. 204].

Remove the posterior part of the right half of the cerebellum by means of a curved, coronal cut through it at right angles to the middle cerebellar peduncle, more or less parallel to the medulla oblongata. Place the edge of the knife against the middle cerebellar peduncle just posterior to the flocculus [FIG. 204] and its tip in the lateral end of the fissura prima. Cut medially along the line of the fissura prima to meet the median section, the inferior part of the edge of the knife passing immediately posterior to the tonsil [FIG. 204]. This section exposes: (1) the folded grey matter of the cortex and the white matter deep to it, which is branched to fit the complex folds of the cortex, the **arbor vitae cerebelli**; (2) a crinkled bag of grey matter buried in the central mass of white matter relatively close to the midline and to the cerebellar recess of the fourth ventricle. this is the **dentate nucleus**, the largest of the deep nuclei of the cerebellum, which extend laterally from the midline along the posterior margin of the cerebellar recess of the fourth ventricle [FIG. 208].

Identify the superior cerebellar artery running along the superior margin of the cerebellum and sending branches into the inferior part of the midbrain and the superior part of the cerebellum. Cut its branches to the cerebellum, and the veins passing superiorly from the cerebellum to the great cerebral vein, the cut end of which may be seen posterior to the superior part of the midbrain. Carefully pull the anterior lobe of the cerebellum postero-inferiorly, and identify the **trochlear nerve** under cover of its margin. Strip the anterior lobe away from the remainder of the cerebellum, tearing its connexions with the white matter, but avoiding damage to the trochlear nerve. This exposes the superior surface of the **middle cerebellar peduncle**, the **superior cerebellar peduncle** passing anterosuperiorly into the midbrain [FIGS. 207, 211] and medial to this peduncle, a thin tongue-like strip of cerebellar cortex (the **lingula**) is fused to the **superior medullary velum**, a delicate white sheet stretched between the superior cerebellar peduncles.

Superior Medullary Velum

This velum extends transversely between the two superior cerebellar peduncles, roofs over the superior part of the fourth ventricle (into which it may be depressed by light pressure from a blunt seeker) and becomes continuous above with the roof of the midbrain. Here a median ridge of white matter (**frenulum veli**) descends on to the velum from the groove between the inferior colliculi [FIG. 211]. It has the fourth cranial nerves emerging on each side of it, these nerves having crossed in the superior part of the velum [FIG. 202].

DISSECTION. With a blunt seeker, lift the surface layer of neuroglia from the superior cerebellar peduncle and expose its fibres running superiorly. Note that as it plunges into the midbrain, its lateral aspect is covered by a low ridge of white matter (the **lateral lemniscus**) passing posteriorly over it into the inferior colliculus. Split the superior medullary velum in the midline and confirm the presence of the fourth ventricle deep to it.

Turn to the inferior surface of the cerebellum and gently lift the right tonsil out of its bed [FIGS. 204, 208]. Avoid damage to the thin roof of the inferior part of the fourth ventricle (**inferior medullary velum**) by dividing the branches of the posterior inferior cerebellar artery which pass into the tonsil from the main stem of the artery on the inferior medullary velum. As the tonsil lifts away from the inferior medullary velum, pull the remainder of the posterior lobe of the cerebellum away with it, but leave the flocculus and nodule intact, and the posterior inferior cerebellar artery on the velum.

If the inferior medullary velum is intact, find the **median aperture of the fourth ventricle** [FIG. 208] and pass a blunt seeker through it to define the extent of the velum and the cavity of the fourth ventricle. Pass the tip of the seeker along the lateral recess of the fourth ventricle to the lateral aperture, and then posteriorly, superior to the nodule, into the cerebellar recess. Follow this recess to its lateral extremity, noting the extent of the inferior medullary velum, and push the tip of the seeker into the cerebellar substance at the posterior apex of the recess; the tip will appear through the substance of the **dentate nucleus** which lies immediately posterior to the lateral part of the recess [FIG. 208].

Note the presence of choroid plexus on the ventricular surface of the inferior medullary velum, and follow it from the median aperture of the fourth ventricle to the lateral recess. The part of the velum to which it is attached is little more than a layer of ependyma and does not appear white like the superior part.

Pull the flocculus away from the middle cerebellar peduncle in a medial direction. A thin strip of white matter will tear away with it (**peduncle of the flocculus**) and this has the inferior medullary velum attached to it [FIG. 204]. Cut through the velum in the midline and turn it downwards to expose the choroid plexus on its anterior aspect, and the lateral recess of the fourth ventricle. Lift

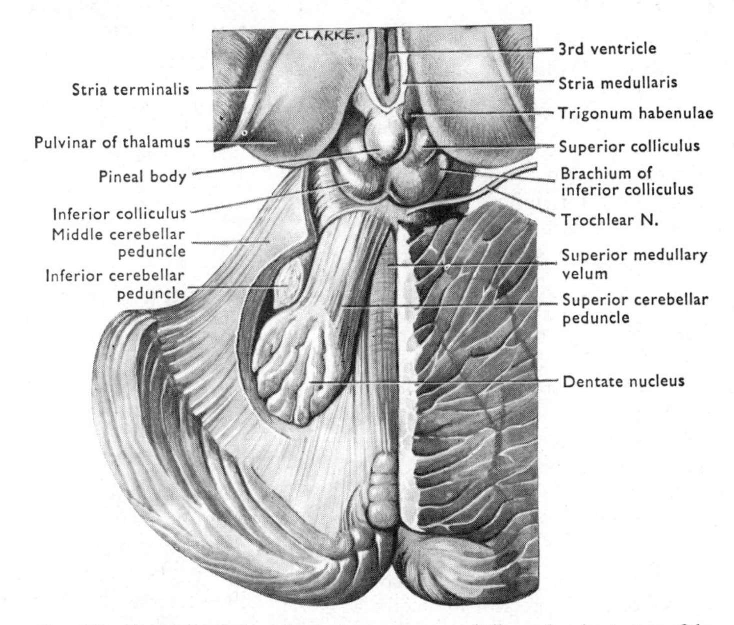

FIG. 207 Dissection to show the dentate nucleus, cerebellar peduncles, tectum of the midbrain, and superior surface of the diencephalon.

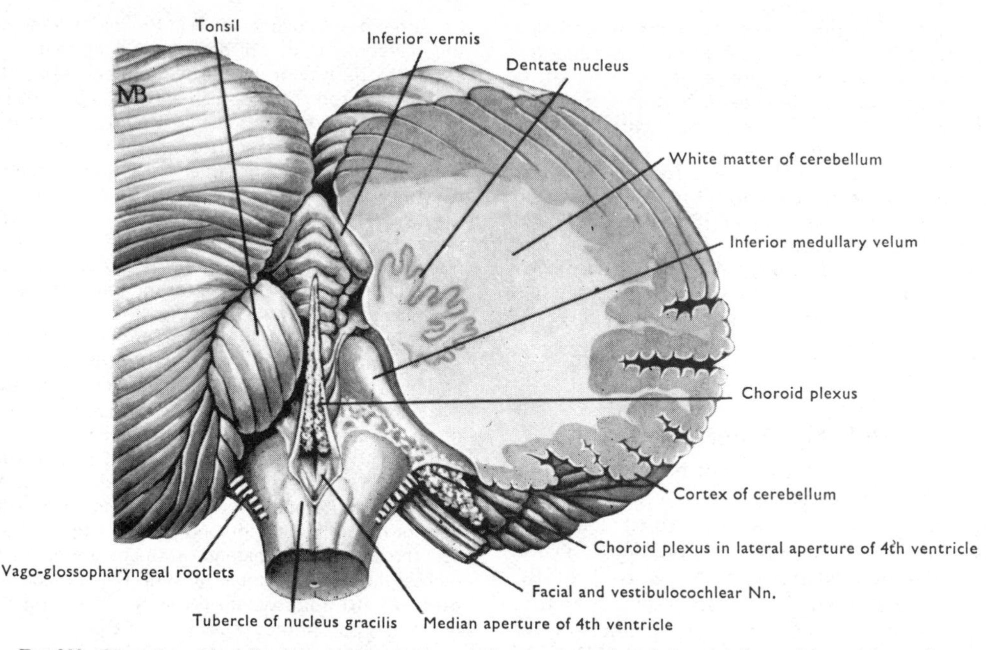

FIG. 208 Dissection of the left inferior surface of the cerebellum to expose the inferior medullary velum and the median and lateral apertures of the fourth ventricle. The medulla oblongata has been displaced anteriorly so as to open up the median aperture. Note the choroid plexus which is visible at both apertures and can also be seen through the thin roof of the fourth ventricle.

the left half of the cerebellum gently backwards and look into the opposite lateral recess from the ventricular aspect.

Inferior Medullary Velum

This velum stretches inferiorly from the nodule and peduncle of the flocculus to (1) the inferior cerebellar peduncles, inferolaterally; (2) the lateral aperture of the fourth ventricle, laterally; (3) the median aperture, inferiorly. The greater part of it forms the bed of the tonsil of the cerebellum (a rounded hollow which contains the posterior inferior cerebellar artery). Its median part lies in contact with the nodule and uvula of the inferior vermis, and curves posteriorly on to the inferior surface of the uvula [FIG. 205] in such a way that the **median aperture of the fourth ventricle** is a V-shaped orifice, which faces inferiorly [FIG. 208]. Thus air introduced into the lumbar subarachnoid space ascends to the cerebellomedullary cistern, and some of it finds its way directly into the fourth ventricle, whence it rises through the cerebral aqueduct to the other ventricles of the brain [FIG. 205].

DISSECTION. Identify the inferior cerebellar peduncle [FIG. 194] on the posterolateral aspect of the medulla oblongata, and note a grey ridge which extends across its posterior aspect from the eighth cranial nerve. This is formed by the **cochlear nuclei**. Strip away these nuclei and expose the upper part of the peduncle. Push a blunt seeker superiorly along the lateral surface of the inferior peduncle between that peduncle and the middle peduncle, and pull it posteriorly through the white matter, thus separating these **peduncles**. Repeat the same process on the medial side of the inferior peduncle, and separate the inferior and superior peduncles. This will show the fibres of the inferior peduncle curving medially over the posterior aspect of the superior peduncle and dentate nucleus. When the inferior peduncle has been completely freed, lift the dentate nucleus posteriorly; the superior peduncle will be raised from its bed since the majority of its fibres arise in the dentate nucleus.

CEREBELLAR PEDUNCLES

These are three compact bundles of nerve fibres on each side, which connect the cerebellum with the brain stem [FIG. 209].

Middle Peduncle

This is the largest and most lateral of the three. It is formed of fibres which arise in the opposite half of the ventral part of the pons, and pass to the cerebellar cortex. They transmit impulses which reach the pons from the cerebral cortex via the crura cerebri. The peduncle enters the cerebellum through the anterior end of the horizontal fissure.

Inferior Peduncle [FIGS. 207, 211]

This peduncle forms on the posterolateral surface of the medulla oblongata, and ascending between the other two peduncles, sweeps posteromedially towards the vermis of the cerebellum, though some of its fibres enter the hemispheres. It consists mainly of afferent fibres to the cerebellum from the spinal medulla, the olive, the reticular formation of the medulla oblongata, and the vestibular nuclei and nerve. It also transmits efferent cerebellar fibres to the medulla oblongata, principally to the vestibular nuclei and the reticular formation.

Superior Peduncle

This is the principal efferent pathway from the cerebellum, and its fibres arise mainly in the dentate nucleus. Each of these peduncles begins in the roof of the fourth ventricle, but passes anteriorly as it ascends, forming the lateral wall of the superior part of the fourth ventricle, and passing towards the opposite peduncle as it enters the midbrain. The two peduncles meet and intermingle in the tegmentum [p. 199] of the inferior part of the midbrain. Here the fibres of each superior peduncle pass to the opposite side (decussate) the majority running superiorly to the thalamus and to the red nucleus in the upper midbrain. Some fibres descend into the dorsal part of the pons as the **descending limb** of the superior cerebellar peduncle mainly in the central tegmental fasciculus [p. 201, FIGS. 200–202, 212–213].

Dentate Nucleus

This is the largest and the most lateral of a group of four nuclei which lie deep in the white matter of each half of the cerebellum, close to the cerebellar recess of the fourth ventricle. It has the shape of a thin, crinkled lamina of grey matter, which is very similar in appearance to the olivary nucleus in section, and, like it, has a wide mouth or hilus which faces anteromedially, and through which the axons of its cells emerge to form the superior cerebellar peduncle. It receives nerve fibres from the Purkinje cells of the greater part of the cerebellar cortex, except the vermis, and collaterals from the afferents to that cortex.

The other three deep nuclei (emboliform, globose, and fastigial) of each half of the cerebellum lie between the dentate nucleus and the midline. They receive nerve fibres from the Purkinje cells of the vermal and paravermal regions of the cerebellar cortex. Most of their efferent fibres leave the cerebellum through the superior peduncle, except for

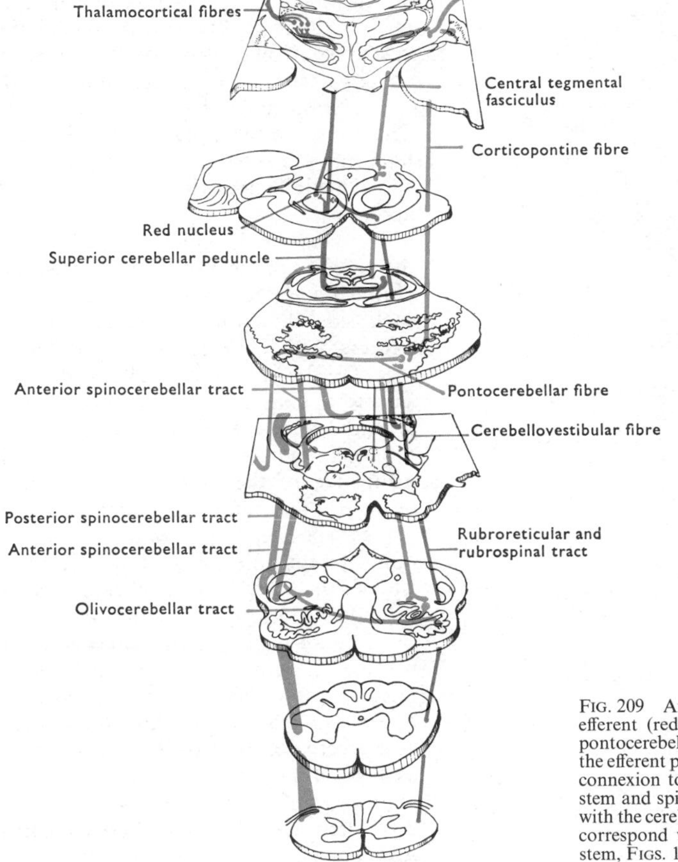

FIG. 209 An outline diagram to show the main afferent (blue) and efferent (red) connexions of the cerebellum. Note the afferent pontocerebellar, olivocerebellar, and spinocerebellar pathways, and the efferent pathway to the red nucleus and thalamus. The efferent connexion to the red nucleus links the cerebellum with the brain stem and spinal medulla; the connexion with the thalamus links it with the cerebral cortex. The outline drawings are not to scale, but correspond with illustrations of transverse sections of the brain stem, FIGS. 195–198, 200–202, and 212–215.

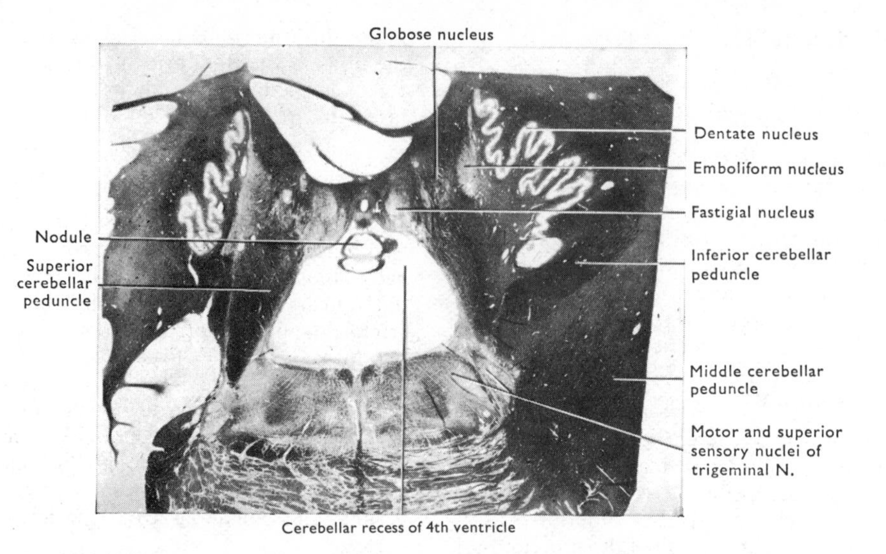

FIG. 210 Oblique transverse section, parallel to the cerebellar recess of the fourth ventricle, to show the cerebellar nuclei.

those of the fastigial nucleus which hook over the superior peduncle and descend in the inferior cerebellar peduncle to the vestibular nuclei and the reticular formation of the medulla oblongata.

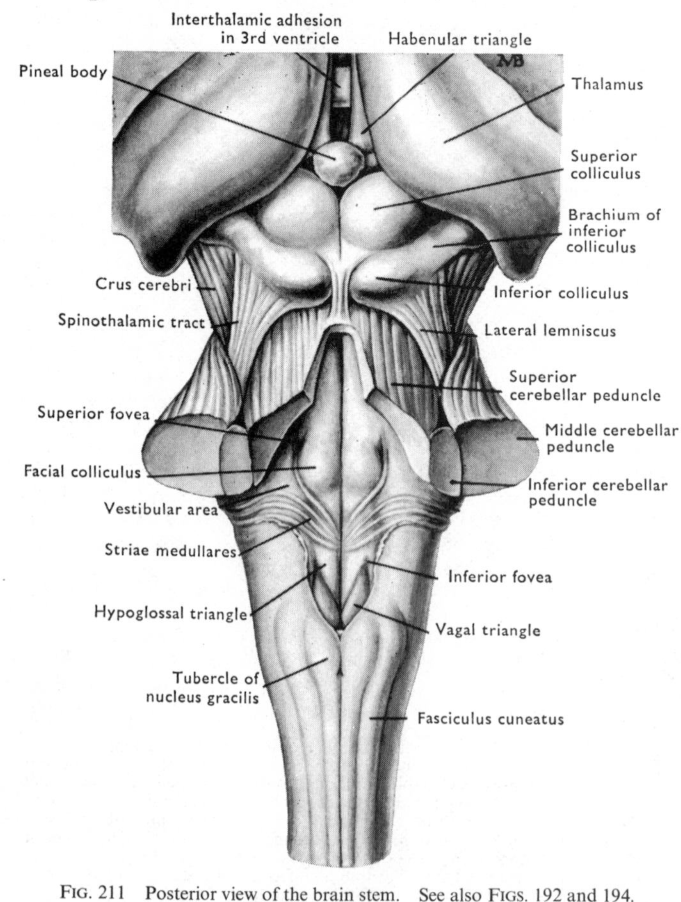

FIG. 211 Posterior view of the brain stem. See also FIGS. 192 and 194.

THE FOURTH VENTRICLE [p. 250]

This is the diamond-shaped cavity of the hindbrain which extends from the superior border of the pons to the middle of the medulla oblongata, and lies behind these structures, and in front of the cerebellum and the medullary vela. It narrows above to the superior angle where it becomes continuous with the cerebral aqueduct in the midbrain. The tiny central canal of the inferior half of the medulla oblongata and spinal medulla opens into it at the inferior angle.

The ventricle is widest at the level of the junction of the pons with the medulla oblongata. Here it extends laterally on each side, to form a tubular **lateral recess**, which curves over the posterior aspect of the inferior cerebellar peduncle. This recess passes as far as the tip of the flocculus, where it opens into the subarachnoid space through the **lateral aperture** of the fourth ventricle, posterior to the ninth and tenth cranial nerves. The deepest part of the ventricle is opposite the inferior part of the pons where the tent-like cerebellar recess extends posteriorly almost to the level of the dentate nuclei. Immediately inferior to this, the inferior medullary velum passes anteriorly and the ventricle becomes a shallow slit, though its depth again increases towards the inferior angle as the velum curves posteriorly on the anterior surface of the uvula of the cerebellum to the margin of the **median aperture** of the fourth ventricle.

ROOF OF THE FOURTH VENTRICLE

Superior to the cerebellar recess, the superior medullary velum with the lingula of the cerebellum fused to its posterior surface, forms the greater part of the roof. The superior medullary velum stretches between the superior cerebellar peduncles, and gradually narrows as it approaches the tectum of the midbrain. Inferior to the cerebellar recess, the thin inferior medullary velum extends in the midline from its cerebellar attachment to the margin of the median aperture [p. 250] on the anterior surfaces of the nodule and uvula. Laterally, it forms the bed of the tonsil and is attached to the posterior surface of the medulla oblongata. These attachments converge towards the median aperture inferiorly, but superiorly they pass laterally over the posterior surface of the inferior cerebellar peduncle to form the inferior limit of the lateral recess [FIG. 208].

Tela Choroidea and Choroid Plexuses of Fourth Ventricle

The pia mater which covers the surface of the brain passes between the inferior medullary velum and the cerebellum with the posterior inferior cerebellar artery. This pia mater (the tela choroidea) forms a tufted, vascular fold which invaginates the ependyma of the roof of the fourth ventricle, on each side of the midline, from the margin of the median aperture to the level of the lateral recess. Here each choroid plexus (vascular pia mater or tela choroidea covered with ependyma) turns laterally and extends into the roof of the corresponding lateral recess as far as the lateral aperture, through which its tufts protrude as a miniature, cauliflower-like excrescence immediately inferior to the flocculus of the cerebellum. Thus each half of the choroid plexus is L-shaped, and the extremities of both can be seen in the intact brain at the median and lateral apertures of the fourth ventricle [FIGS. 192, 208, 285].

FLOOR OF THE FOURTH VENTRICLE

[FIG. 211]

The floor is diamond-shaped and consists of a layer of grey matter which contains the nuclei of various cranial nerves and is separated from the ventricle by a layer of ependyma. The floor is formed by the posterior surfaces of the pons and the superior half of the medulla oblongata. Between the aqueduct superiorly, and the central canal inferiorly, its lateral boundaries are the superior cerebellar peduncles, the inferior cerebellar peduncles, and the cuneate and gracile tubercles.

The floor is divided into right and left halves by a median groove, and a ridge on each side of this is known as the **eminentia medialis**. A slight swelling in each ridge opposite the inferior part of the pons is the **facial colliculus**. This is produced by the nucleus of the abducent nerve with the facial nerve looping round it.

About the middle of each half of the floor a few fine bundles of fibres may be seen through the ependyma. They emerge from the median groove and pass laterally towards the inferior cerebellar peduncle. These are the **medullary striae**, and they are composed of aberrant pontocerebellar fibres which pass from the arcuate nuclei [p. 224] to the cerebellum, thus dividing the floor of the ventricle into pontine and medullary parts. The size and number of these striae is very variable.

In each half of the medullary part, there is a V-shaped depression (the **inferior fovea**) which divides the floor into three triangular areas. The medial area, part of the eminentia medialis, overlies the superior part of the hypoglossal nucleus and is known as the **hypoglossal triangle**. The intermediate part overlies the dorsal nucleus of the vagoglossopharyngeal complex, and is known as the **vagal triangle**. The lateral area is the inferior part of the **vestibular area**—a poorly defined swelling in the lateral part of the floor of the ventricle, extending superiorly into the inferior part of the pons. It is formed by the **vestibular nuclei**, the termination of the vestibular fibres of the vestibulocochlear nerve. Medial to this a slight groove represents the remains of the sulcus limitans of the developing neural tube [pp. 249–250], and is slightly accentuated at the level of the facial colliculus, to form the **superior fovea**. Above the superior fovea, a small area of floor, lateral to the eminentia medialis, has a dark tinge (the locus ceruleus) produced by pigmented cells beneath the ependyma.

DISSECTION. If a satisfactory dissection of the peduncles of the cerebellum has not been obtained on the right half, place your fingers in the medial part of the horizontal fissure of the left half of the cerebellum, and tear off the upper part of this half. This exposes the superior peduncle and parts of the other two. With blunt forceps, pick up thin layers of the white matter near the posterior border of the cerebellum and strip them forwards. In this way the surface of the dentate nucleus is exposed, and its continuity with the superior peduncle can be confirmed.

Remove the remains of the membranes from the surface of the midbrain, avoiding, if possible, damage to the trochlear nerves. Identify the pineal body between the superior colliculi [FIG. 211] and note that it is attached to the dorsal surface of the brain at the junction of the midbrain and diencephalon, inferior to the splenium of the corpus callosum.

The midbrain is the short, thick stalk which connects the hindbrain, in the posterior cranial fossa, with the cerebrum superior to the tentorium cerebelli. It traverses the tentorial notch, and is about 2·5 cm long and slightly more in width. The narrow tubular cavity which traverses it (the **cerebral aqueduct** [FIG. 183]) connects the fourth ventricle inferiorly to the third ventricle superiorly. Identify the inferior end of the aqueduct, and pass a fine probe through it. The tectum is the smaller part of the midbrain posterior to the aqueduct. It consists of the four colliculi. The larger part, anterior to the aqueduct, can be divided into right and left halves, the **cerebral peduncles**. In the undissected brain, the tectum is overlapped by the anterior lobe of the cerebellum and the splenium of the corpus callosum, but the anterior parts of the cerebral peduncles, the **crura cerebri**, can be seen on the base of the brain where they form the posterolateral boundaries of the interpeduncular fossa.

TECTUM

Colliculi

These are four small mounds of grey matter on the posterior surface of the tectum. In the superior pair, the grey matter is on the surface and consists of many layers, in the inferior pair it forms an oval nucleus. The superior cerebellar peduncles enter the substance of the midbrain anterior to the inferior colliculi. Each peduncle is crossed obliquely by a ridge of white matter (the **lateral lemniscus** [FIGS. 194, 202]) which emerges from the superior border of the pons, posterior to the crus cerebri, and passes postero-superiorly into the inferior colliculus. This bundle of nerve fibres transmits impulses from the cochlear nuclei to the inferior colliculus—one of the groups of nerve cells concerned with auditory reflexes. On the side of the midbrain, note a similar ridge (**brachium of the inferior colliculus** [FIGS. 194, 213]) passing anterosuperiorly from the inferior colliculus to the **medial geniculate body**—a small, rounded protuberance lodged between the superior part of the lateral wall of the midbrain and the posterior surface of the crus cerebri. The brachium transmits auditory impulses to this thalamic nucleus for relay to the cerebral cortex where they are consciously appreciated.

The **trochlear nerves** emerge from the dorsal surface of the brain stem immediately inferior to the inferior colliculi. The **pineal body** lies between the two superior colliculi [FIG. 211] each of which receives a bundle of visual fibres in a slight ridge at its superolateral margin—the **brachium of the superior colliculus** [FIG. 250]. This bundle runs over the postero-inferior surface of the thalamus to the colliculus and to the region between it and the root of the pineal body (the **pretectal region**)—both concerned with visual reflexes. The colliculus is connected to the cranial nerve nuclei and to the motor cells of the spinal medulla through the **tectospinal tract** [FIGS. 197, 212, 261] which is responsible for the reflex turning of the head and eyes towards a source of light.

CEREBRAL PEDUNCLES

Each cerebral peduncle consists of: (1) the crus cerebri, anteriorly; (2) the tegmentum, posteriorly; (3) a layer of pigmented cells between these, the substantia nigra [FIGS. 213, 214].

Crus Cerebri

Each of these broad bundles of nerve fibres arises in the cerebral cortex of the corresponding hemisphere, and piercing the hemisphere, converges on its inferior surface, to emerge as the crus where the optic tract crosses its lateral surface. The posterior cerebral artery, the basal vein, and the trochlear nerve also cross the lateral surface of the crus. The two crura are separated by the posterior part of the interpeduncular fossa, which contains the oculomotor nerves and the central branches of the posterior cerebral artery entering the posterior perforated substance. Posteriorly, the crus is separated from the tegmentum of the midbrain by the substantia nigra, and laterally by a shallow groove (the lateral sulcus of the midbrain) in the superior part of which the medial geniculate body is lodged [FIGS. 213, 250].

The nerve fibres in the crus arise from all parts of the cerebral cortex, and converge radially on the crus, so that those from the frontal region lie in its anteromedial part, while those from the occipital and temporal regions lie in its posterolateral part. The majority of its fibres end in the nuclei of the ventral part of the pons and constitute the **corticopontine fibres**. A smaller number traverse the pons and emerge as the pyramid on the medulla oblongata, thus forming the **corticospinal fibres**. A still smaller number leave the crus to cross towards the motor cranial nerve nuclei of the opposite side. These **corticonuclear fibres** correspond to the cortico-spinal fibres but they end in the brain stem and not on the cells of the spinal medulla. The majority of corticospinal and corticonuclear fibres cross to the opposite side before they end. However, the contralateral paralysis of voluntary movement which would be expected when these fibres are destroyed before they cross is not uniform in all muscles. This is usually explained by the assumption that some motor cells are activated by fibres from both sides of the brain. Thus destruction of corticonuclear fibres on one side paralyses voluntary movement in the muscles of the tongue and in the lower part of the face on the other side, but leaves the muscles of the upper part of the face, the muscles of mastication, and those of the pharynx and larynx apparently unaffected. The eye muscles are also spared, though the ability to produce coordinated movements of the two eyes may not be. Similarly in destruction of the corticospinal fibres, the contralateral muscles of the trunk and proximal parts of the limbs are little affected, though those of the distal parts of the limbs are. It seems likely that the explanation of these findings is not entirely correct, but they have considerable value in distinguishing between destruction of the descending fibres from the cerebral cortex (clinically the 'upper motor neuron') from that of the motor nerves or their nuclei of origin (the 'lower motor neuron') which inevitably produces paralysis and rapid wasting of all the muscles innervated by the nerves.

Since most of the corticospinal and the corticonuclear fibres arise in the posterior part of the frontal lobe (*i.e.*, near the middle) of the hemisphere, these fibres lie in the intermediate part of the crus; and of these the corticonuclear fibres are the most medial, while those passing to the lumbosacral part of the spinal medulla are the most lateral [FIG. 199].

In the section which follows, most of the details are not visible in the gross specimen, but refer to stained sections of the brain stem. It is desirable for the student to have these available for study, but, failing this, FIGURES 212, 213, and 214 show the points which are mentioned, and should be consulted.

Substantia Nigra [FIGS. 213, 214]

This curved plate of grey matter lies between the crus cerebri and the tegmentum. It contains many large nerve cells with a considerable amount of melanin in their cytoplasm. These cells have connexions with the tegmentum of the midbrain and with the corpus striatum in the centre of the corresponding hemisphere. They play an important part in the control of muscle tone and activity, and produce dopamine which is passed along their axons to the corpus striatum. The substantia nigra extends from the superior border of the pons into the inferolateral part of the hypothalamus, superior to the midbrain.

Tegmentum

This is a thick column of mixed grey and white matter. It is continuous inferiorly with the dorsal part of the pons, and superiorly with the hypothalamus. It is only visible on the surface of the midbrain in the interpeduncular fossa and on the lateral surface where it is crossed by the brachium of the inferior colliculus and the lateral lemniscus [FIG. 194]. Posteriorly, it is directly continuous with the tectum at the level of the cerebral aqueduct, which is surrounded by a thick tube of grey matter (the **central grey substance**) in which lie the nuclei of the oculomotor and trochlear nerves and the mesencephalic nucleus of the trigeminal nerve [FIGS. 212, 271]. The tegmentum contains a number of important nuclei and tracts:

1. The **red nucleus** is a rounded rod of reddish grey matter which lies in the medial part of each half of the tegmentum at the level of the superior colliculus and the posterior hypothalamus. It is the most obvious structure in a transverse section through the superior part of the midbrain, but is surprisingly white in the fresh tissue because of the mass of fibres from the opposite **superior cerebellar peduncle** which traverse and form a capsule for it. These fibres decussate in the tegmentum at the level of the inferior colliculus, and ascend to the red nucleus where most of them end though some pass directly on to the thalamus. In Man the majority of the cells of the red nucleus send their axons to that part of the thalamus which relays these impulses to the posterior part of the frontal lobe of the cerebral hemisphere, the part from which corticospinal and corticonuclear fibres arise. A small number of cells in the caudal part of the red nucleus give rise to fibres (**rubroreticular and rubrospinal tracts**), which decussate in the ventral part of the tegmentum (**ventral tegmental decussation**), and descend through the dorsal part of the pons to the lateral part of the medulla oblongata and upper spinal medulla.

2. The **medial longitudinal fasciculus** is a small, compact bundle which extends from the upper part of the midbrain into the spinal medulla. It lies adjacent to the midline and the nuclei of the oculomotor and trochlear nerves, and maintains this position through the brain stem. It is an intersegmental bundle which connects the various levels of the brain stem, and receives its major single contribution from the vestibular nuclei. In the spinal medulla it is continuous with the anterior intersegmental tract or fasciculus proprius [FIGS. 196–198, 200–202, 212, 213].

3. The **dorsal longitudinal fasciculus** lies posterior to the nuclei of the third and fourth cranial nerves in the central grey matter. It connects the hypothalamus with the midbrain and pons.

4. The **medial lemniscus** and **spinothalamic tracts** [FIGS. 212, 213, 214] ascend through the tegmentum from the dorsal part of the pons, and pass further laterally as they ascend (the spinothalamic intermingling with the medial lemniscus) so that they come to underlie the brachium of the inferior colliculus in the upper midbrain. At this level the spinothalamic tract gives off fibres to the superior colliculus, the **spinotectal tract**. In addition to the fibres which these tracts contain in the medulla oblongata, trigeminal fibres have joined them when they reach the midbrain level.

5. The **tectospinal tract** consists of fibres which arise in the deeper layer of the superior colliculus. They sweep around the margin of the central grey substance, decussate (the **dorsal tegmental decussation**) anterior to the medial longitudinal fasciculus, and descend through the dorsal part of the pons, the medulla oblongata, and spinal medulla anterior to the fasciculus. Fibres of this tract end in the reticular formation of the brain stem and in the spinal medulla, and produce reflex movements in relation to visual stimuli reaching the superior colliculus; *e.g.*, turning of head and eyes towards a sudden flash of light [FIG. 261].

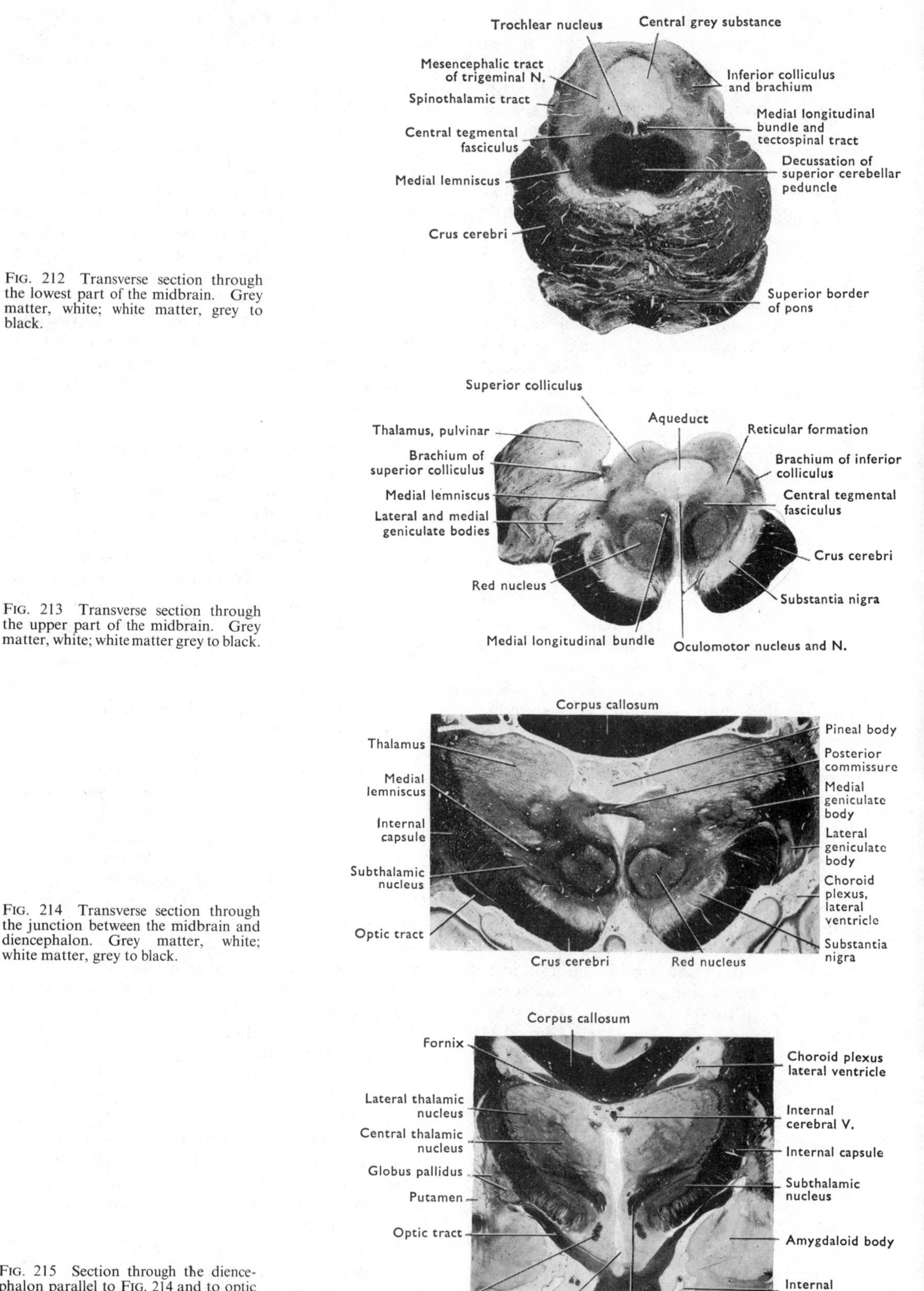

FIG. 212 Transverse section through the lowest part of the midbrain. Grey matter, white; white matter, grey to black.

FIG. 213 Transverse section through the upper part of the midbrain. Grey matter, white; white matter grey to black.

FIG. 214 Transverse section through the junction between the midbrain and diencephalon. Grey matter, white; white matter, grey to black.

FIG. 215 Section through the diencephalon parallel to FIG. 214 and to optic tract. Grey matter, white; white matter, grey to black.

6. The **central tegmental tract** is a large diffuse bundle of fibres, dorsolateral to the red nucleus. It can be traced [FIGS. 198, 200–202, 209, 212] from the upper midbrain, through the dorsal part of the pons to end in the capsule of the olivary nucleus. It probably transmits a number of different fibres, but it almost certainly forms a pathway from the corpus striatum and cerebellum [p. 195] to the olivary nucleus, through which the striatum is connected with the cerebellum by the olivocerebellar fibres.

7. The fibres of the **oculomotor nerve** can be seen in section [FIGS. 213, 272] sweeping ventrally through the medial part of the red nucleus in the superior part of the midbrain. They emerge through the sulcus medial to the crus cerebri into the interpeduncular fossa.

8. In the inferior part of the midbrain, the fibres of the **trochlear nerve** run postero-inferiorly around the central grey substance, to meet and decussate in the most superior part of the superior medullary velum. They emerge posterior to the superior cerebellar peduncles [FIGS. 202, 212, 272].

9. The **reticular formation** of the midbrain occupies the lateral parts of the tegmentum between the central grey substance and the medial lemniscus and spinothalamic tract. It is directly continuous with the reticular formation of the dorsal part of the pons and with the lateral part of the hypothalamus. It is probably intimately concerned with the genesis of complicated reflexes of the righting variety, which may be stimulated by afferent impulses from the vestibular apparatus, or from the proprioceptors of the body, or even by visual impulses from the tectum.

The lateral lemniscus and the auditory pathway has been described already [p. 198].

THE CEREBRUM

The dissectors should begin by reviewing the relation between the skull and brain with the assistance of a dried skull, and should note the relation of the parts of the brain to the folds of dura mater (the falx cerebri and the tentorium cerebelli) if a partially dissected head is available. Reference should also be made to the short account of craniocerebral topography and to FIGURES 277, 296.

The **longitudinal fissure** is the narrow cleft between the two hemispheres. It is occupied by: (1) the falx cerebri; (2) the fold of arachnoid which follows the surface of the falx; (3) the pia mater covering the medial surfaces of the hemispheres; (4) the arteries and veins which lie in the subarachnoid space between the arachnoid and the pia. The falx was removed when the brain was taken from the skull, but the other structures are still *in situ*.

DISSECTION. Separate the two hemispheres at the longitudinal fissure, and expose the mass of white matter which joins them deep in the fissure. This is the **corpus callosum**, the posterior part of which (**splenium**) has already been seen as a thick, rounded mass superior to the pineal body. Identify it in this position and note the **great cerebral vein** issuing from the brain between the pineal and the splenium of the corpus callosum. Clean this vein and identify as many of its tributaries as may be seen filled with blood. These include **superior cerebellar veins**, and the **basal veins** which pass posteriorly around the side of the midbrain to join the great cerebral vein.

Turn to the superior surface of the hemispheres, and, drawing them apart, divide the corpus callosum in the median plane, starting at the splenium. Proceed carefully, noting that the corpus callosum is much thinner immediately anterior to the splenium, and that a thin layer of pia mater extends anteriorly, inferior to the corpus callosum. When about 3 cm of the corpus callosum has been divided, examine the pia mater deep to it and identify two small veins which run posteriorly in it and unite to form the great cerebral vein; these are the **internal cerebral veins** [FIG. 246]. The sheet of pia mater in which they lie separates the corpus callosum from the choroid plexus in the roof of the third ventricle and extends laterally to the choroid plexus of the lateral ventricles within the hemispheres. It is therefore the tela choroidea of the third and lateral ventricles. Divide the pia mater in the midline, between the internal cerebral veins, and so open into the vertical, slit-like cavity of the third ventricle immediately beneath.

Continue the division of the corpus callosum. Keep strictly to the midline and avoid extending the incision beyond its inferior margin. As the incision is carried forwards and the hemispheres are allowed to fall apart, a bridge of tissue will be seen crossing the third ventricle (the interthalamic adhesion [FIG. 249]). Attached to the inferior surface of the corpus callosum further anteriorly are two thin, vertical sheets of white matter (the **laminae of the septum pellucidum**) and these may be pushed apart by gentle blunt dissection to expose a space between them, the **cavum septi pellucidi** [FIG. 239]. As the anterior end of the corpus callosum is reached, it will be seen to turn inferiorly and then posteriorly to form the bend or **genu of the corpus callosum**, which thins out inferiorly to form the **rostrum** [FIG. 222]. Divide the genu and rostrum between the laminae of the septum pellucidum, and separate the two arches (fornices) of white matter, one of which is attached to the inferior margin of each lamina of the septum pellucidum. As this is done, the anterior extremity of the third ventricle will be opened and a round bundle will be seen crossing the midline immediately anterior to the columns of the fornix as they turn vertically downwards. This is the **anterior commissure** [FIGS. 223, 249]. Divide it at the superior margin of the **lamina terminalis** (the anterior wall of the third ventricle) and cut vertically through the thin lamina terminalis.

Turn to the inferior surface of the brain and identify the optic chiasma. Pull it inferiorly and note the divided lamina terminalis above it. Divide the optic chiasma in the midline, and carry the median incision posteriorly through the floor of the third ventricle, dividing the infundibulum and separating the mamillary bodies. Do not carry the incision any further than this, but divide the

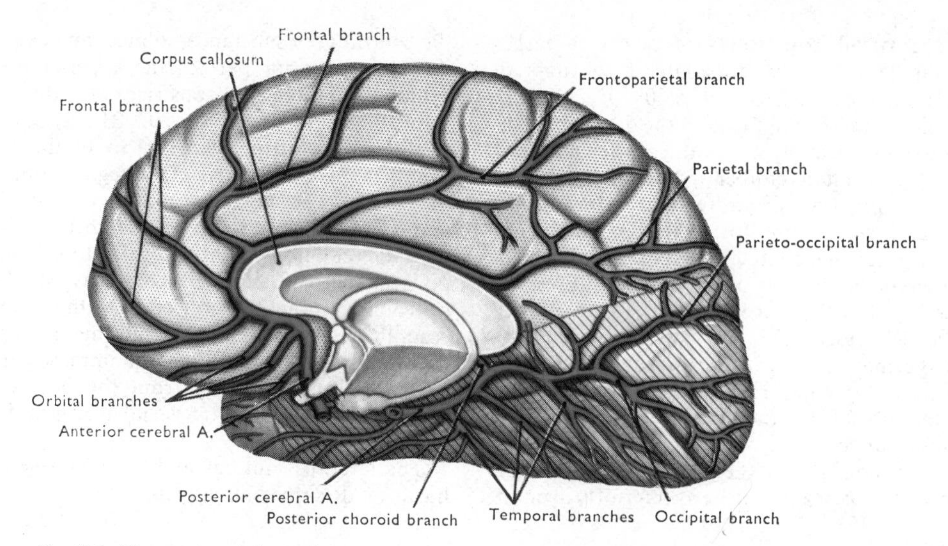

FIG. 216 The arteries of the medial and tentorial surfaces of the right hemisphere. Area supplied by the anterior cerebral artery, red stipple; by the posterior cerebral artery, red cross-hatching; and by the middle cerebral artery, no colour.

right posterior cerebral artery close to its origin from the basilar artery. Turn to the right side of the brain and make a transverse cut through the upper part of the right half of the midbrain so as to join the two ends of the median cut. Now lift away the right hemisphere and examine its medial aspect.

Most of the features of the medial aspect will be examined later, but the dissectors should take this opportunity to pass a fine probe through the cerebral aqueduct from the posterosuperior part of the third ventricle to the fourth ventricle.

Anterior Cerebral Artery [FIG. 216]

The beginning of this vessel has been examined already [p. 179], but it can now be followed from the anterior communicating artery on the surface of the corpus callosum, or in the sulcus cinguli [FIG. 223]. It supplies the medial surface and the adjacent margin of the superolateral surface as far posteriorly as the splenium of the corpus callosum.

Posterior Cerebral Artery

This vessel supplies the remainder of the medial surface of the hemisphere and the greater part of the tentorial surface. Its main branches on the medial surface run in the calcarine and parieto-occipital sulci [FIG. 223].

Follow the branches of these arteries, noting in particular the **posterior choroidal branches** of the posterior cerebral artery. These arise as it runs round the lateral aspect of the midbrain, close to the superior surface of the margin of the tentorium, and they pass deep to the cortical margin. Shortly after its origin the posterior cerebral artery lies on the parahippocampal gyrus [FIG. 230] which ends anteriorly in a rounded swelling, the **uncus**. This is normally grooved by the free edge of the tentorium, and tends to be thrust through the tentorial notch when the supratentorial pressure exceeds that in the infratentorial compartment [FIG. 229].

General Features of Surface of Hemisphere

The **frontal pole** lies opposite the root of the nose and the medial part of the superciliary arch.

The **occipital pole** is more pointed than the frontal. It lies a short distance superolateral to the external occipital protuberance, and its medial aspect may be grooved by the superior sagittal sinus turning into the transverse sinus.

The **temporal pole** fits into the anterior part of the middle cranial fossa, and is overhung by the lesser wing of the sphenoid. The various **borders of the hemisphere** are named: the superomedial, inferolateral, superciliary, medial orbital, medial occipital, and inferomedial [p. 209] borders. The superciliary border is at the junction of the superolateral and orbital surfaces; the medial orbital lies at the junction of the medial and orbital surfaces; the medial occipital border is at the junction of the medial and tentorial surfaces. The superomedial border is self-explanatory. The inferolateral border (at the junction of the superolateral and tentorial surfaces) lies immediately superior to the transverse sinus in its posterior part, and exhibits a notch (**preoccipital notch**) about 3 cm from the occipital pole, where the inferior anastomotic vein enters the transverse sinus. This notch is used as one landmark which artificially separates the occipital and temporal lobes of the brain [FIG. 227].

DISSECTION. Review the vessels on the surface of the hemisphere, and then proceed to strip the meninges from its surface. On the superolateral surface, strip the meninges towards the lateral sulcus [FIG. 217] pulling them off along the line of the other sulci. Note the large branches of the middle cerebral artery which issue from the lateral sulcus.

CEREBRAL SULCI AND GYRI

There are great individual variations in the details of the sulci and gyri, so attention should be given mainly to the major sulci and gyri. These can be identified with the help of FIGURES 217, 218, 223, 225, and 229. From these it will become apparent that the majority of the sulci run longitudinally in the hemisphere with the exception of the central, precentral, postcentral, and parieto-occipital which are more or less transverse to the long axis of the hemisphere.

The following account deals only with some of the important features of the sulci and gyri. The student should confirm, and be conversant with the position of all those mentioned in the above figures.

The **sulci** vary in depth from slight grooves to deep fissures, and some of them (calcarine and collateral, FIGURES 223 and 229) are sufficiently deep to indent the wall of the lateral ventricle in the depths of the hemisphere.

The **gyri** consist of a central core of white matter (nerve fibres running to and from the overlying cortex) covered by a layer of grey matter, the **cerebral cortex**. The cortex extends as an uninterrupted sheet over the whole surface of the hemisphere, and consists of nerve cells which are arranged in six layers parallel to the surface. These layers are differentiated from each other by their different content of cells, but they are difficult to separate in many situations, and tend to be arranged in such a manner that certain areas of the cortex can be recognized easily in a microscopic section, while others can not. Any special features of the different regions of the cortex which are discussed will be mentioned briefly, but the student should refer to the larger textbooks for details of the microscopic structure of the cerebral cortex. It is a great advantage if the student can have microscopic preparations to study, for only very few of the features of the structure of the cerebral cortex are visible to the unaided eye. In general, the cerebral cortex varies both in thickness and microscopic structure in its various parts, but the correlation between structure and function is, as yet, in its infancy. The cortex is thicker on the summits of the gyri than in the depths of the sulci, and is much thicker in the posterior part of the frontal lobe than in the adjacent anterior part of the parietal lobe on the opposite wall of the central sulcus [FIG. 217].

Lateral Sulcus [FIGS. 217, 229]

The various parts of this deep and complex sulcus are formed by the meeting of folds of the surrounding cortex which overgrow a portion of the cortex adherent to the solid central part of the hemisphere (corpus striatum) in such a manner that, during development, that cortex becomes a buried island, **the insula**. [FIG. 284; p. 251].

The **stem** of the sulcus begins in the vallecula and runs laterally between the temporal pole and the posterior part of the orbital surface of the hemisphere. It transmits the middle cerebral artery. The stem ends on the superolateral surface by dividing into anterior, ascending, and posterior **rami** [FIG. 217]. The first two cut into the inferior frontal gyrus, while the horizontal part of the posterior ramus separates the frontal and parietal lobes of the

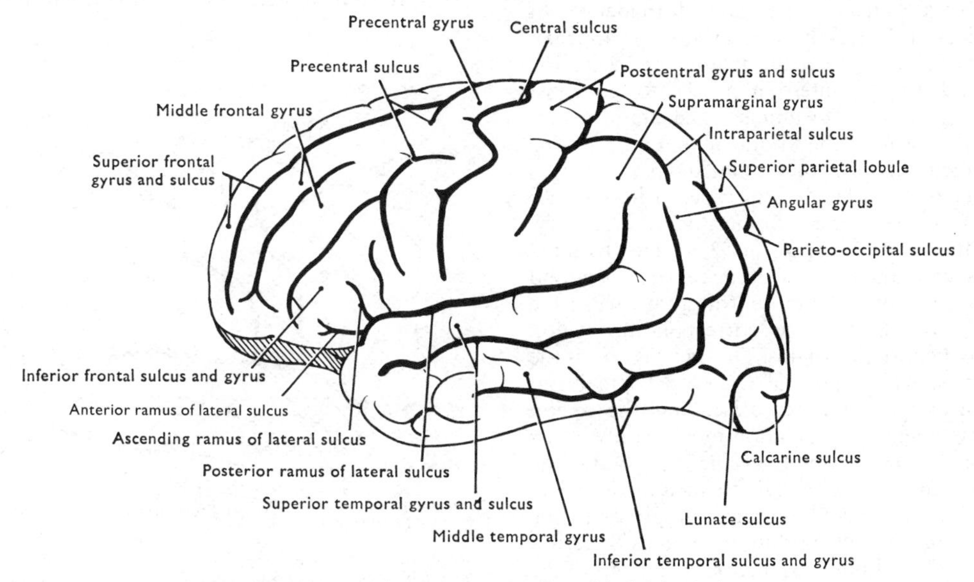

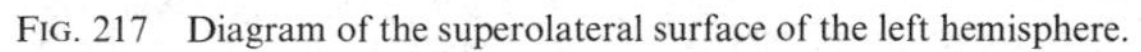
FIG. 217 Diagram of the superolateral surface of the left hemisphere.

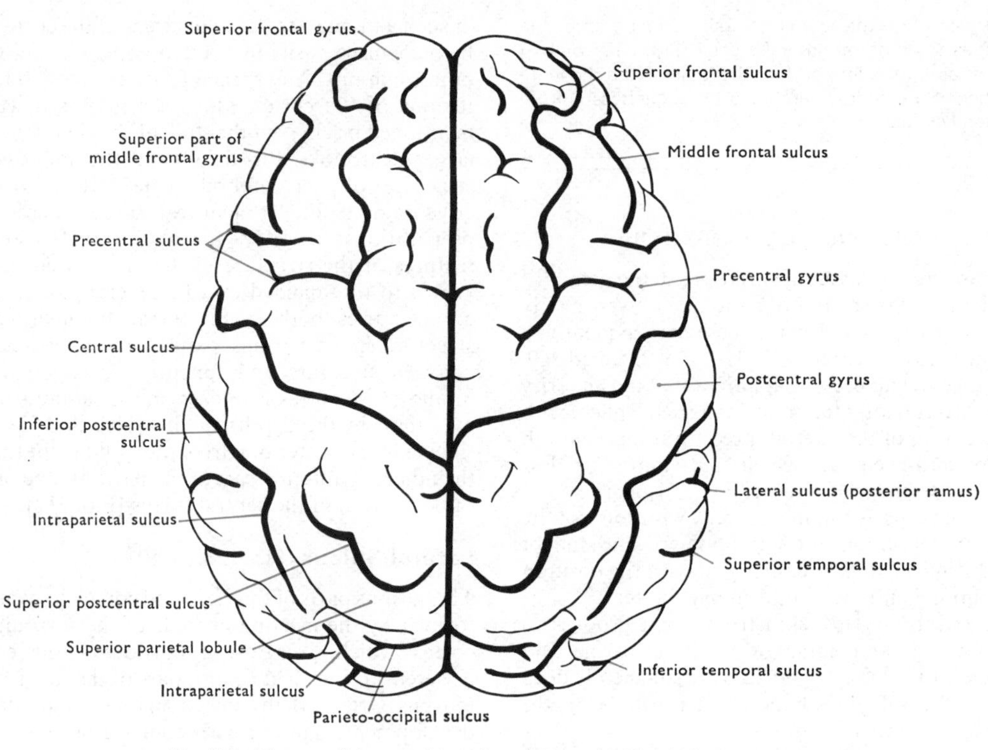

FIG. 218 Diagram of the superior surface of the cerebral hemispheres.

brain superiorly, from the temporal lobe inferiorly. If the lips of this sulcus are pulled apart, the insula is exposed, and the main branches of the middle cerebral artery can be seen running over it before emerging from the sulcus on to the superolateral surface of the hemisphere.

Central Sulcus [FIGS. 217, 223]

This important sulcus is often difficult to identify. It begins superiorly on the medial surface approximately midway between the frontal and occipital poles, and passes antero-inferiorly to end just superior to the posterior ramus of the lateral sulcus, 2–3 cm posterior to the origin of that sulcus. The central sulcus separates the frontal and parietal lobes of the hemisphere, but it is of especial importance because it separates the main 'motor' and 'sensory' areas of the cerebral cortex [FIG. 219]. The **'motor' area** lies in the anterior wall of the central sulcus and the adjacent part of the precentral gyrus, while the 'sensory' area lies in the corresponding region immediately posterior to this. The 'motor' area is the region from which most of the corticospinal and corticonuclear nerve fibres arise, and the region where low intensity stimulation most readily elicits movements of the contralateral side of the body. The **'sensory' area** is the region to which the impulses ascending in the medial lemniscus and spinothalamic tracts are transmitted. Within these areas there is a further functional subdivision, for movements of the opposite side of the head and neck are most easily elicited by stimulation of the inferior precentral region, the upper limb, trunk, and lower limb following in sequence superiorly. A similar arrangement of the sensory fibres is found in the postcentral region; those from the opposite side of the head reaching the inferior postcentral region, with the other regions of the body represented in sequence superiorly.

The frontal branch of the **middle meningeal**

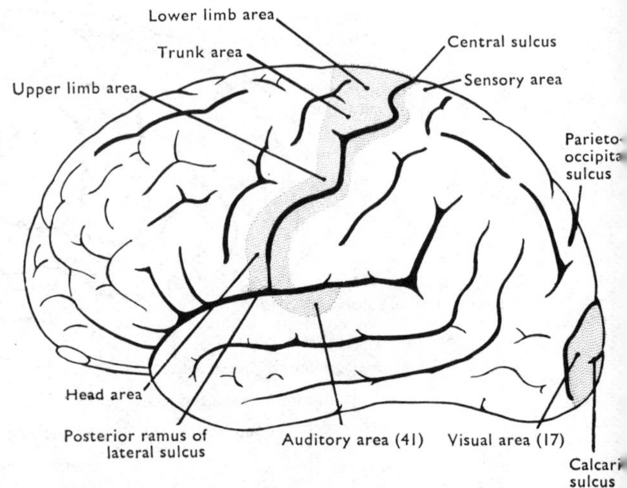

FIG. 219 Diagram of the superolateral surface of the left hemisphere, to show 'motor' (red) and 'sensory' areas (blue).

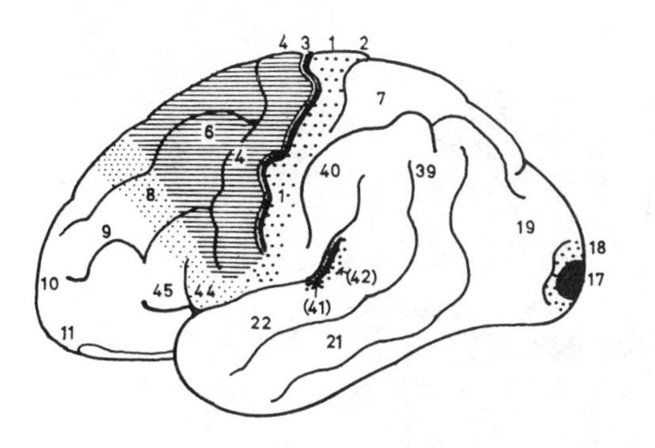

FIG. 220 Diagram of the superolateral surface of the left hemisphere to indicate the position of some of the areas which can be differentiated on the basis of their histological structure. The numbers indicate some of the areas described by Brodmann though the extent of these areas is not shown. Black, koniocortex of the primary receptive areas; large dots, parakoniocortex; small dots, frontal dysgranular cortex; cross-hatching, frontal agranular cortex.

artery runs in the endocranium, parallel, and a short distance anterior to, the central sulcus [FIG. 277]. Thus haemorrhage from this artery is likely to press on this region of the brain and interfere with the voluntary production of movements in the opposite half of the body.

If a small wedge of cortex is cut out across the central sulcus, the greater thickness of the precentral cortex as compared with the postcentral cortex is immediately obvious. This difference in thickness is associated with a marked difference in microscopic structure. The 'sensory' cortex, like the other areas of the cortex to which sensory fibres pass, is composed almost exclusively of small cells, has poorly marked lamination, and a moderately well-defined white line (nerve fibres) which is sometimes visible to the naked eye, running through its fourth layer, *i.e.*, virtually half way through the thickness of the cortex. The precentral, 'motor' cortex is characterized by the presence of giant pyramidal cells in its fifth layer.

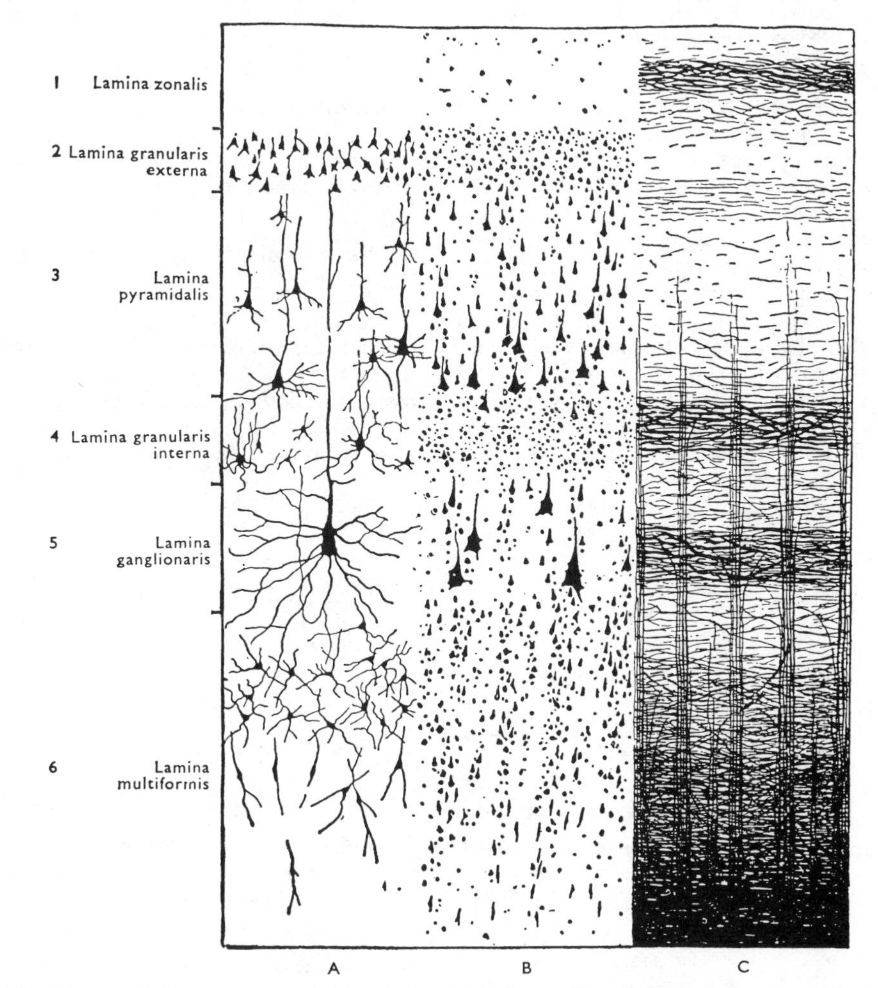

FIG. 221 Diagram of the arrangement of cells and of myelinated nerve fibres in the cerebral cortex. (From C. J. Herrick (1938) *An Introduction to Neurology*.) A, cells shown by Golgi method; B, cells shown by Nissl method; C, nerve fibres shown by Weigert method.

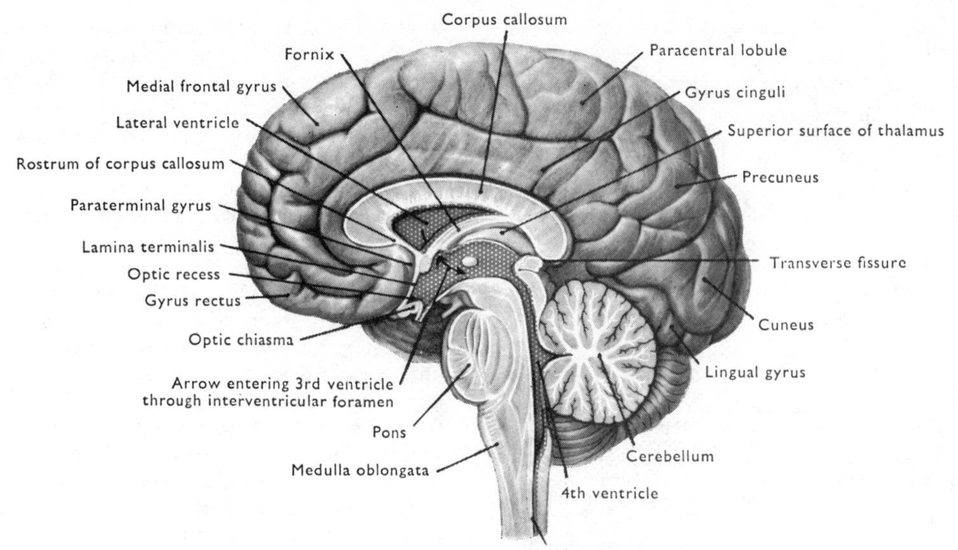

FIG. 222 Medial surface of the right half of bisected brain. The septum pellucidum has been removed to expose the lateral ventricle. The arrow passes through the interventricular foramen from the lateral to the third ventricle, and lies in the hypothalamic sulcus. Ependymal lining of ventricles, blue.

These are among the largest cells in the cerebral cortex, and are largest and most numerous in the superior part of the area. They are known as the **giant pyramidal cells** (of Betz) and they give rise to a small proportion (3 per cent) of the million corticospinal fibres in the pyramid. A few Betz cells are also found in the postcentral gyrus. The 'motor' cortex does not show a white stria such as that in the postcentral cortex, and contains many medium-sized pyramidal-shaped cells in its other layers. Throughout the greater part of the cortex, the fourth layer is composed of small cells, and hence has a granular appearance under the low power of the microscope when only the cell bodies are stained. This granular layer is absent in the posterior part of the frontal lobe, which is, therefore, known as frontal **agranular cortex**—another feature which differentiates it from the parietal lobe.

Parieto-occipital and Calcarine Sulci [FIG. 223]

These sulci form a Y-shaped arrangement on the medial aspect of the posterior part of the hemisphere. Both are very deep sulci. The superior end of the parieto-occipital sulcus reaches the superolateral surface, and the posterior end of the calcarine may extend round the occipital pole on to the

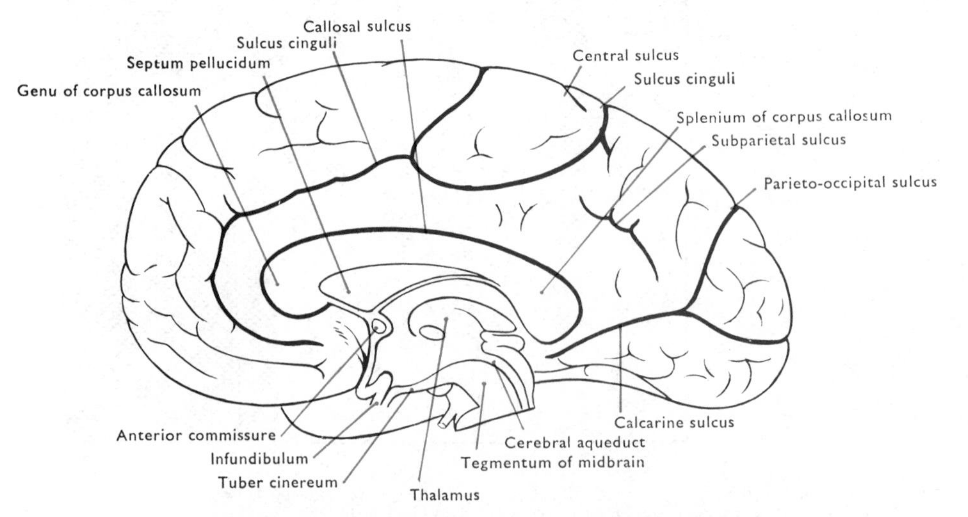

FIG. 223 Diagram of the medial surface of the right cerebral hemisphere.

superolateral surface. In the latter case the end of the calcarine sulcus is surrounded by the **lunate sulcus** [FIG. 225]. An imaginary line joining the superior end of the parieto-occipital sulcus to the pre-occipital notch separates the occipital lobe from the parietal and temporal lobes [FIG. 227].

The **calcarine sulcus** is of considerable functional significance, for it is in the cortex which lines its walls and spills over on to the medial surface for a few millimetres that the nerve fibres conveying visual impulses to the cerebral cortex end [FIGS. 224, 226]. This visual cortex is one of the few regions which can be recognized with the unaided eye, because of a very well defined white stria running through it parallel to the surface. This is equivalent to the stria seen in the postcentral cortex, but is much more marked. The presence of this stria has led to the name **striate cortex** being applied to the visual area in the walls of the calcarine sulcus. At the margin of the visual area the cortex changes its structure abruptly, and the stria ceases to be visible. It is, therefore, possible to map out the visual area with moderate accuracy by cutting a series of thin slices through the occipital lobe, and following the extent of the visual stria in these slices.

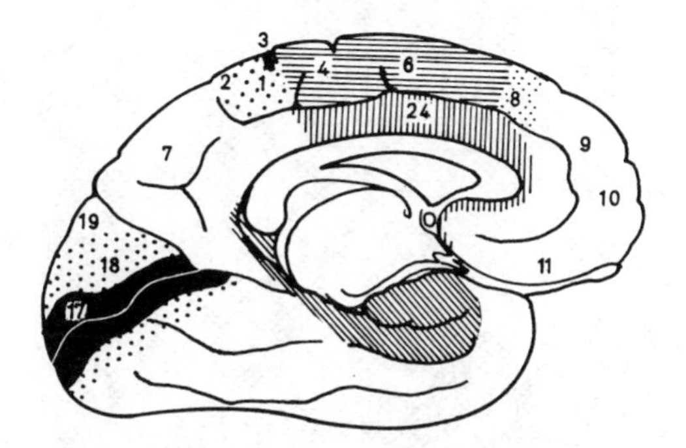

FIG. 224 Diagram of the medial surface of the left hemisphere to indicate the position of some of the areas which can be differentiated on the basis of their histological structure. The numbers indicate some of the areas described by Brodmann though the extent of these areas is not shown. Black, koniocortex of the primary receptive areas; large dots, parakoniocortex; small dots, frontal dysgranular cortex; horizontal cross-hatching, frontal agranular cortex; oblique cross-hatching, allocortex (rhinencephalon); vertical cross-hatching, cingulate agranular cortex.

DISSECTION. Cut a slice through the calcarine sulcus, posterior to its junction with the parieto-occipital sulcus, and examine the cortex for the presence of the stria.

The nerve impulses that reach one calcarine sulcus arise as a result of light stimuli in the opposite half of the field of vision. The arrangement of the nerve fibres carrying these impulses is such that impulses arising as a result of stimuli in the lower half of the field of vision, end in the superior wall of the sulcus, and vice versa, while those from the periphery end far forwards in the sulcus, and those from the centre end posteriorly. The position of small injuries to this region of the brain is, therefore, relatively easy to localize accurately from the defects which they produce in the opposite half of the field of vision.

Lobes of Hemispheres

The arbitrary division of the superolateral surface of the hemisphere into **lobes** is completed by extending the horizontal part of the posterior ramus of the lateral sulcus posteriorly to meet the line joining the parieto-occipital sulcus and the pre-occipital notch, thus separating the parietal and temporal lobes. Although the central sulcus, which separates the

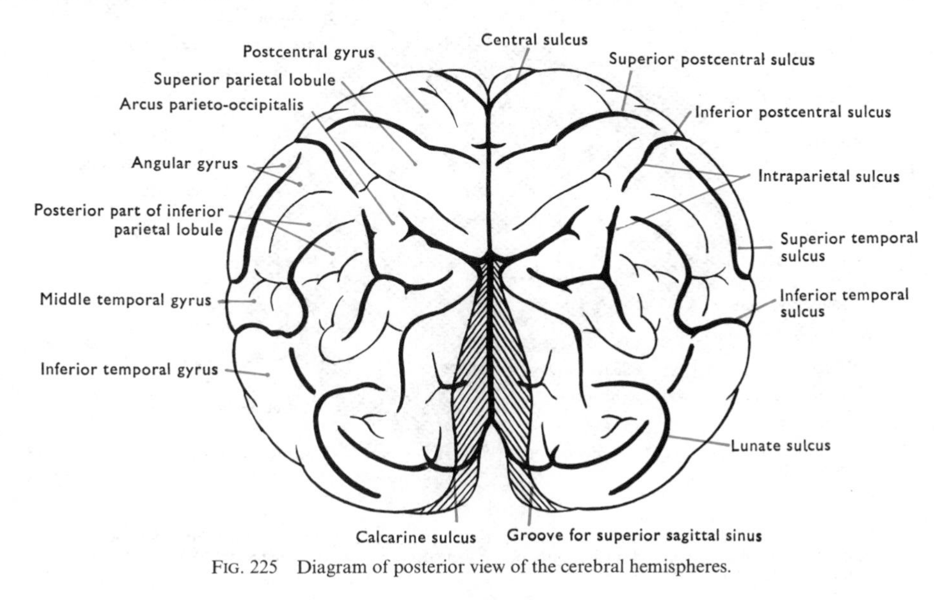

FIG. 225 Diagram of posterior view of the cerebral hemispheres.

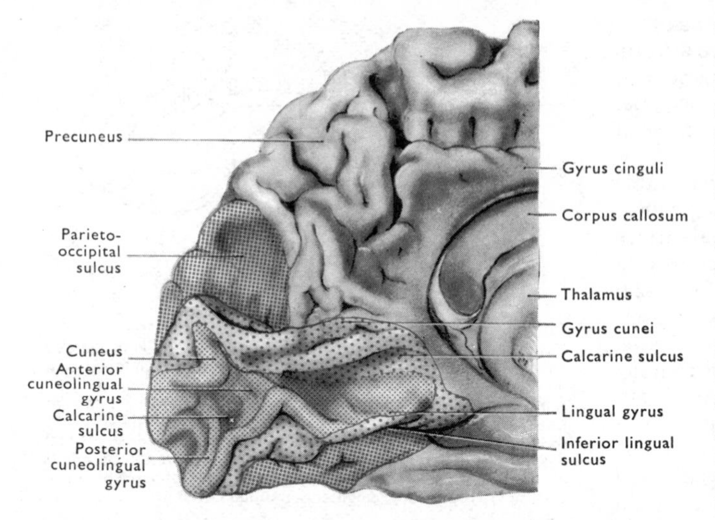

FIG. 226 Posterior part of the medial surface of the left hemisphere. The calcarine sulcus has been forced open to expose the cortex lying in its depths. Fine blue dots, visual receptive area (area 17); coarse blue dots, the area immediately surrounding area 17 to which it sends fibres (area 18); upper and lower dark blue areas (area 19) have long association connexions with other parts of the cortex.

frontal and parietal lobes, also separates parts of the cerebral cortex which are structurally and functionally very different from each other, the other lines separating the lobes of the hemisphere should not be assumed to be equally important; their purpose is purely descriptive, and they are depicted in FIGURE 227.

Other important features of the surface of the hemispheres are mentioned below.

The **'motor speech area'** lies in the region of the anterior and ascending rami of the lateral sulcus [FIG. 217]. This region is intimately connected with the control of movements of the larynx and tongue through the motor area immediately posterior to it, but the common statement that this 'centre' is present in the left hemisphere of right-handed individuals, and vice versa, is not always true. The complex mechanism which subserves speech involves many other regions of the brain, most commonly in the left hemisphere.

The **auditory area**, the cortical area to which auditory impulses are primarily relayed, lies on the middle of the superior surface of the superior temporal gyrus, extending on to the lateral surface of that gyrus, inferior to the postcentral gyrus [FIG. 219]. If the superior temporal gyrus is pulled inferiorly, a transverse gyrus will be seen running across its superior surface. This is the region of the auditory area.

The **inferior parietal lobule** is the part of the parietal lobe inferior to the intraparietal sulcus. The posterior ramus of the lateral sulcus and the superior and inferior temporal sulci sweep superiorly into it and divide it into three parts arranged around the ends of these sulci [FIG. 227]. The two anterior parts are known as the **supramarginal** and **angular gyri** and are believed to be important in the recognition of structures and symbols, and in an extension of this, the recognition of the body image. The part which this region of the brain plays in this activity is not understood, but there are considerable differences in the effects of lesions of this region in the 'dominant' (usually the left) and the 'non-dominant' hemisphere. Disabilities of spatial recognition are most common

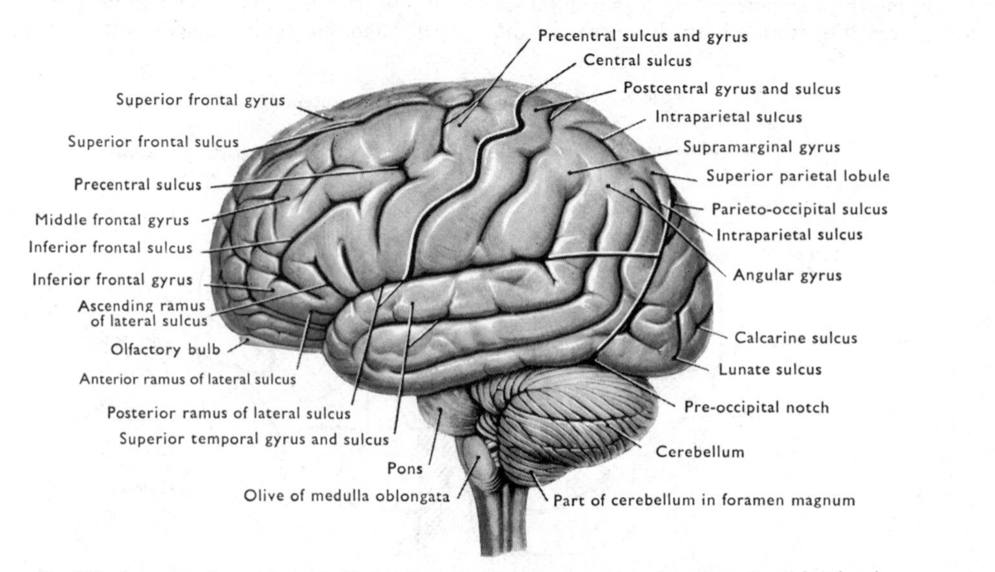

FIG. 227 Lateral surface of the left half of the brain (semi-diagrammatic) to show the main sulci and gyri, and the division of the superolateral surface into lobes by means of the central sulcus and two arbitrary lines shown in black and white.

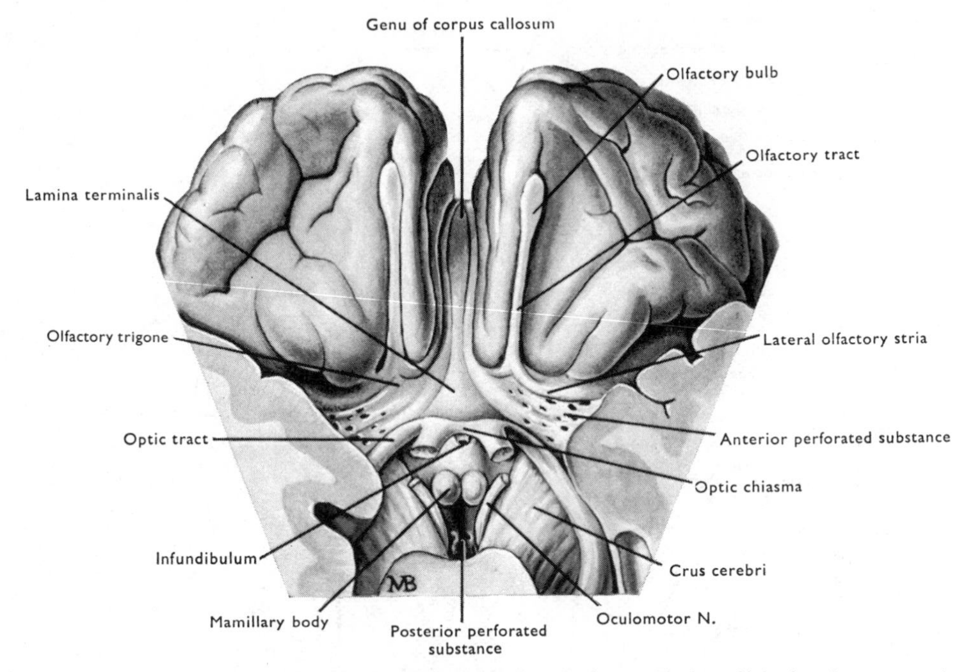

FIG. 228 Orbital surfaces of the hemispheres and the interpeduncular fossa. The frontal lobes have been separated slightly to show the genu of the corpus callosum, and the optic chiasma has been pulled inferiorly to uncover the lamina terminalis. The white ridge, parallel to the optic tract, is the diagonal band.

after right hemisphere damage in this region, while speech defects tend to arise from injuries on the left.

The **olfactory bulb and tract** lie in the **olfactory sulcus** close to the medial margin of the orbital surface of the frontal lobe. The olfactory bulb is a narrow, oval body which lies on the dura mater and endocranium immediately above the cribriform plate of the ethmoid, and these structures alone separate it from the nasal mucous membrane. Lateral to this, the orbital surface lies on the orbital part of the frontal bone. Fractures of the anterior cranial fossa are, therefore, liable to damage the olfactory bulb and to lead to blood and cerebrospinal fluid leaking into the nose and the orbit. A route for infection of the meninges may be opened through such a fracture into the nose, and in those cases where the frontal air sinus invades the orbital part of the frontal bone, infection may extend from that sinus into the cranial cavity.

Follow the olfactory tract posteriorly [FIG. 228]. It splits into a **medial** and a **lateral stria** at the anterior margin of the anterior perforated substance in which there is a small grey elevation, the **olfactory trigone**. The medial stria curves round the posterior end of the gyrus rectus to the medial side of the hemisphere, and disappears in the region of the **paraterminal gyrus** [FIG. 230]. The lateral stria runs posterolaterally on the margin of the anterior perforated substance, curves round the stem of the lateral sulcus and enters the uncus. Grey matter extends from the olfactory trigone to the uncus along the lateral olfactory stria, and these parts, trigone, uncus, and the grey matter of the stria, constitute the **piriform area** in which the olfactory fibres end.

The **uncus**, a raised area on the medial surface of the temporal lobe, is separated from the remainder of the temporal lobe by the **rhinal sulcus** [FIG. 229]. The posterior surface of the uncus is notched and given a hook-like appearance by the anterior extremity of the **hippocampal sulcus**. The uncus is continuous postero-inferiorly with the **parahippocampal gyrus**, which is separated from the remainder of the temporal lobe by the collateral sulcus, which is often continuous with the rhinal sulcus. The uncus and parahippocampal gyrus form the **inferomedial border** of the hemisphere which lies close to the free margin of the tentorium.

Open the hippocampal sulcus on the right hemisphere and find the **dentate gyrus** in its depths. This gyrus is a minute, transversely ridged structure, which is continuous anteriorly with the uncus. Posteriorly, it sweeps upwards towards the splenium of the corpus callosum on the inferior surface of which it becomes continuous with a small ridge of grey matter, the **gyrus fasciolaris**. This gyrus is continuous with a thin layer of grey matter on the superior surface of the corpus callosum (**indusium griseum**) which can be followed over it, in the callosal sulcus, to reach the paraterminal gyrus inferior to the rostrum of the corpus callosum [FIG. 230]. The indusium griseum is continuous with the cortex of the gyrus cinguli in the depths of the callosal sulcus, and contains two delicate, longitudinal bands of fibres buried in it, the medial and

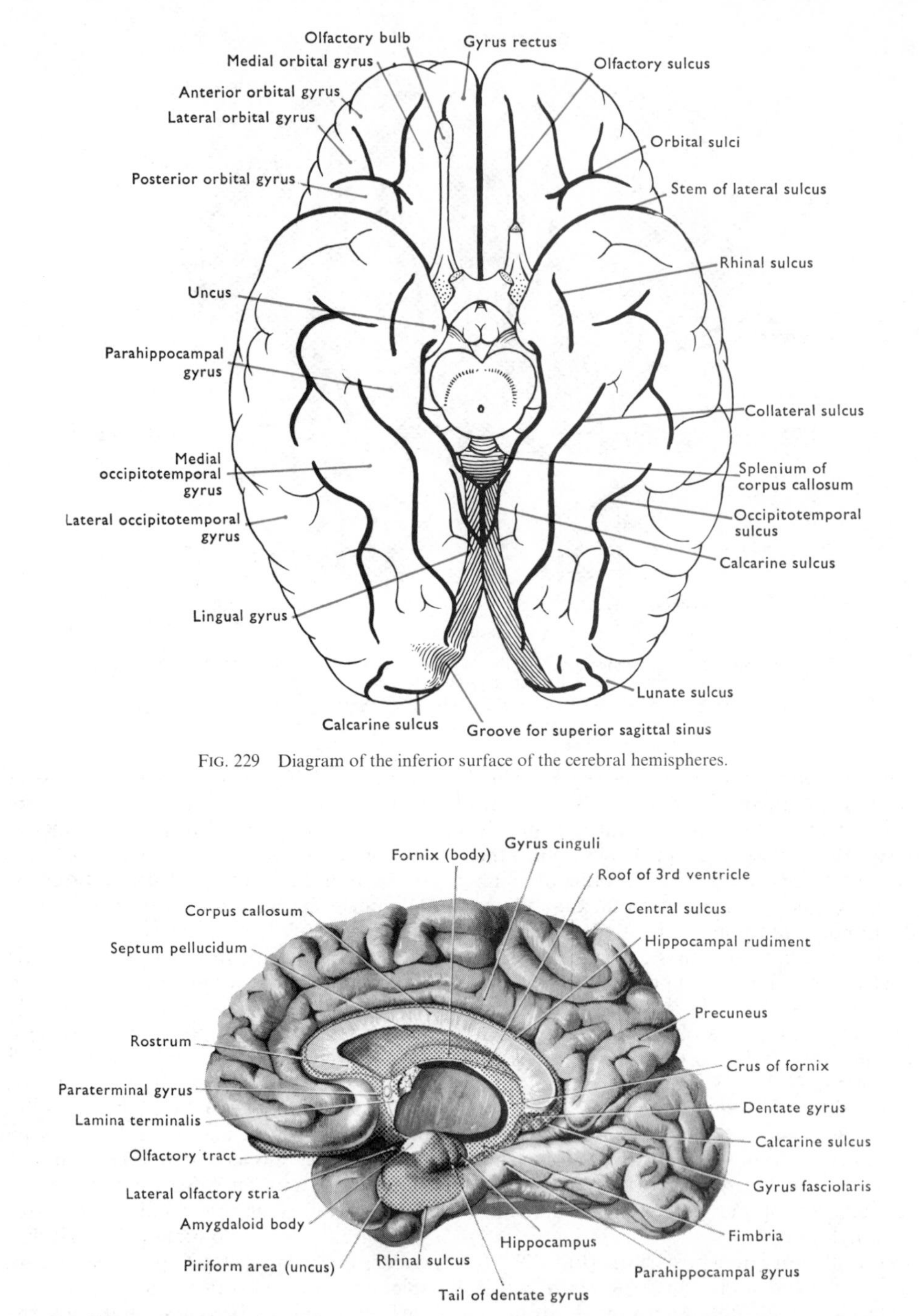

FIG. 229 Diagram of the inferior surface of the cerebral hemispheres.

FIG. 230 Medial aspect of the right cerebral hemisphere with the limbic parts coloured. The olfactory bulb, tract, and lateral stria, the uncus, and the dentate gyrus are shown in red. The hippocampus, supracallosal hippocampal vestige (indusium griseum), paraterminal gyrus, septum pellucidum and medial olfactory stria are shown in blue.

lateral **longitudinal striae**. These run from the hippocampus posteriorly, to the paraterminal gyrus anteriorly.

The olfactory bulb, tract, trigone and striae, together with the uncus, dentate gyrus, hippocampus—which will be seen later—indusium griseum, septum pellucidum, and paraterminal gyrus were called the **rhinencephalon** in the mistaken belief that they were all concerned with olfaction. Only the olfactory trigone and the anterior piriform area are known to receive olfactory fibres direct from the olfactory bulb. This, combined with the large size of the hippocampus in aquatic mammals that have no olfactory bulb, indicates that this ring of cortex, now known as the **limbic system**, has other functions in some mammals at least.

THE WHITE MATTER OF THE CEREBRUM

The white matter of the cerebrum consists of the nerve fibres which lie deep to the cerebral cortex and connect the various parts of the cortex with each other and with the other parts of the central nervous system. In this enormous mass of nerve fibres three different groups may be recognized because of their different connexions. These are: (1) Association fibres, which connect the various parts of the cerebral cortex of one hemisphere with each other. (2) Commissural fibres, which cross the midline and connect the parts of the two hemispheres with each other. (3) Projection fibres, which connect the cerebral cortex with other regions of the central nervous system. All three types are associated with fibres passing in the opposite direction, and all are intermingled in the centre of the hemisphere to form a mass of white matter which underlies the sulci and gyri, but each group forms separate bundles in part of its course, and so may be partly exposed by dissection.

ASSOCIATION FIBRES

These fibres are of all lengths. **Short association fibres** pass from one part of a gyrus to another part of the same gyrus, or they may loop round a sulcus to an adjacent gyrus. Other association fibres run for long distances (*e.g.*, between the frontal and occipital lobes) and it is these long association fibres that form the bundles which can be demonstrated and which will be exposed during the dissection; they are: (1) the cingulum on the medial aspect; (2) the superior longitudinal bundle on the superolateral aspect; (3) the fasciculus uncinatus on the superolateral and orbital aspects; (4) the inferior longitudinal bundle on the tentorial aspect [FIGS. 231, 254].

Cingulum

This is the only association bundle which can be displayed at this stage; the others, together with the projection fibres will be seen later.

DISSECTION. On the right hemisphere, scrape away the grey matter of the gyrus cinguli [FIG. 230] and expose a rounded bundle of white matter lying longitudinally within it. It is simplest to determine the edge of this bundle (the cingulum) by removing the cortex between it and the superior surface of the corpus callosum, and then extend the removal superiorly. The superior surface of the cingulum cannot be defined because nerve fibres enter this surface from the surrounding cortex of the medial aspect of the hemisphere and run in it for variable distances. Expose the medial and inferior surface of the cingulum throughout the length of the corpus callosum and then follow it, anteriorly, towards the anterior perforated substance, and, posteriorly, around the splenium of the corpus callosum and through the length of the parahippocampal gyrus as far as the uncus. In this latter part, the cingulum has a twisted, rope-like appearance. Lift off a portion of the fibres of the cingulum by pushing a blunt instrument between its fibres, and demonstrate its longitudinal nature by tearing these away from the remainder. In this way its continuity into the surrounding cortex will be demonstrated though it is not possible to determine any detailed connexions by this means.

The cingulum is a thick bundle of long association fibres which pass longitudinally between the various parts of the cerebral cortex on the medial aspect of the hemisphere. The cingulum extends from the region of the paraterminal gyrus to the uncus, and thus forms almost a complete circle (cingulum means

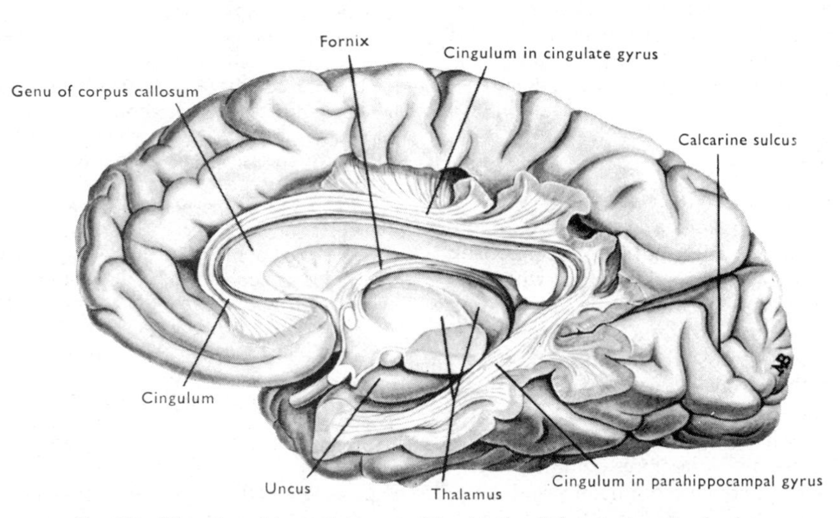

FIG. 231 Dissection of the medial aspect of the right hemisphere to show the cingulum.

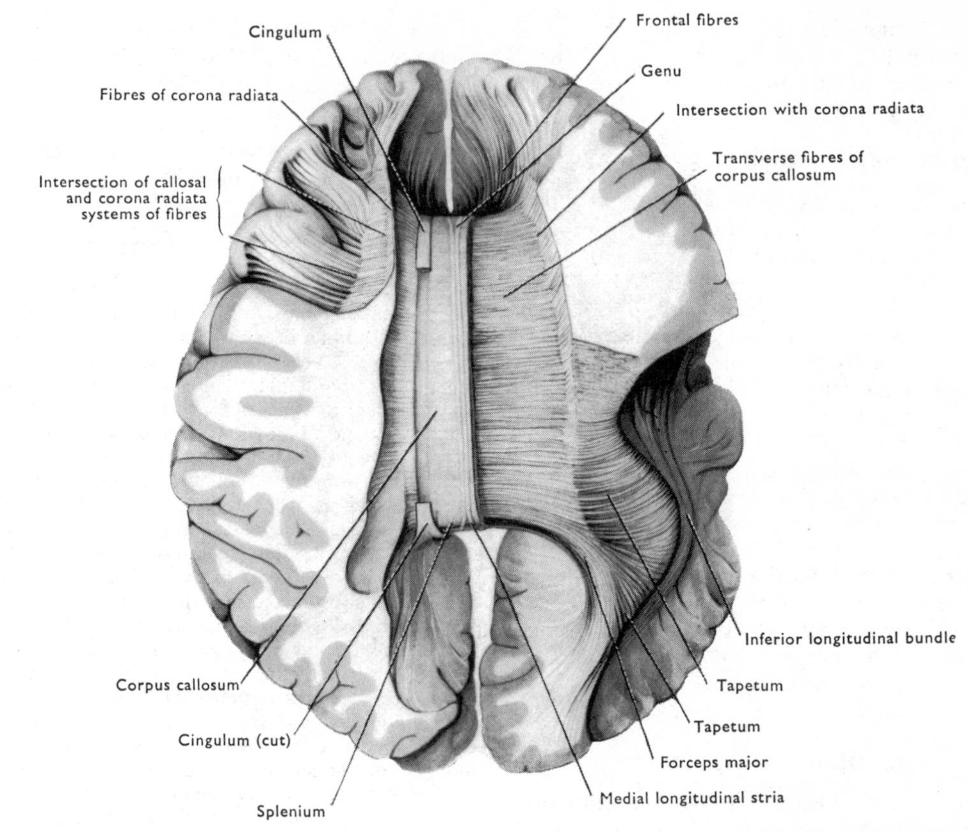

FIG. 232 The corpus callosum exposed from above, with the right half dissected to show the course of some of its fibres.

a girdle) which is capable of connecting all parts of the medial surface of the hemisphere together [FIGS. 231, 232].

COMMISSURAL FIBRES

The greatest number of commissural fibres is found in the corpus callosum, though other commissures of the forebrain are present: (1) the anterior commissure; (2) the optic chiasma; (3) the posterior and habenular commissures in the midbrain-diencephalic junction, at the root of the pineal body; (4) the fornix commissure [p. 219].

Corpus Callosum

This great commissure is formed by nerve fibres which converge on it from the greater part of the cerebral cortex, and then fan out into the opposite hemisphere. The various fibres do not mix with each other, but retain the same relation to the surrounding fibres throughout their course, and are transmitted to the part of the opposite hemisphere which corresponds to that from which they arose. Thus fibres from the medial surface of the hemisphere form U-shaped bundles hooking through the superficial part of the commissure, while fibres from the lateral aspect of the hemisphere lie more deeply and run a more horizontal course [FIG. 284; p. 252].

DISSECTION. On the right hemisphere, lever out the remains of the cingulum from the gyrus cinguli, and pull it away from the remainder of the cortex. This will expose the superficial fibres of the corpus callosum curving into the medial aspect of the hemisphere, and forming, with the corresponding fibres in the opposite hemisphere, two-thirds of a circle, the **forceps minor**. Immediately posterior to the splenium is the deep parieto-occipital sulcus, around which fibres of the splenium have to pass to reach the cuneus [FIG. 222] on the medial aspect of the hemisphere. These fibres should be followed around the parieto-occipital sulcus which forces them further laterally than elsewhere so that they form the **forceps major**. Note that the calcarine sulcus lies immediately inferior to the forceps major, and that it is possible to enter the posterior horn of the lateral ventricle by lifting the cortex of the upper wall of the calcarine sulcus away from the forceps major, for the sulcus and the forceps both indent the medial wall of the posterior horn of the ventricle [FIG. 242]. Make a coronal section through the right occipital lobe immediately posterior to the splenium and confirm the relation of these parts to the lateral ventricle. On the cut surface, a white layer of fibres about 2 mm thick can be seen crossing the roof of the posterior horn of the lateral ventricle and turning down to form its immediate lateral wall. This is the **tapetum**, and it appears whiter than the white matter immediately lateral to it because it is cut parallel to the fibres it contains, while the other fibres are cut transversely. The tapetum consists of those fibres of the splenium which turn inferiorly, as a separate sheet, to pass to the inferior parts of the occipital lobe and to the temporal lobe. Thus the fibres of the splenium which form the forceps major and the tapetum virtually surround the posterior horn of the lateral ventricle [FIGS. 232, 274].

Hold down the splenium of the corpus callosum with one hand, while the other hand pushes the cortex above the splenium laterally. This usually causes the fibres of the white matter to tear along the line of the tapetum. Stop the tearing as soon as the direction of the tapetum is confirmed.

Turn to the genu of the corpus callosum [FIG. 223], push the handle of a knife into it, and lift up a strip of its superficial fibres. Tear these laterally and, if the result is satisfactory, they will be seen to separate from another group of fibres which are passing vertically towards the cerebral cortex, the corona radiata. This appearance is due to the fact that the fibres of the corpus callosum passing horizontally towards the superolateral surface of the hemisphere intersect with the vertically placed fibres of the corona radiata [FIGS. 232, 284].

The corpus callosum unites the medial surfaces of the two cerebral hemispheres for nearly half their anteroposterior length. It lies nearer the anterior than the posterior ends of the hemispheres, and its upper surface forms the floor of the middle part of the longitudinal fissure. The anterior cerebral vessels often lie on the pia mater covering it, and the falx cerebri touches it posteriorly, but does not reach it further forwards where the falx becomes progressively less deep. It is covered on each side by the gyrus cinguli.

The main part or **trunk** of the corpus callosum is thinner than the extremities [FIG. 231]. The posterior end or **splenium** is full and rounded because of the large number of fibres which it transmits from the parietal, occipital, and temporal lobes. It overlies the upper part of the midbrain, and extends posteriorly to the highest part of the cerebellum. Anteriorly the corpus callosum is folded back on itself to form the **genu**; the lower, recurved portion, which thins rapidly as it passes posteriorly, is the **rostrum**. The tip of the rostrum is usually connected by neuroglia to the superior extremity of the lamina terminalis anterior to the anterior commissure [FIG. 223] but occasionally the cavum septi pellucidi is open anteriorly between these two structures.

Every part of the corpus callosum radiates out into the section of the hemisphere which lies opposite it; the rostrum extends inferiorly to the orbital surface; the genu passes to the anterior part of the frontal lobe; the trunk contains fibres to the remainder of the frontal lobe and to the parietal lobe; the splenium carries fibres to the posterior parts of the parietal lobe and to the occipital and temporal lobes. Most of these fibres intermingle with other groups shortly after leaving the corpus callosum and cannot be dissected out separately, but it is possible to define the fibres of the forceps major and minor and of the tapetum [FIG. 232, and p. 252] which remain separate. It is also possible to follow the fibres of the rostrum and inferior part of the genu to the medial and orbital surfaces inferior to it [FIG. 263].

Attached to the inferior surface of the corpus callosum in each hemisphere is the septum pellucidum with the fornix attached to its inferior margin. These two structures form the medial wall and medial part of the floor of the body of the lateral ventricle, while the corpus callosum forms the roof of this part of the lateral ventricle. These parts will be seen when the lateral ventricle is dissected.

In the midline the corpus callosum overlies the thin roof of the median, slit-like third ventricle, but is separated from it by a thin sheet of pia mater, the **tela choroidea of the third ventricle**. This sheet of pia mater extends anteriorly through the **transverse fissure** between the splenium of the corpus callosum and the dorsal surface of midbrain and pineal body. It contains **posterior choroidal** branches of the posterior cerebral arteries, which supply the choroid plexus in the roof of the third ventricle and in the body of the lateral ventricle, and the **internal cerebral veins** which unite to form the great cerebral vein inferior to the splenium [FIG. 246].

Anterior Commissure

This is a round bundle which crosses the median plane in the superior part of the lamina terminals, immediately anterior to the column of the fornix and the interventricular foramen [FIG. 233]. The fibres in this commissure arise mainly in the temporal lobe and will be seen later in the dissection; the remainder unite the olfactory bulbs.

Posterior Commissure [FIGS. 233, 249]

This is a slender bundle which crosses the median plane immediately dorsal to the upper part of the aqueduct and inferior to the root of the pineal body. It is composed mainly of fibres which arise in the midbrain, and it carries fibres from the superior colliculi, the medial longitudinal bundle, and a number of nuclei associated with that bundle in the superior part of the midbrain.

Habenular Commissure

It lies at the root of the pineal body, and is separated from the posterior commissure by the small **pineal recess** of the third ventricle in the root of the pineal. The commissure unites the habenular nuclei in the habenular triangle [FIG. 249] and it has the posterior margin of the thin roof of the third ventricle attached to it [FIG. 233].

PINEAL BODY

This is a small glandular structure which lies between the superior colliculi and is attached to the habenular and posterior commissures by its stalk. It is invaded

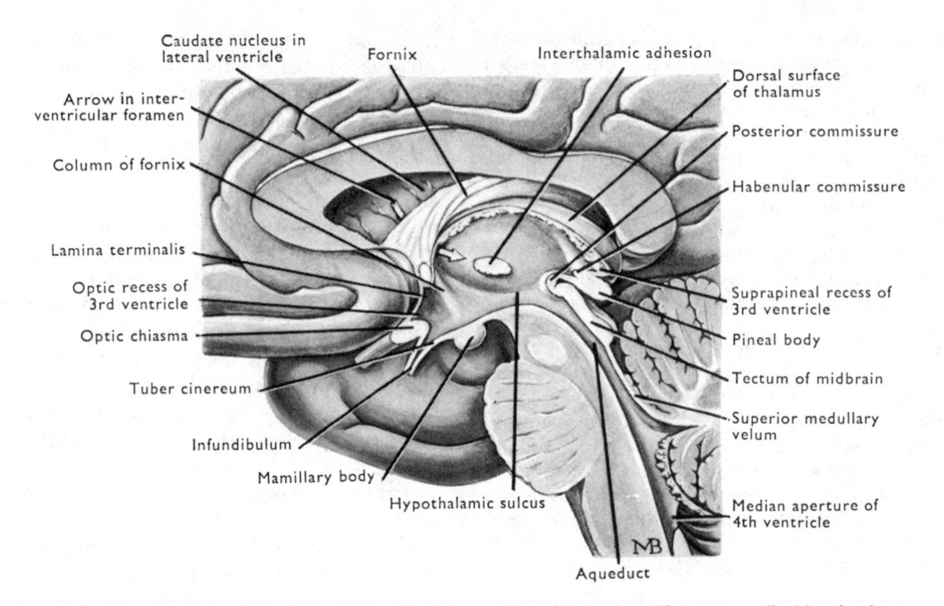

FIG. 233 Median section through the third and fourth ventricles of the brain. The septum pellucidum has been removed to expose the lateral ventricle.

by the pineal recess of the third ventricle, while the thin roof of that ventricle is folded posteriorly over its superior surface to form the **suprapineal recess** of the third ventricle [FIG. 233]. The anterior part of the stalk of the pineal divides into right and left habenulae which pass anteriorly along the medial aspect of the corresponding habenular triangle to become continuous with a narrow, white ridge lying at the junction of the medial and superior surfaces of the thalamus, the **stria medullaris thalami** [FIG. 249]. The lateral edge of the thin roof of the third ventricle is attached to these structures.

The functions of the pineal body are not fully understood, but there is evidence that it produces pharmacologically-active substances, *e.g.*, melatonin and serotonin, and influences gonadal growth in those animals in which this occurs in response to increasing exposure to light. Whatever its functions, the pineal commonly calcifies in middle age, and the ability to visualize it in radiographs may be a simple way of demonstrating a shift of the brain to one side.

The **habenular nuclei** lie in the habenular triangles. These are small triangular areas which lie between the medial aspect of the thalamus, the pineal stalk, and the superior colliculus. The nuclei receive fibres via the stria medullaris thalami, and they give rise to a small bundle of fibres (fasciculus retroflexus) which traverses the upper part of the tegmentum of the midbrain to end in the interpeduncular nucleus, a small collection of cells in the posterior perforated substance.

THE THIRD VENTRICLE
[FIG. 233]

This is the narrow, slit-like cavity which lies in the median plane between the two halves of the diencephalon, and extends from the lamina terminalis anteriorly, to the superior end of the aqueduct and root of the pineal posteriorly.

The **roof** of the ventricle consists of a layer of ependyma which is invaginated on each side by the overlying vascular pia mater to form a minute, linear choroid plexus. The roof is attached on both sides to the stria medullaris thalami, and extends from the interventricular foramen anteriorly, to the habenular commissure posteriorly, looping posteriorly over the superior surface of the pineal body to form the suprapineal recess of the ventricle.

The **floor** extends posteriorly from the **optic recess** on the superior surface of the optic chiasma into the **infundibular recess**, and then passes, above the mamillary bodies and the tegmentum of the midbrain, to the aqueduct.

The anterior commissure indents the **anterior wall** of the ventricle at the superior end of the lamina terminalis. Immediately posterior to this the column of the fornix forms a low ridge in the lateral wall. Posterior to the column of the fornix, and almost hidden by it, is a small, obliquely placed aperture which opens into the lateral ventricle. This is one half of the **interventricular foramen**, and it forms the only communication between the lateral ventricle and the other cavities of the brain. On each side, the narrow strip of choroid plexus in the roof of the third ventricle becomes continuous with the choroid plexus of the corresponding lateral ventricle through the interventricular foramen, and this is often so narrow that it is nearly filled by the choroid plexus. Thus any hypertrophy of the plexus at this situation easily blocks the communication between the two ventricles, causing an increase in pressure in the lateral ventricle. If this is unilateral it may be sufficient to cause compression of the opposite hemisphere and a shift of the midline structures towards that side.

The **lateral wall** of the third ventricle is traversed by a shallow groove (**hypothalamic sulcus**) from the interventricular foramen to the aqueduct. This sulcus separates the thalamus above from the hypothalamus below, and immediately above it the two thalami bulge towards each other so that they frequently meet and fuse, thus forming the **interthalamic adhesion**. This adhesion has no functional significance, is variable in size, and is not a commissure.

THE LATERAL VENTRICLE AND THE CHOROID FISSURE

The lateral ventricle is a C-shaped cavity which extends from its anterior horn in the frontal lobe in a continuous curve posteriorly (central part), then inferiorly, and finally anteriorly, to end in the temporal lobe as the inferior horn. From its convex posterior surface a posterior horn extends backwards to a variable extent into the occipital lobe [FIG. 240]. The size and shape of this ventricle is very variable. In the young, the walls lie almost in apposition, while with increasing age and loss of neural tissue the ventricle expands and may reach a considerable size without any increase in its internal pressure. In young children a prolonged increase of pressure in the ventricles causes the brain and skull to expand (hydrocephalus), but in older children and adults no such expansion is possible and distention of the ventricle can only occur as a result of the loss of nervous tissue.

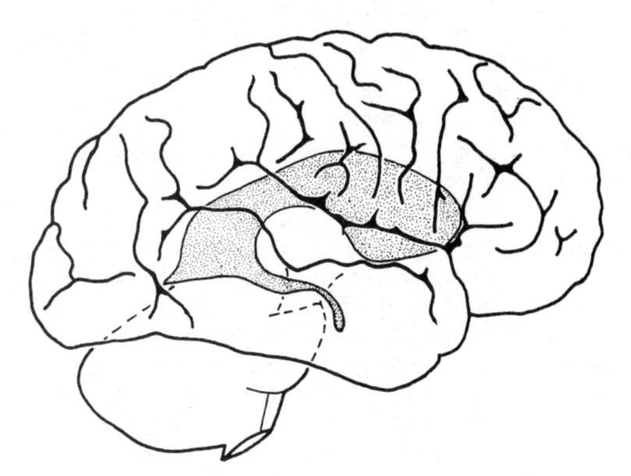

FIG. 234 Drawing to show the position of the lateral ventricle (shaded) in the hemisphere. The posterior horn in this case is particularly small.

The lateral ventricle may be demonstrated radiologically in the living by the introduction of a contrast medium. This is most easily achieved by introducing air into the lumbar subarachnoid space, whence it rises to the cerebellomedullary cistern. Here a considerable amount enters the median aperture of the fourth ventricle, which faces inferiorly. From the fourth ventricle the air rises through the aqueduct to the third ventricle, and passes into the lateral ventricles through the interventricular foramen, which is in the highest part of the third ventricle and lies between the columns of the fornix and the anterior part of the thalamus [FIG. 233].

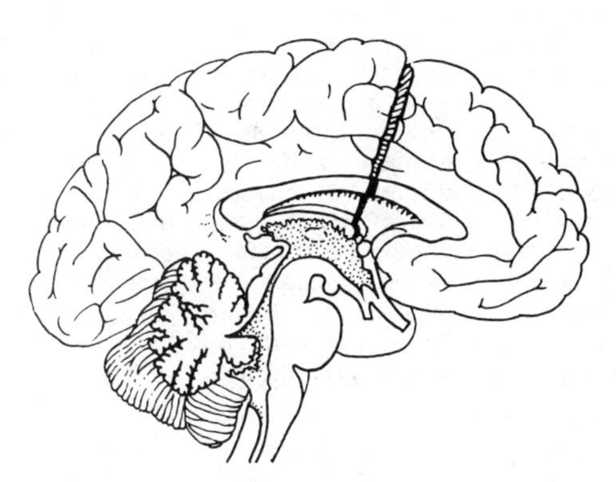

FIG. 235 Drawing to show the first incision to be made in the dissection to expose the lateral ventricle.

The following dissection is illustrated in FIGURES 235, 236, 237, and 238. It has the advantage that it exposes the complete extent of the ventricle by dividing the brain into two parts which may readily be fitted together again so that the form and position of the ventricle and the choroid fissure may be appreciated without doing significant damage to any of the deep structures of the hemisphere. In addition the brain stem remains in continuity with the anterior portion, which greatly facilitates the further dissection of the corpus striatum and the internal capsule. The dissection is, however, difficult and the student should seek the assistance of an experienced demonstrator.

DISSECTION. On the left hemisphere, place the point of a knife in the interventricular foramen and make a vertical cut through the fornix, septum pellucidum, and medial

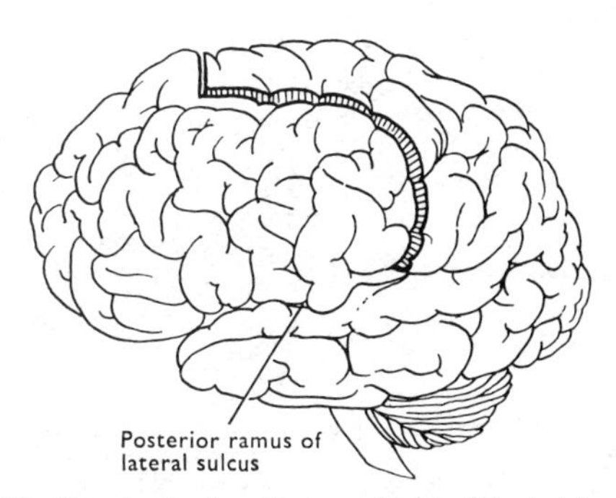

FIG. 236 Drawing to show the second part of the incision to be made in the dissection to expose the lateral ventricle.

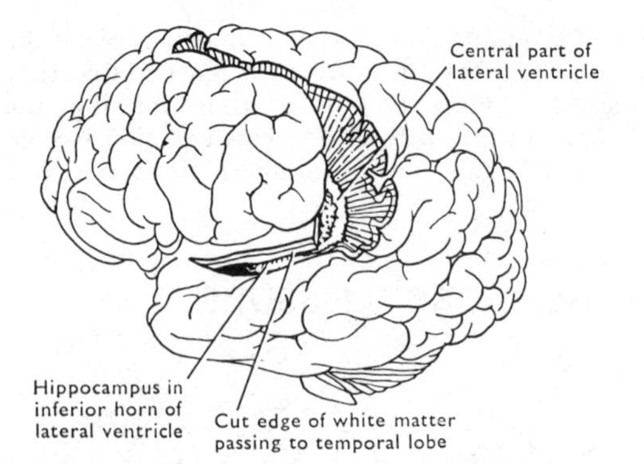

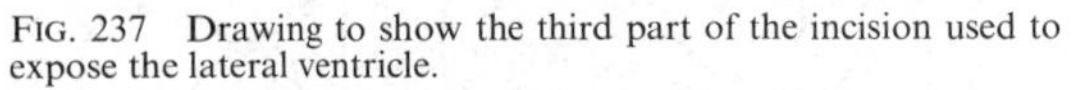

FIG. 237 Drawing to show the third part of the incision used to expose the lateral ventricle.

aspect of the hemisphere as shown in FIGURE 235. Open the cut as it is made, and carry the point of the knife as far as the lateral edge of the lateral ventricle, but avoid cutting into its floor. Now turn the knife so that its edge faces posteriorly and cut backwards and then downwards, following the curve of the lateral edge of the lateral ventricle with the point of the knife as far as the posterior ramus of the lateral sulcus [FIG. 236] and opening the cut as it proceeds. Note the ridge of the choroid plexus protruding into the ventricle from its floor [FIG. 239].

Depress the temporal lobe strongly, exposing the inferior part of the insula. Keeping the knife in the ventricle, and holding it as nearly vertical as possible, cut forwards through the medial part of the transverse temporal gyri on the superior surface of the temporal lobe, and enter the sulcus (circular sulcus) which separates the insula from the temporal lobe. This cut divides the white matter passing horizontally into the temporal lobe and opens into the roof of the inferior horn of the lateral ventricle. Carry the cut anteriorly along the circular sulcus to the stem of the lateral sulcus [FIG. 237] opening the cut and confirming that it enters the inferior horn of the lateral ventricle as you proceed.

Withdraw the knife, and holding the brain with the frontal pole upwards, separate the temporal lobe from the

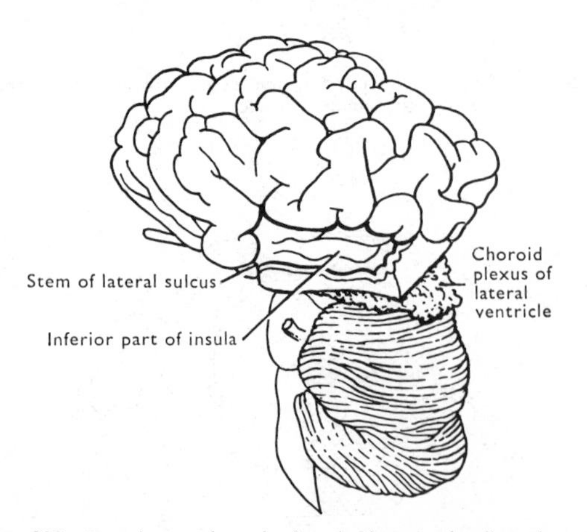

FIG. 238 Drawing to show the fourth phase in the dissection of the lateral ventricle. The third incision has been extended through the uncus into the inferior part of the choroid fissure, and the occipitotemporal part of the brain removed, leaving the choroid plexus of the lateral ventricle attached to the 'frontal' part.

frontal lobe, thus opening the stem of the lateral sulcus. Place the knife in the anterior part of the previous incision, and make a horizontal cut medially through the temporal lobe in the inferior part of the stem of the lateral sulcus, following the anterior edge of the inferior horn of the lateral ventricle. Open this cut as it is made and note the choroid plexus protruding into the inferior horn of the lateral ventricle from its medial wall approximately 1·5 cm from its tip. Cut through the medial wall of the ventricle to the anterior extremity of the choroid plexus in the inferior horn.

The hemisphere is now separated into two parts which are held together by the ependyma which passes over the choroid plexus between them. The two parts are: (1) a 'frontal' part with the brain stem attached to it; (2) an 'occipitotemporal' part which carries the medial parts of the frontal and parietal lobes, the arch of the fornix, fimbria, hippocampus, and the trunk and splenium of the corpus callosum in one piece.

Turn to the medial side, and lifting the fornix away from the superior surface of the thalamus, separate the choroid plexus from the fornix and leave it attached to the thalamus. Slowly separate the occipitotemporal part from the frontal part along this line of the choroid fissure, between the fornix and the thalamus, and note the posterior choroidal branches of the posterior cerebral artery passing to the choroid plexus from the artery as it lies on the inferomedial margin. Divide these to prevent them pulling the choroid plexus from the surface of the thalamus.

When the two parts have been separated, they may be replaced as often as necessary to relate the internal appearances to the surface structures.

CHOROID FISSURE

[p. 252]

This is the fissure through which the two parts of the dissected brain are torn apart. It is the line along which a vascular sheet of pia mater is tucked into the lateral ventricle carrying the ependyma in front of it to form the choroid plexus of that ventricle. No nervous tissue is developed in this narrow strip of the ventricular wall, so that the vascular pia mater is applied directly to the ependymal lining. The choroid fissure follows the line of the C-shaped ventricle. It lies in the concavity of the curve of the **fimbria** and **fornix** [FIG. 244], between the fornix and the thalamus in the central part of the ventricle, and between the fimbria and the roof of the inferior horn of the ventricle in the inferior horn. The choroid fissure is to be distinguished from the transverse fissure [FIG. 222, and p. 213] through which the pia mater (tela choroidea) which forms the choroid plexuses of the third and central parts of the lateral ventricles is continuous with the pia mater on the surface of the midbrain.

Note the position of the choroid plexus of the lateral ventricle, and follow it from the interventricular foramen into the inferior horn of the ventricle. In the central part of each lateral ventricle,

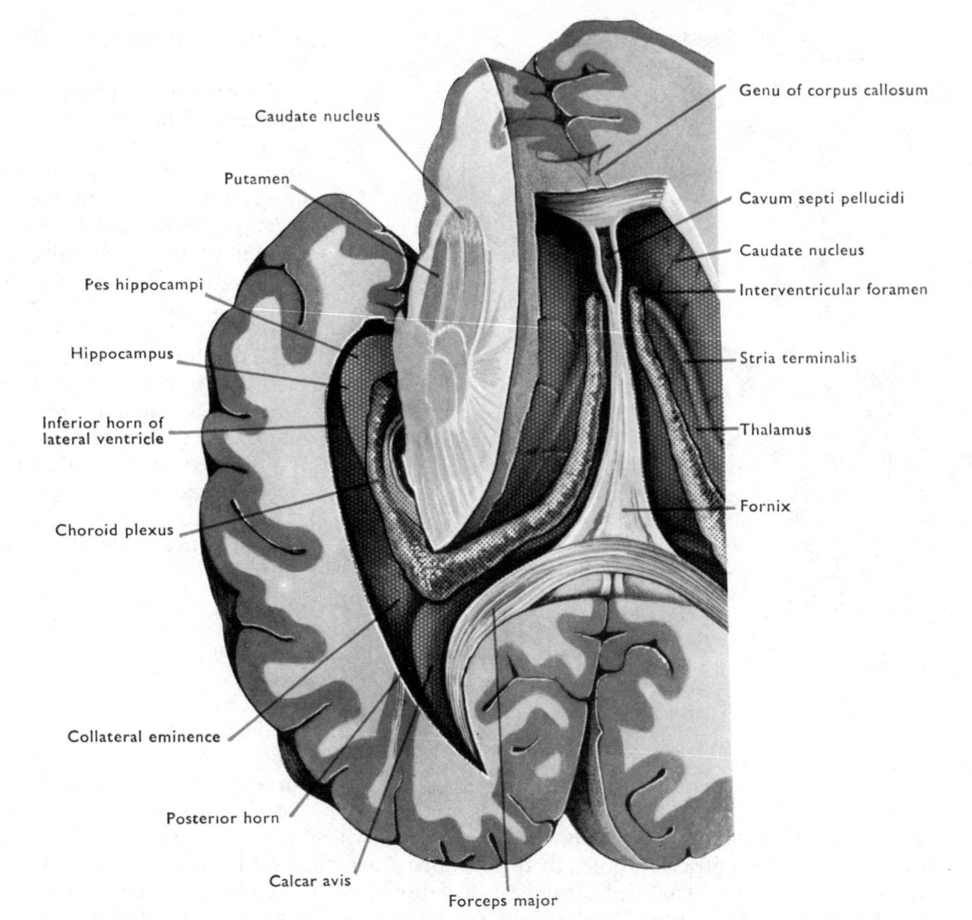

FIG. 239 Dissection from above to show the lateral ventricle. Ependyma, blue; choroid plexus, red. It should be appreciated that the choroid plexus is covered with ependyma on its ventricular aspect, but this is not shown.

the plexus lies at the lateral extremity of a sheet of pia mater—**tela choroidea of the lateral ventricle** [FIG. 246] which covers the dorsal surface of the thalamus and the roof of the third ventricle, and which widens posteriorly as each lateral ventricle and choroid plexus move further from the median plane. As the ventricles turn downwards, this pia mater is continuous with the pia mater on each side of the upper midbrain which also forms tela choroidea where the inferomedial margin of the hemisphere lies against the midbrain. Beyond this, each fissure passes forwards to the posterior margin of the uncus, parallel to the optic tract and between it and the margin of the hemisphere [FIG. 243].

LATERAL VENTRICLE

The **anterior horn** of the lateral ventricle curves inferiorly into the frontal lobe from the interventricular foramen. It is triangular in coronal section. The narrow floor is formed by the rostrum of the corpus callosum; the roof and anterior wall by the trunk and genu of the corpus callosum; the vertical medial wall by the septum pellucidum and column of the fornix; the lateral wall

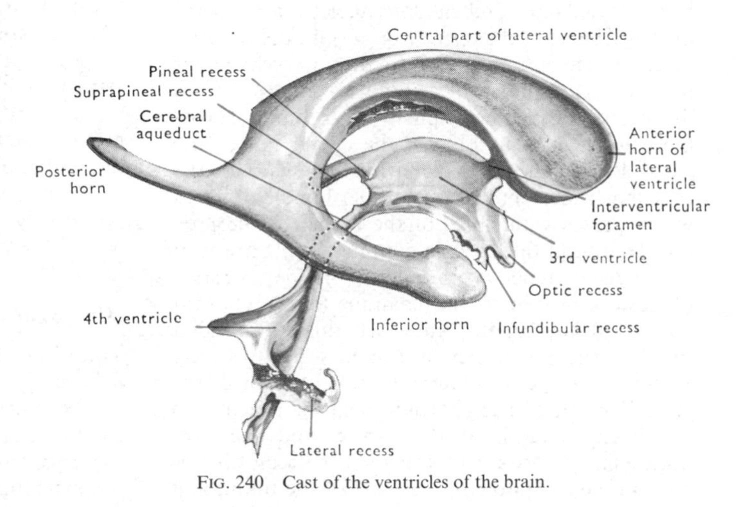

FIG. 240 Cast of the ventricles of the brain.

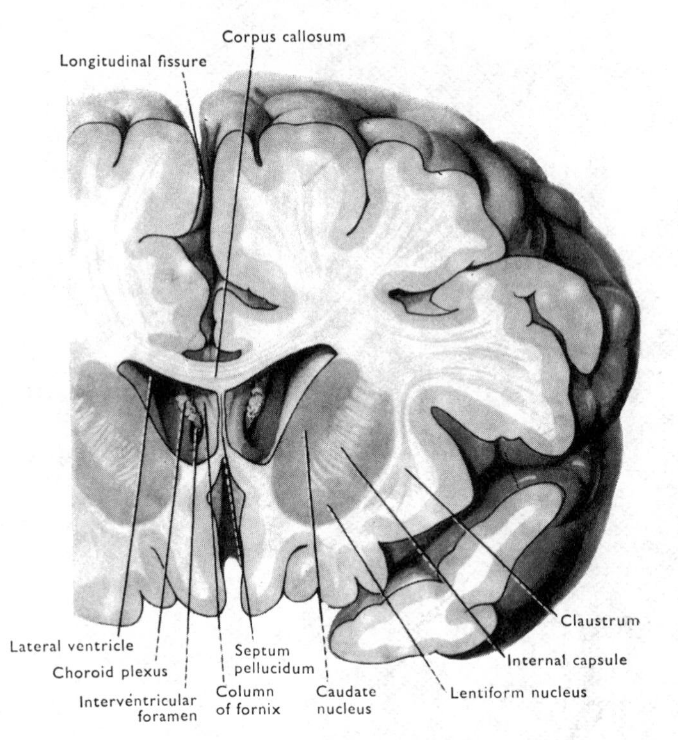

FIG. 241 Coronal section through the hemispheres behind the genu of the corpus callosum, seen from in front. Cf. FIGS. 242, 263.

by the bulging head of the caudate nucleus [FIG. 241].

The **central part** of the ventricle is roofed by the trunk of the corpus callosum. The medial wall is formed by the fornix and septum pellucidum anteriorly, and by the fornix posteriorly, and becomes less extensive as it is followed posteriorly [FIGS. 265, 268].

The floor consists from lateral to medial of the following structures: (1) The **caudate nucleus** which lies in the angle between the floor and the roof, and narrows rapidly as it is traced posteriorly. (2) The **thalamostriate vein** runs anteriorly in the groove between thalamus and caudate nucleus, and passes medially beneath the ependyma to join the internal cerebral vein just posterior to the interventricular foramen. A number of tributaries enter it from the centre of the hemisphere by running across the caudate nucleus outside the ependyma of the ventricle. (3) The **stria terminalis** runs with the thalamostriate vein and is a slender bundle of fibres which passes with fibres of the fornix to the grey matter around the anterior commissure. It arises in the amygdaloid body in the uncus. (4) A narrow strip of the dorsal surface of the thalamus. (5) The choroid plexus. (6) The **fornix**. Anteriorly this is a rounded bundle, but posteriorly it becomes progressively flattened and extends laterally into the floor of the lateral ventricle. The choroid plexus is attached to the lateral margin of the fornix, and the torn ependyma (**taenia fornicis**) can be seen on this edge in the occipitotemporal part of the brain, and on the thalamus of the frontal part.

The **posterior horn** begins at the splenium of the corpus callosum, and extends posteriorly into the occipital lobe, tapering to a point. The roof, lateral wall, and floor are formed by a sheet of fibres (tapetum) from the splenium of the corpus callosum which arches over it and passes inferiorly to the lower parts of the occipital lobe [FIG. 242]. The medial wall is invaginated by two ridges; the upper of these (**bulb of the posterior horn**) is formed by the fibres of the forceps major. The lower ridge (**calcar avis**) is produced by the calcarine sulcus which extends deeply into the medial surface of the occipital lobe.

The **inferior horn** is the direct continuation of the ventricular cavity into the temporal lobe. It runs inferiorly, posterior to the thalamus, and then passes anteriorly, curving medially to end at the uncus. The lateral wall is formed by the tapetum of the corpus callosum. The roof, which can be seen on the inferior surface of the 'frontal' part of the brain, consists of the white matter which passes laterally into the temporal lobe (including sublentiform part of the internal capsule) with the stria terminalis and the tail of the caudate applied to it. Both these structures can be followed from the floor of the central part of the ventricle, and both become continuous with the amygdaloid body at the tip of the inferior horn. The floor is broad posteriorly where the inferior and posterior horns meet, and is often raised (**collateral triangle**) by the collateral sulcus. Anteriorly its floor narrows and its medial part is formed by a convex ridge produced by the **hippocampus** covered by a layer of nerve fibres (the **alveus**) which passes medially from the hippocampus to form a ridge on its medial border (**fimbria** of the hippocampus).

The **amygdaloid body** is an oval mass of grey matter which overlies the tip of the inferior horn of the ventricle in the uncus, and is continuous medially with the cortex of the temporal lobe [FIG. 264], posteriorly with the caudate nucleus, and superiorly with the lentiform nucleus.

Hippocampus

Traced anteriorly the hippocampus turns medially near the tip of the inferior horn of the lateral ventricle and becomes continuous with the uncus. This part of the hippocampus is ridged and known as the **pes hippocampi** [FIG. 244]. Traced posteriorly the hippocampus virtually disappears at the splenium of

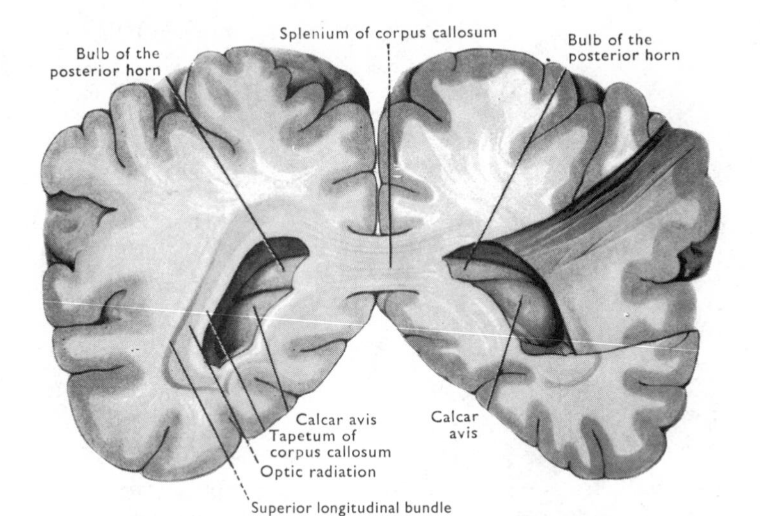

FIG. 242 Coronal section through the posterior horns of the lateral ventricals.

the corpus callosum, but is continuous with a small ridge which curves round the inferior surface of the splenium, the **gyrus fasciolaris** [FIG. 230].

Fimbria of Hippocampus. This narrow strip of white matter on the medial margin of the hippocampus overlies the dentate gyrus. It is formed by the fibres of the alveus which mainly arise in the hippocampus. The medial margin of the fimbria is sharp and has the ependyma reflected off the choroid plexus attached to it, thus forming one margin of the choroid fissure. It is continuous anteriorly with the recurved tip of the uncus, and posteriorly with the **crus of the fornix** [FIG. 244]. The alveus, fimbria, and fornix form the efferent pathway for the cells of the hippocampus, but they also contain fibres which pass from the cells in the region of the anterior commissure (**septal nuclei**) to the hippocampus. The fibres leaving the hippocampus pass: (1) To the opposite hippocampus through the **commissure of the fornix**, which is situated where the crura of the fornix meet inferior to the corpus callosum [FIGS. 239, 246]. (2) To the septal and anterior hypothalamic region. (3) To the mamillary body. In addition, a few of its fibres may pass direct to the anterior nuclei of the thalamus.

Dentate Gyrus

This is a narrow ribbon of grey matter which lies in the medial concavity of the hippocampus, inferior to the fimbria of the hippocampus. It is readily recognized by the transverse notches which divide it into a number of 'teeth'. Posteriorly it is continuous with the gyrus fasciolaris, and through it with the thin layer of grey matter on the superior surface of the corpus callosum, the **indusium griseum**. Anteriorly the dentate gyrus passes into the cleft of the uncus, reappearing as the narrow **tail of the dentate gyrus** which fades out as it passes superiorly over the recurved portion of the uncus [FIG. 244].

THE EDGE OF THE HEMISPHERE

The hippocampus, dentate gyrus, indusium griseum, septum pellucidum, and fornix which form the edge (or limbus) of the hemisphere were long thought to be part of the rhinencephalon or 'smell brain' [p. 210].

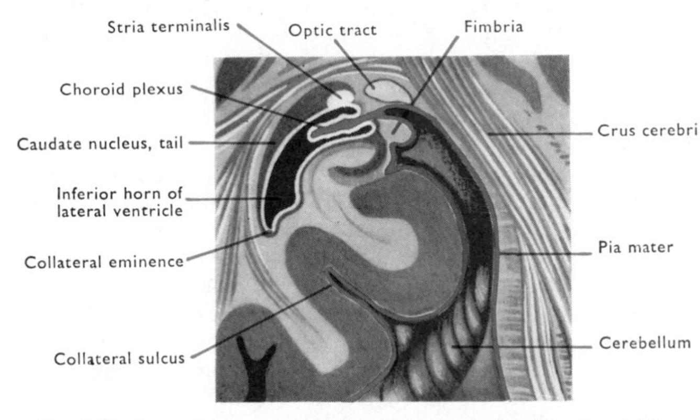

FIG. 243 Coronal section through the inferior horn of the lateral ventricle.

Septum Pellucidum

This thin, vertical partition consists of two parallel laminae, each of which connects the fornix to the corpus callosum and fills the interval in the concavity of its genu. It varies directly in depth and length with the size of the lateral ventricle, and may extend as far posteriorly as the splenium of the corpus callosum or only for a short distance beyond the interventricular foramen. It lies between the anterior horns and central parts of the two lateral ventricles, and its two laminae may be separated by the **cavity of the septum pellucidum.**

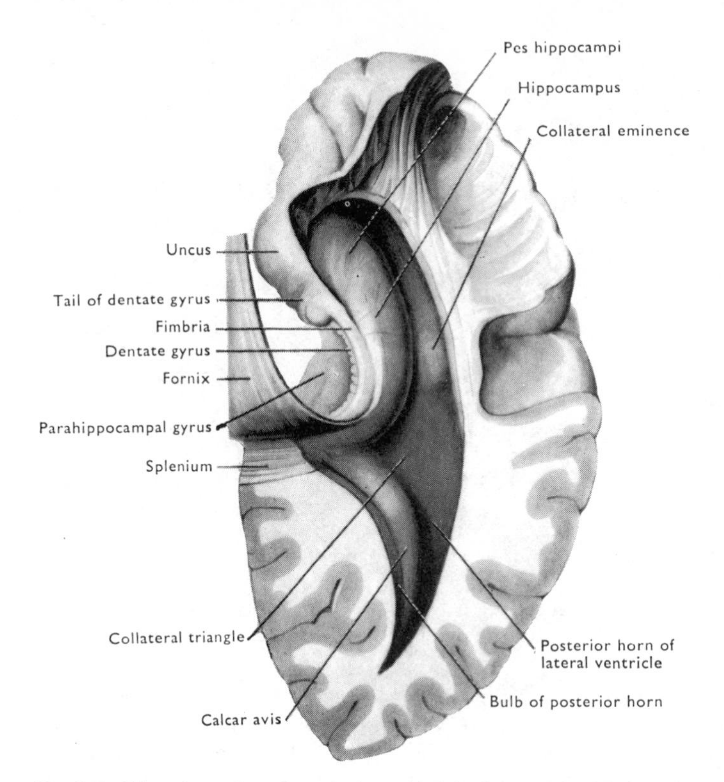

FIG. 244 Dissection to show the posterior and inferior horns of the right lateral ventricle, viewed from above.

In lower mammals it is thicker and contains a number of septal nuclei which are found around the anterior commissure ventral to the septum in Man.

Fornix

The fornix is a paired structure, one of the pair in each hemisphere, but the two are so closely fused beneath the middle of the trunk of the corpus callosum that they are usually described as a single structure. Each half of the fornix is composed of fibres which enter it from the fimbria, and together with that structure the fornix forms almost one complete turn of a spiral. The **fimbria** begins close to the uncus and runs posterosuperiorly to enter the **crus of the fornix**, a flattened strip which arches upwards, medially, and forwards under the splenium of the corpus callosum to become continuous with the body of the fornix. The body is formed where the two crura fuse, but anteriorly they become more cylindrical and separate to form the **columns of the fornix**. These sweep inferiorly between the anterior commissure and the interventricular foramen, and curving posteriorly through the anterior hypothalamus, end in the mamillary body close to the midline and medial to the beginning of the fimbria. This spiral bundle of fibres forms the outer margin of the **choroid fissure** and the edge (limbus) of the telencephalon between the beginning of the fimbria and the interventricular foramen. It therefore has the ependyma of the choroid plexus attached to its sharp lateral margin throughout this length, and it is separated from the inner margin of the choroid fissure (mainly thalamus) by the tela choroidea passing laterally into the choroid plexus.

The crura and body of the fornix groove the posterior and superior surfaces of the thalamus respectively, and they lie in the medial part of the floor of the lateral ventricle. The columns of the fornix form slight ridges on the lateral wall of the third ventricle inferior to the interventricular foramen.

DISSECTION. Make two parallel cuts through the ependyma of the lateral wall of the third ventricle, one on each side of the column of the fornix between the interventricular foramen and the mamillary body, in the 'frontal' part of the left hemisphere. Remove the ependyma between these, and scrape away sufficient of the underlying grey matter to expose the column of the fornix and follow it to the mamillary body.

At the mamillary body avoid injury to another bundle which appears to arise from the anteromedial part of the mamillary body and pass laterally across the fornix into the wall of the third ventricle. Cut through the ependyma on each side of this mamillothalamic tract, and follow it first laterally and then superiorly to the anterior tubercle of the thalamus, which contains the anterior nuclei of the thalamus.

Mamillothalamic Tract [FIG. 245]

If the mamillary body is dissected free from the floor of the third ventricle and only the column of the fornix and the mamillothalamic tract are left connected to it, it gives the appearance of being a loop on the fornix through which it becomes continuous with the mamillothalamic tract. In fact the fibres of the column of the fornix end in the nuclei of the mamillary body, and new fibres arise from its cells to form the mamillothalamic tract. The mamillothalamic tract is a compact bundle which passes laterally and then superiorly to the **anterior nuclei of the thalamus**. On the way its fibres branch to give rise to a diffuse bundle passing inferiorly to the tegmentum of the midbrain (the **mamillotegmental tract**).

The impulses discharged by the hippocampus into the fimbria and fornix reach the anterior nuclei of the thalamus (which project to the **gyrus cinguli**) through the mamillothalamic tract, and can also have an effect on the brain stem mechanisms by way of the mamillotegmental tract. Some fibres which arise in the gyrus cinguli enter the **cingulum** and

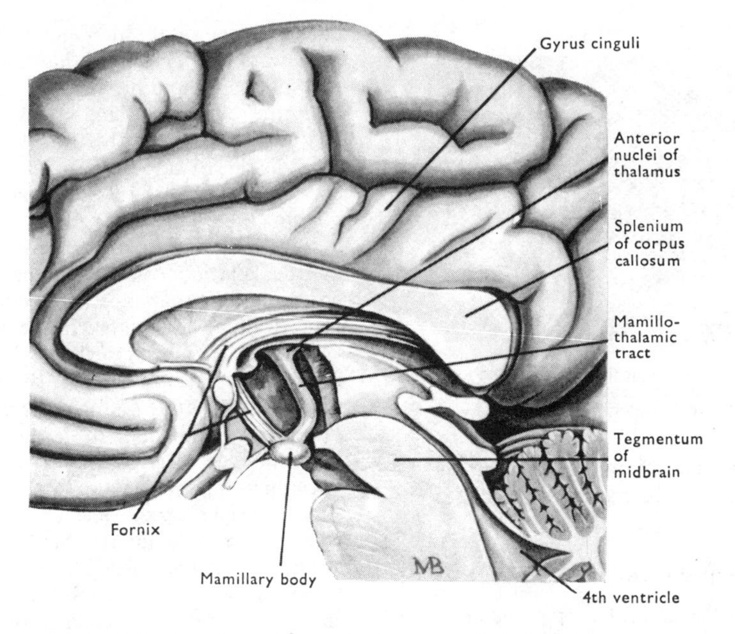

FIG. 245 Dissection to show the fornix and mamillothalamic tract.

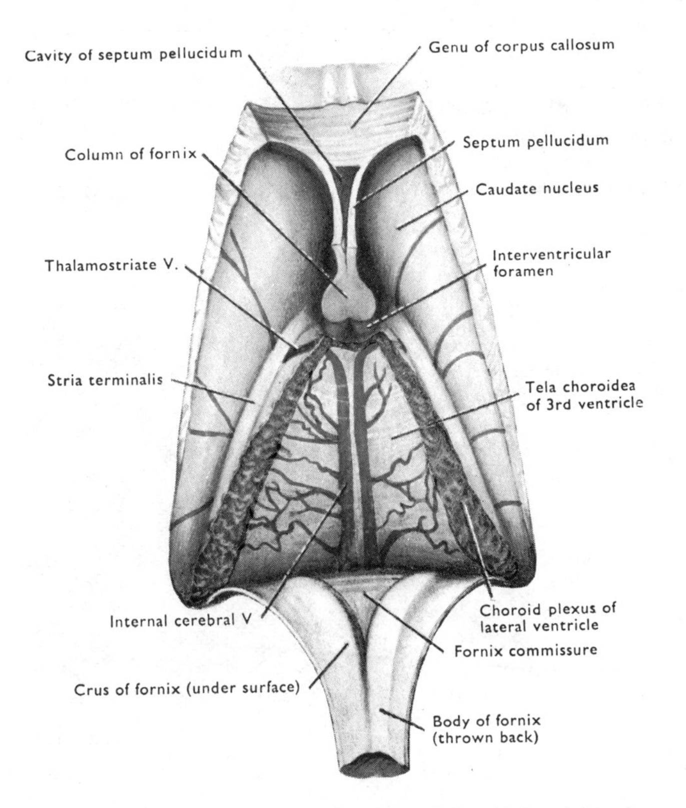

FIG. 246 Dissection of the tela choroidea of the third and lateral ventricles. It is exposed from above by dividing the columns of the fornix and turning them backwards.

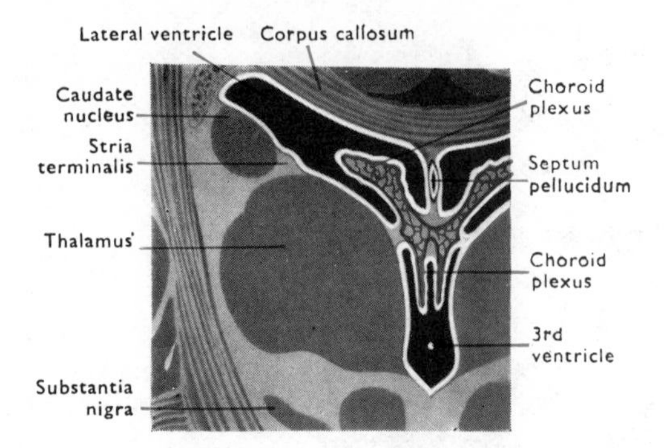

FIG. 247 Coronal section through the thalamus and associated structures just posterior to the interventricular foramen. Note the choroid plexuses of the third and lateral ventricles.

pass through it to the parahippocampal gyrus and the hippocampus, thus producing a feed-back to the hippocampus.

Tela Choroidea and Internal Cerebral Veins

The tela choroidea is a fold of pia mater which extends anteriorly through the transverse fissure [FIG. 222] below the splenium of the corpus callosum. It passes forwards inferior to the fornix and superior to the roof of the third ventricle and the dorsal surface of the thalamus. The fold contains the internal cerebral veins and posterior choroidal branches of the posterior cerebral arteries [FIG. 246]. These vessels supply the vascular choroid plexuses of the central parts of the lateral ventricles, into which the lateral margins of the tela are invaginated beyond the lateral edges of the fornix. The tela choroidea narrows anteriorly as the two lateral ventricles and their choroid plexuses approach one another, and it ends in an apex at the interventricular foramen where the choroid plexus of each lateral ventricle becomes continuous around the anterior extremity of the tela with the corresponding narrow strip of choroid plexus in the roof of the third ventricle. These two strips lie side by side on the inferior surface of the tela choroidea close to the midline, and are supplied by its contained blood vessels.

Each **internal cerebral vein** begins at the apex of the tela choroidea by the junction of the corresponding thalamostriate vein with the choroid vein draining the choroid plexus of the lateral ventricle. The two internal cerebral veins run posteriorly, side by side, in the roof of the third ventricle, and unite to form the great cerebral vein inferior to the splenium of the corpus callosum. This vein emerges through the transverse fissure, and sweeping superiorly on the splenium with the superior layer of the tela choroidea of the third ventricle, joins the inferior sagittal sinus to form the straight sinus. The inferior layer of the tela choroidea becomes continuous with the pia mater covering the surface of the midbrain, and it forms a sheath for the pineal body.

DISSECTION. On the 'frontal' part of the left hemisphere, remove the tela choroidea of the third ventricle by dividing the thalamostriate vein where it joins the internal cerebral vein, and pulling the tela posteriorly. Note that the ependyma covering the choroid plexus has a linear attachment to the superior surface of the thalamus, and identify again its torn margin on the lateral edge of the fornix in the 'occipitotemporal' part of the left hemisphere.

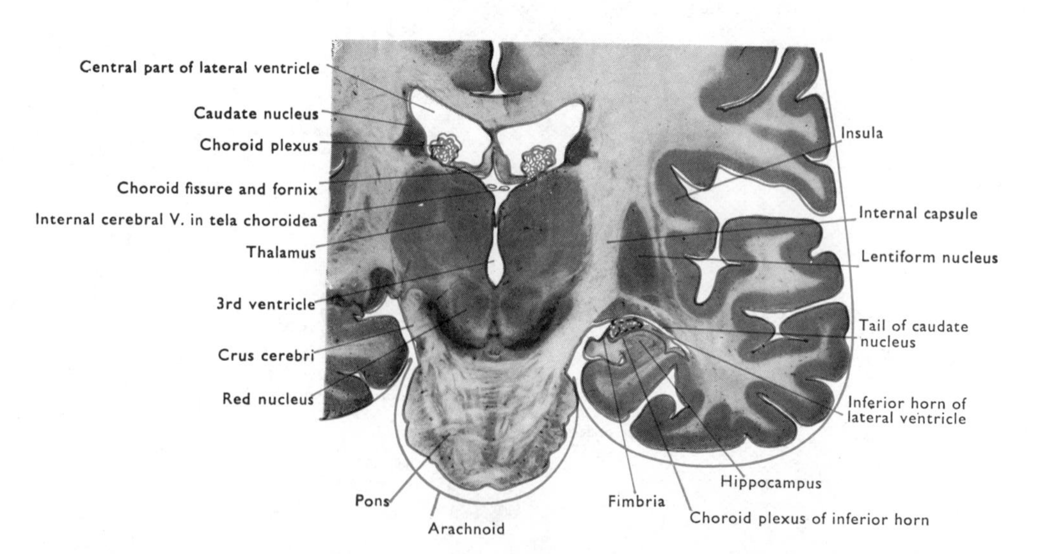

FIG. 248 Coronal section through cerebrum, midbrain, and pons to show the arrangement of the pia mater, and the choroid plexuses of the lateral ventricle in semi-diagrammatic form. Pia mater, red; arachnoid mater, ependyma, and internal cerebral veins, blue.

THE THALAMI AND THE OPTIC TRACTS

THALAMUS

The thalamus is a large mass of grey matter which lies obliquely across the path of the cerebral peduncle as it ascends into the hemisphere. It is the great sensory relay station for the cerebral cortex, and some of its cells transmit ascending sensory impulses to the sensory receptive areas of the cortex. In addition it has reciprocal connexions with most parts of the cerebral cortex and a number of subcortical masses of grey matter.

The **anterior two-thirds** of the medial surfaces of the two thalami are covered with ependyma and form the lateral walls of the third ventricle superior to the hypothalamic sulcus [FIG. 233]. In this part they frequently fuse across the cavity of the ventricle to form the **interthalamic adhesion**. The posterior thirds are more widely separated by the superior part of the midbrain and the pineal body [FIG. 249] and are not covered with ependyma.

The **posterior end** of the thalamus (the **pulvinar**) is wide and overhangs the superior part of the midbrain. Laterally it is grooved by the crus of the fornix, and beyond that it forms a small part of the floor of the lateral ventricle. This part is limited laterally by the slender **stria terminalis** which courses round the lateral margin of the thalamus from the amygdaloid body to the septal nuclei in company with the thalamostriate vein. Lateral to these is the caudate nucleus which forms a ridge in the lateral part of the floor in the central part of the lateral ventricle.

The **superior surface** is limited laterally by the stria terminalis and thalamostriate vein, and it tapers anteriorly to end in the rounded **anterior tubercle** of the thalamus just posterior to the interventricular foramen and the column of the fornix. The tubercle is formed by the **anterior nuclei** of the thalamus which receive the mamillothalamic tract and project to the cingulate gyrus of the cerebral cortex. The groove for the fornix courses obliquely across the superior surface of the thalamus, and the area of this surface lateral to the groove is covered with the ependyma of the floor of the lateral ventricle, while the groove and the area medial to it are covered by the tela choroidea of the lateral ventricle. Where the superior and medial surfaces meet there is a delicate ridge of white matter coursing anteroposteriorly on the thalamus to end in the **habenular triangle** [FIG. 249]. This **stria medullaris thalami** arises in the septal nuclei. It has the thin roof of the third

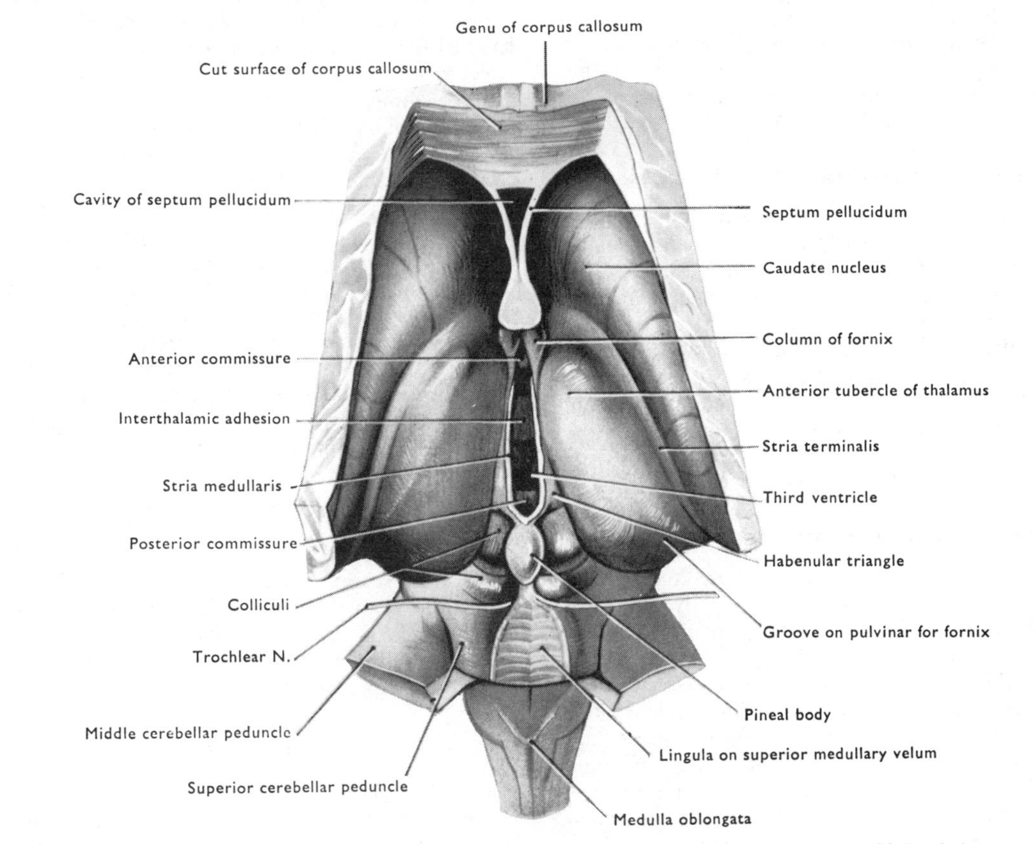

FIG. 249 The thalami and third ventricle seen from above after removal of the overlying tela. Cf. FIG. 246.

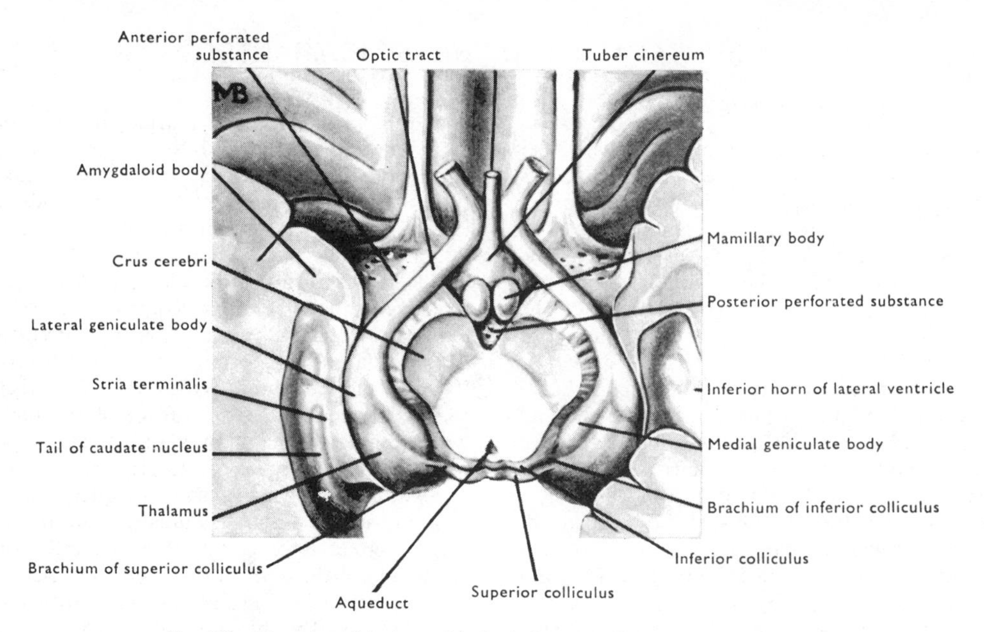

FIG. 250 Dissection of the base of the brain from below to show the optic tracts.

ventricle attached to its medial margin posterior to the interventricular foramen.

The **lateral surface** of the thalamus is hidden at present, but it lies in contact with the internal capsule [FIGS. 266, 268].

The **inferior surface** is mostly hidden for it rests on the hypothalamus. The posterior part of this surface is free and exhibits two swellings, the medial and lateral geniculate bodies [FIG. 250].

Medial Geniculate Body [FIGS. 194, 250]

This is a well defined, oval mass which lies in the angle between the inferior surface of the thalamus, the lateral aspect of the midbrain, and the posterior surface of the crus cerebri. It is the thalamic nucleus which receives the **brachium of the inferior colliculus**. This is a well defined ridge which passes obliquely across the lateral surface of the midbrain from the inferior colliculus to the medial geniculate body, and conveys auditory impulses to the geniculate body for transmission to the superior temporal gyrus of the cerebral cortex.

Lateral Geniculate Body [FIGS. 194, 250]

This body is concave inferiorly [FIG. 268] and so forms a pair of swellings separated by a groove (hilus), anterolateral to the medial geniculate body. It is directly continuous anteriorly with the **optic tract** which appears to split into two roots as it enters the body because of the hilar sulcus. The appearance of one root joining the medial geniculate body is illusory. The body forms the thalamic termination of the fibres of the optic tract, and gives rise to the fibres of the **optic radiation** which convey visual impulses to the occipital lobe of the cerebral cortex. Branches of the fibres in the optic tract bypass the lateral geniculate body, and running over the inferior surface of the thalamus, lateral to the medial geniculate body, enter the superolateral margin of the superior colliculus. These fibres—the **brachium of the superior colliculus**—are concerned with the production of visual reflexes such as the turning of the head and eyes towards a sudden flash of light and the constriction of the pupil when the retina is exposed to light. This brachium differs from the brachium of the inferior colliculus in that it is solely concerned with reflexes, is not part of the pathway concerned with vision, and conducts impulses from the retinae to the midbrain.

The geniculate bodies are, therefore, the thalamic nuclei concerned with the auditory and visual pathways, and are essential in the transfer of the corresponding impulses to the cerebral cortex. They differ from each other in that the lateral geniculate is the only route through which impulses generated by light in the opposite half of the field of vision can reach the cerebral cortex, while each medial geniculate body carries auditory impulses originating in both cochleae. Thus destruction of one lateral geniculate produces blindness in the opposite half of the field of vision, while destruction of one medial geniculate body has little effect on hearing. The colliculi are concerned with reflex activities triggered by auditory (inferior colliculus) and visual (superior colliculus) impulses.

OPTIC TRACTS

Each optic tract begins at the posterolateral corner of the optic chiasma. It passes posterolaterally between the anterior perforated substance and the tuber cinereum, and then lies superior to the medial aspect of the temporal lobe on the lateral aspect of the crus cerebri at its junction with the internal capsule. On the crus cerebri, the optic tract is close to the choroid fissure [FIG. 243] of the inferior horn of the lateral ventricle, and is crossed by the anterior choroid branch [FIG. 188] of the internal carotid artery on its way to the fissure, and by the basal vein passing around the midbrain to the great cerebral vein.

Each optic tract enters the corresponding lateral geniculate body, appearing to divide into two **roots** (see above). The medial root seems to pass to the medial geniculate body, but the only fibres which continue posterior to the lateral geniculate body are those of the brachium of the superior colliculus which run in the interval between the two 'roots' to the superior colliculus.

Each optic tract contains nerve fibres which originate from the temporal half of the ipsilateral retina and the nasal half of the contralateral retina (the fibres from the nasal halves of the retinae crossing in the chiasma). These are the parts of the retinae which receive light from the opposite half of the field of vision. Hence destruction of one optic tract is followed by blindness in the opposite half of the field of vision—an *homonymous hemianopia*. Because fibres from the corresponding parts of the two retinae lie together in the tract, partial damage to the tract leads to partial loss of vision involving the same part of the half field in both eyes (see also optic radiation). By contrast, destruction of one optic nerve causes blindness in that eye, while median destruction of the chiasma produces a bitemporal hemianopia by severing the fibres from the nasal halves of both retinae as they cross—the nasal halves of the retinae receiving light from the temporal half of the field of vision in each eye. Again, because fibres from the parts of the nasal halves of the retinae cross in the corresponding parts of the chiasma, damage to the upper part of the chiasma in the midline interrupts nerve fibres from the upper nasal parts of both retinae—the parts which receive light from the lower temporal fields of vision—producing a lower bitemporal visual defect. Similarly, damage to the lower part of the chiasma produces an upper bitemporal visual defect.

Nerve fibres in the brachium of the superior colliculus reach that colliculus and the **pretectal region** anterior to it. The superior colliculus produces its effects through the **tectoreticular** and **tectospinal tracts** which descend contralaterally from it, anterior to the medial longitudinal bundle [FIGS. 197–202, 212, 213]. Each pretectal region sends fibres to the **accessory nuclei** of both **oculomotor nerves**, whence preganglionic parasympathetic fibres pass in the oculomotor nerves to the ciliary ganglia. These ganglia innervate the sphincter muscles of the **pupils**, causing them to contract even when light shines in only one eye (*consensual response*). Constriction of the pupils when the eyes are converged and focused on a near object (*convergence-accommodation reflex*) probably uses the same peripheral pathway, but other central pathways are involved including the cerebral cortex.

CORTICAL AREAS

For blood supply, see FIGURES 190 and 216.

Before proceeding to the deep dissection of the hemisphere, the areas of the cerebral cortex should be reviewed with special reference to those regions in which particular functions have been localized [pp. 203–211, and FIGS. 219, 220, 226, 230].

THE DEEP DISSECTION OF THE HEMISPHERE

The following dissections are carried out on the left hemisphere on which the lateral ventricle has already been exposed; the right hemisphere is retained so that sections of it can be cut at a later phase. The use of the left hemisphere for these dissections facilitiates the exposure of certain structures, but makes it difficult to see the entire extent of certain of the association bundles and of the corona radiata. This can be overcome by placing the two parts of the hemisphere together and continuing the dissection from the 'frontal' part into the 'occipitotemporal' part of the divided left hemisphere.

INSULA

The insula is that part of the cerebral cortex which is submerged in the lateral sulcus. It is roughly triangular in outline, and its margin is formed by the **circular sulcus** of the insula, where the cortex of the insula becomes continuous with the cortex on the deep surfaces of the folds of the hemisphere (**opercula**) which hide the insula from view. The surface of the insula is marked by a number of sulci and gyri radiating superiorly from the stem of the lateral sulcus, which lies at the apex of the insula known as the **limen insulae** [FIG. 251]. The middle cerebral artery lies on the limen insulae as it passes laterally on to the surface of the insula in the stem of the lateral sulcus. Medial to the limen insulae is the **anterior perforated substance** through which the striate branches of the middle cerebral artery pass into the substance of the brain.

DISSECTION. The inferior part of the insula is already exposed on the 'frontal' part of the left hemisphere. To complete its exposure lift up the frontoparietal operculum, which covers the posterosuperior part of the insula, and tear the operculum upwards and backwards away from the rest of the brain. This exposes the greater part of the insula and the posterior part of the **superior longitudinal bundle** in the root of the operculum

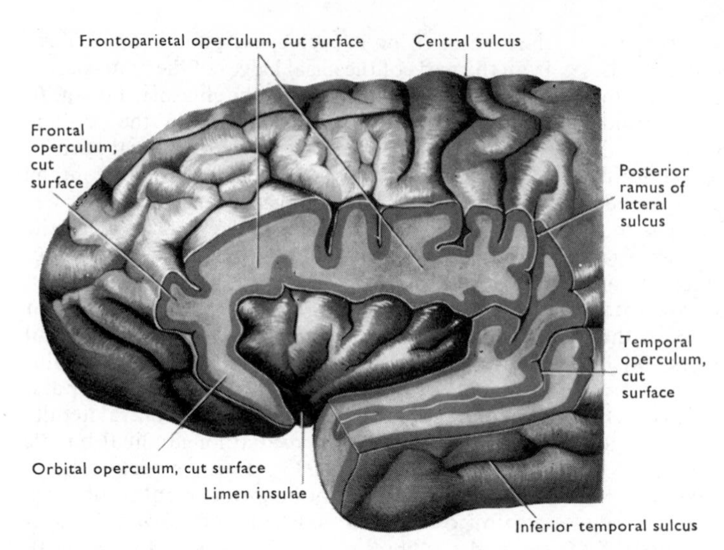

FIG. 251 Dissection of the superolateral surface of the left hemisphere to display the insula.

[FIG. 252]. Tear away the parts of the frontal lobe which still cover the anterior parts of the insula and uncover the anterior part of the superior longitudinal bundle. This part is less clearly shown than the posterior part because it is traversed by many fibres of the corona radiata passing laterally through it.

The function of the insula is unknown, but there is evidence that a secondary sensory area extends over the parietal operculum on to the superior surface of the insula, and a secondary auditory area may also extend over the temporal operculum on to its inferior part. Stimulation of the insula also produces alterations in gastric motility and a number of other actions which may be mediated by way of intracortical connexions.

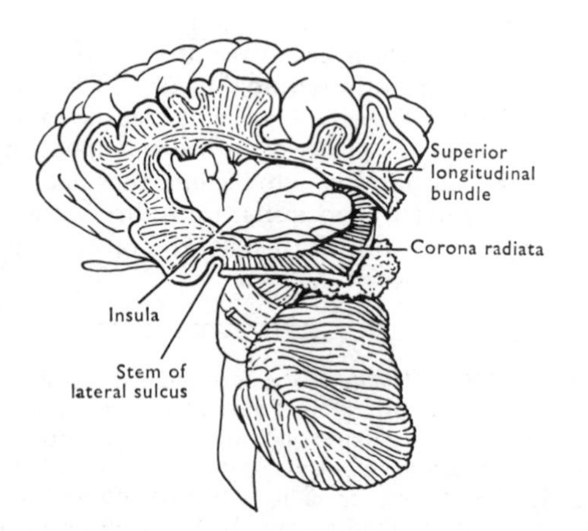

FIG. 252 First phase in the dissection of the 'frontal' part of the brain. The frontal and parietal parts of the operculum have been torn off and the insula and superior longitudinal bundle exposed.

DISSECTION. With a blunt instrument, such as the handle of a knife, raise up the lower border of the insula, which will strip away from the subjacent structures through a thin layer of grey matter which lies deep to the white matter of the insula. This grey matter is the **claustrum**. Lift the whole of the insula away from the rest of the 'frontal' part of the left hemisphere, turning the insula superiorly towards the superior longitudinal bundle. When completely removed it will expose a part of the claustrum overlying a rounded, smooth zone of white matter (the external capsule) and superior to that, a fan-shaped layer of white matter radiating outwards towards the superior longitudinal bundle. This is part of the **corona radiata**.

Using the handle of the knife, scrape away the remains of the claustrum from the external capsule, and note that it passes deep to a ridge of white matter just superior to the stem of the lateral sulcus. This is a frontotemporal association bundle, the **fasciculus uncinatus**, the inferior margin of which hooks round the stem of the lateral sulcus deep to the limen insulae to pass between the frontal lobe and the temporal pole. The superior margin of the fasciculus uncinatus is straight and passes postero-inferiorly towards the temporal lobe from the frontal lobe [FIG. 253].

Pass the handle of the knife downwards between the external capsule and the fasciculus uncinatus towards the lateral margin of the anterior perforated substance. This will free the narrow middle part of the fasciculus from the external capsule, and if the fasciculus is lifted away from its bed its fibres will be seen fanning out into the adjacent parts of the frontal and temporal lobes.

Fasciculus Uncinatus [FIG. 254]

This thick bundle of association fibres passes between the frontal and temporal lobes of the brain. Its inferior fibres hook round the stem of the lateral sulcus into the anterior part of the temporal lobe, while its superior fibres pass directly backwards and downwards towards the posterior part of the temporal lobe. At both ends it fans out, but is narrow in the middle, thus giving it the shape of a sheaf bent inferiorly.

DISSECTION. Place the 'occipitotemporal' part of the left hemisphere in position on the 'frontal' part, and note where the superior longitudinal bundle on the 'frontal' part abuts on the white matter of the 'occipitotemporal' part. Insert the blunt handle of a knife into the white matter of the 'occipitotemporal' part immediately superficial to the point where the superior longitudinal bundle abuts on it, and lever up the surface, tearing it away from the subjacent tissue. This will demonstrate the posterior end of the superior longitudinal bundle turning inferiorly and fanning out into the occipital and temporal lobes, superficial to the fibres of the corona radiata.

Superior Longitudinal Bundle [Fig. 254]

This thick bundle of longitudinal association fibres lies immediately external to the circular sulcus of the insula. It extends around the insula from the frontal pole to the tip of the temporal pole, and is thickest superior to the posterior half of the insula. Fibres enter and leave its external surface throughout its length, and it occupies approximately the same position on the lateral surface of the brain as the cingulum [Fig. 231] does on the medial surface, thus connecting the various parts of the cerebral cortex on the superolateral surface of the hemisphere.

DISSECTION. Turn to the inferior surface of the temporal lobe and insert the handle of the knife obliquely upwards and forwards into its inferior surface close to the occipital pole. Tear the inferior part away from the remainder, and expose a bundle of fibres which runs horizontally forwards into the temporal lobe. This is the inferior longitudinal bundle.

Inferior Longitudinal Bundle

This is a bundle of association fibres which runs horizontally through the temporal and occipital lobes near their inferior surfaces. It is applied to the inferior surface of the corona radiata, and connects the various parts of the cerebral cortex on the tentorial surface of the hemisphere.

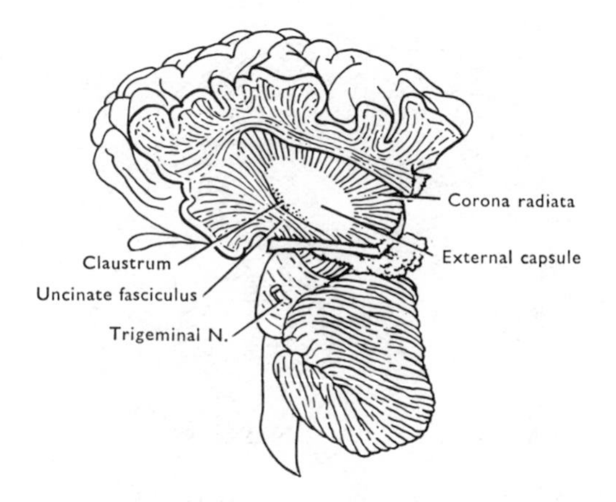

Fig. 253 Second phase in the dissection of the 'frontal' part of the brain. The insula has been removed and the external capsule, claustrum, and fasciculus uncinatus uncovered.

DISSECTION. Remove the external capsule from the lentiform nucleus which lies deep to it. This may be done by removing the external capsule piecemeal, but it is better to lift up a strip of the external capsule and expose the rounded grey surface of the lentiform nucleus; then introduce a blunt instrument beneath the remainder of the external capsule and separate it from the lentiform nucleus by moving the instrument gently back and forth. Then tear the external capsule away in the direction of the corona radiata.

As the external capsule is removed note a number of fine vessels which run in grooves over the lateral surface of the lentiform nucleus. These are the **striate branches of the middle cerebral artery**, and if followed inferiorly, converge on the anterior perforated substance. Follow some of them to the anterior perforated substance.

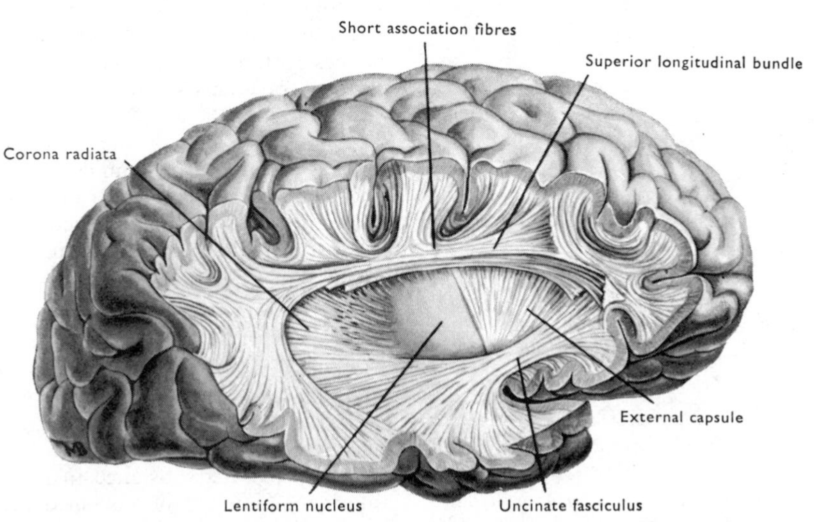

Fig. 254 Dissection of the superolateral surface of the right hemisphere to show the laterally placed association bundles.

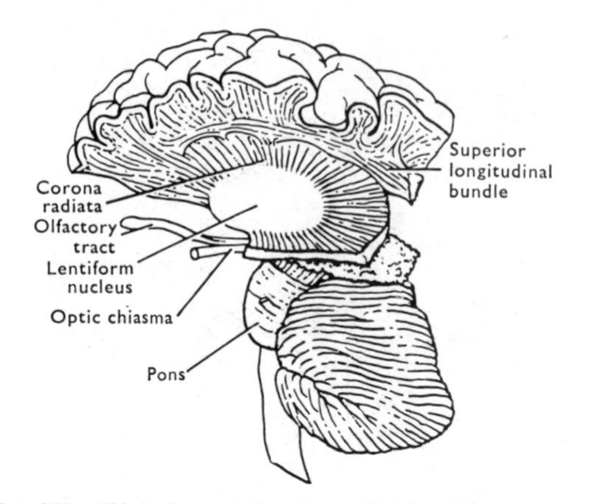

FIG. 255 Third phase of dissection. The fasciculus uncinatus, claustrum, and external capsule have been removed exposing the lentiform nucleus. The rostrum of the corpus callosum, together with the medial orbital part of the frontal lobe, have been torn away from the medial side and the antero-inferior edge of the corona radiata is seen emerging medial to the anterior end of the lentiform nucleus.

External Capsule

This thin layer of white matter lies over the lateral surface of the lentiform nucleus and separates it from the claustrum and the white matter of the insula [FIG. 265]. Since the external capsule strips from the lentiform nucleus with ease it seems unlikely that many of its fibres enter the lentiform nucleus, but a certain number appear to pass to the claustrum, and some are probably association fibres passing between the temporal lobe and the frontal and parietal lobes.

DISSECTION. On the medial side of the 'frontal' part of the left hemisphere, insert the handle of a knife into the anterior horn of the lateral ventricle and tear the genu and rostrum of the corpus callosum forwards, removing them from the rest of the brain. This exposes the head of the caudate nucleus and its continuity with the lentiform nucleus superior to the anterior perforated substance; also the most anterior part of the corona radiata emerging between the caudate and lentiform nuclei as it passes to the frontal lobe. The fibres of the corona which spread laterally and medially in the frontal lobe have been removed, and only the centrally placed fibres remain passing directly forwards into the frontal lobe. [FIG. 255]. The whole extent of the lentiform and caudate nuclei and the corona radiata should now be explored; their position relative to each other and to the internal cover of the lentiform nucleus (the internal capsule) will be clarified by subsequent dissection and the section of the right hemisphere.

Find the anterior commissure on the medial surface of the 'frontal' part of the left hemisphere, and remove the grey matter immediately inferior to it so as to follow it laterally through the inferior parts of the caudate and lentiform nuclei [FIG. 264]. It can be traced, superior to the anterior perforated substance, to enter the shelf of white matter which extends laterally into the temporal lobe above the inferior horn of the lateral ventricle. This is the same shelf of white matter which is cut when the lateral ventricle is exposed, and into which the fibres of the fasciculus uncinatus and the sublentiform part of the corona radiata run to reach the temporal lobe. The fibres of the anterior commissure fan out into this white matter of the temporal lobe.

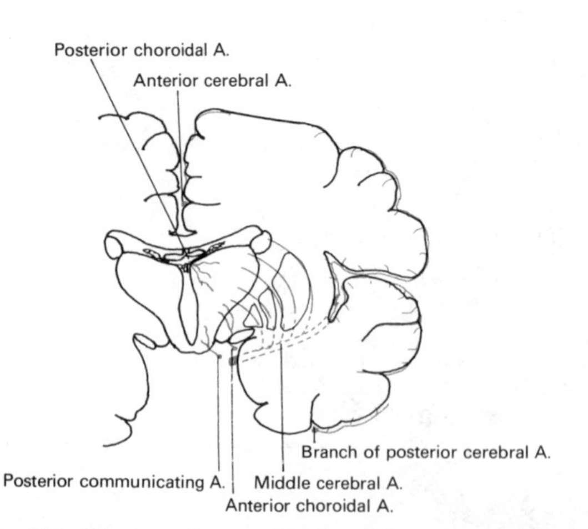

FIG. 256 Diagrammatic coronal section of the cerebrum to show the distribution of some of the arteries to this structure. The middle cerebral artery (broken red lines) and the origins of its striate branches are shown in a more anterior plane.

Anterior Commissure

This commissure is complementary to the corpus callosum, and connects the two temporal lobes. Seen from below the full extent of the commissure has the shape of a cupid's bow, but at its ends it fans out into the temporal lobe. Its fibres have a twisted appearance like a piece of string, and it crosses the midline in the superior part of the lamina terminalis immediately anterior to the columns of the fornix. Laterally it curves slightly forwards, through the head of the caudate nucleus, superior to the anterior perforated substance (olfactory tubercle) and then inclines posteriorly in the lentiform nucleus to enter the temporal lobe. Its fibres fan out posteriorly in the white matter of the temporal lobe.

In lower mammals the anterior commissure conveys a number of commissural fibres between the olfactory bulbs and other olfactory parts of the forebrain, but if such an element is present in the anterior commissure in Man it must be a very small part of it.

Striate Arteries

The lateral striate branches of the middle cerebral artery have already been seen coursing over the lateral surface of the lentiform nucleus and piercing its surface after entering the brain through the anterior perforated substance. In the dissection of the anterior commissure the more medially placed striate arteries have been seen ascending through the anterior perforated substance into the caudate and

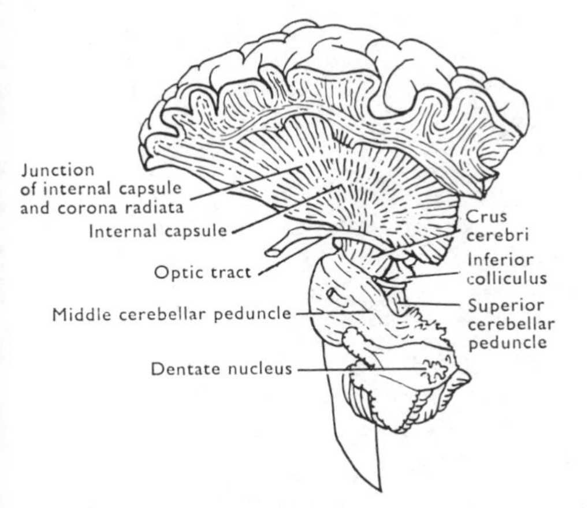

FIG. 257 Fourth phase. The lentiform nucleus has been stripped from the internal capsule to expose the continuity of that structure with the crus cerebri at the level of the optic tract. The cerebellum is also partly dissected to show the superior and middle peduncles.

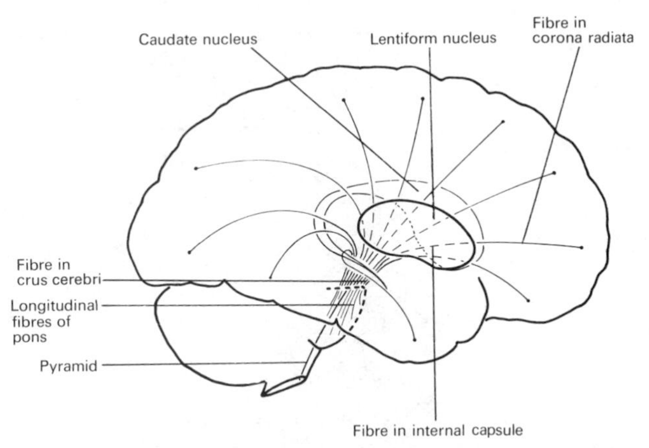

FIG. 258 Diagram to show arrangement of fibres in corona radiata, internal capsule, crus cerebri, pons and pyramid.

lentiform nuclei to supply them and the internal capsule which runs between them [FIG. 265]. The vessels from the lentiform nucleus also pierce the internal capsule and enter the lateral surface of the thalamus on the medial side of the internal capsule. The more medial of the striate arteries arise from the anterior cerebral artery, usually near the anterior communicating artery, and run a recurrent course to the anterior perforated substance [FIG. 187]. These pass to the caudate nucleus and to the anterior hypothalamus. The striate arteries which play a part in the supply of the internal capsule may be thrombosed or ruptured, thus interfering with the pathways to and from the cerebral cortex. This condition, known as a *stroke*, causes paralysis of the opposite side of the body due to the involvement of the corticospinal pathways, and frequently affects sensation because of the injury to thalamocortical fibres in the internal capsule [pp. 240–241].

CORONA RADIATA

This mass consists of the nerve fibres passing between the cerebral cortex and other parts of the central nervous system. It therefore consists of the **projection fibres** of the cerebral cortex; i.e., corticofugal fibres passing inferiorly into the internal capsule, but also **corticopetal fibres** running superiorly, principally from the thalamus. The fibres of the corona radiata are continuous with the internal capsule around the periphery of the lentiform nucleus, and they spread out in a fan-like fashion into the cerebral cortex both anteroposteriorly and transversely. Those which emerge anterior and superior to the lentiform nucleus pass to and from the frontal and parietal lobes; those which emerge posterior to the nucleus (**retrolentiform part**) pass to and from the occipital lobe; while those which emerge beneath the posterior end of the lentiform nucleus (**sublentiform part**) pass to and from the temporal lobe.

DISSECTION. To expose the internal capsule, remove the **lentiform nucleus** from the lateral side of the internal capsule. The lentiform nucleus does not strip easily from the internal capsule because a large number of fibres enter the lentiform nucleus from the internal capsule. Some of these are concentrated into two sheets (the **medullary laminae**) which pass downwards and laterally through the lentiform nucleus and divide it into three parts. The outer, darker part of the lentiform nucleus (the **putamen**) is continuous with the caudate nucleus between the bundles of the internal capsule; the inner, paler segments [FIGS. 262, 265] comprises the **globus pallidus**. The latter segments send the efferent fibres of the lentiform nucleus (the ansa lenticularis) through and around the inferior part of the internal capsule into the hypothalamus.

Bend the corona radiata medially and run a blunt instrument round the margin of the lentiform nucleus to separate it from the base of the corona. Continue stripping the nucleus from the internal capsule downwards towards the stem of the lateral sulcus, noting the small branches of the striate arteries passing from the lentiform nucleus into the capsule. The plane of the lateral surface of the internal capsule runs inferomedially towards the lateral surface of the crus cerebri. When the nucleus is completely separated from the capsule, note that it is still attached to the brain by a loop of fibres which curves round the inferior edge of the internal capsule to enter the hypothalamus, the **ansa lenticularis**.

Fold the lentiform nucleus forwards on the ansa, and note the continuity of corona radiata and the internal capsule, and the optic tract running round the internal capsule at its junction with the crus cerebri [FIG. 259]. Strip the optic tract from the lateral side of the internal capsule and confirm the continuity of the internal capsule with the crus cerebri.

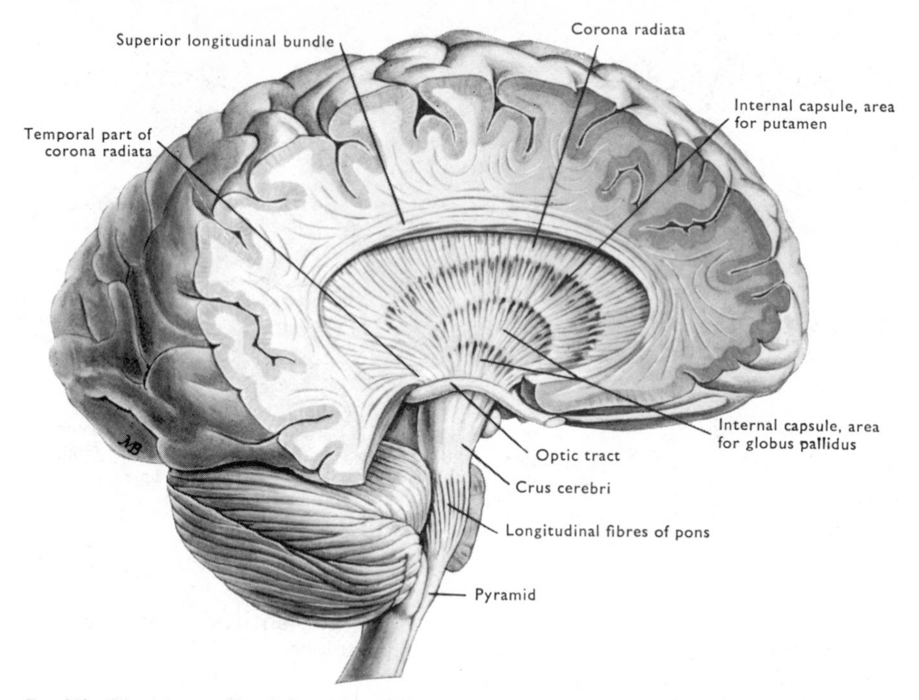

FIG. 259 Dissection to show the continuity of the corona radiata, internal capsule, crus cerebri, longitudinal fibres of the pons, and the pyramid.

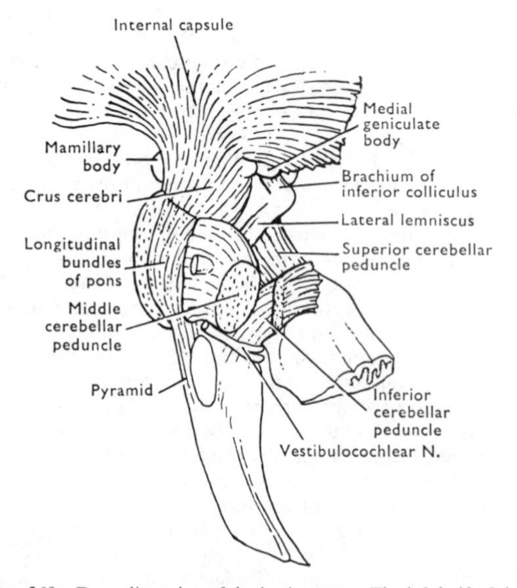

FIG. 260 Deep dissection of the brain stem. The left half of the pons has been torn away between two cuts to display the longitudinal fibres passing through it to become continuous with the pyramid in the medulla oblongata. The three cerebellar peduncles are also displayed.

INTERNAL CAPSULE

Certain gross features of the internal capsule are now obvious. (1) It lies medial to the lentiform nucleus. (2) It is continuous superiorly with the corona radiata, and inferiorly with the crus cerebri. (3) Its fibres radiate in a fan-shaped manner so that the fibres from the frontal lobe enter its anterior part, run a long course in it, and enter the medial part of the crus cerebri. The temporal fibres enter it just superior to the optic tract, run a very short distance in it, and enter the posterolateral part of the crus with the occipital fibres. It follows from this that the position of the various groups of fibres in the internal capsule and crus is determined by their position of origin in the hemisphere, and that the fibres which arise in the precentral gyrus (**corticonuclear** and **corticospinal fibres**), near the midpoint between the frontal and occipital poles, will enter the middle region of the internal capsule and crus. (4) If the dissection has been satisfactory, the strands of grey matter passing between the putamen and the caudate nucleus will be visible near the superior edge of the internal capsule. Also the torn ends of the two medullary laminae will be seen as two rough ridges on the lateral surface of the internal capsule.

DISSECTION. Make two vertical cuts through the transverse fibres of the pons, one in the midline and the other immediately anterior to the entry of the trigeminal nerve. Strip off the superficial transverse fibres of the pons between these, and expose the longitudinal fibres of the pons [FIG. 260]. These are the fibres of the crus cerebri which break up into bundles in the basilar part of the pons, and running between the transverse fibres, mainly end in the pontine nuclei. A small proportion can be followed through the pons to emerge at its inferior border as the pyramid on the anterior surface of the medulla oblongata [FIG. 270].

Follow the pyramid inferiorly. Note that it gradually tapers in the lower part of the medulla oblongata, where the anterior median fissure of the medulla oblongata is filled with interdigitating bundles of fibres. These are the decussating fibres of the pyramids passing towards the lateral funiculus (white column) of the spinal medulla to form the lateral corticospinal tract, which extends into the sacral spinal medulla. A variable proportion of the fibres in the pyramid continues inferiorly, without decussation, into the anterior funiculus of the spinal medulla, to form the anterior corticospinal tract. The fibres of this tract decussate before termination in the grey matter of the spinal medulla, and the tract usually terminates in the upper thoracic region. Other fibres of the pyramid descend in the lateral corticospinal tract of the same side.

The above dissections demonstrate that the corona radiata, internal capsule, crus cerebri, longitudinal fibres of the ventral part of the pons, the pyramid, and the corticospinal tracts form one continuous mass of fibres [FIG. 259] and that only those fibres which enter the corticospinal tracts run through the whole extent of this pathway. It is through the latter, 'motor' pathway that the cerebral cortex can act on the cells of the spinal medulla directly. The majority of the descending cortical fibres end in the thalamus, lentiform and caudate nuclei, and in the pons, and almost all of the fibres ascending to the cortex are found only in the internal capsule and corona radiata, since they enter this system from the thalamus (see below).

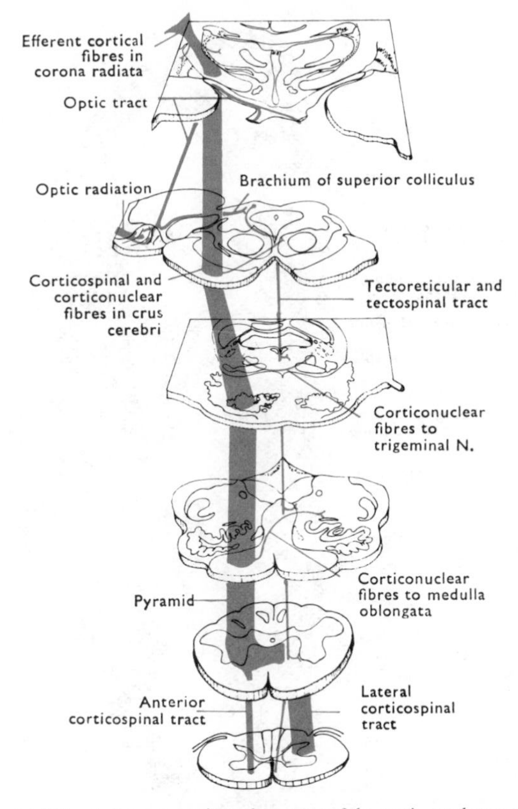

FIG. 261 A diagram to show the course of the corticonuclear and corticospinal (red) fibres (of one side) through the brain stem. In addition, the fibres in one optic tract (blue) are shown passing to the lateral geniculate body and to the superior colliculus, and the descending pathway (tectospinal and tectoreticular, red) for visual reflexes from the superior colliculus. The outline drawings of brain stem sections are not to scale, but correspond to FIGS. 195–198, 200–202, and 212–215. For details of pyramidal decussation see FIGURE 193.

THE DEEP NUCLEI OF THE TELENCEPHALON

The deep nuclei of the telencephalon are: (1) the corpus striatum, consisting of the caudate and lentiform nuclei; (2) the claustrum; and (3) the amygdaloid body. All of these masses of grey matter have been seen in the previous dissections, but their relations are best studied in sections of the hemisphere, which also help to clarify the position and gross connexions of the thalamus and hypothalamus.

Sections of the hemisphere can be carried out on the right side which is nearly intact, but it is desirable to have both horizontal and transverse (coronal) sections for study. For this purpose, groups of students should combine their specimens if it is not possible to have another brain. Specially stained macroscopic sections are also of considerable assistance but the staining masks certain features of the white matter which are visible in freshly cut slices and which help to determine the direction in which particular groups of fibres are running. Thus, when white matter is cut parallel to its fibres it appears whiter than when the section is made across the fibres, and it is possible, for example, to differentiate between the tapetum, the corona radiata, and the longitudinal bundles in sections through the posterior part of the hemisphere [FIG. 242].

DISSECTION. Make a horizontal cut through the right hemisphere which passes through the interventricular foramen and the inferior half of the splenium of the corpus callosum. This is best carried out with a long knife pulled through the brain in a single sweep.

If a second hemisphere is not available for the sections, then the horizontal section should be studied, the two

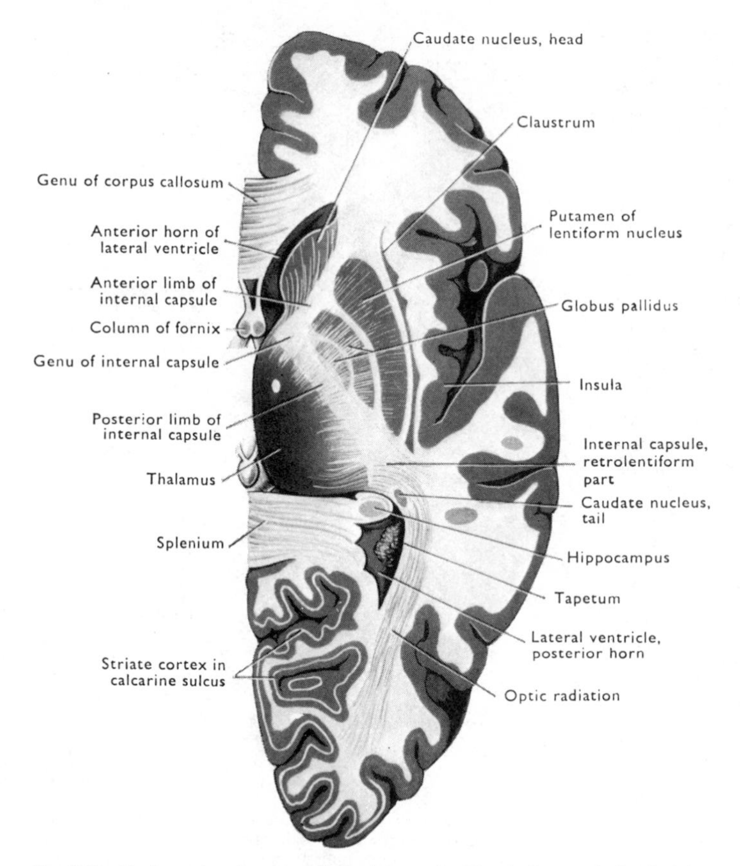

FIG. 262 Horizontal section through the right cerebral hemisphere at the level of the interventricular foramen.

parts placed together, and the transverse sections cut from the same hemisphere. This method is not ideal, and every effort should be made to obtain a separate hemisphere for the transverse sections.

The transverse sections are cut as follows: (1) through the posterior part of the genu of the corpus callosum [FIG. 263]; (2) through the anterior perforated substance [FIG. 264]; (3) immediately anterior to the infundibulum [FIG. 265]; (4) just anterior to the mamillary bodies [FIG. 266]; (5) obliquely downwards through the crus cerebri and pons (if present) [FIG. 268]; (6) through the splenium of the corpus callosum [FIG. 274].

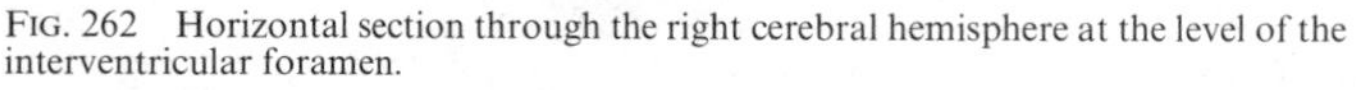

Horizontal Section

The main features of this section are immediately visible in FIGURE 262, but certain special points should be noted:

1. The white and grey matter of the hemisphere, cortex, corpus striatum, claustrum, and thalamus.

2. The buried **insula** with its white matter is lateral to: (1) the **claustrum** (a thin layer of grey matter with a slightly scalloped lateral surface); (2) the **external capsule**; and (3) the **lentiform nucleus**, divided into its three parts by the medial and lateral **medullary laminae**. The outer segment is the **putamen**, the inner two are the **globus pallidus**.

3. The V-shaped **internal capsule** bent round the medial aspect of the lentiform nucleus is divisible, in this view, into anterior and posterior limbs which meet at the **genu**. Medial to the **anterior limb** is the head of the caudate nucleus in the anterior horn of the lateral ventricle, while medial to the **posterior limb** are the thalamus, and further posteriorly, the tail of the caudate nucleus still in the wall of the ventricle. Note that the posterior limb, cut transverse to its fibres, is darker in colour than the anterior limb, cut parallel to its fibres. Similarly, the **optic radiation**, a part of the corona radiata, passing towards the occipital lobe is lighter in colour than the fibres of the **tapetum** which separate it from the posterior horn of the lateral ventricle [*cf.*, FIG. 274].

4. Identify the genu and splenium of the **corpus callosum**, and note the relation of these parts to the lateral ventricle.

5. Note the position and extent of the **striate cortex**. The specimen illustrated in FIGURE 262 has an unusually extensive striate area.

6. Examine the cut surface of the lentiform nucleus and note the cut ends of the striate branches of the middle cerebral artery in it. Note also the cortical branches of the same artery on the insula.

With the assistance of FIGURES 241, and 263–268, identify the structures visible in the transverse sections, and relate their positions to those seen in the horizontal section and to the features seen in the dissection of the left hemisphere. In this manner try to build up a three-dimensional picture of the internal structure of the cerebral hemisphere. This can be achieved most readily by placing the sections together to reconstruct the hemisphere and taking them apart several times.

CORPUS STRIATUM

The corpus striatum consists of the caudate and lentiform nuclei, which are two parts of the same mass partially separated by the fibres of the internal capsule. The periphery of the internal capsule lies between the caudate and the putamen of the lentiform, but centrally it separates the lentiform nucleus from the thalamus and hypothalamus. The **caudate** and **putamen** are not completely separated. The head of the caudate is fused to the putamen, while the body and tail are connected to the putamen by strands of grey matter between the bundles of the internal capsule. This incomplete

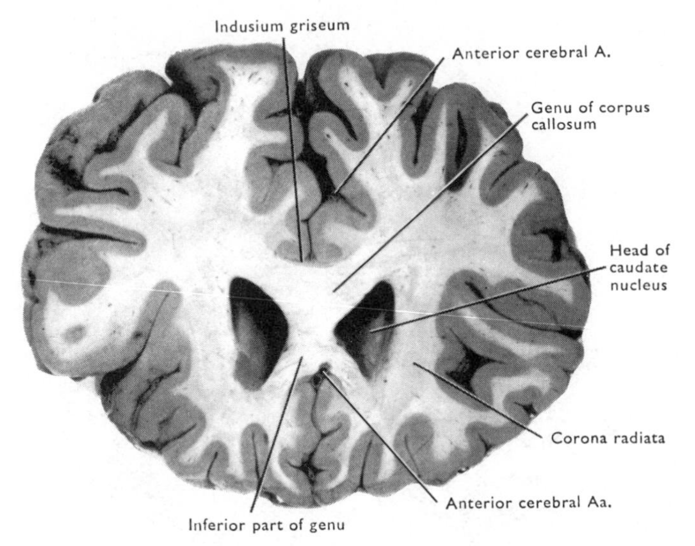

FIG. 263 Coronal section through the genu of the corpus callosum and the anterior horns of the lateral ventricles.

separation does not have the same functional significance as the separation of the putamen from the globus pallidus [FIGS. 262, 265], for the caudate and putamen are identical in histological structure and have similar connexions, while the structure and connexions of the globus pallidus are very different.

Caudate Nucleus

This comma-shaped nucleus has a wide, thick **head** which lies in the lateral wall of the anterior horn of the lateral ventricle, and is fused with the anterior end of the lentiform nucleus inferior to the anterior limb of the internal capsule. Here the common mass is fused with the anterior perforated substance, and the striate branches of the middle and anterior cerebral arteries enter it. The head tapers rapidly to the narrow **body**, which runs posteriorly in the lateral part of the floor of the central portion of the lateral ventricle. The **tail**—narrow and flat—turns inferiorly, and runs anteriorly in the roof of the inferior horn to end in the amygdaloid body. The tail may be discontinuous in places in the roof of the inferior horn.

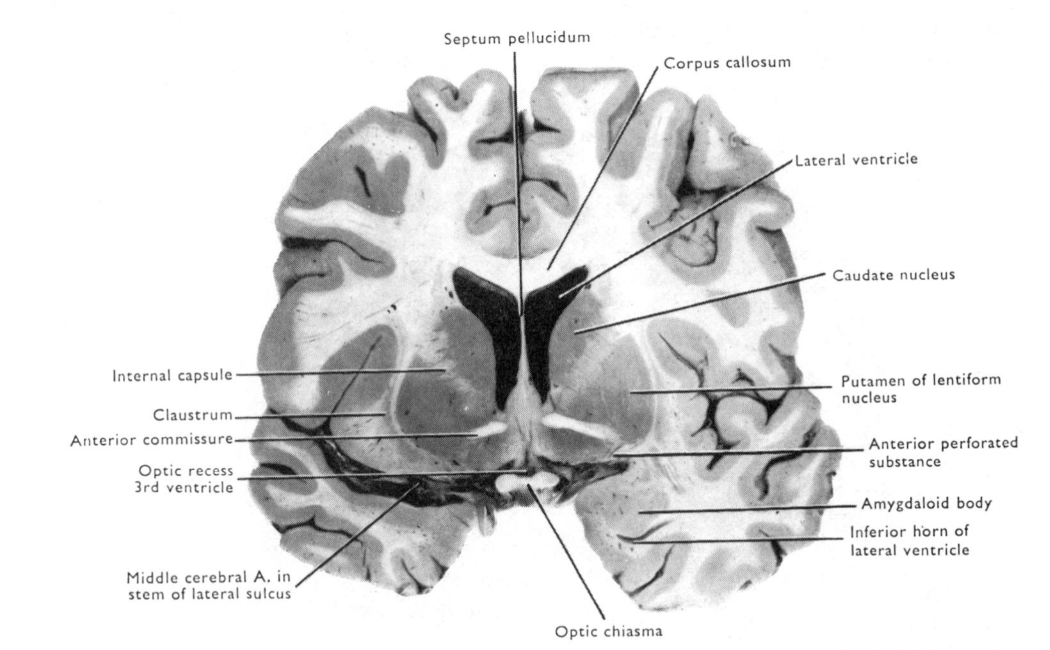

FIG. 264 Coronal section through the brain at the level of the optic chiasma. On the right the section is slightly posterior to that on the left half.

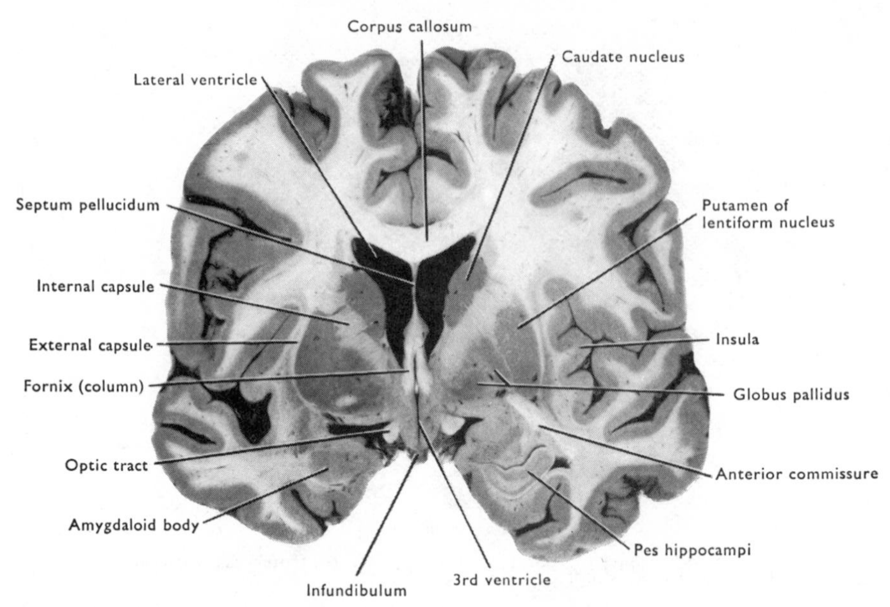

FIG. 265 Coronal section through the brain at the level of the infundibulum.

The ventricular surface of the caudate nucleus is covered with ependyma. The deep surface of the head is fused with the putamen and applied to the anterior limb of the internal capsule, while the deep surfaces of the body and tail lie on the peripheral margin of the internal capsule. A thin strand of association and corticostriate fibres, the fronto-occipital bundle, lies between the convexity of the caudate nucleus and the inferior surface of the corpus callosum.

Lentiform Nucleus

This large, lens-shaped nucleus lies lateral to the internal capsule, and has three surfaces.

The **lateral surface** is smooth and convex, and is grooved by the lateral striate vessels before they sink into its substance. It is separated from the claustrum by the external capsule.

The **medial surface** [FIG. 262] comes to an apex opposite the interventricular foramen. This lies on the genu of the internal capsule, between the head of the caudate nucleus anteriorly and the thalamus posteriorly. The striate arteries traverse and supply the nucleus, and continue through the internal capsule, to supply the caudate nucleus and the lateral part of the thalamus.

The **inferior surface** is fused anteriorly with the

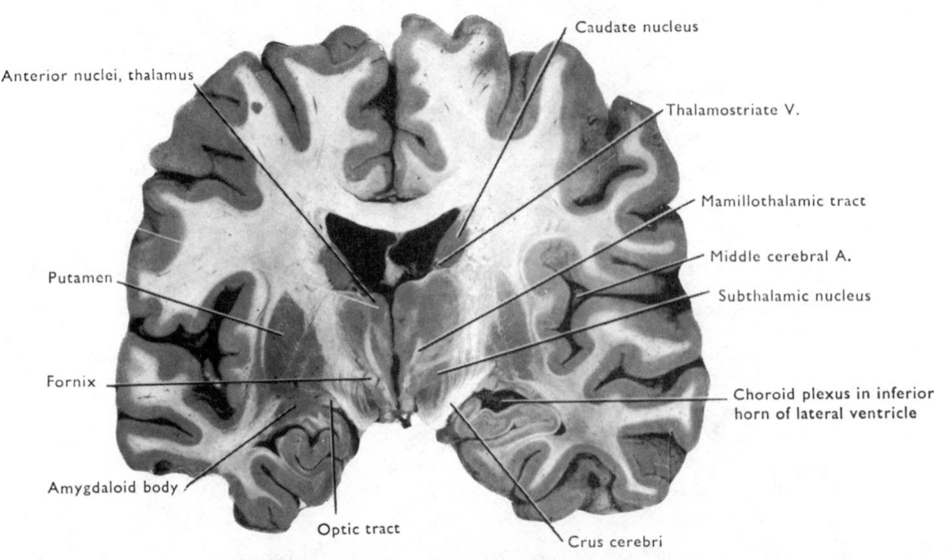

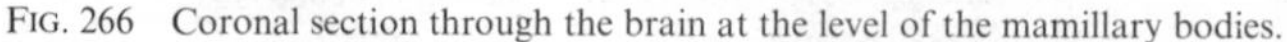
FIG. 266 Coronal section through the brain at the level of the mamillary bodies.

anterior perforated substance and through it is continuous with the amygdaloid body [FIG. 265]. In this situation the lentiform is pierced by the anterior commissure [FIG. 264]. Posteriorly the lentiform nucleus lies on the white matter which passes laterally into the temporal lobe. This shelf of white matter contains the **sublentiform part of the internal capsule** (auditory radiation) and separates the lentiform nucleus from the tail of the caudate nucleus, the optic tract, and the inferior horn of the lateral ventricle [FIG. 268].

The lentiform nucleus is divided by two vertical sheets of white matter (the lateral and medial **medullary laminae**) into three parts. The lateral, darker part is the **putamen**, the two paler, medial segments form the **globus pallidus**.

The **connexions** of the corpus striatum [FIG. 267] are complex and not completely understood. It is clear, however, that the caudate and putamen receive a considerable number of fibres from the cerebral cortex, thalamus, and substantia nigra. The efferent fibres of both these nuclei pass to the substantia nigra and predominantly to the globus pallidus which, unlike the caudate and putamen, contains a number of large nerve cells, and gives rise to most of the nerve fibres which leave the corpus striatum. The efferent fibres of the globus pallidus (**ansa lenticularis**) pass medially through and around the inferior part of the internal capsule to enter the lateral part of the hypothalamus (subthalamus) [FIG. 266]. Here some fibres end on the subthalamic nucleus, while others turn caudally to reach the tegmentum of the midbrain. The majority turn dorsally into the thalamus, whence the impulses they convey may be transmitted to the cerebral cortex or back to the corpus striatum.

The **functions** of the corpus striatum are not clear, but it appears to be concerned with the regulation of muscle tone and the control of automatic movements, as well as playing a part in the control of voluntary movements. The involvement of the corpus striatum in pathological lesions tends to lead to a state of increased muscular tone, the disappearance of associated movements (*e.g.*, swinging the arms in walking), and the development of a tremor which is present during rest. The genesis of these symptoms is not understood, but their disappearance when the internal capsule or the thalamus is subsequently damaged seems to suggest that they may in part be due to the uncontrolled action of the cerebral cortex.

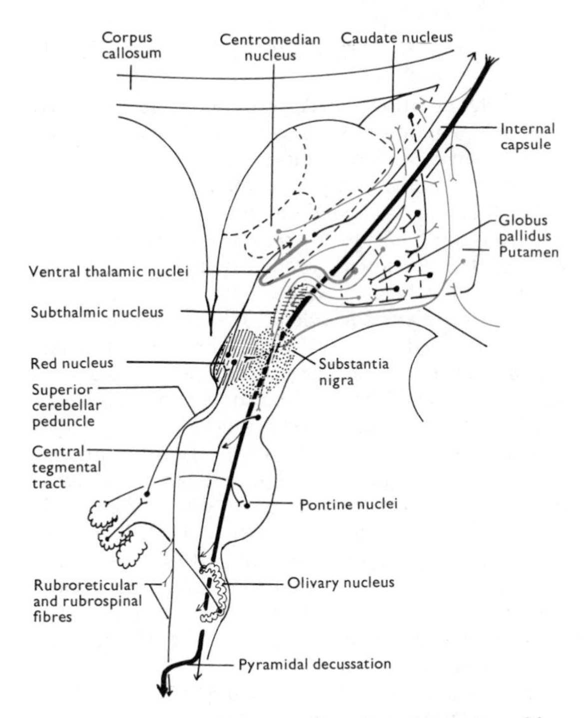

FIG. 267 Diagram to show some of the principal connexions of the corpus striatum. Afferent fibres, blue; efferent fibres, red.

Claustrum [FIGS. 264–268]

This thin plate of grey matter which lies between the external capsule and the white matter of the insula is of unknown function, but appears to belong to the deep nuclei of the telencephalon. Antero-inferiorly the claustrum reaches its greatest thickness and is fused with the anterior perforated substance and the amygdaloid body [FIG. 265]. It seems to receive nerve fibres through the external capsule, but its connexions are unknown.

Amygdaloid Body [FIG. 265]

This complex of nuclei lies over the tip of the inferior horn of the lateral ventricle and is fused with the claustrum and the anterior perforated substance. It is continuous with the tail of the caudate nucleus posteriorly, and with the cortex of the uncus medially. It probably receives afferent fibres from a considerable number of sources, some from the olfactory system. That it is not mainly concerned with olfaction seems clear from its large size in Man and anosmic aquatic mammals. Its efferent fibres pass mainly into the stria terminalis, but some pass by way of the diagonal band [FIG. 228] across the anterior perforated substance to much the same regions, the anterior hypothalamus and septal nuclei.

The **stria terminalis** is a delicate bundle of fibres which arises in the amygdaloid body. It passes posteriorly along the medial side of the tail of the caudate nucleus in the roof of the inferior horn of the lateral ventricle. Curving upwards with the caudate nucleus, it enters the floor of the central part of the lateral ventricle and runs forwards with the thalamostriate vein, between the thalamus and the body of the caudate nucleus. At the interventricular foramen it turns inferiorly with the fornix, and ends in the anterior hypothalamus and septal nuclei.

If the differentiation between grey and white matter is good in the brain slices, the main subdivisions of the thalamus may be seen, but its minor subdivisions are only visible in microscopic sections.

The thalamus is divided into a number of separate cell groups or nuclei by the presence of two layers of white matter (medullary laminae) which run through it approximately in a sagittal plane. The **lateral medullary lamina** lies close to the internal capsule and is separated from it by a thin, broken lamina of grey matter [FIGS. 266, 268] the **reticular nucleus** of the thalamus. The **medial medullary lamina** forms a more complete layer and lies more or less midway between the lateral wall of the third ventricle and the internal capsule. It divides the thalamus into **medial** and **lateral nuclei** [FIG. 268] and inferiorly it sweeps in a medial direction to separate the medial nucleus from the **ventral nuclei** which lie between the medial nucleus and the hypothalamus. There is no clear line of demarcation between the lateral nucleus and the ventral nuclei. Anterosuperiorly the medial medullary lamina splits to enclose the **anterior nuclei** [FIGS. 266, 268] while posteriorly it encloses the **centromedian nucleus** [FIG. 268] and a number of other small **intralaminar nuclei** which are not visible to the unaided eye. These nuclei have connexions mainly with non-cortical parts of the forebrain, *e.g.*, corpus striatum.

Afferent Connexions

The main bundles of afferent fibres to the thalamus are: (1) The **spinothalamic tract** and **medial lemniscus**, which bring impulses arising in sensory endings in the opposite side of the body, together with some uncrossed fibres in the spinothalamic tract and from the trigeminal nuclei; also ascending fibres from the reticular formation of the brain stem, which carry similar, bilateral impulses arising from activity in the spinoreticular tract [p. 238]. (2) Visual fibres of the **optic tract**, which end in the lateral geniculate body. (3) **Acoustic fibres**, which pass to the medial geniculate body in the brachium of the inferior colliculus. (4) The **mamillothalamic tract**, which conveys impulses from the mamillary body through the plane of the medial medullary lamina to the anterior nuclei of the thalamus. Some of the latter impulses arise as a result of impulses which reach the mamillary body through the fornix.

In addition to these, most of the thalamic nuclei receive fibres from the area of cerebral cortex to which they project, and many intrathalamic connexions.

Thalamic nucleus	Area of cortex sending fibres to thalamic nucleus.
Ventral	Posterior frontal and anterior parietal lobes.
Medial	Anterior part of frontal lobe.
Lateral (including pulvinar)	Parietal and occipital lobes.
Lateral geniculate body	Striate cortex of occipital lobe.
Medial geniculate body	Superior temporal gyrus.
Reticular	The greater part of the cortex.

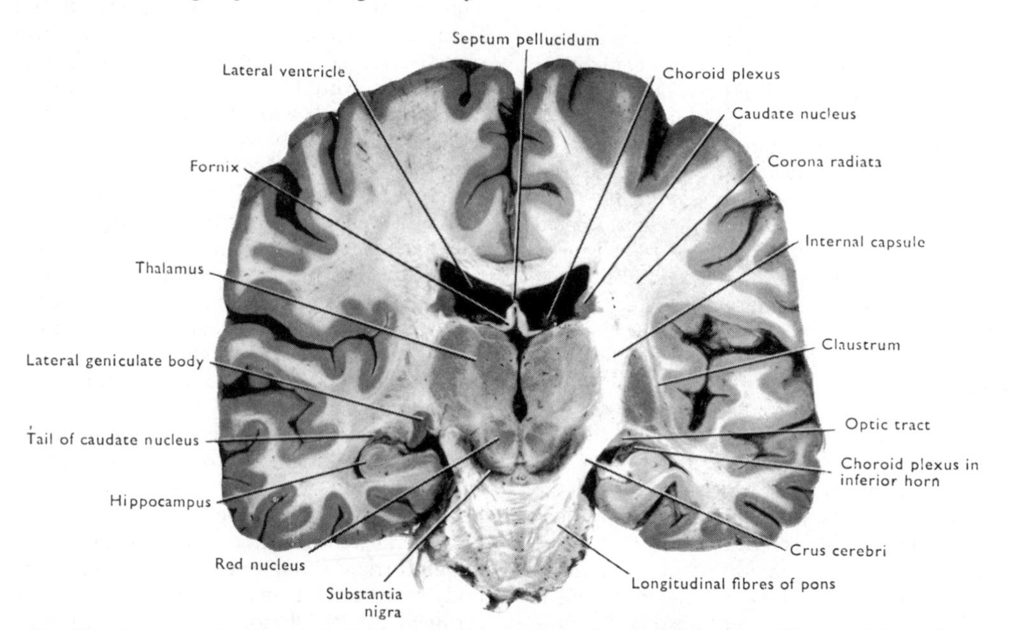

FIG. 268 Coronal section through the brain passing through the basilar part of the pons. Note the division of the thalamus into anterior, medial, lateral, and central nuclei on the left.

All these fibres reach the thalamus via the corona radiata and internal capsule.

Anterior — Hippocampus via fornix directly, and mamillary body indirectly.

DISSECTION. To expose the ascending pathways from the brain stem and spinal medulla, make a deep median incision into the pons and deepen the cut which was made through the left half of the pons immediately anterior to the trigeminal nerve to expose the longitudinal fibres of the pons. With the handle of a knife, free the left pyramid from the anterior surface of the medulla oblongata and pull it upwards, tearing away all the longitudinal and transverse fibres of the ventral part of the pons between the median and lateral cuts. As the superior border of the pons is reached pull the crus cerebri away with it in a superior direction as far as the inferior part of the internal capsule [FIG. 269].

The removal of this ventral mass of tissue from the brain stem exposes a flat sheet of fibres running longitudinally, and in the midbrain a sheet of pigmented grey matter, the substantia nigra [FIG. 268]. The longitudinal sheet of fibres in the pons and midbrain [FIG. 270] consists of the **medial lemniscus**, the **spinothalamic tract**, and the **lateral lemniscus** at the lateral border. Follow the lateral margin of the sheet superiorly on to the exposed lateral surface of the midbrain and note its continuity with the lateral lemniscus previously exposed there. Also note how the medial lemniscus sweeps laterally as it ascends, and comes to lie on the lateral aspect of the midbrain at the level of the superior colliculus. Here some of the fibres of the spinothalamic tract pass into the superior colliculus (**spinotectal tract**). Follow the medial lemniscus superiorly till it disappears into the thalamus on the medial aspect of the internal capsule.

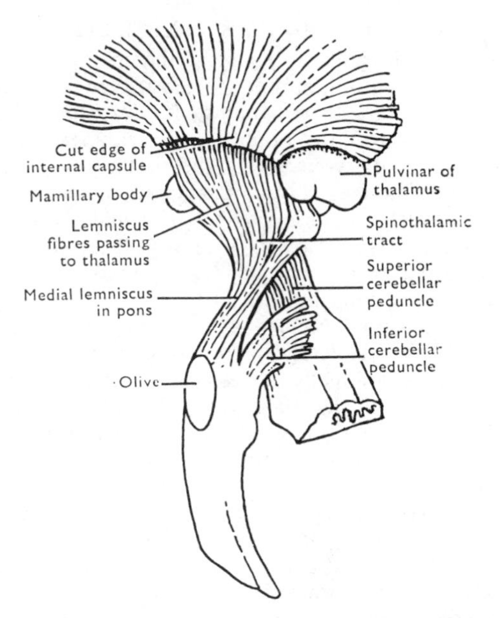

FIG. 269 Deep dissection of brain stem, see also FIG. 260. The pons, middle cerebellar peduncle, crus cerebri, and pyramid have been torn away, leaving the superior and inferior cerebellar peduncles and the lemniscus system exposed. Note the continuity of the dentate nucleus with the superior cerebellar peduncle.

ASCENDING TRACTS IN THE BRAIN STEM

The sheet of fibres now exposed consists of the medial lemniscus, spinothalamic tract, and lateral lemniscus from medial to lateral.

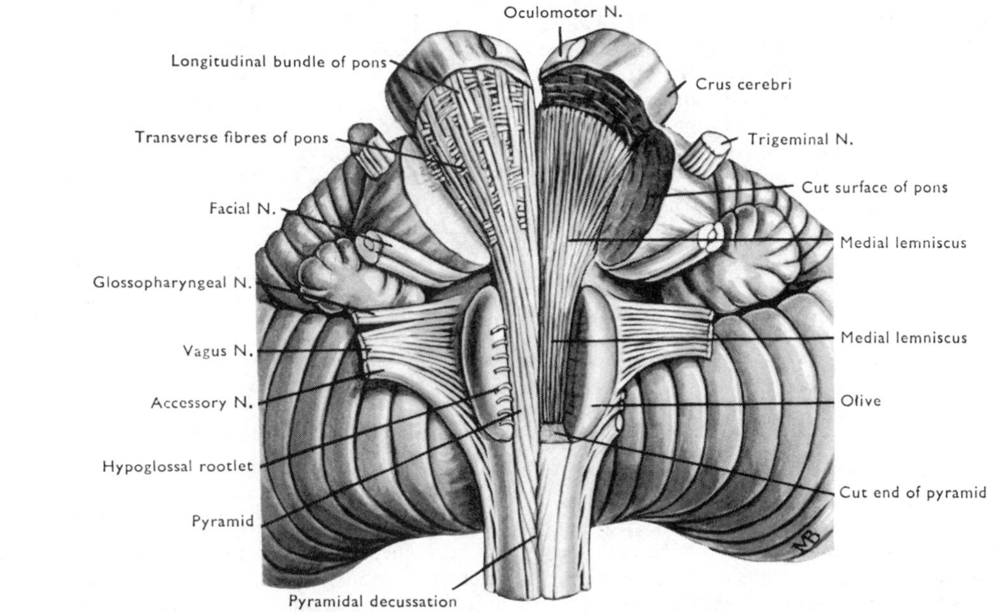

FIG. 270 Dissection of the pons and medulla oblongata. On the left of the illustration, the longitudinal fibres of the pons and pyramid have been exposed, but they have been removed on the right to expose the medial lemniscus lying posterior to them.

Medial Lemniscus [FIGS. 197, 201, 212, 271]

The medial lemniscus is one of the great ascending tracts. It transmits impulses which arise in tactile and proprioceptive sensory endings. It is formed by fibres which arise in the opposite nuclei cuneatus and gracilis, and sweep across the midline to ascend through the medulla oblongata as a vertical sheet immediately posterior to the pyramid. It is the anterior surface of this mass of fibres which is exposed when the pyramid is raised out of its bed. As it enters the pons the medial lemniscus turns through a right angle and becomes a horizontal sheet of fibres which extends laterally from the midline. As it ascends, it gradually moves away from the midline and reaches the lateral aspect of the superior part of the midbrain, ascending vertically from this position to end in the thalamus. Thus the fibres of the two sides which transmit these impulses through the spinal medulla (fasciculi gracilis and cuneatus) and medulla oblongata (medial lemniscus) lie side by side adjacent to the midline, and may therefore be injured bilaterally by a single lesion in the midline. In the pons and midbrain this is no longer possible because the medial lemnisci diverge from each other.

Spinothalamic Tracts [FIGS. 175, 197, 271]

The spinothalamic tracts which ascend through the anterolateral parts of the white matter of the spinal medulla and the lateral parts of the medulla oblongata, are widely separated from the posterior funiculi and medial lemnisci respectively in these two situations. But in the pons, the medial lemnisci extend laterally to join the spinothalamic tracts, and continue superiorly fused with them. Thus the spinothalamic tracts, which are principally concerned with evoking the sensations of pain, temperature, and some touch, may be injured on one side of the spinal medulla and medulla oblongata without injury to the posterior funiculi or medial lemniscus, but in the pons and midbrain the medial lemniscus and corresponding spinothalamic tract may be injured by a single localized lesion.

Again, because the spinothalamic tracts are pierced by the ventral roots of the spinal nerves in the spinal medulla, lesions of these tracts are frequently associated with ventral root injury. The position of the spinothalamic tracts in the lateral part of the medulla oblongata places them in association with the laterally attached cranial nerves (facial, vestibulocochlear, glossopharyngeal, vagus, and cranial part of the accessory) which may be damaged in association with injuries of the spinothalamic tracts in the medulla oblongata. **Spinoreticular fibres** in the spinal part of the spinothalamic tract end in the lateral part of the reticular formation of the medulla oblongata. In addition to their functions in modifying reticular activity, such fibres may be responsible for stimulating cells of the bilateral ascending **reticulothalamic system**—probably concerned with alerting the brain to unpleasant stimuli affecting any part of the body.

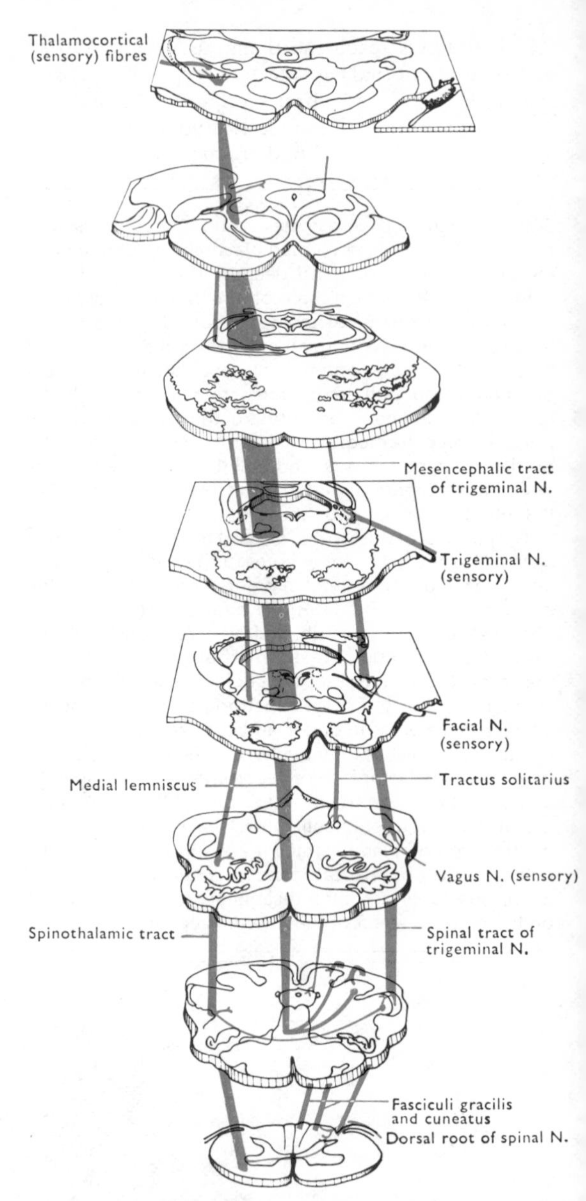

FIG. 271 An outline diagram to show the course taken by the primary sensory fibres (red) in the spinal and cranial nerves, and the secondary ascending pathways (blue) through which the impulses they carry are transmitted to the thalamus and cerebral cortex. The pathways shown are those concerned with the conscious appreciation of the various sensory impulses. The outline drawings of the brain stem slices are not to scale, but they correspond to FIGS. 195–198, 200–202 and 212–215. Fibres cross the spinothalamic tract obliquely along the medial lemniscus and not transversely as shown.)

Lateral Lemniscus [FIGS. 194, 202, 211, 272]

The lateral lemniscus, which is formed by fibres arising in the **cochlear nuclei** of the vestibulocochlear nerves of both sides, joins the lateral edge of the combined medial lemniscus and spinothalamic tracts in the pons, and together these three groups of fibres form the ascending sheet which may be exposed by removal of the ventral part of the pons. In this dissection it is possible to demonstrate the continuity of the lateral lemniscus with the cochlear nuclei by displacing the middle and inferior cerebellar peduncles laterally. The fibres of the lateral lemniscus on the lateral edge of the sheet turn posteriorly across the superior cerebellar peduncle to enter the inferior colliculus, while some of the fibres of the spinothalamic tract on the dorsal aspect of the medial lemniscus run posteriorly into the superior colliculus (**spinotectal tract**), the remainder passing superiorly into the thalamus.

The fibres of the lateral lemniscus reach it from the cochlear nuclei partly as decussating fibres which pass through the medial lemniscus in the pons (the **trapezoid body**) and partly as direct fibres from the cochlear nuclei of the same side. Thus each lateral lemniscus, and the other parts of the auditory pathway superior to it, carry impulses from both ears. This permits accurate measurement of differences in the time of arrival of sounds at the two ears (sound location), and explains why damage to this pathway on one side does not prevent acoustic impulses from either ear reaching the thalamus and cerebral cortex, and hence does not cause unilateral deafness such as would arise from injury to the vestibulocochlear nerve on one side.

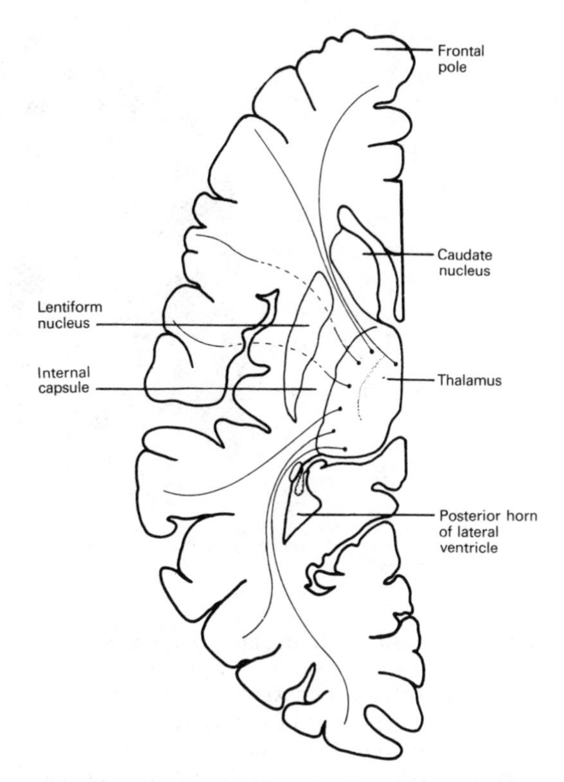

FIG. 273 Diagram of the arrangement of the thalamocortical radiations to the upper part of the cerebral hemisphere. The broken lines indicate fibres arching superiorly out of the section over the lentiform nucleus.

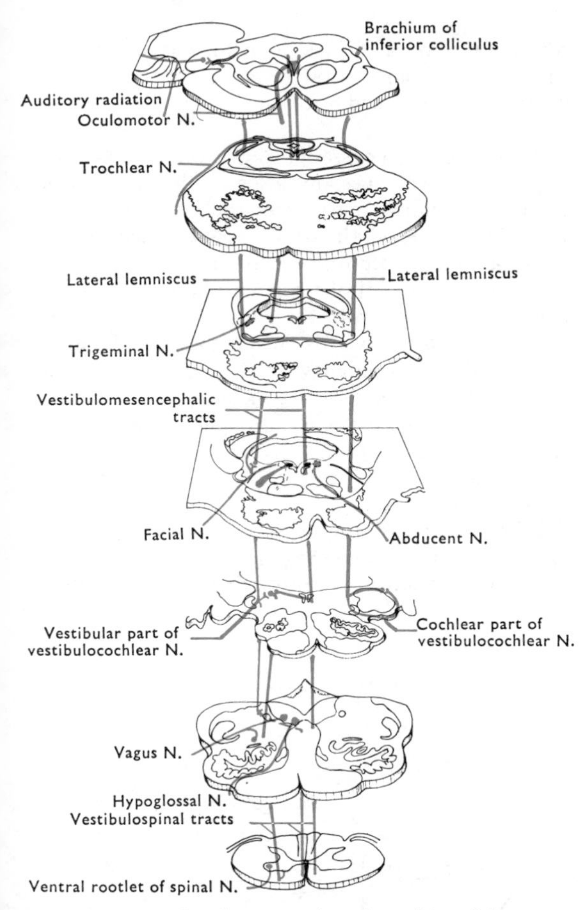

FIG. 272 An outline diagram to show the position of the motor cranial nerve nuclei and the course taken by their fibres passing through the brain stem to their points of exit (red). On the right the ascending connexions of the cochlear part of the vestibulocochlear nerve are shown (blue); on the left, the principal connexions of the vestibular part of the same nerve are indicated (blue). The outline drawings are not to scale, but correspond with FIGS. 195–198, 200–202 and 212–215.

The lateral lemniscus ends in the inferior colliculus, but the impulses which it carries are transmitted to the thalamus (medial geniculate body) via the **brachium of the inferior colliculus**.

Thalamic Termination of Sensory Pathways

It has been seen already that the fibres of the optic tract end in the lateral geniculate body, and that the acoustic fibres in the brachium of the inferior colliculus end in the medial geniculate body. The fibres of the medial lemniscus and spinothalamic tracts terminate in the **posterior part of the ventral nucleus of the thalamus**.

EFFERENT FIBRES OF THE THALAMUS

The table on page 287 which shows the afferent connexions of the thalamus from the cerebral cortex is equally appropriate to show the efferent fibres from the thalamus to the cerebral cortex (with the exception of the anterior nuclear complex which projects to the cingulate gyrus, and the reticular

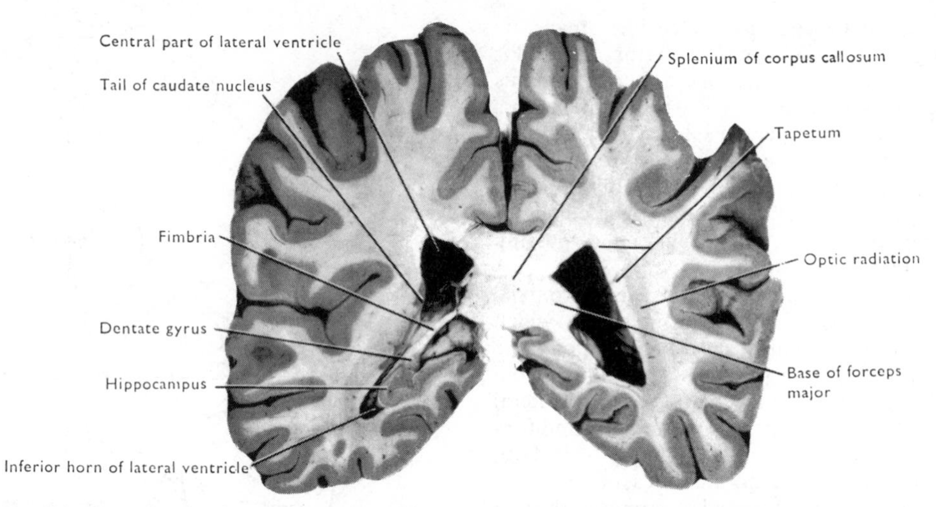

FIG. 274 Coronal section through the splenium of the corpus callosum to show the continuity of the central part and inferior horn of the lateral ventricle. Note how the fibres of the splenium and its tapetum virtually surround the posterior part of the lateral ventricle, as the fibres of the genu surround the anterior horn. Cf. FIG. 263.

nucleus which projects to the midbrain tegmentum) for most of these connexions are reciprocal.

In addition to these connexions the thalamus sends fibres to the hypothalamus and to the corpus striatum. In particular the medial nucleus sends fibres to the hypothalamus, and the centromedian and intralaminar nuclei are said to project to the striatum. There is physiological evidence that the intralaminar nuclei may project diffusely to the cerebral cortex, but there is no anatomical confirmation of this.

Within the general mass of the thalamocortical fibres are those which transmit the sensory information carried to the thalamus in the optic tract, the acoustic pathway, and the medial lemniscus and spinothalamic tracts. These are known as the sensory radiations, and they reach the cerebral cortex through the internal capsule and corona radiata in the same fashion as all the other thalamocortical fibres.

Optic Radiation

The cells of the **lateral geniculate body** form the fibres of the radiation. They transmit the impulses which reach the geniculate from the parts of the retinae concerned with the contralateral half of the field of vision [p. 225]. The fibres sweep forwards into the retrolentiform part of the internal capsule, spreading out into a broad sheet which turns round the concave surface of the lateral ventricle, and passes towards the occipital lobe, lateral to the tapetum and the posterior horn of the lateral ventricle [FIG. 262]. On the concave surface of the ventricle, the inferior fibres sweep far forwards on the superior surface of the inferior horn of the lateral ventricle (close to the optic tract) before turning backwards. The fibres in the upper, middle, and lower parts of the sheet respectively transmit impulses from the same regions of the corresponding halves of the two retinae; the fibres at the edges of the sheet being concerned with the peripheral parts of the retina, while those in the centre are concerned with the central or macular parts.

As the radiation passes posteriorly, the fibres at its upper and lower edges, one after another, hook over the corresponding margins of the lateral ventricle [FIG. 275] and reach the striate cortex respectively in the superior and inferior walls of the calcarine sulcus on the medial aspect of the hemisphere. Thus fibres carrying impulses from the peripheral parts of the retinae end far forwards, while the central fibres concerned with macular vision end in the posterior part of the calcarine sulcus, including the occipital pole. Because the lower fibres of the radiation pass further forwards than the upper fibres, their terminations in the calcarine cortex also extend further forwards.

Fibres carrying impulses from corresponding points of the two retinae lie together in the radiation. Hence a small injury in the radiation produces a blind

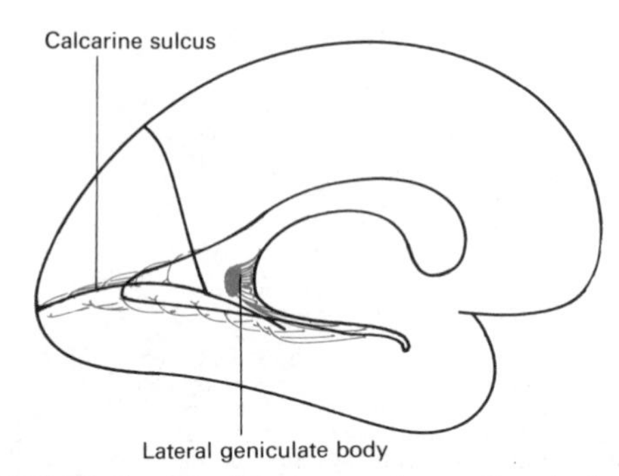

FIG. 275 A diagram of the course of nerve fibres in the optic radiation. Medial view of left cerebral hemisphere showing relation of radiation to lateral ventricle and calcarine sulcus.

spot (*scotoma*) in the contralateral half of the field of vision in the corresponding area of both eyes. It is therefore more obvious to the individual than a similar scotoma produced by a small retinal lesion in one eye which is covered by the intact retina in the other eye, as in the case of the physiological blind spot—the optic disc. Scotomata produced by injuries of the upper part of the radiation are seen in the lower half of the field of vision and vice versa, but both are in the contralateral field.

In addition to the fibres passing to the cerebral cortex, the optic radiation contains fibres passing in the opposite direction which run to the lateral geniculate body and to the superior colliculus. The latter are concerned with the production of eye movements through the descending tectal pathways.

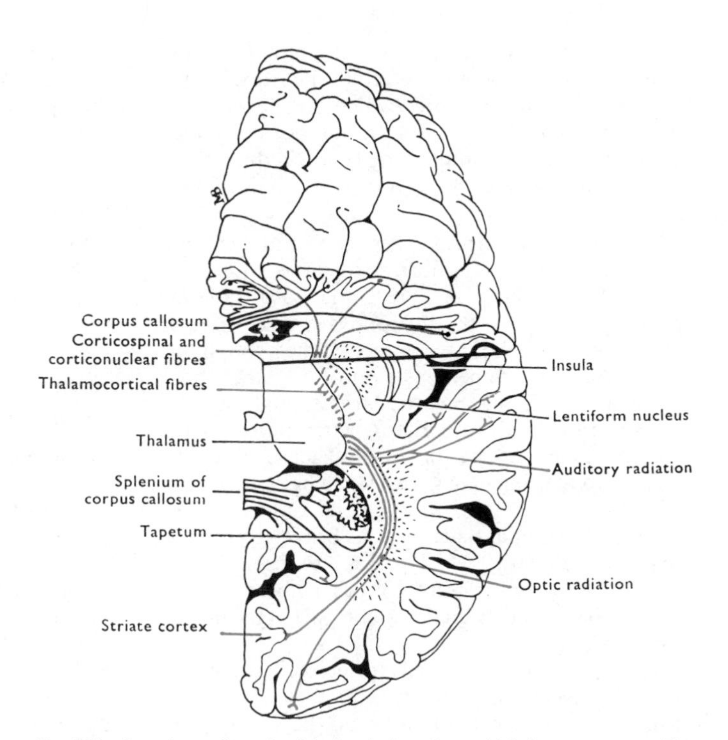

FIG. 276 Superior surface of a right hemisphere from which the upper parts of the parietal and occipital lobes have been removed. On the cut surfaces the various groups of fibres are shown diagrammatically. Red, descending /ascending fibres; blue, sensory fibres; black, commissural fibres of corpus callosum.

Acoustic Radiation

The fibres of this radiation arise in the medial geniculate body and pass anterolaterally into the sublentiform part of the internal capsule, between the roof of the inferior horn of the lateral ventricle and the lentiform nucleus. They then turn superiorly to end on the superior surface of the superior temporal gyrus (transverse temporal gyrus) and to a small extent on its lateral surface [FIGS. 219, 268, 276].

General Sensory Radiation

The cells of the posterior ventral nucleus of the thalamus, which receive impulses through the medial lemniscus and spinothalamic tracts, send fibres through the lateral aspect of the thalamus into the posterior limb of the internal capsule. These fibres ascend through the internal capsule and fan out to end in the posterior wall of the central sulcus and the immediately adjacent part of the postcentral gyrus [FIG. 219]. They are so arranged that fibres conveying sensory information from the head end inferiorly, and those carrying impulses from the lower limb and perineum end superiorly, even extending on to the medial surface of the hemisphere. The fibres which transmit impulses from the medial lemniscus retain the discrete point to point localization which is characteristic of that tract and of the posterior funiculi from which it receives the impulses. Therefore they convey information which allows accurate localization of the part of the body stimulated. The spinothalamic system does not have the same discrete point to point localization and is also partially bilateral in the information which it carries. It follows from this that lesions of this sensory radiation or of the postcentral gyrus produce profound disturbances of proprioception and tactile sense on the opposite side of the body, while pain and temperature sensations are only slightly affected though their localization (a function of the medial lemniscus system) is greatly disturbed.

HYPOTHALAMUS

This part of the brain forms the lateral wall of the third ventricle inferior to the hypothalamic sulcus, and laterally extends to the lower part of the internal capsule. It appears on the inferior surface of the brain surrounded by the optic tracts and crura cerebri, and has the mamillary bodies and infundibulum attached to this surface. In coronal section it is possible to divide the hypothalamus into lateral (subthalamus) and medial parts, only the medial appearing on the inferior surface, while the former overlies the inferior part of the internal capsule. Both parts are directly continuous inferiorly with the tegmentum of the midbrain without any line of demarcation. The medial part (the hypothalamus of the physiologist) is particularly concerned with visceral activity.

The connexions of the hypothalamus are complex and incompletely understood, and details are available in larger textbooks, but a few of the major connexions are mentioned here.

Medial Part

Afferent Connexions. These reach the medial hypothalamus from: (1) the hippocampus via the fornix; (2) the thalamus, conveying impulses from many systems including the cerebral cortex; (3) the olfactory system, by fibres which pass posteriorly into the hypothalamus from the anterior perforated substance; (4) the brain stem by ascending fibres passing through the tegmentum of the midbrain.

Efferent Connexions. These can be divided into three main groups: (1) fibre connexions with other parts of the brain; (2) fibre connexions with the posterior lobe of the hypophysis; (3) vascular connexions with the anterior lobe of the hypophysis.

(1) The fibre connexions with other parts of the brain are numerous but the largest single tract is the **mamillothalamic tract** which passes to the anterior nuclei of the thalamus. Thence fibres are relayed to the cingulate gyrus which has connexions with the hippocampus through the cingulum. The fibres in the mamillothalamic tract branch to give rise to a descending tract which passes into the midbrain, and a considerable number of other fibres descend into the brain stem from the lateral part of the medial hypothalamus.

(2) Certain groups of cells (principally the supraoptic, paraventricular, and tuberal nuclei) send their axons into the posterior lobe of the hypophysis. These do not innervate the posterior lobe in the usual manner, but act as ducts for the secretory materials synthesized in cells of the nuclei, and these materials are stored in the posterior lobe.

(3) The vascular connexion consists of capillaries of the hypothalamus (medial part) which form veins running inferiorly with the infundibulum. These veins break up into capillaries in the anterior lobe of the hypophysis, and are believed to carry hormones (releasing factors) from the hypothalamus which can stimulate the anterior lobe to specific activity.

The nervous and vascular connexions with the hypophysis are concerned with the slow alterations in hormonal concentrations in the blood stream, by means of which the hypothalamus may produce slow, prolonged alterations in the activity of other organs. The connexions with the brain stem are presumably concerned with rapid, short duration alterations in visceral activity.

Lateral Part

The lateral part of the hypothalamus (subthalamus) consists of a number of groups of cells among which the **subthalamic nucleus** is the large oval structure seen in coronal sections of the brain [FIG. 266]. It lies superior to the upper extremity of the substantia nigra, and its inferior part is close to the red nucleus.

The subthalamus receives fibres from the thalamus and a considerable number of the fibres of the ansa lenticularis which enter it from the globus pallidus of the lentiform nucleus. Most of its efferent fibres pass back to the globus pallidus, but some probably descend into the midbrain. Lesions of the subthalamic nucleus give rise to hemiballismus, a violent or writhing movement which principally affects the opposite upper limb, but its part in the normal functioning of the brain is unclear.

CRANIOCEREBRAL TOPOGRAPHY

After the dissection of the brain is completed, the dissectors should review the relation of the brain to the skull and meninges. This is a most important topic since the effects of head injuries in different regions are dependent on the association of the parts of the brain to the skull and to the structures in or on the surfaces of the skull. Thus fractures of the base of the skull may allow blood and cerebrospinal fluid to escape into the nose, orbit, or external acoustic meatus, and fractures which cross the grooves for the middle meningeal artery may lead to extensive haemorrhage between the dura and periosteum with compression of the brain. Depressed fractures of the skull may irritate the underlying nervous tissue or interfere with its function, thus producing convulsions or paralysis if over the motor area of the cerebral cortex (precentral gyrus) and visual disturbances if the occipital pole is involved, the central area of vision being particularly affected in such a posterior injury.

Some of the main features of cranial topography have been described already [pp. 171–2] but the following points should be noted again. It should always be remembered that there are many different shapes of skull and that there is no such thing as an exact relationship between the parts of the brain and the surface features of the skull; the points mentioned below are approximate only:

1. A point two fingers' breadth posterior and one finger's breadth superior to the notch on the posterior margin of the junction of the frontal and zygomatic bones (which is easily palpated) marks (a) the approximate position of the inferior end of the **coronal suture**, (b) the anterior end of the **posterior ramus of the lateral sulcus**, (c) the point where the lesser wing of the sphenoid meets the lateral wall of the skull, and (d) the situation where the **frontal branch of the middle meningeal artery** is frequently buried in the bone of the skull.

2. The coronal suture passes superiorly and slightly backwards from this point to its meeting with the sagittal suture at the **bregma**. The point of meeting of these sutures can be felt as a slight depression, and represents the position of the **anterior fonticulus** in the infant.

3. The frontal branch of the middle meningeal artery ascends more or less parallel to the coronal suture and frequently about a finger's breadth posterior to it.

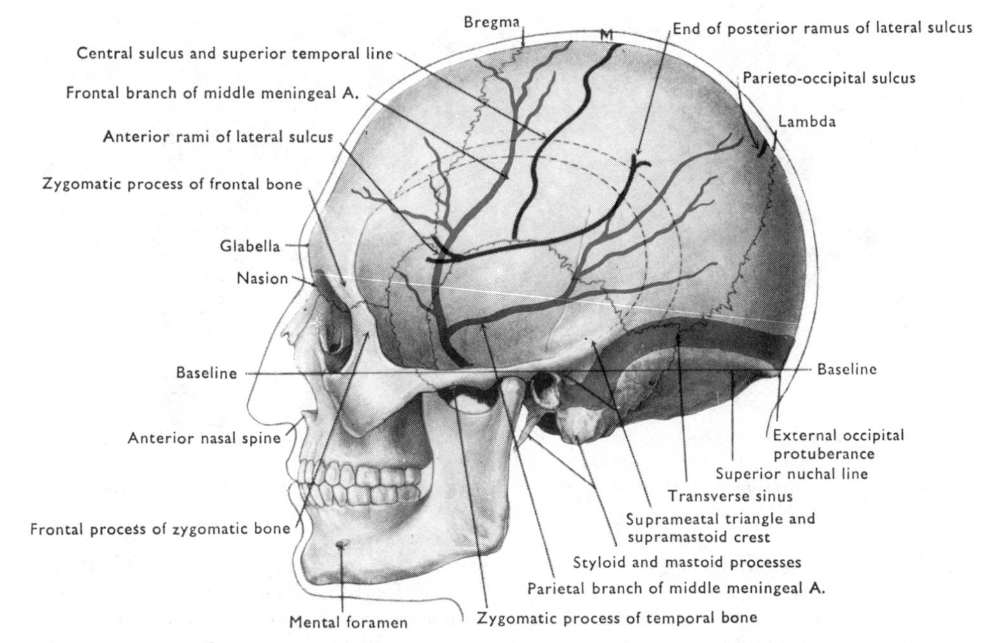

FIG. 277 Lateral view of the skull to show the position of certain important intracranial structures. M = midpoint between nasion and inion.

4. The **central sulcus** also lies parallel to the coronal suture, but approximately two fingers' breadths posterior to it. The superior extremity of this sulcus lies 1 cm posterior to the midpoint on a median line joining the root of the nose to the external occipital protuberance.

5. The posterior ramus of the lateral sulcus passes posteriorly and slightly superiorly. At first it lies on the line of the squamoparietal suture, but then passes deep to the parietal bone; its terminal part ascends for a short distance inferior to the parietal eminence.

6. If the line of the posterior ramus of the lateral sulcus is extended anteriorly to pass a short distance superior to the orbital margin, it marks the inferior margin of the frontal lobe.

7. A line, convex anteriorly, drawn from the point mentioned in (1) to meet the superior margin of the zygomatic arch approximately at its middle, marks the anterior extremity of the temporal lobe. The **inferolateral margin of the cerebral hemisphere** is indicated by the extension of the same line posteriorly along the superior margin of the zygomatic arch (immediately superior to the external acoustic meatus) and then horizontally to a point 1 cm superolateral to the external occcipital protuberance.

THE DEVELOPMENT OF THE HEAD, NECK, AND BRAIN

The following brief accounts give an introduction to some of the more important elements in the development of this region. They should prove sufficient to explain some of the more obscure elements of the gross anatomy and form a basis for an understanding of the common congenital abnormalities.

THE DEVELOPMENT OF THE FACE

The face develops by the growth of the tissue (**mesoderm**) which lies between the surface **ectoderm** and the lining of the gut tube (**entoderm**) around the anterior extremity of that tube. This growth of mesoderm causes the **buccopharyngeal membrane** (the meeting of ectoderm and entoderm at the anterior extremity of the gut tube) to be displaced from the surface so that it lies in the depths of an ectodermal pit—the **stomodaeum**. At the upper margin of the stomodaeum, and between it and the brain, the proliferating mesoderm forms the **frontonasal process**. This is innervated by the ophthalmic nerve and supplied by the ophthalmic artery. Its mesoderm does not grow uniformly, but splits into two medial and two **lateral nasal processes** which partly surround two ectodermal pits, the **olfactory pits**, continuous with the stomodaeum ventrally. Lateral and inferior to the stomodaeum, the mesoderm forms the first of a series of U-shaped bars in the wall of the cephalic extremity of the gut tube—the first pharyngeal or **mandibular arch** [FIG. 278]. On each side, the eye protrudes from the brain to lie

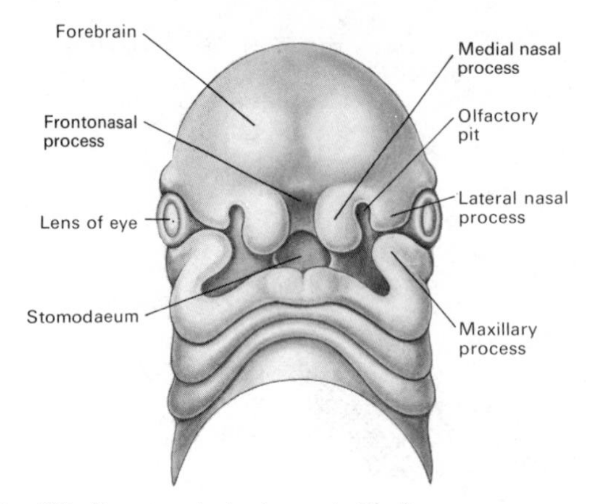

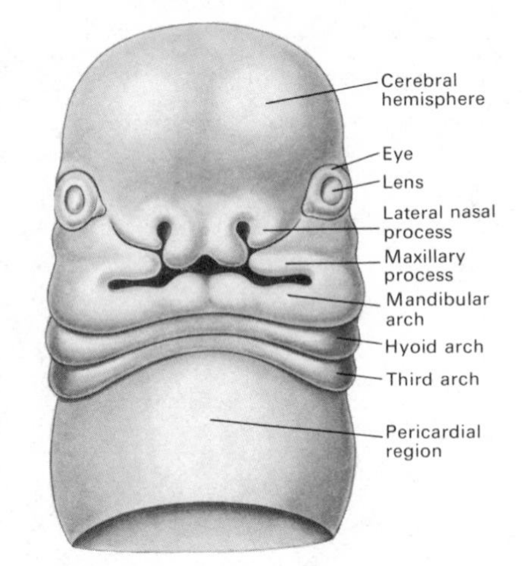

FIG. 278 Two stages in development of the face.

A. Boundaries of stomodeum before completion of primitive upper lip.

B. Completion of upper lip. Note union of maxillary process with lateral and medial nasal processes to form cheek and upper lip.

beneath the skin between the dorsal end of this arch and the frontonasal process.

The mesoderm in the dorsal end of the first pharyngeal arch proliferates and bulges forwards as the **maxillary process** (supplied by the maxillary nerve) below the corresponding eye to fuse with the **lateral nasal process** between the eye and the olfactory pit [FIG. 278]. This is the line of formation of the **nasolacrimal duct** either from ectoderm buried at the time of fusion or as a secondary downgrowth from the ectoderm covering the eye (conjunctiva). The maxillary processes continue their growth anteromedially to meet and fuse with the **medial nasal processes**, thus cutting off the olfactory pits from the stomodaeum and meeting each other below the frontonasal process to form the upper lip. Posterior to these maxillary processes, the deep, blind ends of the **olfactory pits** subsequently break through into the stomodaeum leaving the processes as a **primitive palate** beneath each pit. The anterior openings of the pits are the **nares** or nasal openings.

Failure of either maxillary process to fuse with the medial nasal process leads to a vertical defect in the upper lip which extends into the naris (**hare-lip**). This may be unilateral or bilateral and is often associated with a cleft palate—a feature which is also due to improper formation of the maxillary processes (see below). Failure of the maxillary processes to fuse with each other, or failure of the frontonasal process to grow properly may lead to a median defect in the lip—a much less common abnormality.

THE DEVELOPMENT OF THE MOUTH AND NASAL CAVITIES

The nasal cavities and the anterosuperior part of the oral cavity are developed from the **stomodaeum** (see above). This ectodermal pit leads to the **buccopharyngeal membrane** which separates it from the entodermal pharynx in the earliest stages. Initially the anterior end of the neural tube lies on the roof of the stomodaeum, and these two ectodermal derivatives remain in contact immediately in front of the buccopharyngeal membrane, even after the ingrowth of mesoderm separates them elsewhere. As a result, the contact area of the stomodaeal roof is pulled upwards to form a pocket (**Rathke's pouch**) which subsequently separates from the stomodaeum and forms the anterior and intermediate parts of the **hypophysis**. The original site of this contact corresponds to the junction of the posterior edge of the nasal septum with the roof of the nasal part of the pharynx. The position of the ventral edge of the buccopharyngeal membrane has no such defined position, but must lie anterior to the foramen caecum of the tongue.

The **margins of the stomodaeum** are: (1) anterosuperiorly, the **frontonasal process** which is overlapped inferiorly by the anterior extensions of the maxillary processes forming the primitive palate (see above); (2) posterosuperiorly, the **maxillary processes** [FIG. 278]; (3) laterally and inferiorly, the mandibular arches. Mesoderm which migrates into the floor of the stomodaeum from the occipital myotomes forms the **muscles of the tongue**. This overlies the ventral ends of the first and second

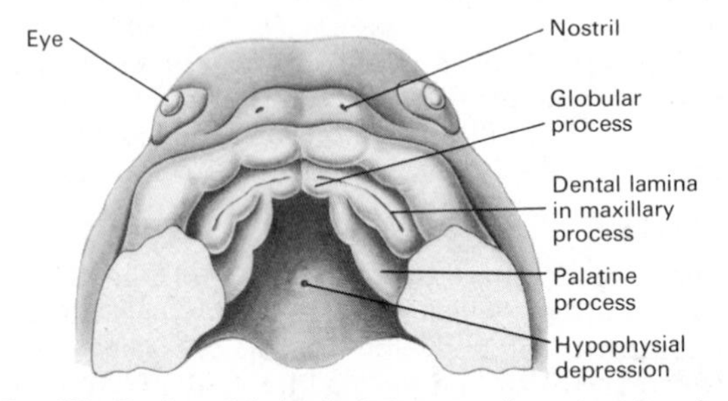

FIG. 279 Portion of the head of a human embryo about 8 weeks old. The lips are separated from the gums, and the line of the dental lamina is visible in the gums. The palatine processes are growing inwards from the maxillary process.

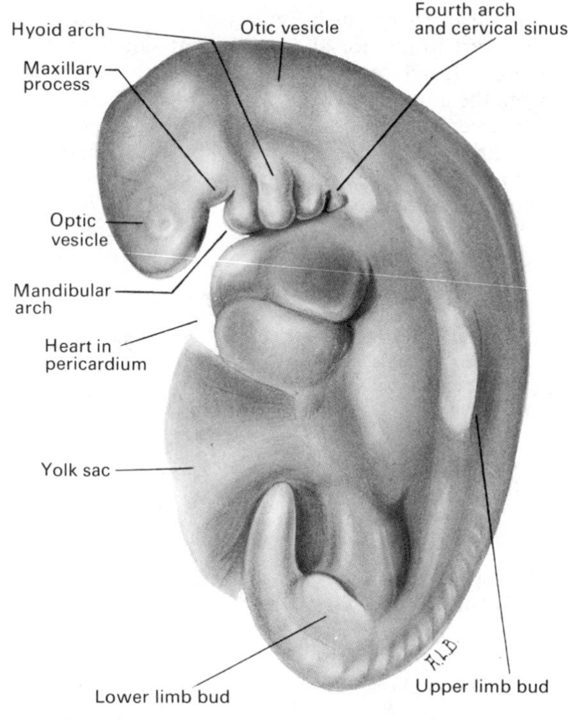

FIG. 280 Human embryo, 5·3 mm long, to show the development of the limb buds. The rudiment of the upper limb, more advanced than that of the lower limb, appears as a longitudinal ridge on the side of the embryo. (After G. L. Streeter, *Contr. Embryol. Carneg. Instn.*, 1945; embryo no. 8066.)

pharyngeal arches [FIG. 278] and bulges the mucous membrane of this part of the floor upwards to form the mass of the tongue.

Mesoderm of each maxillary process grows forwards to fuse with the frontonasal process (see above) and extends medially behind this to separate the roof of the stomodaeum from the brain. On each side, the inferior part of this medial growth produces a shelf-like protrusion of the stomodaeal lining which extends downwards, lateral to the upgrowing tongue, as a **palatal process**.

As the volume of mesoderm (frontonasal anteriorly, and maxillary posteriorly) between the brain and the stomodaeum increases, it is excavated from below on both sides of the midline. These excavations form the nasal parts of the common oronasal cavity, and leave the **nasal septum** between them with its inferior margin resting on the dorsum of the tongue. As the depth of the oral cavity increases, the **palatal processes** are elevated above the tongue, fuse anteriorly with the primitive palate (maxillary processes) and frontonasal process and, behind this, with each other and the inferior edge of the nasal septum to form the **definitive palate** [FIG. 279]. Thus the walls of the nasal cavities are formed anterosuperiorly by the frontonasal process (supplied by the ophthalmic nerve and artery) and posteriorly and inferiorly by the maxillary process (supplied by the maxillary nerve and artery) with a small contribution from the third arch [p. 247] posteriorly.

Failure of proper formation or fusion of the palatal processes produces varying degrees of **cleft palate** ranging from a bifid uvula to a complete anteroposterior defect on one or both sides. Such defects are paramedian behind the incisor teeth in the frontonasal process, but pass forwards lateral to these teeth to become continuous with a **hare-lip** when this is also present [p. 244]. This anterior separation of the right and left defects is due to the development of the incisive part of the maxilla from the frontonasal process.

Cleft palate is an important defect because it prevents alterations of pressure in the oral cavity independent of that in the nasal cavity. Hence it interferes with sucking, blowing, and swallowing [pp. 118, 255] and permits the passage of fluids from the oral into the nasal cavities from which they may be inhaled into the lungs—a problem from the time of birth. It also interferes with the separation of the oral and nasal elements of speech, giving a slurred 'nasal' quality to the voice. This cannot be overcome by the considerable hypertrophy of the horizontal fibres of palatopharyngeus and of the superior constrictor which is present in these cases and produces an unusually powerful closure of the opening between the oral and nasal parts of the pharynx (pharyngeal isthmus).

THE DEVELOPMENT OF THE PHARYNX

The mesoderm in the lateral and inferior walls of the anterior end of the gut tube (primitive **pharynx**) proliferates to form a series of U-shaped, rod-like thickenings which partly surround it. These thickenings, reminiscent of the gill arches of fishes, are known as the **branchial** (gill) or **pharyngeal arches** [FIG. 283]. Four such arches are visible externally [FIG. 280] though there are probably two others of which there is no external evidence (see below). Between the arches, the mesoderm does not proliferate and so leaves the skin ectoderm, externally, and the pharyngeal entoderm, internally, virtually in contact—a situation akin to the gill slits of fishes but normally without a breakthrough from the pharynx to the exterior. Thus on both sides of the pharynx, the **ectoderm** forms a **groove** and the **entoderm** a **pouch** between each arch.

The arches and the intervening grooves and pouches lie ventral to a longitudinal ridge of branchial mesoderm which is the primordium of the **sternocleidomastoid muscle**, and they extend from the stomodaeum to the pericardium which lies in the floor of the pharynx at this stage. Hence the arches will be situated in the **anterior triangle of the neck** when the head grows forwards away from the pericardium. Inferior to the ventral surfaces of the pharyngeal arches, and between them and the pericardium, is a strip of muscle which migrates from the occipital and upper cervical somites caudal to the arches, and extends cranially ventral to them. This strip of muscle forms the **muscles of the tongue** at its cranial extremity and the **infrahyoid muscles** caudal to this. The nerves which supply this muscle (hypoglossal and ansa cervicalis) initially follow the line of migration of the muscle and loop caudal to the pharyngeal arches.

In each arch, the mesoderm forms (1) a cartilage rod; (2) an artery which connects the outflow tract of the heart to the dorsal aorta of that side; (3) some muscle. In addition, a nerve grows into each arch from the adjacent part of the brain—the hindbrain. These nerves supply the structures in the corresponding arches and the entoderm on their internal surfaces: only the trigeminal having a significant cutaneous distribution.

FATE OF THE ELEMENTS IN EACH PHARYNGEAL ARCH

Arch	Cartilage	Artery	Muscles	Nerve
First or mandibular	Malleus, its ant. ligament, sphenomandibular ligament, cartilage on medial surface of mandible	Disappears	Of mastication, tensors tympani and palati, digastric (ant. belly), mylohyoid	Trigeminal, mandibular
Second or hyoid	Stapes, styloid process, stylohyoid lig., lesser cornu and upper part of body of hyoid bone	Disappears	Of facial expression, stylohyoid, digastric (post. belly), stapedius	Facial
Third	Hyoid bone (lower part of body and greater cornu)	Common and internal carotid	Stylopharyngeus and pharyngeal	Glossopharyngeal
Fourth	Thyroid cartilage	Aortic arch (L), subclavian (R)	Cricothyroid and pharyngeal	Vagus, superior laryngeal
Fifth	Probably only represented by a transitory artery			
Sixth	Cricoid (?)	Pulmonary, ductus arteriosus	Laryngeal and pharyngeal	Vagus, recurrent laryngeal

Cartilages [Fig. 281]

With the development of the mandible on the lateral side of the first arch cartilage and the articulation of the mandible with the skull, the articulation of the primitive jaw (**first arch cartilage**) with the skull is released to form the malleus and its joint with the incus. It may also form the incus, though this probably arises from the part of the skull with which both the first and second arch cartilages articulate. The inferior part of the first arch cartilage fuses with the medial side of the mandible at the mylohyoid line. The rest of the cartilage becomes ligamentous [Fig. 281].

The origin of the **thyroid** and **cricoid cartilages** is uncertain because the fourth arch does not have a well-formed cartilage in the embryo, and the sixth appears to have none.

Arteries

See Volume 2.

Muscles

These mostly lie in the position of the arch in which they are formed, but some of the muscle of the second arch migrates to the surface of the face where it forms the muscles of facial expression. Hence the nerve supply of these muscles is the facial nerve. Sternocleidomastoid and trapezius muscles are formed from branchial mesoderm.

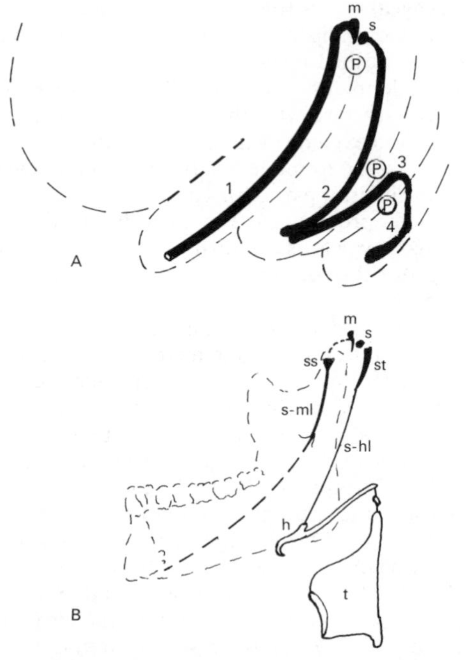

Fig. 281 Diagrams to show the fate of the pharyngeal arch cartilages. A. Cartilages in the embryo. B. Derivatives of the cartilages in the adult. h. Hyoid bone. m. Malleus. P. Position of pharyngeal pouches. s. Stapes. to s-hl. Stylohyoid ligament. s-ml. Sphenomandibular ligament. ss. Spine of sphenoid. st. Styloid process. t. Thyroid cartilage.

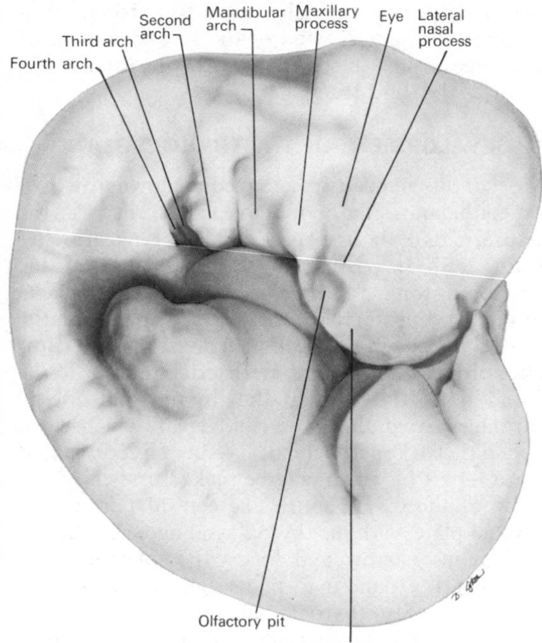

FIG. 282 Human embryo, 6·5 mm. long, to show the development of the pharyngeal arches. The second arch has extended backwards partly covering the third and fourth which lie in a deep cervical sinus.

Changes in the Arches [FIG. 282]

Each **second arch** grows laterally and caudally so as partly to cover the third and fourth arches. This is the same type of growth which occurs in the bony fishes and produces the operculum which covers the gills caudal to it. This growth of the second arch results in several changes.

1. It forms a depression (**cervical sinus** [FIG. 283]) caudal to the second arch with the third and fourth arches in its medial wall. This sinus usually fills up later, but it may persist as a blind pocket opening on the surface of the neck anterior to sternocleidomastoid. If one of the entodermal pouches at the level of the sinus breaks through to the exterior, such a cervical sinus may open into the pharynx and so form a **cervical fistula** between the pharynx and the skin. Remnants of the ectoderm of the sinus may persist in the neck without an external opening. These form small pockets of epithelium adjacent to thyroid gland, and may later distend and produce swellings in the neck (**branchial** or **pharyngeal cysts**) often in young adults.

2. The **facial nerve** is carried caudally with the arch—a feature which explains the angled course of the nerve in the temporal bone.

3. The **second arch** is withdrawn from the pharyngeal wall, allowing the first and third arches to come together—a feature which explains the supply of adjacent areas of the mucous membrane of the pharynx by the trigeminal (first arch) and glossopharyngeal (third arch) nerves, *e.g.*, in the tongue.

4. On each side, the first pouch and the dorsal part of the second pouch are drawn caudally with the second arch and compounded into a single, elongated pouch (**tubotympanic recess**) which forms the cavities of the **auditory tube** and **middle ear** [FIG. 283]. This places the first arch and its contents anterolateral to the middle ear cavity while the second is posterior to it, and the third medial to it. This explains the presence of the **styloid process** and the vertical part of the **facial nerve** posterior to the middle ear cavity and lateral to the nerves of the more caudal arches (IX and X in the jugular foramen). It also explains the presence of the **internal carotid artery** (third arch artery) in the carotid canal medial to the middle ear, and the fact that the medial wall of the middle ear receives its

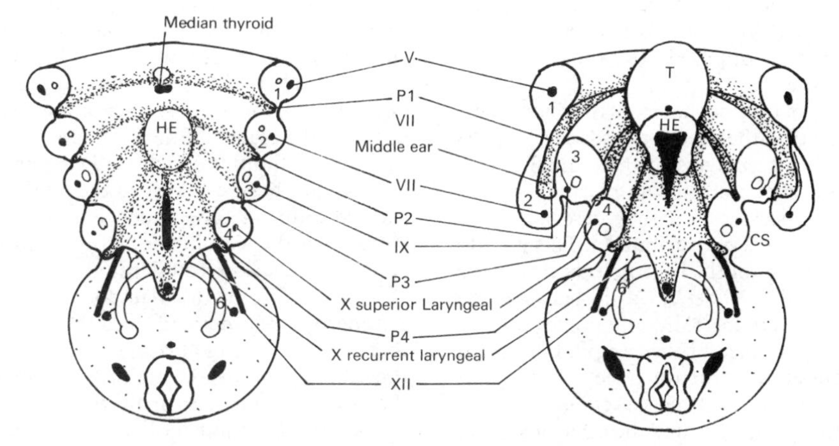

FIG. 283 Diagrammatic views of the ventral and lateral walls of the pharynx at two stages in the development of the pharyngeal arches. The roof of the pharynx has been removed by a curved, horizontal slice through the arches. 1–6 the pharyngeal arches. CS, the cervical sinus. HE, hypobranchial eminence. P1–P4, entodermal pharyngeal pouches. T, tongue. Roman numerals indicate the cranial nerves or their branches.

nerve supply (**tympanic plexus**) from the glossopharyngeal nerve (third arch).

Entodermal Pouches

The **first pouch** is simple, but the others are bifid with dorsal and ventral wings. The first pouch and the dorsal part of the second are involved in the production of the auditory tube and middle ear cavity (see above). The ventral wing of the second pouch forms a bed for the **palatine tonsil**.

The **third** and **fourth pouches** elongate and separate from the ectoderm as the mesodermal proliferation spreads around them. The epithelium of each wing proliferates to form a solid body, while the neck of the pouch, elongating and narrowing, disappears leaving the pouch derivatives as bilobed structures buried in the neck. The ventral wing of each third pouch complex becomes adherent to the pericardium ventral to it and, as the neck elongates, the entire pouch complex on each side is drawn caudally with the pericardium. The ventral part on each side elongates to form a lobe of the **thymus** which usually becomes a thoracic organ attached to the pericardium. The derivatives of the dorsal wings are the **inferior parathyroids**. These are usually released from the complexes as they reach the caudal part of the median thyroid (see below) in their descent, though one or both of them may be drawn down into the thorax with the thymus. Commonly a strand of tissue unites the cranial pole of each lobe of the thymus to the corresponding lobe of the thyroid gland in the region of the inferior parathyroid—a remnant of this embryological connexion. The strand may consist entirely of thymic tissue.

The **fourth pouch** complex, which may have contributions from more caudal, rudimentary pouches (**ultimobranchial body**) becomes adherent to the posterior surface of the corresponding lateral lobe of the thyroid gland. The dorsal wing of the pouch forms the **superior parathyroid**, the ventral wing contributes to the median thyroid. It may give rise to the pale cells of the thyroid which are believed to secrete **thyrocalcitonin**—a substance which has actions opposed to those of the parathyroid hormone.

The separation of the pouches from the pharyngeal wall and the growth of the head forwards carrying the brain and, with it, the origins of the cranial nerves which enter the arches, pulls the more caudal nerves into line with the neck so that they descend in the adult to reach their area of distribution. This growth also carries the origins of the **hypoglossal nerves** cranially so that they slide upwards between the pharynx and the overlying skin, and come to hook round the inferior surface of the occipital artery. If there is a **cervical fistula** (see above) the corresponding hypoglossal nerve cannot ascend above it, but always curves round the inferior surface of such a fistula.

This method of development results in a cervical fistula or *persistent pharyngeal pouch* having an internal opening into the **tonsillar fossa** if it involves the ventral wing of the second pouch, or opening through the **thyrohyoid membrane** if it involves the third [FIG. 281].

DEVELOPMENT OF THE THYROID GLAND

Most of this gland arises as a median downgrowth of the epithelium of the floor of the pharynx between the ventral ends of the first and second pharyngeal arches [FIG. 283]. It passes into the substance of the developing tongue and becomes adherent to the roots of the two third-arch arteries as they arise from the ventral aorta. As the neck elongates, this **median thyroid** descends with these arteries anterior to the hyoid bone and then to the thyroid and cricoid cartilages, expanding into a curved plate which leaves an epithelial tract behind it—the **thyroglossal duct**—part of which may be tucked up behind the body of the hyoid bone. The central part of the thyroid plate, with the thyroglossal duct attached to it, remains narrow and forms the **isthmus**; the lateral parts expand to form the right and left **lobes**.

Any part of the thyroglossal duct may persist. This is least common in the tongue where the **foramen caecum** marks the approximate point of origin of the gland, but **cysts** may form from isolated remnants in the region of the body of the hyoid bone and anywhere between this point and the isthmus of the gland. The lower part of the duct may persist as an upwards extension of the isthmus, **the pyramidal lobe** of the thyroid gland, or connective tissue surrounding this part of the duct may remain as a fibrous or muscular strand (**levator glandulae thyroideae**) extending upwards from the isthmus or pyramidal lobe to the hyoid bone. Very rarely the thyroid gland may not descend. It is then situated in the substance of the tongue.

THE DEVELOPMENT OF THE CENTRAL NERVOUS SYSTEM

The tubular central nervous system develops from a thickened plate of ectoderm on the amniotic (dorsal) surface of the embryonic disc. This **neural plate** extends from the buccopharyngeal membrane cranially to the cloacal membrane caudally. The edges of the plate and the skin ectoderm with which they are continuous fold dorsally towards each other, so that they meet and fuse in the midline. This turns the plate into a tube which is separated from the surface by the simultaneous fusion of the adjacent edges of the skin ectoderm dorsal to it, and by subsequent ingrowth of mesoderm between them. Ectodermal cells derived from the edges of the plate form a ridge (**neural crest**) on the dorsal surface of the tube. From this the majority of the nerve cells of the **spinal** and **autonomic ganglia** are produced together with a number of other structures, including the **sheath cells** of peripheral nerves.

The closure of the tube begins at the middle of the plate and spreads cranially and caudally so that the two ends are the last to be completed, the tube remaining open for a while through the anterior and posterior **neuropores**. It is here that varying degrees of failure of closure of the tube, or of its separation from the overlying skin by ingrowth of mesoderm most commonly occur. This results in **anencephaly** at the cranial end or **spina bifida** at the caudal end—the commonest malformations of the neural tube and the commonest developmental defects in the earliest stages of development.

Initially the neural tube consists of a thick layer of epithelium (**ependyma**) which soon forms four quadrants—two dorsal, **alar laminae** and two ventral, **basal laminae**, separated by a **sulcus limitans** on each side. This ependyma proliferates to form both nerve cells (grey matter) and neuroglia on its external surface. In the part forming the **spinal medulla**, this is principally on the anterolateral (basal laminae) and posterolateral (alar laminae) surfaces, to form the anterior and posterior **horns of the grey matter**. The nerve cells spin out processes or axons. A minority (**efferent cells**) in the basal laminae send their axons transversely outwards into the surrounding tissue to form the **ventral roots** of the spinal nerves. The majority in both laminae are **interneurones** which either communicate with adjacent cells or, more usually, send their axons on to the surface of the grey matter where they run longitudinally forming an *external layer of nerve fibres*. Some of these processes cross the midline, producing a **commissure** ventral to the ependyma, before running longitudinally on the opposite side. Relatively few of these longitudinal, interneuronal fibres are formed on the posterior surface of the spinal medulla. Here processes of the spinal ganglion cells (**afferent fibres** forming the **dorsal roots**) pass on to the surface of the tube and also extend longitudinally forming the **posterior funiculi**.

All these longitudinal fibres send many branches into the grey matter. Thus those that arise in the cells of the spinal medulla form connexions between its segments (**intersegmental tracts** or **fasciculi proprii**), while those from the spinal ganglia carry

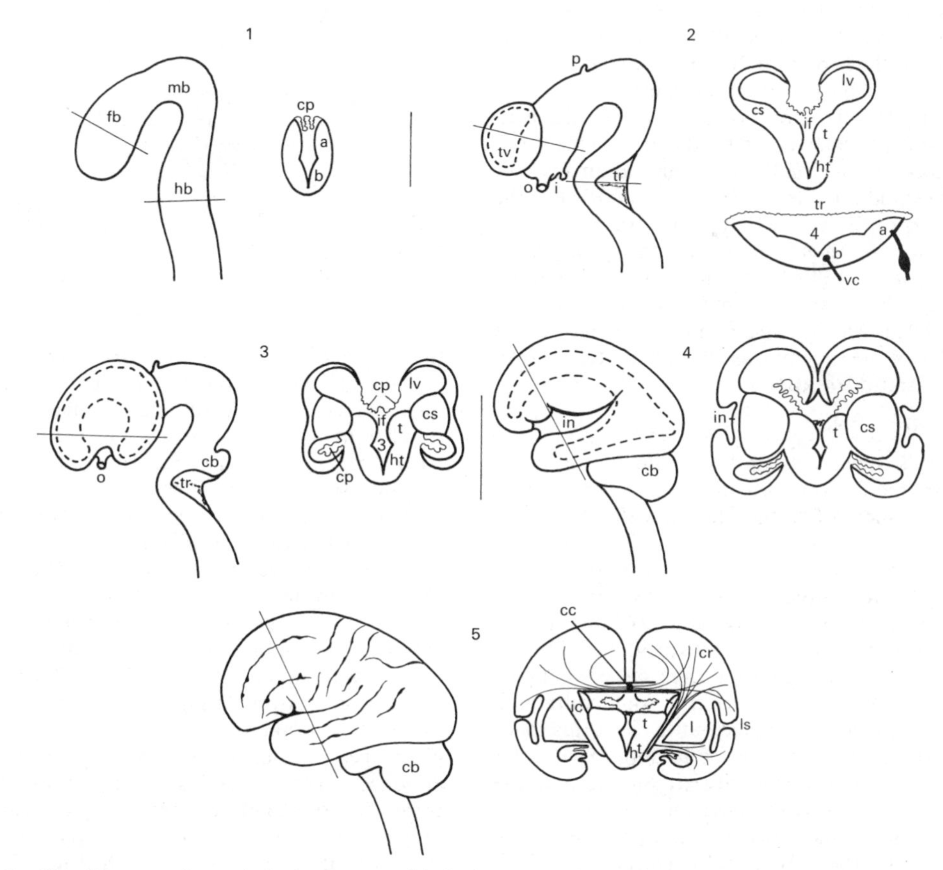

FIG. 284 Diagrams of stages in the development of the brain, not to scale. Each diagram contains a lateral view of the brain and transverse section(s) corresponding to the line(s) drawn on the lateral view. a. Alar lamina. b. Basal lamina. c. Caudate nucleus. cb. Cerebellum. cc. Corpus callosum. cp. Choroid plexus. cr. Corona radiata. cs. Corpus striatum. fb. Forebrain. hb. Hindbrain. ht. Hypothalamus. i. Infundibulum. ic. Internal capsule. if. Interventricular foramen. in. Insula. l. Lentiform nucleus. lc. Lateral cranial nerve. ls. Lateral sulcus. lv. Lateral ventricle. mb. Midbrain. o. Optic outgrowth (cut). p. pineal body. t. Thalamus. tr. thin roof of fourth ventricle. tv. telencephalic vesicle. vc. Ventral cranial nerve. 3. Third ventricle. 4. Fourth ventricle.

sensory information to many levels of the spinal medulla and to the brain. Subsequently, on the lateral and anterior surfaces and superficial to the intersegmental tracts, longitudinal nerve fibres descend from nerve cells in the brain to the spinal medulla, and others ascend to the brain from cells in the spinal medulla. These complete the **anterior** and **lateral funiculi** of the spinal medulla which are separated by the emerging ventral roots. When all the longitudinal fibres myelinate, all the funiculi become white in colour forming the **white matter** of the spinal medulla. As a result of this method of development, **efferent cells** lie at the level of emergence of their axons from the central nervous system, while **afferent axons** extend considerable distances within it so that they have the potential of numerous connexions through a multiplicity of branches (**collaterals**). The interneurones further disseminate the impulses in such a manner that those entering at very different levels of the central nervous system may be integrated rapidly.

BRAIN

At an early stage, the part of the neural tube within the developing skull is bent inferiorly so that its rostral part, the forebrain, lies parallel and anterior to its caudal part (the hindbrain), the flexure (**cephalic flexure**) occurring at the **midbrain**. Subsequently this flexure is partly straightened, but the long axis of the forebrain remains at an angle to that of the hindbrain, leaving the dorsal surface of the midbrain more extensive than its ventral surface.

In the *forebrain and the hindbrain*, the ependyma of the **roof plate** (the line of closure of the neural tube) does not proliferate, but remains a single layer of epithelial cells which is invaginated on each side into the cavity of the tube by vascular pia mater to form linear **choroid plexuses**.

Hindbrain

As the brain tube grows, the hindbrain is folded at its middle so that it forms an acute angle ventrally. This changes the shape of the tube dramatically [FIG. 285]. The cavity becomes a diamond-shaped space, the **fourth ventricle**, which is widest at the line of folding (junction of the two parts of the hindbrain—the pons and medulla oblongata) and tapers superiorly to the narrow **aqueduct** in the midbrain and inferiorly to the **central canal** in the lower part of the medulla oblongata. The thin roof, containing the choroid plexuses, is pulled out to cover this space posteriorly, and at the line of folding extends far laterally as the roof of the **lateral recesses** of the ventricle. At the tips of these recesses and at the inferior angle of the ventricle, the thin roof breaks down forming the only **apertures** through which the cavity of the neural tube communicates with the surrounding **subarachnoid space**, and through which protrude the ends of the linear choroid plexuses, each now L-shaped [FIGS. 284, 285].

The flattening of the hindbrain which results from

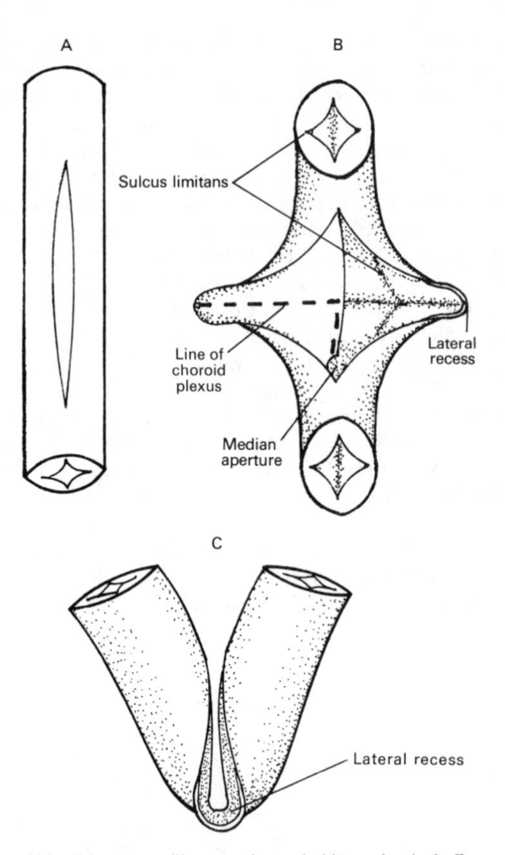

FIG. 285 Diagram to illustrate the probable mechanical effects of the pontine flexure in the formation of the fourth ventricle. A, before folding. B, posterior view after folding; the thin roof has been removed on the right side. C, lateral view after folding with the thin roof removed.

the folding displaces the **alar laminae** so that they lie lateral to the **basal laminae**. Thus the receptive nuclei which arise from the alar laminae are lateral to the efferent nuclei which arise from the basal laminae. Also the ganglionated dorsal roots are attached to the lateral aspect of the pons and medulla oblongata, though the ventral roots emerge ventrally near the midline. These two sets of roots, equivalent to those of the spinal nerves, fail to unite, and thus form two separate groups of **cranial nerves**—ventrally emergent **motor** nerves, the abducent and hypoglossal (the oculomotor and trochlear nerves in the midbrain are similar, though the trochlear emerges dorsally after decussating), and laterally attached nerves, trigeminal, facial, vestibulocochlear, glossopharyngeal, vagus, and accessory. The laterally attached nerves contain **sensory** fibres, and most transmit **visceral efferent** (parasympathetic preganglionic) and/or modified visceral efferent (**branchial motor**) fibres—a feature which they share with primitive spinal nerves in which the dorsal and ventral roots do not unite.

The increased size of the pons and medulla oblongata compared with the spinal medulla is the result of the development of the **reticular**

formation and of groups of cells primarily related to the cerebellum (*e.g.*, **pontine** and **olivary nuclei**).

Cerebellum

This organ arises as an extension of the alar laminae into a narrow, transverse strip of the thin roof of the fourth ventricle at the pontine level. This divides the thin roof into a small cranial part between the cerebellum and the midbrain (**superior medullary velum**) and a larger caudal part which connects the cerebellum to the medulla oblongata (**inferior medullary velum**) and contains the choroid plexuses. The cerebellum does not extend its attachment as it grows. Hence the nerve fibres which enter and leave it are clustered together in a small area at the pontine level—the compact **cerebellar peduncles** [FIG. 211]. The cerebellum is formed by cells derived from the ependyma of the roof of the ventricle. At first they lie close to the roof, but many subsequently migrate to the surface of the cerebellum to form the **cerebellar cortex**, those remaining near the roof of the ventricle form the **cerebellar nuclei**.

The enormous expansion of the cerebellar surface, while it retains a limited attachment to the brain stem, causes the cerebellum to become markedly convex posteriorly and to draw the **fourth ventricle** posteriorly into itself as a tent-like **cerebellar recess**. The cerebellar nuclei lie along the ridge of the tent.

Midbrain

This part of the brain retains a narrow, tubular cavity—the **cerebral aqueduct**, continuous below with the fourth ventricle and above with the cavity of the forebrain. Posterior to the aqueduct, the alar laminae form the **tectum** consisting of two superior and two inferior **colliculi** receiving respectively visual and auditory impulses, and concerned with reflexes involving these senses. Anterior to the aqueduct, the basal laminae give rise to the **tegmentum** of the midbrain (reticular formation, red nuclei, third and fourth cranial nerve nuclei) and the **substantia nigra**. These are mostly concerned with the motor apparatus of the brain.

Forebrain

At first this is a simple, median tube with a thin roof containing two linear choroid plexuses. The wall of the part adjacent to the midbrain differentiates into the **thalamus** dorsally and the **hypothalamus** ventrally. The remainder gives rise to a number of outgrowths of the full thickness of its wall. There are two median outgrowths. (1) The **pineal body** grows posteriorly from the roof at its junction with the midbrain, and lies on the dorsal surface of the midbrain between the two superior colliculi. (2) The **neural lobe of the hypophysis** is formed by a downgrowth (the **infundibulum**) from the floor of the anterior hypothalamus. Two pairs of outgrowths also arise from the forebrain. (a) Immediately anterior to the infundibulum, an **optic outgrowth** arises on each side. The expanded ends of these outgrowths form the retina and its pigment layer, the ciliary part of the retina, and both epithelial layers of the iris. The stalks form the **optic nerves** which meet in the **chiasma** anterior to the infundibulum. The original cavities of the optic outgrowths are reduced to mere slits between the two retinal layers when the distended end is invaginated upon itself to form a double layered cup. (b) Immediately anterior to the thalamus, the dorsal wall bulges outwards on each side to form the telencephalic vesicles, leaving the remainder of the forebrain as the **diencephalon**, composed of the thalamus and hypothalamus.

Each **telencephalon** begins as a small, hollow vesicle. Its cavity is one of the **lateral ventricles** which communicates with the median cavity of the diencephalon (**third ventricle**) and with the other lateral ventricle through the common **interventricular foramen** at the anterosuperior extremity of the third ventricle. This is immediately behind the anterior wall of that ventricle—the **lamina terminalis** which now unites the two telencephalic vesicles across the median plane. When the telencephalic vesicles grow out from the neural tube, each carries part of the ependymal roof of the third ventricle and the corresponding **choroid plexus** into the medial wall of the vesicle [FIG. 284].

The wall of each vesicle is thick inferiorly (**corpus striatum**) but thin superiorly. The thin, superior part grows rapidly in area and so overgrows the corpus striatum which increases in volume rather than area. As a result the thin walled part extends backwards and then downwards on the corpus striatum, carrying the ependymal wall and choroid plexus with it. It then extends forwards below the corpus striatum as the **temporal lobe** of the hemisphere. The anterior thin walled part extends forwards and downwards on the corpus striatum as the **frontal lobe** of the brain. In this way the thin walled part of the vesicle and the contained lateral ventricle form a C-shaped structure surrounding the solid, central corpus striatum which only appears on the surface where the temporal lobe abuts on the frontal lobe at the **stem of the lateral sulcus**. It is this initial growth which determines the shape of the hemisphere and the arrangement of all the structures within it. When the temporal lobe abuts on the frontal lobe, further expansion of the vesicle in the sagittal plane occurs by backward growth of its convex surface to form the **occipital lobe** and the pointed **posterior horn of the lateral ventricle** [FIG. 284].

Lateral expansion of the thin walled part of the vesicle causes it to overgrow and hide from view the cerebral cortex which is adherent to the lateral surface of the corpus striatum [FIG. 284]. This process forms the **lateral sulcus** when the temporal lobe, rolling upwards over this cortex from below, meets the frontoparietal lobes descending in the same way from above. The buried cerebral cortex is the **insula** which may be exposed by pulling apart the lips of the lateral sulcus.

Medial expansion of the vesicle brings the corpus striatum into closer contact with the diencephalon, and carries the adjacent choroid plexus and the medial part of the thin walled vesicle over the surface of the diencephalon. Thus the **choroid plexus**, invaginated through the **choroid fissure**, is applied to the thalamus and hypothalamus, and the adjacent edge (**limbus**) of the thin walled vesicle lies medial to the plexus and encircles the corresponding half of the diencephalon. This is the edge of the hemisphere in which the **fornix** develops [FIG. 230]. Superiorly, the medial extension brings these parts of the hemispheres into contact above the ependymal roof of the third ventricle and the vessels which overlie it (**internal cerebral veins** and **posterior choroidal arteries**) and permits the growth of nerve fibres between the two hemispheres to form the **corpus callosum**. Posteriorly the two hemispheres are held apart by the intervening midbrain, hence the corpus callosum extends no further posteriorly than the pineal body on the dorsal surface of the midbrain, compelling callosal fibres from the temporal and lower occipital lobes to ascend to its caudal part and form the massive splenium [FIG. 232]. It is here that the internal cerebral veins emerge from beneath the corpus callosum uniting to form the **great cerebral vein**.

Thus the *general structure of the hemisphere* is that of a centrally placed corpus striatum applied to the lateral surface of the diencephalon. Wrapped around this is the remainder of the hemisphere containing the lateral ventricle with the choroid plexus invaginated into it. All these structure follow the same C-shaped curve around the corpus striatum and diencephalon, failing to cover the corpus striatum only in the stem of the lateral sulcus where it lies close to the surface.

The **cerebral cortex** rapidly thickens by the growth of its cells which have migrated to the surface from the ependyma of the ventricle, thus the relative size of the ventricle diminishes rapidly. The cells of the cortex send fibres medially into the **corpus callosum**. They converge on it without mixing, and diverge again when they enter the opposite hemisphere. Hence they unite the corresponding parts of the two hemispheres [FIG. 284]. **Projection fibres** grow radially inwards (**corona radiata**) from the cortex towards the corpus striatum [FIG. 258]. These pierce the corpus striatum near its outer margin, partly separating the rim of the corpus striatum (**caudate nucleus**) from the remainder (**lentiform nucleus**) which lies in the concavity of the C-shaped sheet of fibres (**internal capsule** [FIG. 258]) formed by the projection fibres as they pierce the corpus striatum. Thence the internal capsule passes towards the base of the brain between the diencephalon and the lentiform nucleus [FIG. 266], condensing and emerging from it as the **crus cerebri** which separates the inferior part of the medial edge of the cerebral cortex and the corresponding part of the choroid fissure from the hypothalamus. Each crus, now lying on the anterolateral surface of the median brain tube, descends on that surface of the midbrain to enter the pons where most of its fibres end. The **pyramid** is the sole remaining part which emerges from the caudal extremity of the pons on to the anterior surface of the medulla oblongata.

Of the **projection fibres** which enter the **internal capsule**, a considerable number end in the corpus striatum and diencephalon, others pass to the midbrain, pons, and medulla oblongata (including **corticonuclear** and **corticoreticular** fibres), while the remainder enter the spinal medulla through the pyramid. In addition, **ascending fibres** pass from the thalamus to the cerebral cortex in the internal capsule and corona radiata, while those which unite the parts of the corpus striatum to each other and to the diencephalon pierce the internal capsule or hook round its inferior surface (ansa lenticularis).

The simple convergence of fibres from the cerebral cortex into the internal capsule and vice versa makes it possible to predict the position of any particular group of fibres in this pathway provided it is known from which part of the cerebral cortex the fibres come or to which part they are going [FIGS. 258, 284].

TABLE 1
Movements of the Neck and of the Head on the Neck

Movement	Muscles	Nerve supply
Flexion	Sternocleidomastoid	Accessory
	Longus colli	Cervical ventral rami
	Longus capitis	Cervical ventral rami
	Rectus capitis anterior*	C. 1. ventral ramus
Extension	Splenius cervicis and capitis	Cervical dorsal rami
	Erector spinae	Cervical dorsal rami
	Rectus capitis posterior major and minor*	C. l. dorsal ramus
	Obliquus capitis superior*	C. l. dorsal ramus
	Trapezius	Accessory
Lateral flexion and rotation	Sternocleidomastoid‡	Accessory
	Scalenes	Cervical ventral rami
	Longus colli	Cervical ventral rami
	Rectus capitis lateralis*	C. 1. ventral ramus
	Levator scapulae (shoulder fixed)	Cervical ventral rami
	Splenius†	Cervical dorsal rami
	Longissimus	Cervical dorsal rami
	Obliquus capitis superior‡ and inferior	C.1. dorsal ramus†

* These muscles act mainly as ligaments of adjustable length and tension.
‡ In rotation turns the face towards the opposite side.
† In rotation turns the face to the same side.

TABLE 2
Movements of the Tongue

These are complex interactions of the intrinsic* and extrinsic muscles. The actions of individual muscles should not be learnt, but the general arrangements can be obtained from this table.

Movement	Muscles	Nerve supply
Tip		
Elevation	Superior longitudinal*	Hypoglossal
Depression	Inferior longitudinal* and genioglossus	Hypoglossal
Retraction	Genioglossus (anterior fibres) and longitudinal muscles*	Hypoglossal
Turning to one side	Longitudinal* of that side with protrusors of opposite side	Hypoglossal and C. 1.
Body		
Widening	Longitudinal* and vertical*	Hypoglossal
Heightening	Longitudinal* and transverse*	Hypoglossal
Shortening	Longitudinal*	Hypoglossal
Elongation	Transverse* and Vertical*	Hypoglossal
Depression		
of median part	Genioglossus	Hypoglossal
of edges	Hyoglossus	Hypoglossal
of all	Lowering of hyoid bone (*q.v.*)	
Elevation	Elevation of hyoid bone (*q.v.*)	
	Styloglossus	Hypoglossal
	Mylohyoid	Trigeminal
	Palatoglossus	Pharyngeal plexus
Protrusion in the midline	Geniohyoids†	Ventral ramus C. 1.
	Genioglossus (posterior fibres)†	Hypoglossal
	Vertical*	Hypoglossal
	(Transverse*)	Hypoglossal
to one side	Action of above muscles on opposite side ± retractors of same side	
Retraction	Longitudinal*	Hypoglossal
	Styloglossus	Hypoglossal

† These muscles help to maintain the patency of the airway when lying supine: the reason for holding the jaw forwards in the unconscious patient in this position.

TABLE 3
Movements of the Hyoid Bone

MOVEMENT	MUSCLES	NERVE SUPPLY
Elevation	Digastric	Trigeminal and facial
	Stylohyoid	Facial
	Mylohyoid	Trigeminal
Depression	Sternohyoid	Ansa cervicalis
	Omohyoid	Ansa cervicalis
	Thyrohyoid and sternothyroid	C. 1 and ansa cervicalis
Protraction	Geniohyoid	C. 1 ventral ramus
	Genioglossus (posterior fibres)	Hypoglossal
Retraction	Middle constrictor of pharynx	Pharyngeal plexus
	Stylohyoid	Facial

TABLE 4
Movements of the Larynx

MOVEMENT	MUSCLES	NERVE SUPPLY
As a whole		
Elevation	Thyrohyoid (hyoid fixed)	C. 1 ventral ramus
closes laryngeal	Stylopharyngeus	Glossopharyngeal
orifice	Palatopharyngeus	Pharyngeal plexus
Depression	Sternothyroid	Ansa cervicalis
opens laryngeal orifice		
Of vocal cords		
Tightening		
all	Cricothyroid	Superior laryngeal
part	Vocalis	Recurrent laryngeal
Slackening		
all	Thyro-arytenoid	Recurrent laryngeal
part	Vocalis	Recurrent laryngeal
Abduction	Posterior crico-arytenoid	Recurrent laryngeal
Adduction	Lateral crico-arytenoid	Recurrent laryngeal
	Oblique and transverse arytenoid	Recurrent laryngeal

TABLE 5
Movements of the Mandible

MOVEMENT	MUSCLES	NERVE SUPPLY
Elevation of chin (closing mouth)	Masseter	Mandibular
	Medial pterygoid	Mandibular
	Temporalis (anterior part)	Mandibular
Depression of chin (opening mouth)	Lateral pterygoid	Mandibular
	Digastric (hyoid fixed)	Mandibular and facial (ansa cervicalis)
	Geniohyoid and mylohyoid with infrahyoid muscles	C. 1 ventral ramus Mandibular Ansa cervicalis
Protraction	Lateral pterygoid	Mandibular
	Medial pterygoid	Mandibular
	Masseter, superficial part	Mandibular
Retraction	Temporalis (posterior part)	Mandibular
	Digastric (hyoid fixed)	Mandibular and facial (ansa cervicalis)
Chewing—a grinding movement; alternation of,		
one side	Medial and lateral pterygoid and masseter	Mandibular
other side	Temporalis (all parts) masseter (deep part)	Mandibular

TABLE 6
Swallowing

This complex movement is not entirely understood, but an approximate sequence of events is given below. More muscles are used than those given, and the phases merge into one another.

Movement	Muscles	Nerve supply
1. Mouth shut	See above	
2. Respiration stopped		
3. Tip of tongue elevated in front of bolus	Superior longitudinal	Hypoglossal
4. Bolus moved back by elevating tongue against hard and tensed soft palate	See above Tensor palati	 Mandibular
5. Hyoid displaced forwards to open pharynx	Geniohyoid	C. 1 ventral ramus
6. Larynx raised (closes laryngeal orifice) with epiglottis which is tipped back by the posterior surface of tongue, and bent down above laryngeal orifice. Palate raised to close pharyngeal isthmus.	Thyrohyoid Stylopharyngeus Aryepiglotticus Levator palati Superior constrictor Palatopharyngeus (horizontal fibres)	C. 1 ventral ramus Glossopharyngeal Recurrent laryngeal Pharyngeal plexus Pharyngeal plexus Pharyngeal plexus
7. Bolus grasped by then displaced downwards with larynx (opening laryngeal orifice) hyoid and tongue. Hyoid pulled backwards. Then soft palate relaxed to re-establish continuity of oral and nasal parts of pharynx. Respiration starts.	Superior then middle constrictor Infrahyoid muscles Middle constrictor	Pharyngeal plexus Ansa cervicalis Pharyngeal plexus

TABLE 7
Sucking

This is the process of lowering the pressure in the isolated oral cavity. The muscles involved vary with the intensity of the effort, but respiration continues through the nose until swallowing begins.

Movement	Muscles	Nerve supply
1. Lips closed around object to be sucked, or lips closed if object is in mouth. Teeth closed as far as possible to support cheeks. Palate depressed against posterior part of tongue to separate mouth from pharynx	Orbicularis oris See movements of mandible Palatopharyngeus Palatoglossus	Facial Mandibular Pharyngeal plexus Pharyngeal plexus
2. Tongue lowered	Lowering of hyoid bone (*q.v.*) Genioglossus Hyoglossus Vertical muscles of tongue	 Hypoglossal Hypoglossal Hypoglossal

TABLE 8
Sneezing and Coughing

The first part of the sequence is the same in both activities.

Movement	Muscles	Nerve supply
1. A breath is taken in	Inspiratory—see Volume 2	
2. Intrathoracic pressure is raised by:		
a. closing glottis, vocal cords tensed and adducted	Cricothyroid	Superior laryngeal
	Oblique and transverse arytenoids, lateral crico-arytenoid	Recurrent laryngeal
b. Contracting expiratory muscles	Oblique and transverse abdominal	Intercostal and subcostal
	Internal intercostal, transversus thoracis, levator ani and sphincter urethrae	Intercostal Pudendal
3. Sneezing		
Palate depressed against posterior part of tongue to close mouth from pharynx	Palatopharyngeus	Pharyngeal plexus
	Palatoglossus	Pharyngeal plexus
Coughing		
Palate raised,	Levator palati	Pharyngeal plexus
tensed,	Tensor palati	Mandibular
and pulled against posterior pharyngeal wall to separate oral and nasal parts of pharynx.	Superior constrictor	Pharyngeal plexus
	Palatopharyngeus (horizontal part)	Pharyngeal plexus
4. Vocal cords suddenly abducted to release intrathoracic air pressure through nose or mouth.	Posterior crico-arytenoid and loss of tension in adductors	Recurrent laryngeal

In blowing a wind instrument or other activity of this kind, the same mechanism is used as for coughing except that the intrathoracic pressure is built up against an external resistance and the glottis is never closed. Blowing is maintained during the next inspiration by distending the cheeks with expired air, separating the oral cavity from the pharynx with the soft palate, as in sucking or sneezing, and contracting buccinator during inspiration through the nose.

THE BONES OF THE CERVICAL VERTEBRAE AND SKULL

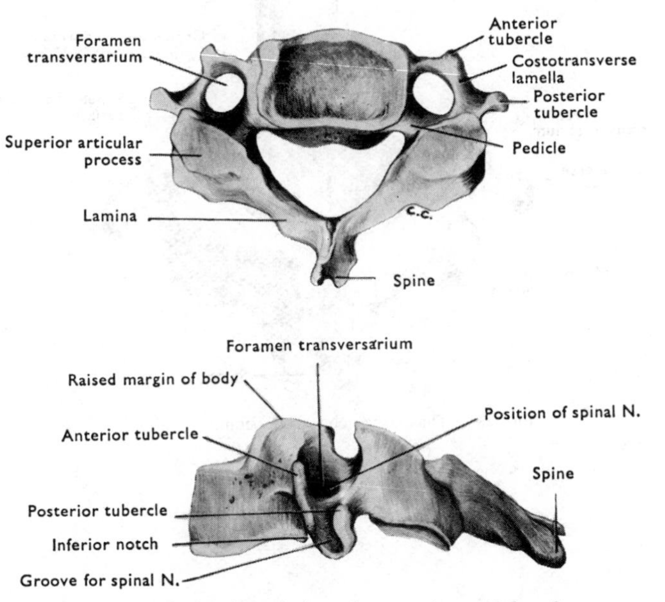

FIG. 286 The fourth cervical vertebra, superior and left surfaces.

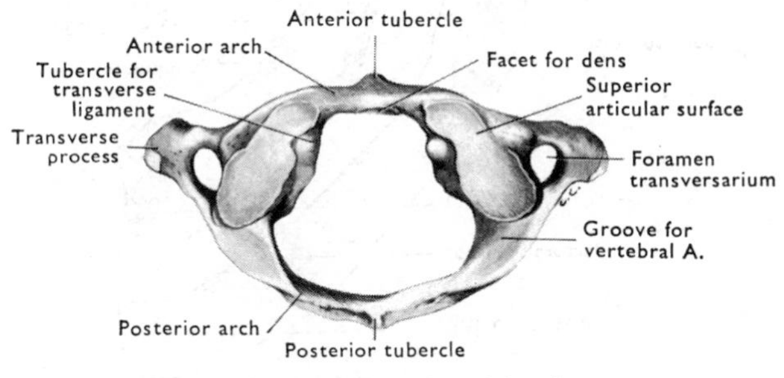

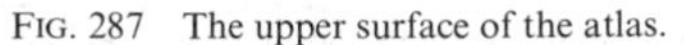
FIG. 287 The upper surface of the atlas.

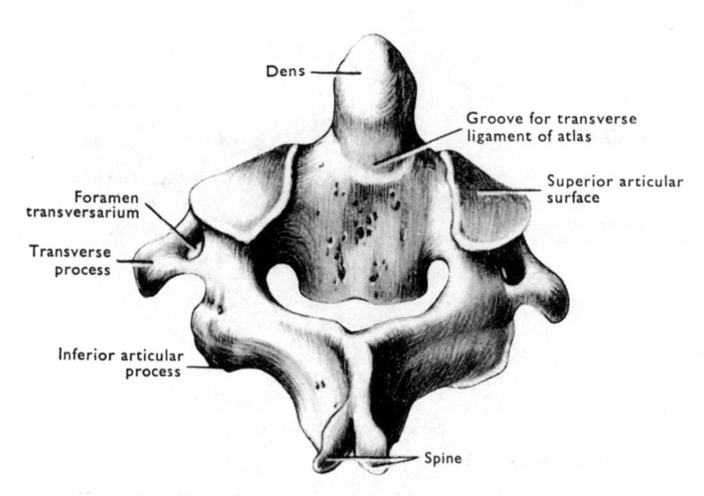

FIG. 288 The axis vertebra from behind and above.

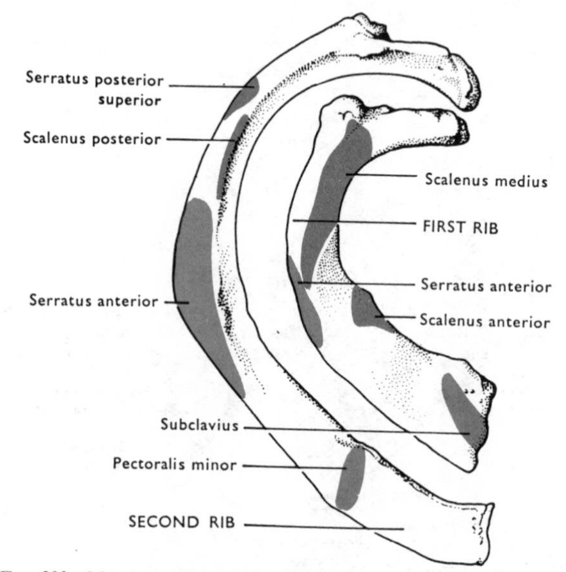

FIG. 289 Muscle attachments to the first two ribs, excluding the intercostals. Origins, red; insertions, blue.

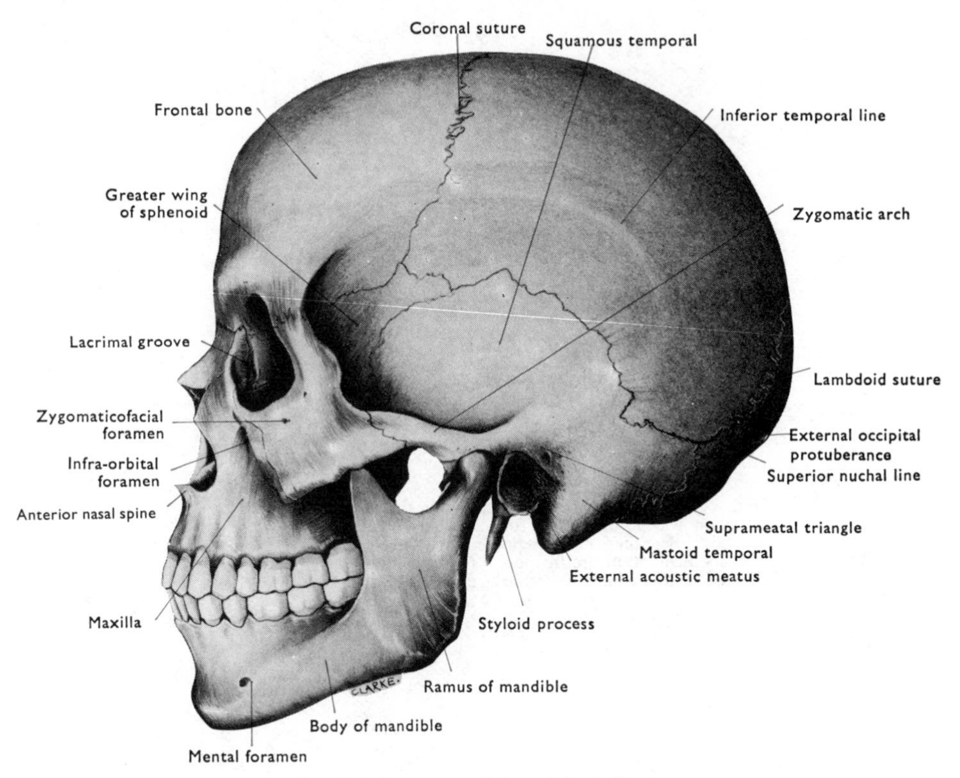

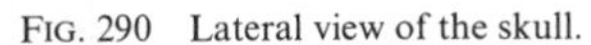

FIG. 290 Lateral view of the skull.

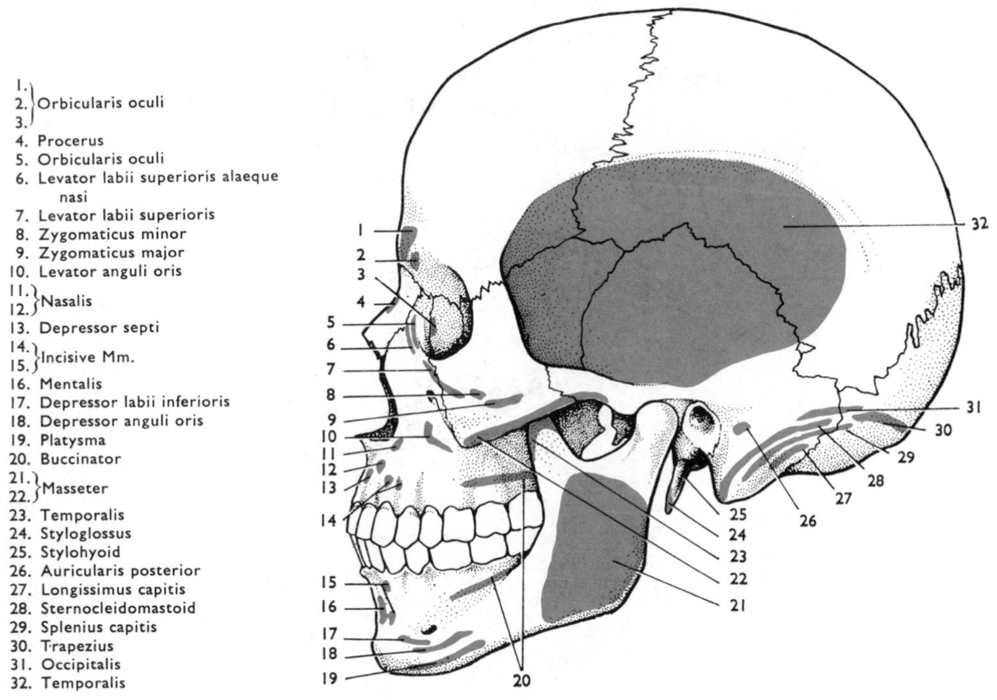

FIG. 291 Lateral view of the skull showing the muscle attachments.

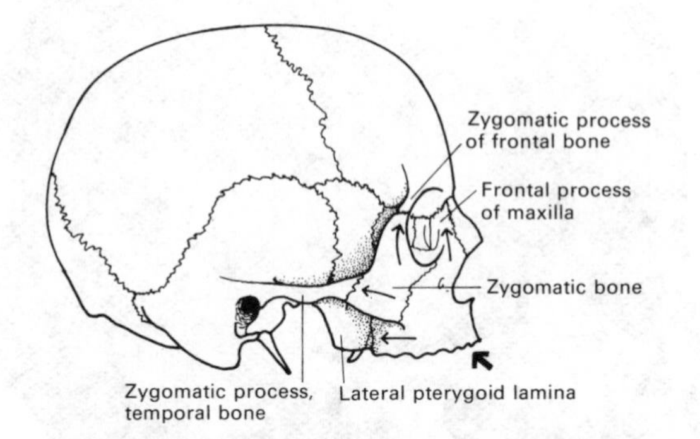

FIG. 292 Lateral view of skull to show the parts of the facial skeleton which transmit to the cranium forces applied to the maxilla. The thick arrow represents the force applied – the thin arrows the lines of transmission. These are the parts of the skull liable to fracture in blows to the face.

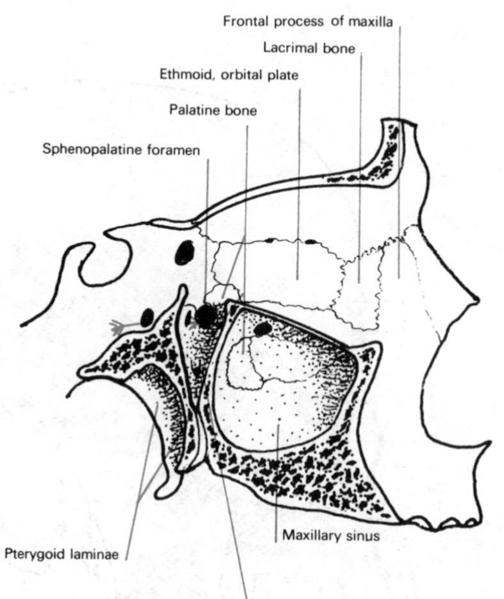

FIG. 293 Lateral view of sagittal section through anterior part of skull to show pterygopalatine fossa, and the construction of the medial wall of the orbit. Arrow in foramen rotundum.

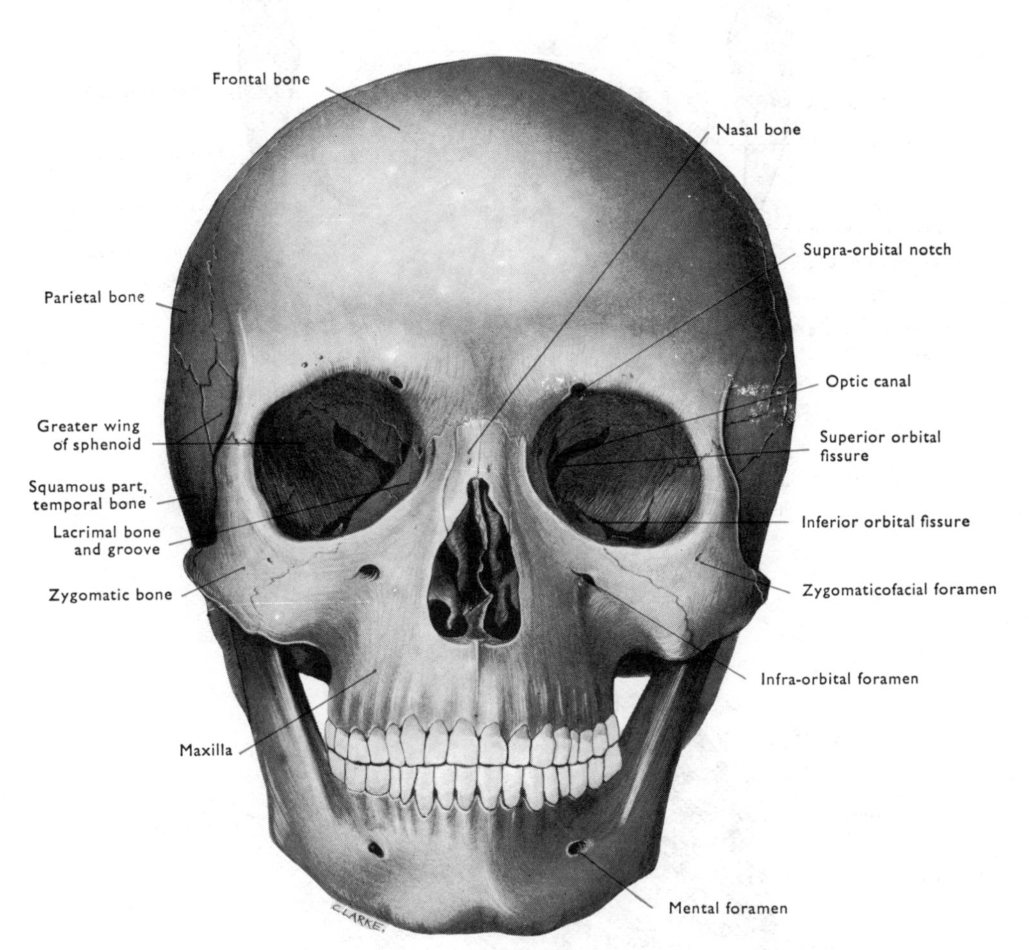

FIG. 294 Anterior view of the skull.

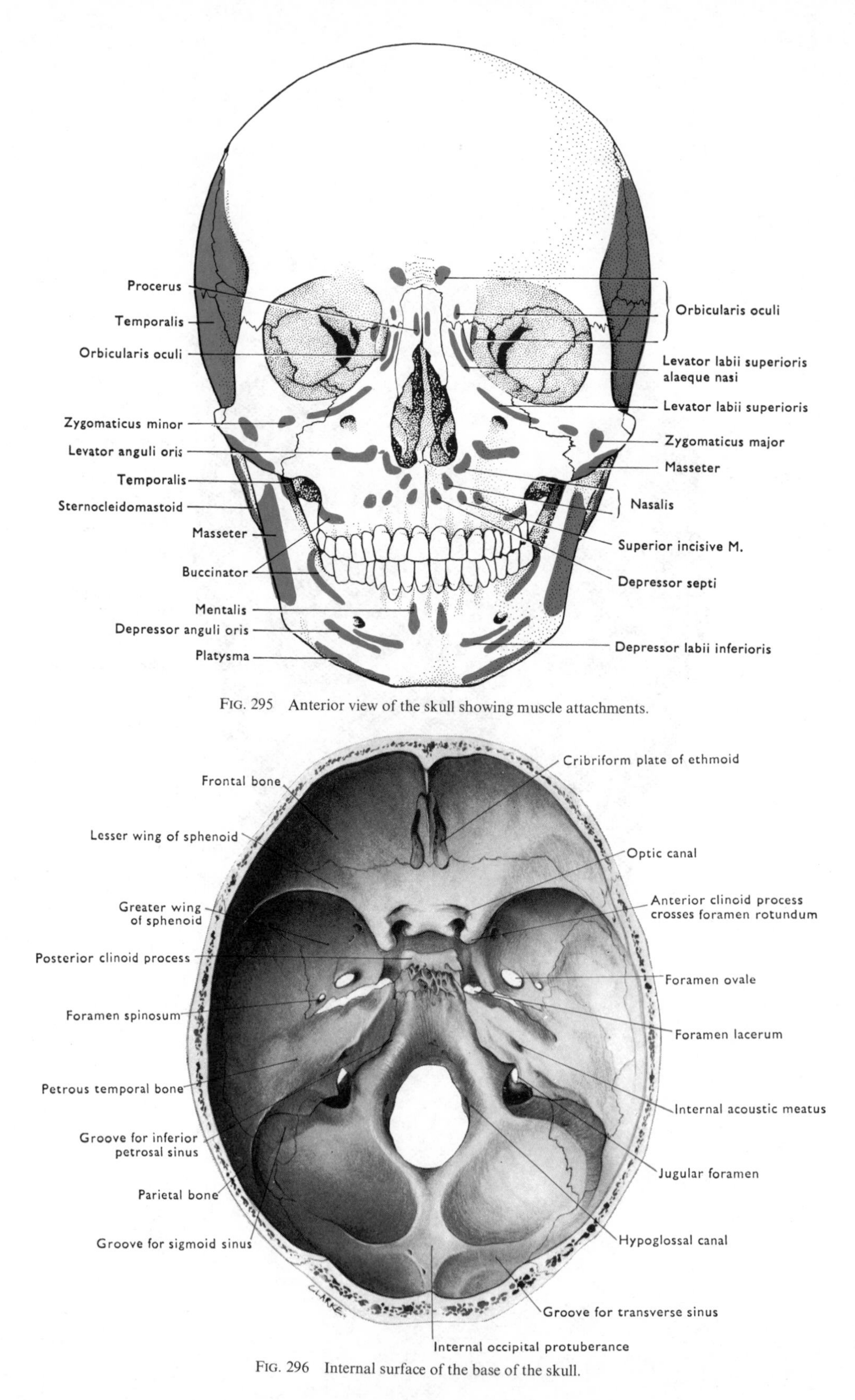

FIG. 295 Anterior view of the skull showing muscle attachments.

FIG. 296 Internal surface of the base of the skull.

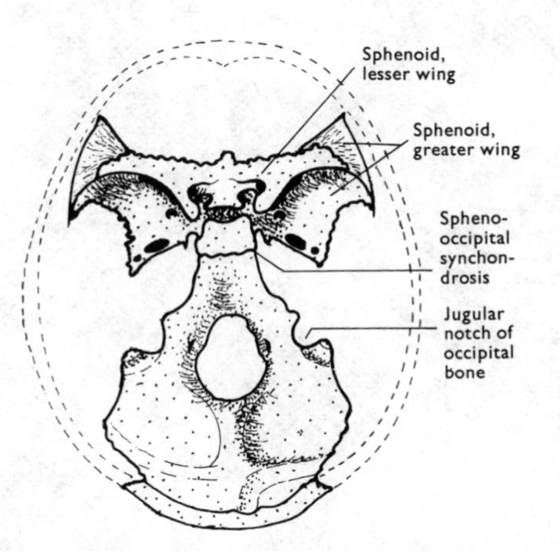

FIG. 297 The part of the internal surface of the base of the skull formed by the sphenoid and occipital bones. On each side a temporal bone is wedged between them. The frontal and ethmoid bones complete the anterior part.

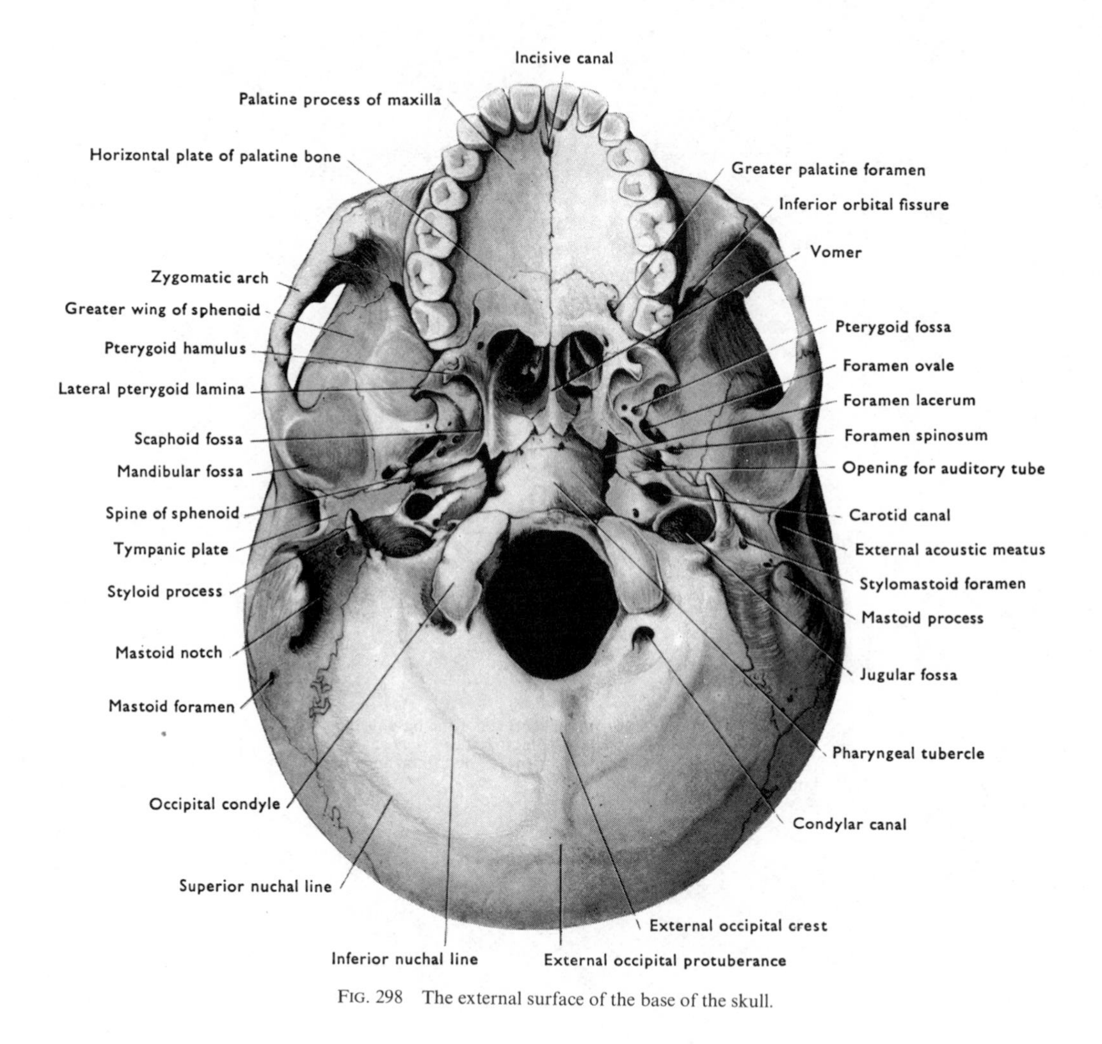

FIG. 298 The external surface of the base of the skull.

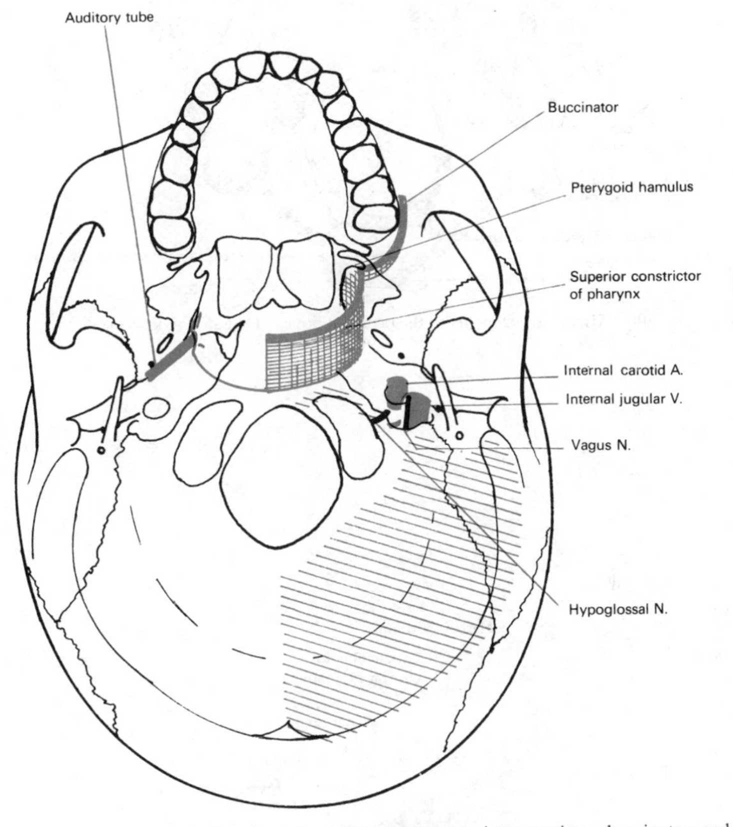

FIG. 299 External surface of base of skull to show the position of the superior constrictor, buccinator, and the auditory tube.

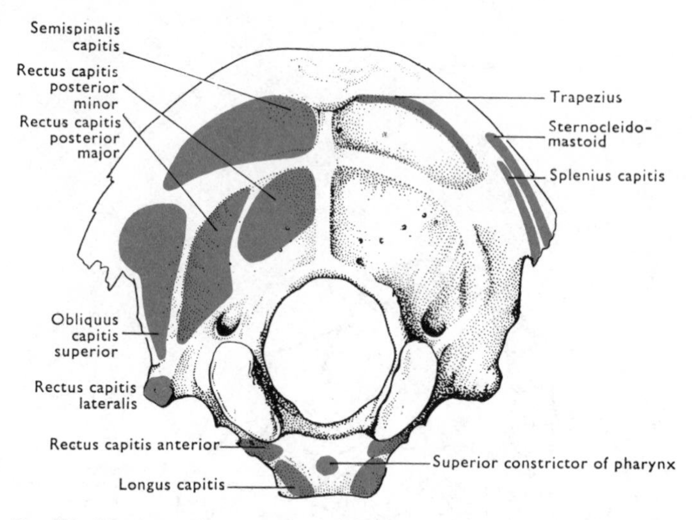

FIG. 300 Muscle attachments to the occipital bone. Origin, red; insertions, blue.

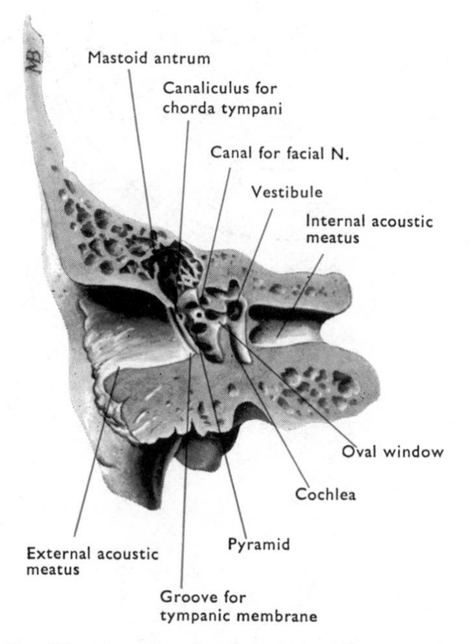

FIG. 301 Coronal section through the right temporal bone seen from in front.

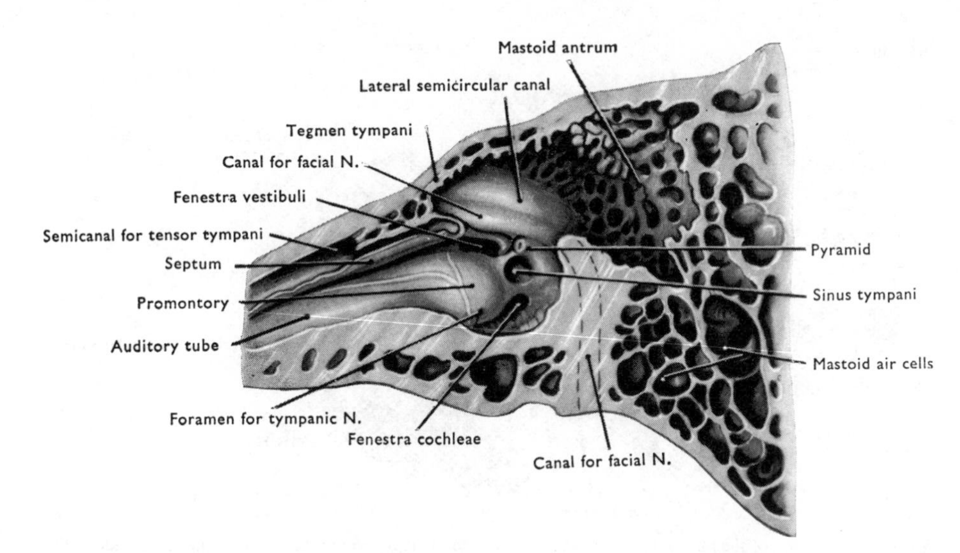

FIG. 302 Vertical section through the left middle ear and auditory tube to expose the medial or labyrinthine wall of the middle ear. The auditory ossicles have been removed.

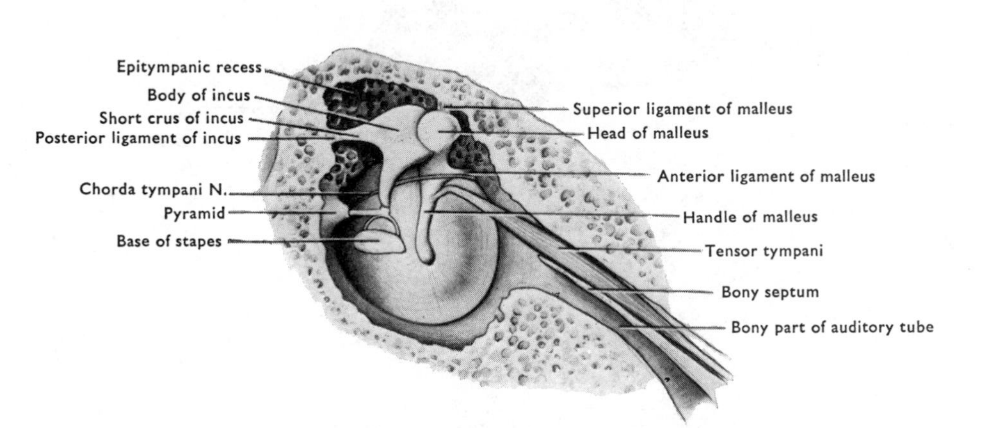

FIG. 303 Left tympanic membrane and auditory ossicles seen from the medial side. Note the auditory tube, and superior to it, tensor tympani and its semicanal.

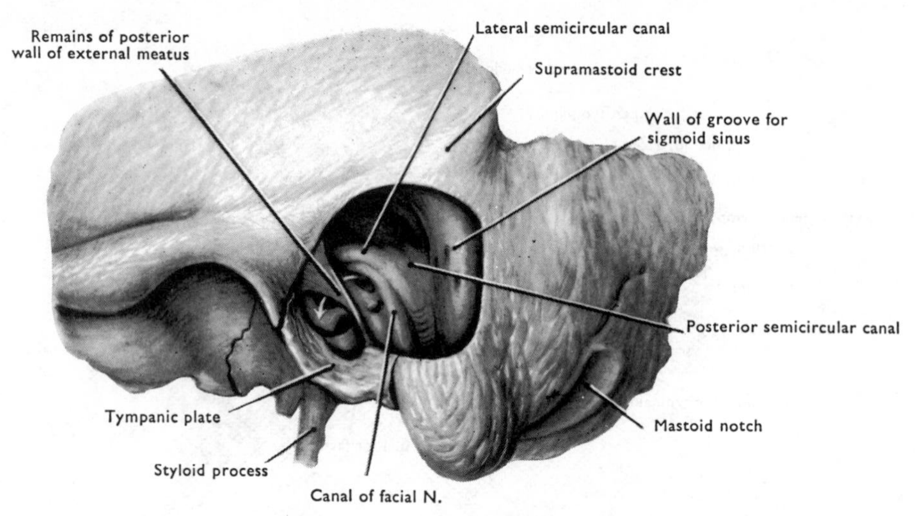

FIG. 304 Dissection of the mastoid antrum and petromastoid part of temporal bone from the lateral side. The arrow passes from the mastoid antrum into the tympanic cavity through the aditus.

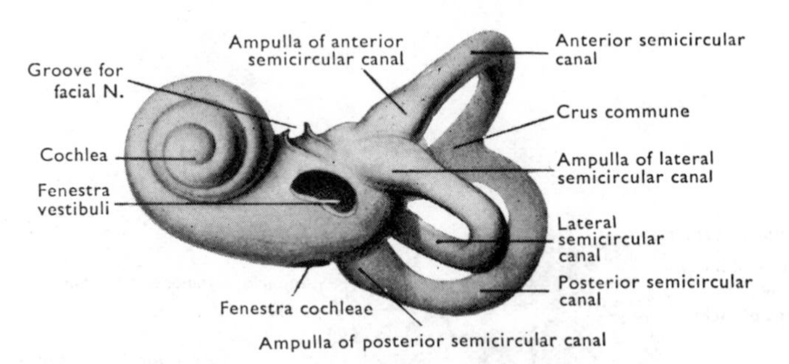

FIG. 305 Left bony labyrinth viewed from the lateral side.

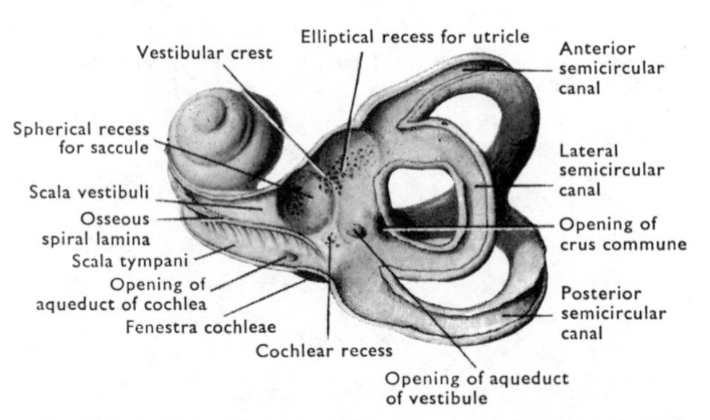

FIG. 306 Interior of the left bony labyrinth viewed from the lateral side.

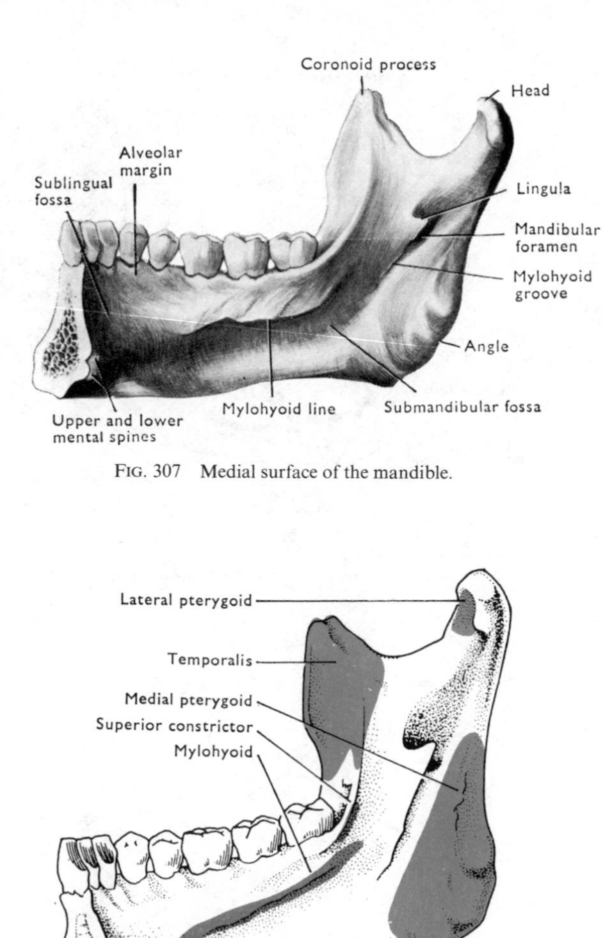

FIG. 307 Medial surface of the mandible.

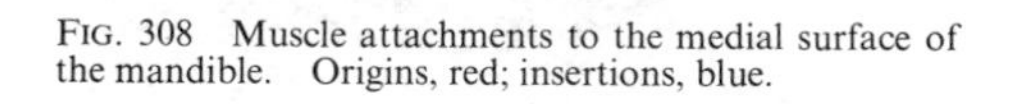

FIG. 308 Muscle attachments to the medial surface of the mandible. Origins, red; insertions, blue.

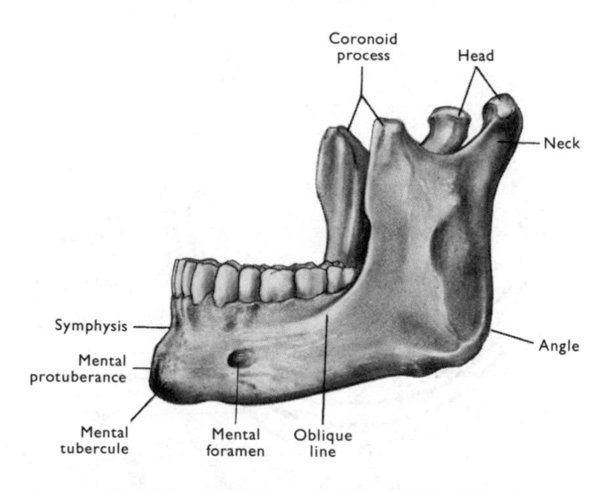

FIG. 309 Mandible as seen from the left side.

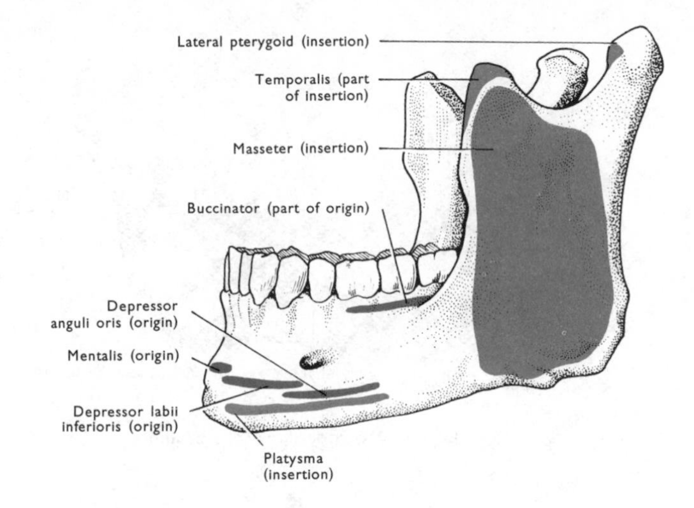

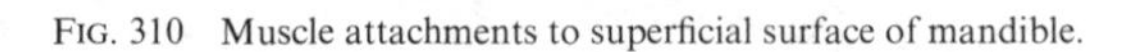

FIG. 310 Muscle attachments to superficial surface of mandible.

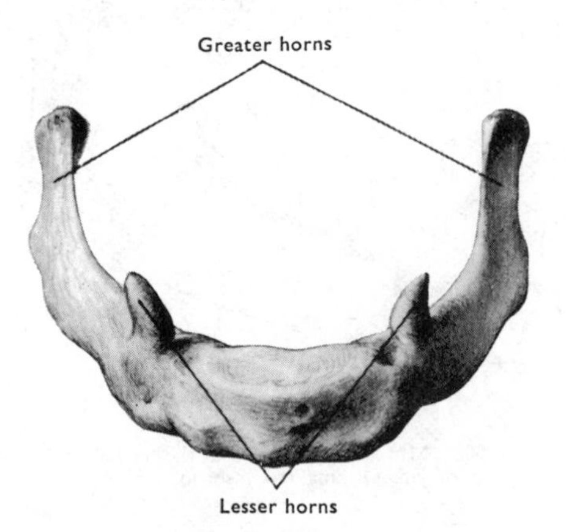

FIG. 311 Anterior view of the hyoid bone.

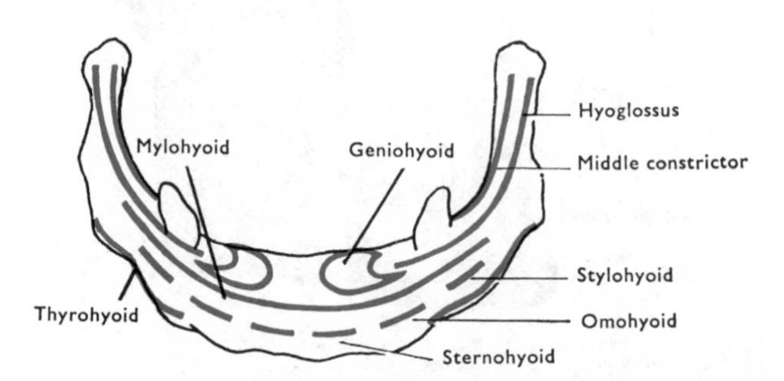

FIG. 312 Anterior view of hyoid bone to show muscle attachments.

INDEX TO VOLUME THREE